DISCRETE MATHEMATICS

Anuranjan Misra
Professor and Dean Academics
Bhagwant Institute of Technology
Ghaziabad, U.P., India

Copyright © 2010, Author

All rights reserved. No part of this book may be reproduced, stored in a retrieval system, or transmitted in any form or by any means, electronic, mechanical, photocopying, recording or otherwise, without the prior written permission of the publisher. Making copies of this book or any portion thereof for any purpose other than your own is a violation of copyright law.

ISBN: 978-93-80408-14-9

First Edition: New Delhi, 2010

Published by
ACME LEARNING PRIVATE LIMITED
2/8, Ansari Road, Daryaganj
New Delhi-110 002

Printed at Rajive Book Binding, Delhi-110040

DEDICATED

To
My Grand Parents Late Shri & Smt. Nand Kishore Misra

Anuranjan Misra

Detailed Contents

Preface .... *xiii*
Acknowledgements ...... *xv*

Chapter 1: Set Theory ... 1-38

1.1	Introduction	 1
1.2	Type of Sets	 2
1.3	Methods of Representing a Set	 3
1.4	Combination of SETs	 3
1.5	Multisets	 6
1.6	Laws of Operations (General Identities on Sets)	 12
1.7	Venn Diagrams	 21
1.8	Ordered Pairs	 33

Chapter 2: Relation . 39-80

2.1	Relation	 39
2.2	Range and Image	 39
2.3	Operations on Relations	 40
2.4	Composite Relation	 41
2.5	Properties of Binary Relation	 41
2.6	Congruence Modulo Relation on Set of all Integers	 43
2.7	Equivalence Relation	 46
2.8	Equivalence Classes	 54
2.9	Partition of a Set	 55
2.10	Quotient Set	 56
2.11	Ordering Relations and Ordered Set	 57
2.12	Partial Order Relation	 58
2.13	Partially Ordered Set	 58
2.14	Totally Ordered Set	 59
2.15	Geometrical Representation of Ordering on Set	 59

2.16	Representation of Relations	 62
2.17	Matrix Representation of Composition	 67
2.18	Warshall's Algorithm	 76

Chapter 3: Function — 81-102

3.1	Correspondence	 81
3.2	Types of Correspondence	 81
3.3	Mapping (Function)	 83
3.4	Types of Mapping (Functions)	 83
3.5	Classification of Function	 85
3.6	Operations on Functions	 91
3.7	Growth of Functions	 93
3.8	Recursively Defined Functions	 94
3.9	Product or Compositions of Mappings (Functions)	 97

Chapter 4: Natural Numbers — 103-120

4.1	Natural Numbers	 103
4.2	Notation	 104
4.3	Notion of Proof	 108
4.4	Fallacies	 110
4.5	Mathematical Induction	 111
4.6	Proof By Counter Examples	 111

Chapter 5: Algebraic Structure — 121-148

5.1	Binary Operation	 121
5.2	Types of Binary Operations	 122
5.3	Composition Table	 124
5.4	Formation of Composition Table	 124
5.5	Algebraic Structures	 125
5.6	Some Definitions	 127
5.7	Some Properties of Groups	 139
5.8	Alternative Definition of a Group	 144

Chapter 6: Subgroups — 149-172

6.1	Subgroups and Subgroup Tests	 149
6.2	Cyclic Groups	 149
6.3	Homomorphisms and Normal Subgroups	 152
6.4	Normal Subgroups	 154

6.5	Quotient Groups	 155
6.6	Permutation Group	 158
6.7	Symmetric Groups and Cayley's Theorem	 159
6.8	Conjugacy in Symmetric Groups	 161
6.9	The Alternating Groups	 162
6.10	Some Special Groups	 163
6.11	Dihedral Groups	 166
6.12	Small Groups	 167
6.13	Polyhedral Groups	 169

Chapter 7: Rings and Fields   173-208

7.1	Introduction	 173
7.2	Subrings	 182
7.3	Homomorphisms and Quotient Rings	 184
7.4	Factorisation	 191
7.5	Fields	 201

Chapter 8: Posets, Hasse Diagram and Lattices   209-227

8.1	Introduction	 209
8.2	Totally Ordered (or Linearly ordered) Set	 210
8.3	Properties of Posets	 211
8.4	Hasse Diagrams	 212
8.6	The Möbius Function	 214
8.7	Lattices	 215

Chapter 9: Boolean Algebra   229-250

9.1	Introduction	 229
9.2	Boolean Algebra and Applications	 229
9.3	Theorems of Boolean Algebra	 231
9.4	Logic (Digital) Circuits and Boolean Algebra	 233
9.5	Boolean Functions, Applications	 242

Chapter 10: Graph Theory   251-312

10.1	Introduction	 251
10.2	Graphs—Basics	 251
10.3	Eulerian and Hamiltonian Graphs	 261
10.4	Isomorphism	 265
10.5	Isomorphic Digraphs	 267

10.6	Some Properties of Isomorphic Graphs	 267
10.7	Some Observation on Isomorphism	 268
10.8	Homeomorphism Graphs	 268
10.9	Digraphs	 268
10.10	Matrix Representations	 275
10.11	Tree Structures	 284
10.12	Binary Trees	 288
10.13	Binary Search Trees	 288
10.14	Traversal of a Binary Tree	 289
10.15	Planar Graphs	 292
10.16	The Travelling Salesman Problem	 302
10.17	Kruskal's Algorithm	 302
10.18	Prim's Algorithm	 303

Chapter 11: Mathematical Logic   313-336

11.1	Proposition Logic	 313
11.2	Operators of Proposition	 313
11.3	Truth Table	 314
11.4	Laws	 315
11.5	Types of Statements	 315
11.6	Logical Equivalence	 317
11.7	Predicated Logic	 318
11.8	Inference Theory	 320
11.9	Algebra of Propositions	 323

Chapter 12: Combinatorics   337-370

12.1	Basic Counting Techniques	 337
12.2	Combinatorics	 338
12.3	The Pigeonhole Principle	 342
12.4	Recurrence Relations	 346
12.5	Generating Functions	 351
12.6	Deriving Generating Functions From Recurrences	 355
12.7	Recurrence Relations	 361

Chapter 13: Karnaugh Map   371-420

13.1	Introduction	 371
13.2	Venn Diagrams and Sets	 372
13.3	Boolean Relationships on Venn Diagrams	 374

13.4	Making a Venn Diagram Look Like a Karnaugh Map	 379
13.6	Logic Simplification with Karnaugh Maps	 388
13.7	Larger 4-variable Karnaugh Maps	 394
13.8	Minterm vs Maxterm Solution	 399
13.9	Sum and Product Notation	 410
13.10	Don't Care Cells in the Karnaugh Map	 412
13.11	Larger 5 and 6-variables Karnaugh Maps	 415

Chapter 14: Graph Colouring   421-430

14.1	Introduction	 421
14.2	Definition and Terminology	 421
14.3	Properties	 424
14.4	Algorithms	 426
14.5	Applications	 429

Chapter 15: PÓlya Counting Theorem   431-440

15.1	PÓlya enumeration theorem	 431
15.2	Informal PET Statement	 431
15.3	Definitions for the Univariate Case	 432
15.4	Theorem Statement	 432
15.5	Example Computations	 432
15.6	Formal Statement of the Theorem	 437

Solved Exercises .. 441-465
Appendix .. 467-475
Glossary .. 477-486
Index .. 487-488

Preface

Discrete mathematics usually means the mathematics of discrete sets. You will appreciate the role of discrete mathematics in computer science as you go along—not only for the duration of studying this subject but for subjects like data structures, design and analysis of algorithms, logic, automata, languages and computation which you will study in coming semesters.

The distinction between discrete mathematics and other mathematics is somewhat artificial as analytic methods are often used to study discrete problems and vice versa. Number theory in particular sits on the boundary between discrete and continuous mathematics, as does finite topology (the study of finite topological spaces) which is literally the intersection of combinatorics and topology.

Though the scope of discrete mathematics is broad, we will try to focus on every aspects of discrete mathematics in this book. The book has been written in very easy language and with daily life example so that student can correlate it easily. This book consists of 15 chapters as well as Examples, Questions, Multiple Choice Questions, Glossary etc.

The book also contains one Appendix, Appendix I: Automata, Grammars and Languages. Since rapid changes can be witnessed in the every subject and Discrete Mathematics is not left untouched. I have taken my best efforts to incorporate the latest updates taking in the Discrete Mathematics. Furthermore your views, suggestions, comments and recommendations are invited and will be highly appreciable.

We have tried to bring the book free from errors. However, it is not unusual that certain flaws might have crept into the body of the text. I take personal responsibility for the same and request to let know of those misprints or mistakes. I assure the readers to remove them in the next addition of this book.

—**Author**

Acknowledgements

We are indebted to a number of individuals in academic as well as in social circles who have contributed in different way in the preparation of this book. We realize that the acknowledgement is something that really comes from the bottom of the heart of every writer. We take this opportunity to express my gratitude to all those who have helped and encouraged us to prepare this book.

We are highly thankful to Er. Anil Singh, Chairman, Dr. Asha Singh, Vice Chairman Bhagwant Group of Institutions, Prof. V.K. Sharma, Principal, Dr. V. Satyanathan, Vice Principal, Prof. R.C. Marwaha, Bhagwant Institute of Technology, Muzaffarnagar, Mr. Neeraj Goel (General Secretary), Prof. Surendra Kumar, Director Administration (Formally of Senior Scientist CBRI, Roorkee), Prof. R.G. Mendiratta, Director Academics (Formally of Dy. Director, IIT Delhi), Prof. U.S. Gupta HOD of MCA (Formally of Professor at IIT, Roorkee), Prof. C.L. Mittal HOD of EC dept., Prof. A.K. Sinha, HOD of CSE/IT, Prof. H.C. Mathur, HOD of MCA Dept. (Formally of Banaras Hindu University), Prof. P.N. Koul, Director (CCDP), Prof. Dinesh (Dy Director) Prof. Hariom Upadhyay, Director of MCA, (Formally senior Programmer at New Zercy, US) of ABES Engineering College, Ghaziabad, Prof. Raj Kishore from BN College, Patna, Mr. Vishal Kumar, Project leader, Accenture, Hyderabad, Mr. Vinod Kulkarni, Project Leader, Optimal Network Resource Ltd. Hyderabad, Mr. Amit Goel, Project Leader, TCS, Noida, Mr. Vijay Rastogi, Senior Manager, CSC, Noida, Mr S.C. Tyagi, Executive Director, IDC Foundation, Mr. Anilji Garg, Director, Doodle Powel, Noida, Prof. Abhay Bansal, HOD of IT, ITS Ghaziabad, Prof Yogesh Mittal, HOD of IT, AKGEC, Ghaziabad, Prof V.P. Gupta, Professor, KIET, Ghaziabad, Prof. K.K. Gupta, Dean IT, Ideal Institute, Ghaziabad. Mr. Umesh Gautam, Chairman, Invertis Group of Institutions, Bareilly, Prof. K.D. Misra, Ex- Dean, IT-BHU, Banaras, Prof. R.K. Sukla, Prof. J.N. Srivastava, Prof. Santosh Kumar, Prof. Ravi Shankar Sukla of Invertis Group of Institutions Bareilly.

Our special thanks to Mr. Ashok Kumar Agarwal (B.Tech, IIT Kanpur) Managing Director M.S. System. Ltd. SIIC IIT Kanpur, Praveen Kumar Katiyar (Team Leader RSS Global Kanpur), Mr. Anil Tiwari (Team Leader MSSPL SIIC, IIT Kanpur), Mr. Jitendra Prajapati (Software Developer), Mr. Jitendra Parihar (Senior Software Developer)Mr. Raj Kishor Rathour (Director of CIIT Hardoi), for his support, encouragement, invaluable suggestions and inspirations.

It is our duty for giving special acknowledgements of to the following for their notable contributions to this guide:

Mr. Pankaj Singh Chauhan, Software Engineer, Brainsoft Pvt Ltd, Delhi.
Miss. Ranjna Sharma, HR UG Soft Solution, Delhi.
Mr. Kumar Sagar Gupta, Mindfire Solution, Delhi.
Mr. Hari Mohan Sharma, BSN Info Tech Ltd, Lucknow.
Mr. Yogash Srivastva, Data Base Administrator Ebix Pvt Ltd, Delhi
Mr. Amit Kumar, Genpact Pvt Ltd, Gurgaun (Hariyana).
Mr. Sishu Prajapati, Project Leader Real Time System Ltd, Delhi.

We thank to the Director of Acme Learning for his kind cooperation through the project.
We thanks to our family members for their cooperation throughout the project. At last we thank all the persons who helped us in this project directly and indirectly.

CHAPTER 1

SET THEORY

1.1 INTRODUCTION

A set is a collection of definite distinguishable objects such that, given a set and an object, we can ascertain whether or not the specified object is included in the set.

Sets

A set is a well-defined collection of distinct objects.

By a 'well defined' collection of objects, we mean that there is a rule (s) by means of which it is possible to say that without ambiguity, whether a particular object belongs to the collection or not.

By 'distinct' we mean that we do not repeat an object over and over again in a set.

Elements

Each object belonging to a set is called an element (or a member) of the set. Sets are usually denoted by capital letters A, B, C, N, Q, R, S, etc. and the elements by lower case letters. a, b, c, x, y, etc.

The relationship of an object to a set of which it is an element, is called a relation of belonging.

Examples:
(i) *Set of natural numbers*:
$$N = \{1, 2, 3, 4, 5, 6, ... \}$$
(ii) *Set of integers*:
$$I = \{0, \pm 1, \pm 2, \pm 3, \pm 4, ...\}$$
(iii) *Set of even integers*: Any integers of the form $2n$, where n is an integer is called an even integer. Thus $0, \pm 2, \pm 4, \pm 6, ...$ are even integers.
(iv) *Set of odd integers*: Any integer of the form $2n-1$ or $2n+1$, where n is an integer is called an odd integer. Thus $\pm 1, \pm 3, \pm 5, ...$ are called odd integers.
(v) *Set of non-negative integers*: It contains all positive integers
$$0, 1, 2, 3, 4, ...$$

(vii) Set of all rational numbers: Any number of the form p/q, where p and q are integers and $q \neq 0$ is called a rational numbers. The set of all rational numbers is denoted by Q.

(viii) Set of all complex numbers: Any number of the form $a + ib$ where $i = \sqrt{-1}$.

1.2 TYPE OF SETS

(*i*) **Singleton Set:** If a set consists of only one element, it is called a singleton set.
For example:
$$\{1\}, \{0\}, \{a\}, \{b\}, \text{etc.}$$

(*ii*) **Finite set:** A set consisting of a natural number of objects, *i.e.*, in which the number of elements is finite, is called finite set.
For example:
$$A = \{5, 7, 9, 11\}$$
and
$$B = \{4, 8, 16, 32, 64, 91\}$$
Since A contains 4 elements and B contains 6 elements, so both are finite sets.

(*iii*) **Infinite set:** If number of elements in a set is infinite, the set is called infinite set.
For Example Set of natural numbers.
$$N = \{1, 2, 3, 4, \ldots\} \text{ is an infinite set.}$$

(*iv*) **Equal set:** Two sets A and B consisting of the same elements are called equal set.
For example:
$$A = \{1, 5, 9\}$$
$$B = \{1, 5, 9\}$$
So Here A = B.

(*v*) **Pair set:** A set having two elements is called pair set.
For example:
$$\{1, 2\}, \{0,3\}, (4,9\}, \{3,1\}, \text{etc.}$$

(*vi*) **Empty set:** If a set consists of no elements, it is called the empty sector null set or void set and is represented by ϕ.

Here a point is to be noted that ϕ is a null set but $\{\phi\}$ is singleton set.

Example: Which of the following are valid set definitions.
$$A = \{4, 5, 8\}$$
$$B = \{a, b, a, c\}$$
$$C = \{a, b\}$$
$$D = \{1, 2, 3, 4, 5, 6\}$$

S: A is a valid set, containing three elements 4, 5, and 8.

B looks to be equally valid except for the occurring of two *a*'s. We can certainly check for inclusion within the set and this is surely the most important element. Hence we might regard this as valid and equal to $\{a, b, c\}$. However, there are problems that original definition of B and remove one of the *a*'s then we apparently have $a \in B$ and $a \in B$. This conclusion is not allowed since it is inconsistent, hence we shall regard repetition within a set as referring to the same element and its duplication as being an oversight; the removal of duplicates forms the basis of several mathematical arguments later on. C is a valid set, containing 2 elements, *a* and *b*.

D is a valid set, containing 6 elements 1, 2, 3, 4, 5 and 6.

1.3 METHODS OF REPRESENTING A SET

The most common method of describing the sets are as follows:
  (i) Roster method or listing method or Tabular method.
  (ii) Set Builder method or property method or Rule method.
  (i) **Roster Method:** In this method, a set is described by listing all its elements, separating them by commons and enclosing them within brackets (only brackets).

Example:
  (i) If A is the set of even natural numbers less than 8.
$$A = \{2, 4, 6\}$$
  (ii) If B is the set of odd and numbers less than 17.
$$B = \{1, 3, 5, 7, 9, 11, 13, 15\}$$
  (iii) **Set Builder Method:** Listing elements of a set is sometimes difficult and sometimes impossible.

In this method, a set is described by means of some property which is shared by all the elements of the set.

Example:
  (i) If A is the set of all prime numbers then,
  A = { $x : x$ is a prime numbers}
  (ii) If A is the set of all natural numbers between 10 and 1000 then,
  A = { $x : x \in N$ and $10 < x < 1000$}.

1.4 COMBINATION OF SETS

  (i) **Subsets of a given set:** Let A be a given set. Any set B, each of whose element is also an element of A, is said to be contained in A and is called a subset of A. Subset is represented by $\subset$.

Example:
(1) Let A = {a, b, c}, B is a subset of A represented by B $\subset$ A then subsets of A are.
$$\{a\}, \{b\}, \{c\}, \{a, b\}, \{b, c\}, \{c, a\}, \{a, b, c\} \text{ and } \phi.$$

Example:
Proper subset: If B is subset of A and B $\neq$ A then B is called proper subset of A. In other words, if each elements of A, which is element of B, then B is called a proper subset of A.

In example (i) {a} {b} {c} {a, b} {b, c} {c, a} and ϕ are proper subset (S) of A. and in Example (ii)

{1}, {2}, {3}, {4}, {1, 2}, {1, 33, 21, 43, {2, 4} {3, 4}, {1, 2, 3}, {1, 2, 4}, {2, 3, 4} and φ are proper subset of A.

(ii) **Universal set:** Any set which is super set of all the sets under consideration is known as the universal set and is denoted by Ω or S.

(iii) **Intersection of sets:** If P and Q are any two sets then all the elements which are common in P and Q, set of these elements is called intersection of P and Q and is denoted by P ∩ Q.

In Symbols:
$$P \cap Q = \{x | \in P \text{ and } x \in P \text{ and } x \in Q\}$$

Example 1. Let P = {1, 2, 3, 4} and Q = {2, 4, 6, 8}. then
$$P \cap Q = \{2, 4\}$$

Example 2. Let P = {a, b, c, d, e, f} and Q = {a, b, c, d, g, h, i, j}
$$P \cap Q = \{a, b, c, d\}.$$

(iv) **Union of sets:** If P and Q are any two sets then all the elements which are either in P or in Q, set of these elements is called union of P and Q and is denoted by P ∪ Q.

In symbols:
$$P \cup Q = \{x | x \in P \text{ or } x \in Q\}$$

Example 1. Let P = {7, 8, 9, 10} and Q = {9, 11, 12, 14}, then
$$P \cup Q = \{7, 8, 9, 10, 11, 12, 14\}$$

Example 2. Let P = {a, b, c, d, e, f} and Q = {a, b, c, d, g, h, i, j}, then
$$P \cup Q = \{a, b, c, d, e, f, g, h, i, j\}$$

(v) **Disjoint sets:** Two sets are said to be disjoint. If they have no elements in common. Let A and B are any two sets with no common element.

or
$$A \cap B = \phi$$

i.e. two sets called disjoint if there intersection is a null set.

Example 1. Let A = {1, 2, 3, 4} and B = {5, 6, 7, 8}
$$A \cap B = \phi$$

Since no element is common in set A and B.

Example 2. Let A = {a, b, c} and B = {e, f, g}
So A ∩ B = φ
Since no element in A and B is common.

(vi) **Difference of two sets:** The difference of two sets A and B is the set of all elements which are in A but not in B and is denoted by A − B.

Or symbolically: $A - B = \{x | x \in A; x \notin B\}$.

Similarly: The difference of two sets B and is the set of all elements which are in B but not in A and is denoted by B − A.

or

Symbolically:
$$B - A = \{x \mid x \in B; x \notin A\}$$

Example 1. Let $A = \{1, 2, 3, 4\}$ and $B = \{2, 4, 6, 8\}$, then find $A - B$ and $B - A$

Solution: $A - B = \{1, 3\}$
$$B - A = \{6, 8\}$$

Example 2. Let $P = \{a, b, c, d\}$ and $Q = \{c, d, e, f, g, h\}$, then find $P - Q$ and $Q - P$

Solution: $P - Q = \{a, b\}$.
$$Q - P = \{e, f, g, h\}.$$

(vii) **Symmetric difference of two sets:** The symmetric difference of two sets. A and B is the set of the elements which are in the union of $(A - B)$ and $(B - A)$ and is represented by Δ.

So,
$$\Delta_{(A,B)} = (A - B) \cup (B - A)$$

Example 1. Let $A = \{1, 2, 3, 4, 5, 6\}$ and $B = \{4, 5, 6, 7, 8, 9, 10\}$ then, find symmetric difference of sets A and B.

Sol.: Here $A - B = \{1, 2, 3\}$
$$B - A = \{7, 8, 9, 10\}$$
So, $\Delta_{(A,B)} = (A - B) \cup (B - A)$
$$= \{1, 2, 3, 7, 8, 9, 10\}$$

(viii) **Super set:** If A and B are two sets and A is the subset of B then B is a super set of A. It can be written as,
$$B \supset A$$

Example 1. $A = \{1, 2, 3\}, B = \{1, 2, 3, 4\}$
Since $A \subset B$
then here, we can say that,
$$B \supset A$$

(ix) **Universal set:** Any set which is super set of all the sets under consideration is known as the universal set and is denoted by Ω or S.

Example: Let us take some sets
$$A = \{1, 2, 3\}, B = \{4, 5, 6\}, C = \{6, 7, 8, 9\}$$
$$D = \{10, 11, 12\}, E = \{9, 10, 11\} \; F = \{6, 7, 9\}, \text{ then}$$
Universal set,
$$S = \{1, 2, 3, 4, 5, 6\; 7, 8, 9, 10, 11, \text{ or any super set of S}\}$$

(x) **Complement of A set:** Let U be the universal set and $A \subset U$. Then, the complement of A is the set of those elements of U which are not in A. The complement of A is denoted by A^C.

$$A^C = U - A$$
$$= \{x : x \cup \in \text{ and } x \notin A\}$$
$$= \{x : x \notin A\}$$

Complement of a set is set itself.

i.e., $(A^C)^C = A$

Some Symbols

Symbol	Meaning
$\in$	'belongs to' or 'is a member of'
$\notin$	does not belong to
$\subset$	is a subset of
$\supset$	is a super set of
$\cup$	union of sets
$\cap$	intersection of sets
A^C or A'	complement of A
iff	If and only if
: or δ.t. or |	Such as
$\forall$	for all
$\exists$	there exists
$\Rightarrow$	implies

(xi) **Partition of a set:** A set {A, B, C, ...} of non-empty subset A, B, C, ... of a set S, is called a partition of S if.

(1) $A \cup B \cup C \cup ... S$

(2) The intersection of every pair of distinct subsets is the empty set.

(3) Clearly shows no two of A,B,C. ... have any element in common.

Example: Consider the set S = {1, 3, 5, 7, 9, 11} and its.

Subsets A, B and C such

$$A = \{1, 5, 9\}, B = \{3, 7\}, C = \{11\}$$

Clearly (i) A, B, C are non empty.

(ii) $A \cup B \cup C = \{1, 3, 5, 9, 11\} = S$.

(iii) $A \cap B = \phi, B \cap C = \phi$ and $C \cap A = \phi$

Hence the set {1, 8, 9}, {3, 7} and {11} is a partition of the set S.

1.5 MULTISETS

Multisets are unordered collections of elements where an element can occur as a member more than once. The notation $\{m_1 \times a_1, m_2 \times a_2, ..., m_r \times a_r\}$ denotes the multiset with element a_1 occurring m_1 times, element a_2 occurring m_2 times, etc. The numbers m_i where $i = 1, 2,$

..., r, are called the **multiplicities** of the elements a_i, where $i = 1, 2, ..., r$. Alternatively, a multiset can simply be denoted as a set where repeated elements simply appear multiple times in the representation, such as {a, a, b, a}.

Let P and Q be multisets. The **union** of $P \cup Q$ is a multiset such that the multiplicity of an element in $P \cup Q$ is equal to the *maximum* of the multiplicities of the element in P and in Q. For example,

P = {a, a, a, c, d, d} Q = {a, a, b, c, c} $P \cup Q$ = {a, a, a, b, c, c, d, d}

Note that some definitions of the union of $P \cup Q$ of two multisets P and Q state that the multiplicity of each element in the union is the *sum* (rather than the maximum) of the multiplicities of the element in P and Q. Such a definition is also useful in some contexts, but for the purposes of this class, we'll simply define the union using the *maximum* of the multiplicities of the element in the original sets, and define a separate operation (defined later) for the sum. With that said, when dealing with multisets, make sure you know which definition of *union* you are dealing with!

Now, let set multiset P consist of {EE, EE, EE, ME, MA, MA, CS} — these represent the majors of the personnel needed to complete phase 1 of a project. Multiset Q consists of {EE, ME, ME, MA, CS, CS} – these are the majors of the personnel needed for phase 2 of the project. The multiset $P \cup Q$ gives us the exact number of personnel we need to hire for the project.

The intersection of $P \cap Q$ is a multiset where the multiplicity of an element in $P \cap Q$ is the *minimum* of the multiplicities of the element in P and in Q. For example,

P = {a, a, a, c, d, d} Q = {a, a, b, c, c} $P \cap Q$ = {a, a, c}

The intersection could be used in the example dealing with the project above to tell us the personnel involved in both phases of the project.

The **difference** of two multisets $P - Q$ is the multiset such that the multiplicity of an element in the difference $P - Q$ is equal to the multiplicity of the element in P minus the multiplicity of the element in Q if the difference is positive, and 0 otherwise. For example,

P = {a, a, a, b, b, c, d, d, e} Q = {a, a, b, b, b, c, c, d, d, f} $P - Q$ = {a, e}

This would give the personnel that have to be reassigned after the first phase of the project (they are not involved in the second phase).

We define a fourth operation on multisets that does not apply to regular sets. The **sum** of two multisets $P + Q$ is the multiset where the multiplicity of an element in $P + Q$ is equal to the *sum* of the multiplicities of the element in P and in Q. For example,

P = {a, a, b, c, c} Q = {a, b, b, d} $P + Q$ = {a, a, a, b, b, b. c, c, d}

As an example, think of P as the multiset of all account numbers of transactions made at a bank on a particular day. If Q is the multiset of all account numbers of transactions made the next day, then $P + Q$ is a combined record of account numbers of transactions for the two days. You would want the sum because one account number may have made several transactions in the two days.

Multisets can be useful for a certain class of problems (such as the engineering project example).

Example

P is a multiset that has as its elements the types of computer equipment needed by one department of a university where the multiplicities are the number of items required. Q is the same type of multiset but for a different department.

P = {107 × PCs, 44 × routers, 6 × servers}

Q = {14 × PCs, 6 × routers, 2 × Macs}

Problem 1: What combination of P and Q represents that the university must buy if there is no sharing of equipment?

Solution: Since no equipment is shared, each department needs to have its own version of the equipment. Thus, the combination of equipment needed is :

P + Q = {121 × PC's, 50 × routers, 6 × servers, 2 × Macs}

Problem 2: What combination of P and Q represents, what must be bought if the two departments share?

Solution: If the two departments share, then the number of each item is simply the maximum of the needs of both departments. Thus, the combination of equipment needed is:

P ∪ Q = {107 × PC's, 44 × routers, 6 × servers, 2 × Macs}

Problem 3: What combination of P and Q represents the actual equipment that both departments share?

Solution: The combination of equipment shared by the two departments is simply the overlap in the equipment needed by each department. Thus, the combination of equipment shared is:

P ∩ Q = {14 × PC's, 6 × routers}.

SOLVED EXAMPLES

Example 1: Find the smallest set X such as,
$$X \cup \{1, 2\} = \{1, 2, 3, 5, 9\}$$

Solution: Since
$$X \cup \{1, 2\} = \{1, 2, 3, 5, 9\}$$

So,

Smallest set X contains all the elements except {1, 2}

So,
$$X = \{3, 5, 9\}$$

Example 2: Let A = {1, 2}, B = {2, 4, 6, 8} and C = {3, 4, 5, 6}

Find. A ∪ B, B ∩ C, A − B, B − A, A ∩ (B ∪ C), A ∪ (B ∩ C), and (A ∪ B ∪ C)

Solution: Here given,
$$A = \{1, 2, 3\}, B = \{2, 4, 6, 8\} \text{ and } C = \{3, 4, 5, 6\}$$
$$A \cup B = \{1, 2, 3, 4, 6, 8\}$$
$$B \cap C = \{4, 6\}$$
$$A - B = \{1, 3\}$$

$$B - A = \{4, 6, 8\}$$
$$A \cap (B \cup C) = \{1, 2, 3\} \cap \{2, 4, 6, 8\} \cup \{3, 4, 5, 6\}$$
$$= \{1, 2, 3\} \cap \{2, 3, 4, 5, 6, 8\}$$
$$= \{2, 3\}$$
$$A \cup (B \cap C) = \{1, 2, 3\} \cup \{2, 4, 6, 8\} \cap \{3, 4, 5, 6\}$$
$$= \{1, 2, 3\} \cup \{4, 6\}$$
$$= \{1, 2, 3, 4, 6\}$$
$$A \cup B \cup C = \{1, 2, 3\} \cup \{2, 4, 6, 8\} \cup \{3, 4, 5, 6\}$$
$$= \{1, 2, 3, 4, 5, 6, 8\}$$

Example 3: If $U = \{1, 2, 3, 4, 5, 6, 7, 8, 9\}$
$$A = \{1, 2, 3, 4\}, B = \{2, 4, 6, 8\} \text{ and } C = \{3, 4, 5, 6\}$$
Find A', $(A \cup B)'$, $(A \cap C)'$ and $(B - C)'$

Solution : Here given
$$U = \{1, 2, 3, 4, 5, 6, 7, 8, 9\}, A = \{1, 2, 3, 4\}, B = \{2, 4, 6, 8\} \text{ and }$$
$$C = \{3, 4, 5, 6\}$$

now,
$$A' = U - A$$
$$A' = \{5, 6, 7, 8, 9\}$$
$$A \cup B = \{1, 2, 3, 4, 6, 8\}$$
$$(A \cup B)' = U - (A \cup B)$$
$$= \{5, 7, 9\}$$
$$(A \cap C) = \{3, 4\}$$
$$(A \cap C)' = U - (A \cap C)$$
$$= \{1, 2, 5, 6, 7, 8, 9\}$$
$$B - C = \{2, 8\}$$
$$(B - C)' = U - (B - C)$$
$$= \{1, 3, 4, 5, 6, 7, 9\}$$

Example 4: Let $U = R$ the set of all real numbers. If $A = \{x : x \in R, 0 < x < 2\}$, $B = \{x : x \in R, 1 < x \leq 3\}$ then,

Find A', B', $A \cup B$, $A \cap B$, $A - B$ and $B - A$.

Solution: $A' = R - A = \{x : x \in R \text{ and } x \notin A\}$
$$= \{x : (x \in R \text{ and } x > 2) \text{ or } (x \in R \text{ and } x \leq 0)\}$$
$$= \{x : x \in R \text{ and } x > R\} \cup \{x : x \in R \text{ and } x \leq 0\}$$
$$B' = \{x : x \in R \text{ and } x < 1\} \cup \{x : x \in R \text{ and } x > 3\}$$

Similarly:
$$A \cup B = \{x : x \in R \text{ and } 0 < x \leq 3\}$$
$$A \cap B = \{x : x \in R \text{ and } 1 < x < 2\}$$
$$A - B = \{x : x \in R \text{ and } 0 \leq 1\}$$

Example 5: If $A = \{2, 3, 4, 8, 10\}$, $B = \{1, 3, 4, 10, 12\}$ and $C = \{4, 5, 6, 12, 14, 16\}$ Find $(A \cup B) \cap (A \cup C)$ and $(A \cap B) \cup (A \cap C)$

Solution: Here given $A = \{2, 3, 4, 8, 10\}$, $B = \{1, 3, 4, 10, 12\}$ and $C = \{4, 5, 6, 12, 14, 16\}$

then,
$$A \cup B = \{1, 2, 3, 4, 8, 10, 12\}$$
$$A \cup C = \{2, 3, 4, 5, 6, 8, 10, 12, 14, 16\}$$
$$(A \cup B) \cap (A \cup C) = \{2, 3, 4, 8, 10, 12\}$$
$$A \cap B = \{3, 4, 10\}$$
$$A \cap C = \{4\}$$
$$(A \cap B) \cup (A \cap C) = \{3, 4, 10\}$$

Example 6: Solve $3x^2 - 12x = 0$ where,
(i) $x \in N$, (ii) $x \in I$, (iii) $D \in S = \{a + ib; b \# 0, a, b \in R\}$

Solution:
we have $3x^2 - 12x = 0$
$$3x(x - 4) = 0$$
$$3x(x - 4) = 0$$
$$x = 0, 4$$

(i) If $x \in N$ then,
$$x = \{4\}$$

(ii) If $x = I$ then,
$$x = \{0, 4\}$$

(iii) If $x \in S$, then since there is no root of the form $a + ib$ where a and b are constant and $b \# 0$ the solution set in this case is ϕ.

EXERCISE 1.1

1. Find the smallest set A such as,
$$A \cup \{1, 2, 3\} = \{1, 2, 3, 4, 5, 6, 7, 8\}$$
2. Let $A = \{a, b, c, d\}$, $B = \{b, c, d, e, f\}$, $C = \{e, f, g, h, i\}$
 Find $A \cup B$, $B \cap C$, $A - B$, $B - A$, $A \cap (B \cup C)$, $A \cup (B \cap C)$ and $A \cup B \cup C$.
3. If $x = \{2, 4, 6, 8, 10\}$ $y = \{6, 8, 10, 12, 14\}$ $z = \{10, 12, 14, 15, 16\}$ and $U = \{1, 2, 3, 4, 6, 8, 10, 12, 14, 16\}$ then find
 x', $(x \cup y)'$, $(x \cap y)'$ and $(x - y)'$
4. If $x = \{a, b, c, d, e, f\}$, $y = \{e, f, g, h\}$ and $z = \{g, h, i, j\}$
 Find $(x \cup y) \cap (x \cup z)$ and $(x \cap y) \cup (x \cap z)$.
5. If $x = \{1, 2, 3\}$ then find all subsets of x.

6. If $x = \{a, b, c, d, e, f\}$ and $y = \{d, e, t, g, h\}$ find.
   Symmetric difference of sets x and y.
7. If $x = \{a, b, c\}$ then find all the proper subsets of x.
8. Define finite and infinite sets.
9. Let $A = \{a, b, c\}$, $B = \{b, c\}$ and $C = \{a, d, e\}$. Determine which of the following statements are true give reasons.

   (i) $\{b\} \in A$ (ii) $\{c\} \in A$ (iii) $\phi \subseteq C$
   (iv) $\phi \in B$ (v) $\{c\} \in B$ (vi) $B \subseteq A$
   (vii) $B \cap A \subseteq C$ (viii) $\{d\} \in C$ (viii) $\{d\} \in C$
   (ix) $A \cap C \subseteq B$ (x) $B \subset C$

Example 7: Let $A = \{\theta : 2\cos^2\theta + \sin\theta < 2\}$ and $B = \{Q : \pi/2 < \theta < 3\pi/2\}$ then find $A \cap B$.

Solution: As given sets, B consists of all value of θ in the interval $\pi/2 \leq \theta \leq 3\pi/2$ and set A consists of all value of θ which satisfy the inequality

$$2\cos^2\theta + \sin\theta \leq 2 \qquad \ldots(i)$$

Hence $A \cap B$ will consist of all those values of θ in the interval $\pi/2 < \theta < 3\pi/2$ which satisfy the inequality (1),

The inequality (1) is equivalent to the inequality

$$2 - 2\sin^2\theta + \sin\theta \leq 2$$

i.e., $\qquad \sin\theta(1 - 2\sin\theta) \leq 0 \ldots > (2)$

The inequality (2) is satisfied by all those values of θ which satisfy $\sin\theta < 0$ or $\sin\theta > 1/2$

Now the values of θ which lie in the interval $\frac{\pi}{2} \leq \theta \leq 3\pi/2$.

and satisfy $\sin\theta \leq 0$ are given by $\frac{\pi}{2} \leq \pi \leq 3\pi/2$. And then values of θ which lie in the interval $\frac{\pi}{2} \leq \theta \leq 3\pi/2$ and satisfy $\sin\theta \geq \frac{1}{2}$ are given by $\frac{\pi}{2} \leq \theta \leq 5\pi/6$.

Thus the solution set of inequality (1) consists of all values of θ in the intervals $\pi/2 \leq \theta \leq 5\pi/6$ and $\pi/2 \leq \theta \leq 3\pi/2$,

Hence, $A \cap B = \{\theta : \pi/2 < \theta < 5\pi/6 \text{ or } \pi < \theta < 3\pi/2\}$

Example 8: If $X = \{4^n - 3n - 1 : n \in N\}$ and $y = \{9(n-1) : n \in N\}$ prove that $x \subset y$.

Solution: Let $x_n = 4^n - 3n - 1$ $(n \in N)$, then,

$$x_1 = 4 - 3 - 1 = 0$$

and for $n > 2$, we have

$$x_n = (1+3)^n - 3n - 1$$

$$= 1 + {}^nc_1 3 + {}^nc_2 3^2 + \ldots + 3^n - 3n - 1$$
$$= 1 + 3n + {}^nc_2 3^2 + \ldots + 3^n - 3n - 1$$
$$= {}^nc^2 3^2 + \ldots + 3n = 9 \, [{}^nc_2 + \ldots + 3^{n-2}]$$

Thus xn is some five integral of 9 for $n > 2$.

Hence x consists of all positive integral multiple of 9 of the form $9a_n$ where $a_n = nc_2 + nc_3.3 + \ldots + 3^{n-2}$ together with 0.

Also y consists of elements of the form $9(n-1)$ $(n \in N)$ that is, y consists of all positive integral multiple of 9 together with 0.

It follows that $X \subset Y$.

So,
$$x = \{0, 9, 54, 243 \ldots\}$$
and
$$y = \{0, 9, 18, 27, 36, 45, 54 \ldots\}$$

Hence X is a proper subset of Y.

or $X \subset Y$

Example 9: If $9N = \{ax : x \in N\}$. Describe the set $3N \cap 7N$.

Solution: According to the given nation,
$$N = \{3x : x \in N\} = \{3, 6, 9, 12, \ldots\}$$
and
$$7N = \{7x : x \in N\} = \{7, 14, 21, 28, 35, 42, \ldots\}$$
Hence
$$3N \cap 7N = \{21, 42, 63 \ldots\}$$
$$= \{21x : x \in N\} = 21N$$

1.6 LAWS OF OPERATIONS (General Identities on Sets)

If A, B, C are any subset of a universal set $\cup$ then,

(1) Idempotent Laws:
  (i) $A \cup A = A$
  (ii) $A \cap A = A$

(2) Identity Laws:
  (i) $A \cup \phi = A$
  (ii) $A \cap \phi = A$

(3) Commutative Laws:
  (i) $A \cup B = B \cup A$
  (ii) $A \cap B = B \cap A$

(4) Associative Laws:
  (i) $(A \cap B) \cup C = A \cap (B \cup C)$
  (ii) $(A \cap B) \cap C = A \cup (B \cap C)$

(5) Distributive Laws:
   (i) $A \cup (B \cap C) = (A \cup B) \cap (A \cup C)$
   (ii) $A \cap (B \cup C) = (A \cap B) \cup (A \cap C)$

(6) De-Morgan's Law:
   (i) $(A \cup B)' = A' \cap B'$
   (ii) $(A \cap B)' = A' \cup B'$

   i.e.,(i) Complement of union of two sets is the intersection of their complements.
   (ii) Complement of intersection of two sets is the union of their complements.

More generally, If $A_1, A_2 \ldots A_n$ are any sets,

$(A_1 \cup A_2 \cup \ldots \cup A_n)' = A_1' \cap A_2' \cap A_3' \ldots \cap A_n'$

$(A_1 \cap A_2 \ldots \cap A_n)' = A_1' \cup A_2' \ldots \cap A_n'$

(7) Complement Law:
   (i) $A \cup A' = S$
   (ii) $A \cap A' = \phi$
   (iii) $(A')' = A$

(8) Law of inclusion:
   (i) $A \subset (A \cup B)$ and $B \subset (A \cup B)$
   (ii) $(A \cap B) \subset A$ and $(A \cap B) \subset B$

(9) Law of difference:
   (i) $A - \phi = A$
   (ii) $A - A = \phi$

Proof of operation on sets:

1. Properties of subset,

   (i) To prove $A \cup A = A$

Let, $x \in A \cup A \Rightarrow x \in A$ or $x \in A$
$\Rightarrow \quad x \in A$
i.e., $\quad A \cup A \subseteq A$ ...(1)

Conversely,

Let, $x \in A \Rightarrow x \in A$ or $x \in A$
$\Rightarrow \quad x \in A \cup A$
i.e., $\quad A \subseteq A \cup A$ ...(2)

By equations (1) and (2) we get,

$A \cup A = A$

(ii) To prove $A \cap A = A$

Here, $\quad x \in A \cap A \Rightarrow x \in A$ and $x \in A$

$\Rightarrow \quad x \in A$

i.e., $\quad A \cap A \subseteq A$ ...(3)

Conversely,
$$x \in A \Rightarrow x \in A \text{ and } x \in A$$
$$\Rightarrow x \in A \cap A$$
$$A \subseteq A \cap A \qquad ...(4)$$

By equations (3) and (4) we get,
$$A \cap A = A$$

To prove $A \cup B = A$, Iff $B \subseteq A$

$\therefore \quad x \in B \Rightarrow x \in A$ ...(5)

Now $x \in A \cup B \Rightarrow x \in A$ or $x \in B$
$\Rightarrow x \in A$ by equation (5)

i.e., $\quad A \cup B \subseteq A$ ...(6)

Conversely,
$$x \in A \Rightarrow x \in A \text{ or } x \in B$$
$$\Rightarrow x \in A \cup B$$
$$A \subseteq A \cup B \qquad ...(7)$$

So by equations (6) and (7) we get,
$$A \cup B = A$$

Now let $A \cup B = A$

i.e., $\quad x \in A$ or $x \in B \Rightarrow x \in A$
$\quad x \in B \Rightarrow x \in A$

Hence, $\quad B \subseteq A$

i.e., $\quad A \cup B = A \quad \Rightarrow \quad B \subseteq A$

or $\quad A \cup B = A$, Iff $B \subseteq A$

(iv) To prove $A \cap B = A$, Iff $A \subseteq B$

Let, $\quad A \subseteq B$

$\therefore \quad x \in A \Rightarrow x \in B$ ...(8)

Now $x \in A \cap B \Rightarrow x \in A$ and $x \in B$,
$\Rightarrow x \in A$

i.e., $\quad A \cap B \subseteq A$ ...(9)

and $x \in A \Rightarrow x \in A$ and $x \in B$ by equation (8)
$\Rightarrow x \in A \cap B$
$$A \subseteq A \cap B \qquad ...(10)$$

By equations (9) and (10) we get,
$$A \cap B = A \qquad ...(11)$$

Now let $A \cap B = A$

i.e., $x \in A \to x \in A$ and $x \in B$
∴ $x \in A \Rightarrow x \in B$
i.e., $A \subseteq B$ ...(12)

By using equations (11) and (12) we get,
$$A \cap B = A \text{ iff } A \subseteq B$$

Properties of Intersection operation on sets:

(i) Intersection operation is commutative
i.e., $A \cap B = B \cap A$

Proof: Let $x \in A \cap B$
by the definition of intersection:

$x \in A \cap B$
$\Rightarrow x \in A$ and $x \in B$
$\Rightarrow x \in B$ and $x \in A$
$\Rightarrow x \in B \cap A$

i.e., $A \cap B \subseteq B \cap A$ ...(1)

Conversely

$y \in B \cap A$

By the definition of intersection

$x \in B \cap A$
$\Rightarrow x \in B$ and $x \in A$
$\Rightarrow x \in A$ and $x \in B$
$\Rightarrow x \in A \in B$

i.e., $B \cap A \subseteq A \cap B$ ...(2)

By using equation (1) and (2) we get,
$$A \cup B = B \cap A$$

(ii) Associative law holds in case of intersection of three equal sets.
$$A, B \text{ and } C : (A \cap B) \cap C = A \cap (B \cap C)$$

Proof: Let $x \in (A \cap B) \cap C$,
So,
$x \in (A \cap B) \cap C$
$\Rightarrow x \in (A \cap B)$ and $x \in C$
$\Rightarrow x \in A$ and $x \in B$ and $x \in C$
$\Rightarrow x \in A$ and $x \in (B \cap C)$
$\Rightarrow x \in A \cap (B \cap C)$.

i.e., $(A \cap B) \cap C \subseteq A \cap (B \cap C)$ ...(1)

Conversely let,

$b \in A \cap (B \cap C)$
$\Rightarrow b \in A$ and $b \in B \cap C$

$\Rightarrow \quad b \in A$ and $b \in B$ and $b \in C$
$\Rightarrow \quad (b \in A$ and $b \in B)$ and $b \in C$
$\Rightarrow \quad b \in (A \cap B)$ and $b \in C$
$\Rightarrow \quad b \in (A \cap B) \cap C$

i.e., $\quad (A \cap B) \cap C \subseteq (A \cap B) \cap C$ ...(2)

By using equations (1) and (2) we get,
$$(A \cap B) \cap C = A \cap (B \cap C)$$

Properties of Union operation:

(i) Commutative Law: To prove that
$$A \cup B = B \cup A$$

Proof: Let $\quad x \in A \cup B$
$\therefore \quad x \in A \cup B$
$\Rightarrow \quad x \in A$ or $x \in B$
$\Rightarrow \quad x \in B$ or $x \in A$
$\Rightarrow \quad x \in B \cup A$
$\therefore \quad A \cup B \subseteq B \cup A$ ...(1)

Conversely let, $\quad y \in B \cup A$
But $\quad y \in B \cup A$
$\Rightarrow \quad y \in B$ or $y \in A$
$\Rightarrow \quad y \in A$ or $y \in B$
$\Rightarrow \quad y \in A \cup B$
$\therefore \quad B \cup A \subseteq A \cup B$ ...(2)

By using equation (1) and (2) we have equations:
$$A \cup B = B \cap A$$

(ii) Associative Law: To prove that,
$$A \cup (B \cup C) = (A \cup B) C$$

Proof: Let, $\quad x \in A \cup (B \cup C)$
Hence, $\quad x \in A \cup B \cup C$
$\Rightarrow \quad x \in A$ or $x \in B$ or x
$\Rightarrow \quad (x \in A$ or $x \in B)$ or $x \in C$
$\Rightarrow \quad x \in (A \cup B)$ or $x \in C$
$\Rightarrow \quad x \in (A \cup B) \cup C$
$\therefore \quad A \cup (B \cup C) \subseteq (A \cup B) \cup C$ ...(1)

Conversely, let $x \in (A \cup B) \cup C$,
Here,
$\quad y \in (A \cup B) \cup C$
$\Rightarrow \quad (x \in A$ or $x \in B)$ or $x \in C$

$\Rightarrow$ $\quad x \in$ A or $x \in$ B or $x \in$ C
$\Rightarrow$ $\quad x \in$ A or $(x \in$ B or $x \in$ C)
$\Rightarrow$ $\quad x \in$ A or $x \in (B \cup C)$
$\Rightarrow$ $\quad x \in$ A $\cup (B \cup C)$
$\Rightarrow$ $\quad (A \cup B) \cup C \subseteq A \cup (B \cup C)$ ...(2)

By using equations (1) and (2) we get,
$$A \cup (B \cup C) = (A \cup B) \cup C$$

(iii) **Distributive laws:** For three given sets A, B and C to prove that,
(a) $A \cap (B \cup C) = (A \cap B) \cup (A \cap C)$
(b) $A \cup (B \cap C) = (A \cup B) \cap (A \cup C)$

Proof : Let $a \in A \cap (B \cup C)$,

Here $\quad a \in A \cap (B \cup C)$
$\Rightarrow \quad a \in$ A and $a \in B \cup C$
$\Rightarrow \quad a \in$ A and $(a \in$ B or $a \in$ C)
$\Rightarrow \quad (a \in$ A and $x \in$ B) or $(a \in$ A) or $a \in$ C)
$\Rightarrow \quad a \in A \cap B$ or $a \in A \cap C$
$\Rightarrow \quad a \in (A \cap B) \cup (A \cap C)$

Hence,
$$A \cup (B \cup C) \subset (A \cap B) \cup (A \cap C) \quad \text{...(1)}$$

Conversely let $\quad b \in (A \cap B) \cup (A \cap C)$

Here, $\quad b \in (A \cap B) \cup (A \cap C) \Rightarrow b \in A \cap B$ or $b \in A \cap C$
$\Rightarrow \quad (b \in$ A and $b \in$ B) or $(b \in$ A and $b \in$ C)
$\Rightarrow \quad (b \in$ A and $b \in$ B) or $(b \in$ C)
$\Rightarrow \quad (b \in$ A and $b \in B \cup C)$
$\Rightarrow \quad (b \in A \cap (B \cup C)$
$\therefore \quad (A \cap B) \cup (A \cap C) \subseteq A \cap (B \cup C)$ ...(2)

By using equations (1) and (2) we get,
Hence, $\quad A \cap (B \cup C) = (A \cap B) \cup (A \cap C)$

Number of elements in the union of two or more sets:

Let A, B and C be three finite sets and let $n(A)$, $n(B)$, $n(C)$ respectively denote the number of elements in these sets. Then we see that,
$$n(A \cup B) = n(A) + n(B) - n(A \cap B) \quad \text{...(1)}$$

In case A and B are disjoint sets then, $A \cap B = \phi$ and $n(A \cap B) = n(\phi) = 0$

i.e., for disjoint sets A and B
$$n(A \cup B) = n(A) + n(B)$$

Similarly we can show that,
$$n(A \cup B \cup C) = n(A) + n(B) + n(C) - n(A \cap B) - n(B \cap C)$$
$$- n(A \cap C) + n(A \cap B \cap C) \quad \text{...(2)}$$

Example 10: In a group of 1000 people, there are 750 who can speak Hindi and 400 who can speak Bengali. How many people can speak Hindi only? How many people can speak Bengali only? How many people can speak both Hindi and Bengali?

Solution: Let H and B denote the sets of those people. Who can speak Hindi and Bengali respectively.

Given:

$$n(H \cup B) = 1000, n(H) = 750, n(B) = 400$$

Now

$$n(H \cap B) = n(H) + n(B) - n(H \cup B)$$
$$= 750 + 400 - 1000$$
$$\Rightarrow n(H \cap B) = 150$$

Hence 150 people can speak both Hindi and Bengali. Now we have to find those people who can speak Hindi only, that is, $n(H \cap B')$

But $\quad n(H \cap B') = n(H) - n(H \cap B) = 750 - 150 = 600$

So, 600 people can speak Hindi only.

Similarly:

$$n(B \cap H') = n(B) - n(B \cap H) = 400 - 150 = 250$$

So, 250 people can speak Bengali only.

Example 11: In a survey of 200 students of a higher secondary school, it was found that 120 studied mathematics; 90 studied physics; and 70 studied chemistry; 40 studied mathematics and physics; 30 studied physics and chemistry; 50 studied chemistry and mathematics, and 20 studied none of these subjects. Find the number of students who studied all the three subjects.

Solution: Given,

$$n(U) = 200, n(M) = 120, n(P) = 90, n(C) = 70$$
$$n(M \cap P) = 40, n(P \cap C) = 30, n(C \cap M) = 50, n(M \cup P \cup C)' = 20$$

Now

$$n(M \cup P \cup C)' = n(U) - n(M \cup P \cup C)$$
$$\Rightarrow 20 = 200 - n(M \cup P \cup C)$$
$$\Rightarrow n(M \cup P \cup C) = 200 - 20 = 180$$

and;

$$= n(M) + n(P) + n(C) - n(M \cap P) - n(P \cap C)$$
$$n(M \cup P \cup C) - n(C \cap M) + n(M \cap P \cap C)$$
$$180 = 120 + 90 + 70\ 40 - 30 - 50 + n(M \cap P \cap C)$$
$$\Rightarrow n(M \cap P \cap C) = 180 + 120 - 280 = 20$$

Hence 20 students studied all the three subjects.

Example 12: In a survey of population of 450 people, It is found that 205 can speak English, 210 people can speak Hindi and 120 people can speak Tamil. If 100 people can speak both English and Hindi, 80 can speak both English and Tamil, 35 people can speak Hindi and

Tamil, and 20 people all the three languages. Find the numbers of people who can speak. English but not Hindi or Tamil. Find also the number of people who can speak neither English nor Hindi nor Tamil.

Solution: Let F, H and T denote the sets of those people who can speak English, Hindi and Tamil respectively.

Given,
$$n(U) = 450, n(E) = 205, n(H) = 210, \text{ and } n(T) = 120$$
$$n(H \cap T) \ 35, n(H \cap E) = 100, n(E \cap T) = 80,$$
$$n(H \cap T \cap E) = 20$$

Now, we have to find the human of people who can English but not Hindi or Tamil.
$$n(H \cap E' \cap T') = n(H) - n(H \cap E) - H(H \cap T) + n(H \cap E \cap T)$$
$$= 210 - 100 - 35 + 20$$
$$= 95$$

Now,
$$n(H \cup E \cup T) = n(H) + n(E) + n(T) - n(H \cap E) - n(E \cap T)$$
$$- n(H \cap T) + n(H \cap E \cap T)$$
$$= 250 + 210 + 120 - 35 - 100 - 80$$
$$= 320$$

Now we have to find the number of people who can speak neither English nor Hindi nor Tamil.
$$n(H \cup E \cup T)' = n(U) - (H \cup E \cup T)$$
$$= 450 - 320$$
$$= 130$$

So, 130 people cannot speak Hindi, English or Tamil.

Example 13: At a certain conference of 100 people there are 29 Indian women and 23 Indian men of these Indian people 4 are doctors and 24 are either men or doctors. There are no foreign doctors. How many foreigners are attending the conference? How many women doctors are attending the conference?

Solution: Let E denote the set of all Indian people attending the conference. Then E' will denote the set of all foreigners attending the conference. Also let W, M and D denote the set of Indian women, Indian men and Indian doctors respectively.

Given:
$$n(E \cup E') = 100, n(W) = 29, n(M) = 23, n(B) = 4, (M \cup D) = 243$$

Clearly
$$W \cap H = \phi \text{ and } W \cup M = E$$

So,
$$n(E) = n(W) + n(M) = 29 + 23 = 52$$

Also $n(E \cup E') \ 100$, or $n(E) + n(E') = 100$

∴
$$52 + n(E') = 100$$

or
$$n(E') = 100 - 52 = 48$$

So, 48 foreigners are attending the conference since there are no foreign doctors there can be no foreign women doctors. Hence in order to find out women doctors, we simply find Indian women doctors,

i.e., $n(W \cap D)$

Now we have,
$$n(M \cup D) = n(M) + n(D) - n(M \cap D)$$
or $\qquad 24 = 23 + 4 - n(M \cap D)$

i.e., $\qquad n(M \cap D) = 3$

Thus the number of male Indian doctors is 3 since there are all 4 Indian doctors.

The number of Indian women doctors = 4 − 3 = 1

Hence
$$n(W \cap D) = 1$$

Example 14: An analysis of 100 personnel injury claims made upon a motor insurance company revealed that loss or injury in respect of an eye, an arm or a leg occurred in 30, 50 and 70 cases respectively, claims involved this loss or injury to two of these member numbered 44. How many claims involved loss or injury to all the three, we must assume that one or another of the three members was mentioned in each of the 100 claims.

Solution: Let E, A and L denote the sets of people having injuries in eyes arms or legs respectively.

Given,
$$n(E \cup A \cup L) = 100, n(E) = 30, n(A) = 50, n(L) = 70$$

Also,
$$n(E \cap A \cap L') \cup (A \cap L \cap E') \cup (L \cap E \cap A')] = 44 \qquad \ldots(1)$$

Now,
$$n(E \cup A \cup L) = n(E) + n(A) + n(L) - n(E \cap A) - n(A \cap L)$$
$$- n(L \cap E) + n(E \cap A \cap L)$$
$$100 = 30 + 50 + 70 - n(E \cap A) - n(A \cap L) - n(L \cap E)$$
$$+ n(E \cap A \cap L)$$

$\Rightarrow n(E \cap A) + n(A \cap L) + n(L \cap E) - n(E \cap A \cap L)$
$= \qquad 30 + 50 + 70 - 100 = 50 \qquad \ldots(2)$

Since: $(E \cap A \cap L'), (A \cap L \cap E'), (L \cap E \cap A')$ are disjoints

So by equation (1), we get

$n(E \cap A \cap L') + n(A \cap L \cap E) + n(L \cap E \cap A') = 44$

or $(E \cap A) - n(E \cap A \cap L) + n(A \cap L) - n(A \cap L \cap E) + n(L \cap E) - n(L \cap E \cap A) = 44$

or $n(E \cap A) + n(A \cap L) + n(L \cap E) - 3n(E \cap A \cap L) = 44 \qquad \ldots(3)$

$\Rightarrow$ (2) − (3) We get,
$$2n(E \cap A \cap L) = 6$$
or $\qquad n(E \cap A \cap L) = 3$

1.7 VENN DIAGRAMS

We can also represent sets in diagrammatic form. Diagrammatic representation cannot replace a proof but may be useful in convincing as either that a particular statement is true and hence that a proof is possible or that it is false, in which case it may indicate how to construct an example to prove that it is false. The diagram we use to represent a set(s) are called Venn Diagrams.

In these diagrams, first we draw rectangle to represent universal set S. Second draw circles or (any other appropriate closed curve) within the rectangle to represent the sets. These must intersect in the most general way required by the problem and should be suitably labelled as shown in the following figure points. Which lie within the various regions of the diagram can now be considered to represent the elements of the respective sets. If the number of elements in the sets is small then the individual elements may be written within the appropriate regions.

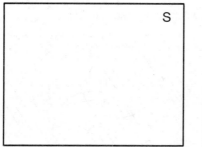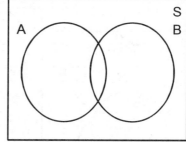

Figure 1.1

Let A = {1, 2, 3, 4}
B = {4, 5, 6, 7, 8, 9, 10}

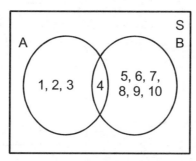

Figure 1.2

Example 15: Represent A ∪ B and A ∩ B by using venn diagram where A and B are any two subsets of S.

Solution: A ∪ B

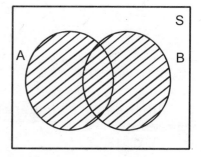

Figure 1.3

In the above figure shaded region represents A ∪ B.
A ∩ B

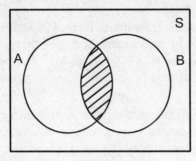

Figure 1.4

In the above figure shaded region represents A ∩ B.

Example 16: Represent the set by venn diagram.

A ∪ (B ′ ∩ C)

Sol.: Let A, B and C be any three subsets of a universal set S.

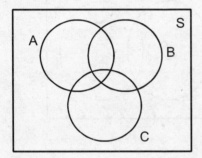 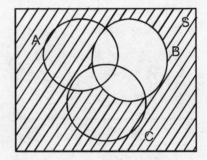

Figure 1.5

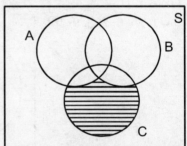

In the above figure shaded region represents B′.

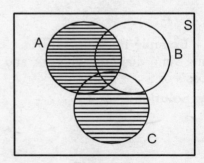

In the above figure shaded region represents B ′ ∩ C

Figure 1.6

Now we add whole region of set A in figure then we get (*iv*)
So clearly figure (*iv*) represents
$$A \cup (B' \cap C)$$

Example 17: If A and B are any two disjoint.
i.e., $A \cap B = \phi$ then represent it.
Solution:

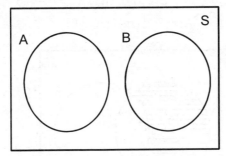

Figure 1.7

Above figure represents $A \cap B = \phi$

Example 18: If $A \subset B$ then, represent it by venn diagram.
Solution:

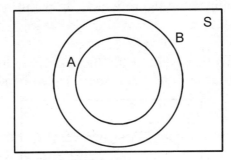

Figure 1.8

The above figure represents $A \subset B$.

Example 19: Let A = {1, 2, 4, 5, 6, 7, 8} and B = { 7, 8, 9, 10, 11, 12} then find $A \cup B$ and $A \cap B$ by using venn diagram.
Solution:

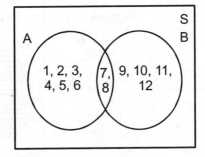

Figure 1.9

As it is clean from the above figure
A ∩ B = { 7, 8}
A ∪ B = {1, 2, 3, 4, 5, 6, 7, 8, 9, 10, 11, 12}

Example 20: Let A, B be any two subsets of a universal set S. Then represent the following by using venn diagram

(i) A – B
(ii) B – A
(iii) (A, B), i.e., (A – B) ∪ (B – A)
(iv) (A – B)′

Solution: (i) A – B

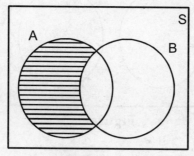

Figure 1.10

(A – B) means that paint of the figure which are A but not in B. As represented by shaded figure in the above figure.

(ii) A – B

Similarly (B – A) means that paint of the fig. which are in set B but not in set A as represented by are in set B but not in set A as represented by shaded region in the below given figure

B – A

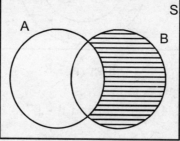

Figure 1.11

(iii) (A – B) ∪ (B – A) = Δ (A, B)

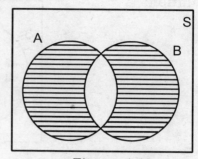

Figure 1.12

(iv) (A – B)′

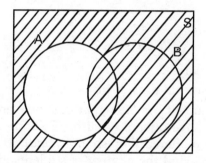

Figure 1.13

Example 21: Let us take set of computers with the following specifications. Find how many of them have all three handworks?

Computer	CD Written	Sound Car	Speeker
1.	Yes	Yes	No
2.	Yes	No	Yes
3.	Yes	Yes	Yes
4.	No	No	No
5.	No	Yes	Yes
6.	No	No	Yes
7.	Yes	No	No
8.	No	No	Yes

Solution: Let A, B and C be the sets of computers with a CD written sound card and speeker represently now we have.

$|A| = 4$ $|B| = 3$ $|C| = 5$
$|A \cap B| = 2$ $|A \cap C| = 2$ $|B \cap C| = 2$
$|A \cap B \cap C| = 1$

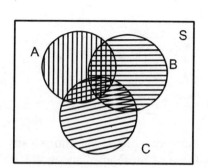

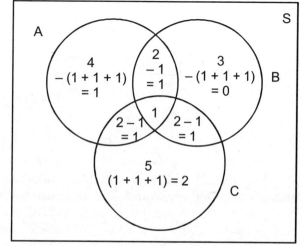

Figure 1.14

i.e., $|A \cup B \cup C| = ?$
So by figure
$$A \cup B \cup C = 1+1+1+0+2 = 5$$
Since $(A \cup B \cup C) = 5$
So, five computers out of 8 have all three handworms.

Example 22: In a class each student has offered at least one language paper of English, Hindi and Sanskrit. If 23 of them have offered English; 15 Hindi and 21 Sanskrit; 10 have offered Hindi and Sanskrit 13 have offered English and Hindi; 21 English and Sanskrit; 8 have offered all the three language. Find the number of students in the class.

Solution: Let language papers English, Hindi and Sanskrit can be represented by the Set E, H and K respectively

$|E| = 23$ $H = 15$ $|K| = 21$
$|H \cap K| = 10$ $|E \cap H| = 13$ $|E \cap K| = 21$
$|H \cap K \cap E| = 8$
$|H \cap K \cap E| = ?$

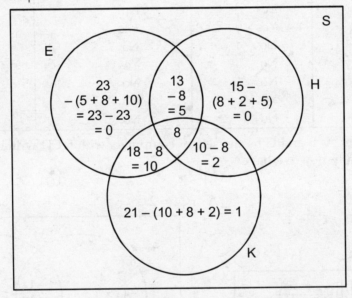

Figure 1.15

So
$$|H \cup S \cup E| = 0+5+8+10+0+2+1 = 26$$
So, Total No. of students in the class = 26

Example 23: In a survey of 1000 households, washing machines, vacuum cleanerness and refrigerators were counted. Each house had at least one of these appliances, 400 had no refrigerator, 380 no vacuum cleaner and 542 no washing machine 294 had both a vacuum cleaner and a washing machine, 277 both a refrigerator and a vacuum cleaner, 190 both a refrigerator and a washing machine. How many households and all the three applications? How many households had only a vacuum cleaner.

Solution: Let P, Q and R denote the sets of house holds housing a washing machine, a vacuum cleaner and a refrigerator respectively.

$|P \cup Q \cup R| = 1000$; $|R'| = 400$;
$|Q'| = 380$, $|P'| = 542$. Also $|P \cup Q R| = |U|$

As we know that,

$$|P| + |P'| = |U|$$

So,
$$|P| = |U| - |P'| = 1000 - 542 = 458$$

Similarly,
$$|Q| = |U| - |Q'| = 1000 - 380 = 620$$
$$|R| = |U| - |R'| = 1000 - 400 = 600$$

also given,
$$|P \cap Q| = 294; |Q \cap R| = 277; |P \cup R| = 190$$

Let,
$$|P \cap Q \cap R| = x$$

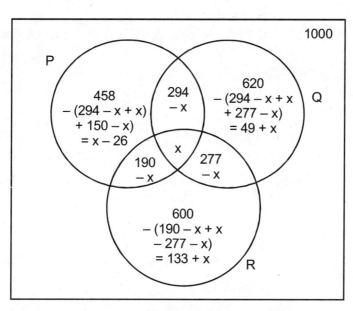

Figure 1.16

So, $|P \cup Q \cup R| = x - 26 + 294 - x + 190 - x + 49 + x + 277 - x + 133 + x = x + 917$
So,
$$x + 917 = 1000$$
$$\Rightarrow \qquad x = 1000 - 917 = 83$$

So, 83 households had all the three applications.
Households having only vacuum cleaner.

$$= 49 + x$$
$$= 49 + 83 = 132$$

Theorem: Prove that $A - B = A \cap B' = B' - A'$

Proof: Let $x \in (A - B)$

$\therefore \qquad x \in (A - B) \Rightarrow x \in A, x \notin B$

$\Rightarrow \qquad x \in A, x \in B'$

$\Rightarrow \qquad x \in A \cap B'$

i.e., $\qquad A - B \subset A \cap B'$ ...(1)

Conversely,

$$x \in A \cap B'$$

$\Rightarrow \qquad x \in A, x \in B'$

$\Rightarrow \qquad x \in A; x \notin B$

$\Rightarrow \qquad x \in (A - B)$

i.e., $\qquad A \cap B' \subseteq (A - B)$ ...(2)

By using equations (1) and (2) we get,

$$A - B = A \cap B' \qquad ...(3)$$

Now $B' - A' = B' \cap (A')' = B' \cap A$ Because $A - B = A \cap B'$ and $(A')' = A$

i.e., $\qquad B' - A' \ A \cap B'$ ...(4)

By equation, (3) and (4) we get,

$$A - B = A \cap B' = B' - A'$$

Example 24: De Morgan's Laws: To prove that

(a) $\qquad (A \cup B)' = A' \cap B'$

(b) $\qquad (A \cap B)' = A' \cup B'$

(c) $\qquad A - (B \cup C) = (A - B) \cap (A - C)$

(d) $\qquad A - (B \cap C) = (A - B) \cup (A - C)$

Solution: (a) Let $x \in (A \cup B)'$

But $x \in (A \cup B)' \Rightarrow x \in A \cup B$

$\Rightarrow x \in A$ and $x \notin B$

$\Rightarrow x \in A'$ and $x \notin B'$

$\Rightarrow x \in A' \cap B'$

i.e., $\qquad (A \cup B)' \subseteq A' \cap B'$ ...(1)

Conversely, let

$$y \in A' \cap B'$$

$\Rightarrow \qquad y \in A'$ and $\in B'$

$\Rightarrow \qquad y \notin A'$ and $y \notin B$

$\Rightarrow \qquad y \notin A \cup B$

$\Rightarrow \qquad y \in (A \cup B)'$

i.e., $\qquad A' \cap B' \subseteq (A \cup B)'$ ...(2)

By using equations (1) and (2) we get,
$$(A \cup B)' = A' \cap B'$$

(b) Let $a \in (A \cup B)'$

But $a \in (A \cap B)' \Rightarrow a \notin (A \cap B)$

$\Rightarrow \qquad a \notin A$ or $a \notin B$

$\Rightarrow \qquad a \in A'$ or $a \in B'$

$\Rightarrow \qquad a \in A' \cup B'$

i.e., $\qquad (A \cap B)' \subseteq A' \cup B'$ ...(1)

Conversely, let $b \in A' \cup B'$

But $b \in A' \cup B' \Rightarrow b \in A'$ or $b \in B'$

$\Rightarrow \qquad b \notin A$ or $b \notin B$

$\Rightarrow \qquad b \notin A \cap B$

$\Rightarrow \qquad b \in (A \cap B)'$

i.e., $\qquad A' \cup B' \subseteq (A \cap B)'$ ...(2)

By using equations (1) and (2) we get,
$$(A \cap B)' = A' \cup B' \qquad ...(b)$$

(c) Let $x \in A - (B \cup C)$

But $x \in A - (B \cup C) \Rightarrow x \in A$ and $x \notin B \cup C$

$\Rightarrow x \in A$ and $[x \notin B$ and $x \notin C]$

$\Rightarrow [x \in A$ and $x \notin B]$ and $[x \in A$ and $x \notin C]$

$\Rightarrow x \in (A - B)$ and $x \in (A - C)$

$\Rightarrow x \in (A - B) \cap (A - C)$

i.e., $\qquad A - (B \cup C) \subseteq (A - B) \cap (A - C)$ ...(1)

Conversely let

$b \in (A - B) \cap (A - C) \Rightarrow b \in (A - B)$ and $b \in (A - C)$

$\Rightarrow b \in A$ but $b \in B$ and $b \in A$ but $b \in C$

$\Rightarrow b \in A$ but $b \notin B$ and $b \notin C$

$\Rightarrow b \in A$ but $b \in B \cup C$

$\Rightarrow b \in A - (B \cup C)$

i.e., $\qquad (A - B) \cap (A - C) \subseteq A - (B \cup C)$ ...(2)

By using equations (1) and (2) we get,
$$A - (B \cup C) = (A - B) \cap (A - C) \qquad ...(C)$$

(c) Here, $A - (B \cap C) = A \cap (B \cap C)'$

$\qquad = A \cap (B' \cup C')$ by equation *(b)*

$\qquad = (A \cap B') \cup (A \cap C')$ by distributive law

$\qquad = (A - B) \cup (A - C)$

Hence, $A - (B \cap C) = (A - B) \cap (A - C)$

Example 25: Prove that,

(i) $A \cap (B - C) = (A \cap B) - (A \cap C)$
(ii) $(A - B) \cup B = A$ iff $B \subset A$
(i) $A \cap (B - C) = A \cap B - A \cap C$

Solution : Let $a \in A \cap (B - C)$

But
$$a \in A \cap (B - C) \Rightarrow a \in A \text{ and } a \in (B - C)$$
$$\Rightarrow a \in A \text{ and } a \in B \text{ but } a \notin C$$
$$\Rightarrow (a \in A \text{ and } a \in B) \text{ but } a \notin A \cap C$$
$$\Rightarrow a \in (A \cap B) \text{ but } a \notin A \cap C$$
$$\Rightarrow a \in (A \cap B) - (A \cap C)$$

i.e., $A \cap (B - C) \subseteq (A \cap B) - (A \cap C)$ ...(1)

Now let $b \in (A \cap B) - (A \cap C)$

$\therefore \quad b \in (A \cap B) - (A \cap C) \Rightarrow b \in A \cap B, b \notin A \cap C$
$$\Rightarrow b \in A \text{ and } b \in B \text{ and } b \in A \text{ but } b \notin C$$
$$\Rightarrow b \in A \text{ and } b \in B \text{ but } b \notin C$$
$$\Rightarrow b \in A, b \in B - C$$
$$\Rightarrow b \in A \cap (B - C) \quad\quad ...(2)$$

By using equations (1) and (2) we get,
$$A \cap (B - C) = (A \cap B) - (A \cap C)$$

(ii) $(A - B) \cup B$ A iff $B \subset A$

$(A-B) \cup B = A \Leftrightarrow (A \cap B') \cup B = A$

$\Leftrightarrow (A \cup B) \cap (B' \cup B) = A$ (by distributive law)

$\Rightarrow \quad\quad \Leftrightarrow (A \cup B) \cap S = A \ (\therefore B' \cup B = S \text{ universal law})$

$\Rightarrow \quad\quad \Leftrightarrow A \cup B \ A \ (\therefore A \cap S = A) \Rightarrow$ all elements of B are in A

$\Rightarrow B \subset A$

Hence $(A - B) \cup B \Rightarrow B \subseteq A$

Example 26: Prove that,

(a) $(A - B) \cup (B - A) = (A \cup B) - A \cap B)$
(b) $(A - B) = \phi$ iff $A \subseteq B$

Solution: (a) We have to prove that $(A - B) \cup (B - A) = (A \cup B) - (A \cap B)$

(a) Hence, $(A - B) \cup (B - A) = (A \cap B') \cup (B \cap A')$

$\therefore \quad\quad A - B = A \cap B'$

$= [(A \cap B') \cup B] \cap [(A \cap B') \cup A']$ ...by distributive law

$= [(A \cup B) \cap (B' \cup B)] \cap [(A \cup A') \cap (B' \cup A')]$

...by distributive law

$= [(A \cup B) \cap S] \cap [S \cap (B' \cup A')]$ ... $\therefore B' \cup B = S$

$$= (A \cup B) \cap (B' \cup A')$$
$$= (A \cup B) \cap (B \cap A)'$$
$$= (A \cup B) - (B \cup A)$$

... $\therefore A \cap S = S$
... by De Morgan's law
... $\therefore A \cap B' = A - B$

Hence,
$$(A - B) \cup (B - A) = (A \cap B') \cup (B \cup A')$$

(b) Here, $\quad A - B = \phi \Rightarrow A \cap B' = \phi \quad\quad \therefore (A - B) = A \cap B'$

$\Rightarrow$ A and B' are disjoint

$\Rightarrow$ all elements of A belongs to B

$\Rightarrow A \subseteq B$

Conversely,

$A \subseteq B$ there is no elements in A which does not belongs to B

$\Rightarrow A - B = \phi$

Hence $A - B = \phi \Rightarrow A \subset B$

Example 27: If A = {1, 2, 3}, B = {2, 4, 6}, C = {4, 6, 8, 10} and U = {1, 2, 3, 4, 5, 6, 7, 8, 9, 10} find A' B' and C' prove that,

(i) $(A \cup B)' = A' \cap B'$

(ii) $(A \cap B)' = A' \cup B'$ where U is taken as universal set.

Solution: A = {1, 2, 3} B = {2, 4, 6} and C = {4, 6, 8, 10}

$$U = \{1, 2, 3, 4, 5, 6, 7, 8, 9, 10\}$$
$$A' = \{4, 5, 6, 7, 8, 9, 10\}$$
$$B' = \{1, 3, 5, 7, 8, 9, 10\}$$
$$C' = \{1, 2, 3, 5, 7, 9\}$$

Now (i) $\quad A \cup B = \{1, 2, 3, 4, 6\}$

$(A \cup B)' = \{7, 8, 9, 10\}$

$A' \cap B' = \{5, 7, 8, 9, 10\}$

Hence,
$$(A \cup B)' = A' \cap B'$$

(ii) $A \cap B = \{2\}$

$(A \cap B)' = \{1, 3, 4, 5, 6, 7, 8, 9, 10\}$

$A' \cup B' = \{1, 3, 4, 5, 6, 7, 8, 9, 10\}$

Hence, $\quad (A \cap B)' = A' \cup B'$

Example 28 : Of the members of there athletic teams a certain schools, 21 are on the basketball team, 26 on hockey team and 29 on the football team 14 play hockey and basketball, 15 play hockey and football, 12 play football and basketball and 8 play all the three games. How many members are there in all.

Solution: We can solve this problem by using formula and venn diagram.

By Formula: Let B, H, and F denote the sets of members who are on the basket-ball team, hockey team and football team respectively.

So, we are given
$$n(B) = 21, n(H) 26, n(F) = 29,$$
$$n(H \cap F) = 15, n(F \cap B) = 12$$
and $n(B \cap H \cap F) = 8$
$n(B \cup H \cap F) = ?$

We know that,
$n(B \cup H F) = n(B) + n(H) + n(F) - n(B \cap H) - (H \cap F) - n(F \cap B) + n(B \cap H \cap F)$
$= 21 + 26 + 29 - 14 - 15 - 12 + 8 = 43$

Hence, there are 43 members in all.

By venn diagrams: 8 play all the three games.

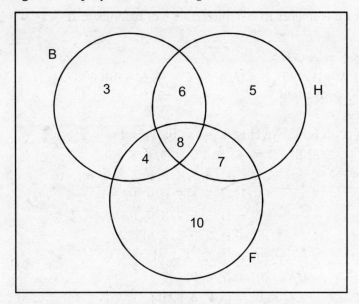

Figure 1.17

Since 14 play hockey and basketball of which 8 are already written in $B \cap H \cap F$, we write 6 in the remaining part of $H \cap B$. Similarly we write 7 and 4 in the remaining parts of $H \cap F$ and $F \cap B$. Finally there are 21 members on the basketball of which $6 + 8 + 4 = 18$ have already been written, So we write 3 in the remaining part of B. Similarly we write 5 and 10 in the remaining parts of H and F respectively.

So,

Total Number = $3 + 6 + 5 + 4 + 8 + 7 + 10 = 43$.

Example 29: An investigator interviewed 100 students to determine their preferences for the three drinks: milk (M), coffee (C) and tea (T). He reported the following: 10 students and all the three drinks M, C and T; 20 had M and C; had C and T; 25 had M and T; 12 had M only; 5 had C only and 8 had T only using a venn diagram. Find how many did not take any of the three drinkss.

Solution: We first draw a venn-diagram representing the set U of 100 students by a rectangle and its three subsets M, C and T by closed curves inside ∪.

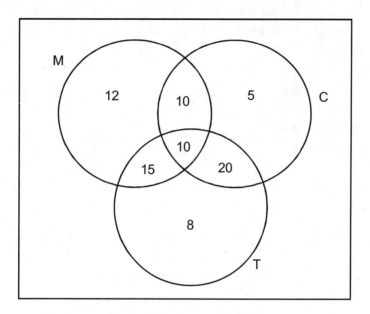

Figure 1.18

Hence, (M ∩ C ∩ T) = 10, We write it first since n (M ∩ C) = 20 out of which 10 have already by counted in h (M ∩ C ∩ T). So we write 10 in the remaining part of M ∩ C.

Similarly, we write 20 and 15 in the remaining parts of C ∩ T and M ∩ C, so that the total number in there parts are 30 and 25 respectively.

Finally we put 12 in 'M only', 5 in 'C only' and 8 in 'T only'

Thus the total No. of students taking drinks M or C or T.

$$= n (M \cup C \cup T) = 12 + 10 + 5 + 15 + 10 + 20 + 8 = 80$$

Hence, number of students taking none of the drinks

$$= n (M \cap C \cap T)' = n (\cup) - n (M \cup C \cup T) = 100 - 80 = 20$$

1.8 ORDERED PAIRS

Let the set P consists of n distinct elements. Any element can form a pair either with other distinct element or it may form the pair with an equal element. Hence each element of the set can form n distinct pairs.

So total No. of distinct ordered pair = $n \times n = n^2$

Example: (*i*) Set (*a, b*) has 2 elements so it contains $2^2 = 4$ ordered pairs.

i.e., (*a, a*) (*a, b*) (*b, a*) (*b, b*)

(*ii*) Set (1, 2, 3) has 3 element so it contains $3^2 = 9$ ordered pairs.

i.e., (1, 2) (1, 1) (1, 3) (2, 1) (2, 2) (2, 3) (3, 1) (3, 2) (3, 3)

Cartesian Product of sets: The cartesian product of set of all distinct ordered pairs (a, b) in which the first component a is an element of A, while the second component b is an element of B and is denoted by the symbol A × B in that order.

Let A × B is the set of all ordered pairs (a, b) such that $a \in A, b \in B$ them.

A × B = $\{(a, b) : a \in A, b \in B\}$

A × B is read as 'A cross B'

Given n sets $A_1 \ldots A_n$, the set of all n-tuples $(x_1 \ldots x_n)$ that $x_1 \in A, x_2 \in A_2 \ldots, x_n \in A_n$ is called the Cartesian. Product of A_1 to A_n denoted by $A_1 \times A_2 \ldots A_n$ and road 'A_1 cross A_2 cross ... cross An' using indexed notation, this product can be written as

Example: Find the cartesian products A × B and B × A If
$$A = \{1, 2, 3\}, B = \{5, 6\}$$
So, $\quad$ A × B = $\{(1, 5), (2, 5), (3, 5), (1, 6) (2, 6) (3, 6)\}$
and,
$$B \times A = \{(5, 1), (6, 1) (5, 2) (6, 2) (5, 3), (6, 3)\}$$

Example 30: If X = $\{a, b, c\}$ and Y = (a, b, e, f) compute X × Y and Y^2

Solution: Given X = $\{a, b, c\}$, Y = (a, b, e, f)

X × Y = $\{(a, a), (b, a), (c, a), (a, b), (b, b), (c, b)$
$\qquad (a, e), (b, e), (c, e) (a, f), (b, f), (c, f)\}$

and

$Y^2 = Y \times Y = \{a, b, e, f\} \times \{a, b, e, f\}$
$\quad = \{(a, a) (b, a) (e, a), (f, a), (a, b) (b, b) (e, b), (f, b)$
$\qquad (a, e) (b, e) (e, e) (f,e), (a, f), (b, f), (e, f) (f, f_1)\}$

Example 31: Find the cartesian product of A = $\{a, b, c\}$ and B = $\{b, c, d\}$ in that order. Also find (A × B) – (B × A) and (A × B) ∩ (B × A)

Solution: Here A = $\{a, b, c\}$ and B = $\{b, c, d\}$

So,
$\qquad$ A × B = $\{c , b), (b, b), (c, b) (a, c), (b, c), (c, c), (a, d), (b, d) (c, d)\}$
and $\qquad$ B × A = $\{(b, a), (c, a), (d, a), (b, b) (c, b), (d, b), (b, c), (c, c) (d, c)\}$

Now : (A × B) – (B × A)
$\qquad = \{ (a, b), (a, c), (a, d) (b, d), (c, d)\}$
and $\quad$ (A × B) ∩ (B × A) = $\{(b, b) (c, c), (c, b) (b, c)\}$

Example 32: Prove that $[(A \times B) = \phi]$ either $[A = \phi$ or $B = \phi]$
Solution: Given
$[A \times B = \phi] \Rightarrow A \times B$ consists of no ordered pairs
$\Rightarrow$ either there is no element in A or no element in B
$\Rightarrow$ either $A = \phi$ or $B = \phi$
Conversely,

If either $A = \phi$ or $B = \phi$ and if ordered pairs are formed so that first component belongs to A and second component belongs to B, the ordered pair cannot be formed for either they will have no antecedent or no successor.

Thus, $A \times B$ will have no elements.
or $\qquad A \times B = \phi$
Hence,
$[A \times B = \phi] \Rightarrow$ either $[A = \phi$ or $B = \phi]$

Example 33: If $A \subset B$, prove that,
$$A \times A = (A \times B) \cap (B \times A)$$
Let A consists of m distinct elements and B consist of n distinct elements.
Since $A \subset B \Rightarrow m < n$

Now, the set $A \times B$ consists of $m \times n$ distinct elements and $B \times A$ also consists of $m \times n$ distinct elements. Since all the m elements of A belongs to B therefore they have m elements of A belong to B. Therefore they have m element in common. Hence, $A \times B$ and $B \times A$ have m^2 elements in common which have been formed by pairing up the common element of A and B. Hence it is the set $A \times A$.

Also $(A \times B) \cap (B \times A)$ is the set of all elements common to $A \times B$ and $(B \times A)$.
Hence,
$$A \times A = (A \times B) \cap (B \times A)$$

Example 34: Prove the following
$$(A \cup B) \times C = (A \times C) \cup (B \times C)$$
Solution: Let (a, b) be an element of $(A \cup B) \times C$ then,
$\qquad (a, b) \in [(A \cup B) \times C] \Rightarrow a \in (A \cup B)$ and $b \in C$
$\Rightarrow \qquad (a \in A$ or $a \in B)$ and $b \in C$
$\Rightarrow \qquad (a \in A$ and $b \in C)$ or $(a \in B$ and $b \in C)$
$\Rightarrow \qquad (a, b) \in (A \times C)$ or $(a, b) \in (B \times C)$
$\Rightarrow \qquad (a, b) \in [(A \times C) \cup (B \times C)]$
i.e., $\qquad (A \cup B) \times C \subseteq (A \times C) \cup (B \times C) \qquad ...(1)$
Conversely,
$\qquad (a, b) \in (A \times C) \cup (B \times C)$
$\Rightarrow \qquad (a, b) \in (A \times C)$ or $(a, b) \in (B \times C)$
$\Rightarrow \qquad (a \in A$ and $b \in C)$ or $(a \in B$ and $b \in C)$

$\Rightarrow$ $\qquad$ (a, $\in$ A or $a \in$ B) and $b \in$ C

$\Rightarrow$ $\qquad$ a, $\in$ A $\cup$ B and $b \in$ C

$\Rightarrow$ $\qquad$ $(a, b) \in$ (A $\cup$ B) $\times$ C

i.e., $\quad$ (A $\times$ C) $\cup$ (B $\times$ C) $\subseteq$ (A $\cup$ B) $\times$ C $\qquad\qquad$ ...(2)

By using equations (1) and (2) we get,

$$(A \cup B) \times C = (A \times C) \cup (B \times C)$$

Example 35: Prove that,

$$(A \cap B) \times (C \cap D) = (A \times C) \cap (B \times D)$$

Solution: Let $(x, y) \in (A \cap B) \times (C \cap D)$; then,

$\qquad (x, y) \in (A \cap B) \times (C \cap D) = x \in (A \cap B), y \in (C \cap D)$

$\Rightarrow \qquad x \in$ A and $x \in$ B while $y \in$ C and $y \in$ D

$\Rightarrow \qquad x \in$ A, $y \in$ C and $x \in$ B, $y \in$ D

$\Rightarrow \qquad (x, y) \in (A \times C)$ and $(x, y) \in (B \times D)$

$\Rightarrow \qquad (x, y) \in (A \times B) \cap (B \times D)$

$\qquad (A \cap B) \times (C \cap D) \subset (A \times C) \cap (B \times D) \qquad\qquad$ ...(1)

Conversely,

$\qquad (x, y) \in (A \times C) \cup (B \times C)$

$\Rightarrow \qquad (x, y) \in (A \times C)$ or $(x, y) \in (B \times C)$

$\Rightarrow \qquad x \in$ A, $y \in$ C or $x \in$ B, $y \in$ C

$\Rightarrow \qquad x \in$ A or $x \in$ B and $y \in$ C

$\Rightarrow \qquad x \in (A \cup B)$ and $y \in$ C

$\Rightarrow \qquad (x, y) \in (A \cup B) \times$ C

$\therefore \quad$ (A $\times$ C) $\cup$ (B $\times$ C) $\subseteq$ (A $\cup$ B) $\times$ C $\qquad\qquad$ ...(2)

By using equation (1) and (2) we get,

$$(A \cup B) \times C = (A \times C) \cup (B \times C)$$

Example 36: Prove that $(A \cup B) \times (C \cap D) = (A \times C) \cap (B \times D)$

Solution: Let $(a, b) \in (A \cap B) \times (C \cap D)$

But $(a, b) \in (A \cap B) \times (C \cap D) \Rightarrow a \in (A \cap B), b \in (C \cap D)$

$\Rightarrow a \in$ A and $a \in$ B while $b \in$ C and $b \in$ D

$\Rightarrow a \in$ A, $b \in$ C and $a \in$ B, $b \in$ D

$\Rightarrow (a, b) \in (A \times C)$ and $(a, b) \in (B \times D)$

$\Rightarrow (a, b) \in (A \times C) \cap (B \times D)$

i.e., $(A \cap B) \times (C \cap D) \subseteq (A \times C) \cap (B \times D)$ $\qquad\qquad$ ...(1)

Conversely let,

$\qquad (a, b) \in (A \times C) \cap (B \times D)$

$\Rightarrow \qquad (a, b) \in (A \times C)$ and $(a, b) \in (B \times D)$

$\Rightarrow \qquad a \in$ A, $b \in$ C and $a \in$ B, $b \in$ D

$\Rightarrow$ $\quad\quad\quad (a \in$ A and $a \in$ B) and $(b \in$ C and $b \in$ D)

$\Rightarrow$ $\quad\quad\quad a \in (A \cap B)$ and $b \in (C \cap D)$

$\Rightarrow$ $\quad\quad\quad (a, b) \in (A \cap B) \times (C \cap D)$

i.e., $\quad (A \times C) \cap (B \times D) \subseteq (A \cap B) \times (C \cap D)$ ...(2)

By using equations (1) and (2) we get,

$\quad (A \cap B) \times (C \cap D) = (A \times C) \cap (B \times D)$

Symbolic logic: George Boole (1855 – 1886) was first to reduce ordinary logical reasoning to an algebraic symbolism. This subject is known as symbolic logic.

Example 37: Prove the following statement by set theory:

(i) All planets revolve sound the sun.

(ii) Earth is plant; therefore earth revolves round the sun.

Solution: Let P be the set of all planets and θ be the set of all objects revolving round the sun. Let element x denote Earth.

Hence by (ii) $x \in$ P and by (i) $P \subset \theta$

$\therefore x \in \theta$, i.e., Earth revolves round the sun.

Example 38: Prove the following by symbolic logic:

(i) All girls are figure – conscious

(ii) Rashmi is girl; therefore Rashmi is figure– conscious.

Solution: Let P be the set of all girls that figure– conscious and θ be the singleton set containing Rashmi let x element denote Rashmi.

by (ii) $x \in \theta$ and by (i) $\theta \subset P$.

$\therefore x \in$ p.

i.e., Rashmi is figure conscious.

EXERCISE 1.2

1. For what values of x, y the following ordered pairs are equal.

(a) $(5x - 6, 6) = (24, y + 1)$ (b) $(2x + 1, 2y + 1) = (7, 9)$

(c) $(8x + 1, 2x) = (9, y)$ (d) $(3x - 1, y - 2) = (8, 2)$

(e) $(x + 2, 2y, -3) = (6, 2)$ (f) $(x, 2) = (2y, 2x)$

2. If $x = \{a, b\}$ and $y = \{a, c, d, e, f\}$ then find out the following values,

(a) $x \times y$ (b) $y \times x$

(c) x^2 (d) y^2

3. Prove that If $A \subseteq X$ and $B \subseteq Y$ then,

$\quad A \times B \subseteq X \times Y$

4. If $P = \{a, b\}, Q = \{b, c, d\}, R = \{c, d, e\}$ then find.

(i) $(A \times B) \cup (A \times C)$

(ii) $(A \times B) \cap (A \times C)$

and Hence verify that,
$(A \times B) \cap (A \times C) = (A \cap A) \times (B \cap C)$

5. Prove that,
   (i) $(A \cap B) \times C = (A \times C) \cap (B \times C)$
   (ii) $(A - B) \times C = (A \times C) - (B \times C)$

6. On the cartesian plane, draw the graph of,
   $\{(x, y) : x \geq 2, y \leq -1\}$

7. If P, Q are not empty sets, prove that,
   $P \times Q = Q \times P \Rightarrow P = Q$

8. Prove that,
   (a) $(A \times C) \cap (D \times B) = A \times B$;
   (b) $(A \times C)' = A' \, C'$
   (c) $(A \times B)' = (A' \times C) \cup (D \times B')$

◻◻

CHAPTER 2

RELATION

2.1 RELATION

A relation between two sets A and B is a subset of A × B. Symbolically, we can write

R is a relation from A to B iff $R \subset A \times B$.

If A = B, then we say that R is a relation on A. We can write aRb iff $(a, b) \in$ R and say that a is R-related to b or that b is a R-relation of a.

We can write $a(\sim R)b$. If a is not R-related to b. We can say that A relation, binds together two-objects of a partition class according to some rule.

When we say that, an ordered pair (x, y) satisfies or belongs to a relation R, we write : $(x, y) \in$ R.

Example. Let X = {2, 3, 6, 9, 18, 27} and R stand for "is thrice of".

Here 6R2, 9R3, 18R6, 27R9. Hence obtained ordered pairs are (6, 2), (9, 3), (18, 6), (27, 9).

Hence, R is defined as the set of ordered pairs :
$$= \{(6, 2), (9, 3), (18, 6), (27, 9)\}$$

2.2 RANGE AND IMAGE

As usual, the pair (a, b) of the relation R has two element. First element a and second element b and $a, b \in$ R. If collect first elements from each pairs. i.e. all first element (a) together to form a set and also collect all second elements (b) to form another set, we get two sets each being a subset of A.

The set of all first elements a for which there is a. Corresponding b given by $(a, b) \in$ R and $a, b \in$ A is called the Domain of R.

The set of all b's to which there is some corresponding a such that $(a, b) \in$ R and $a, b \in$ A is called the range or image of R.

Example. Consider the relation R = $\{(x, y) : y = 4x^2, x \in N\}$

Here, N = {1, 2, 3, 4, ...}.

∴ $y = 4x^2$, $x \in N$ gives the domain of R = $\{x : x \in N\}$ = N and the range of R = {4, 16, 36, ...}.

So, $\qquad$ R = {(1, 4), (2, 16), (3, 36)....}.

Binary Relation

A relation R between pairs of elements of a given set is called a binary relation.

2.3 OPERATIONS ON RELATIONS

Inverse of a Relation

The inverse of a relation R is the set of all reversed pairs of R and is denoted by R^{-1}.

So if $\qquad$ R = {(x, y) : xRy and $x, y \in R$}

then $\qquad$ R^{-1} = {(y, x) : (x, y) ∈ R}

i.e. $xRy \Leftrightarrow yR^{-1}x$

Example. (i) If A = {a, b, c}, B = {1, 2, 3}

Let a relation R such that

$\qquad$ R = {(a, 1), (b, 2), (c, 3)}

So here domain of $\qquad$ R = (a, b, c)

and Range of $\qquad$ R = (1, 2, 3)

and $\qquad$ R^{-1} = {(1, a), (2, b), (3, c)}

Domain of $\qquad$ R^{-1} = (1, 2, 3)

and Range of $\qquad$ R^{-1} = (a, b, c).

Example. (ii) The inverse of relation "is greater than" is the relation "is less than" i.e. $x > y \Leftrightarrow y < x$.

Identity Relation

A relation R in a set A is said to be identity relation denoted by I_A if

$$I_A = \{(a, b) : (a, b) \in A \text{ and } a = b\}$$

This relation is also called Equality relation.

Example. Let A = {a, b, c}

I_A = {(a, a), (b, b), (c, c)} is an identity or equality relation in A.

Void Relation

A relation R in a set A is said to be a void relation. If R = φ.

Example. Let A = (1, 2, 3) and R is a relation defined by aRb. Iff a divides b then R = $\phi \subset [A \times A]$ is a void relation.

Universal Relation

A relation R in a set A is said to be universal relation. If R coincide with A × A.
i.e. R is universal relation. Iff R = A × A.

Example. A = {1, 2, 3} then
R = A × A = {(1, 1), (1, 2), (1, 3), (2, 1), (2, 2), (2, 3), (3, 1), (3, 2), (3, 3)} is a universal relation.

2.4 COMPOSITE RELATION

Let R_1 and R_2 be the relations from the sets A to B and B to C respectively then the composite of R_1 and R_2 is a relation from A to C denoted by $R_2 \circ R_1$

$$R_2 \circ R_1 = [(a, c) \exists b \in B \text{ such that } (a, b) \in R_1 \text{ and } (b, c) \in R_2]$$

$\Rightarrow (a, b) \in R_1, (b, c) \in R_2$

$\Rightarrow (a, c) \in R_2 \circ R_1$

Example. If two relations R_1 and R_2 are such that :

$$R_1 = [(a, c), (b, a), (c, d)]$$
and
$$R_2 = [(a, b), (b, c), (c, d)]$$

then calculate $R_2 \circ R_1$

Domain $(R_2 \circ R_1) = \text{dom}(R_1)$

$(a, c) \in R_1$ and $(c, d) \in R_2 \Rightarrow (a, d) \in R_2 \circ R_1$

$(b, a) \in R_1$ and $(a, b) \in R_2 \Rightarrow (b, b) \in R_2 \circ R_1$

$(c, d) \in R_1$ and $(d, c) \in R_2 \Rightarrow (c, c) \in R_2 \circ R_1$

Hence, $R_2 \circ R_1 = [(a, d), (b, b), (c, c)]$

2.5 PROPERTIES OF BINARY RELATION

(i) Reflexive

A relation R on a set A is said to be reflexive. If each member $a \in A$ is R-related to itself.

i.e. $aRa \ \forall \ a \in A$

or a relation R in a set A is said to be reflexive. If every element of A is related to itself *i.e.* aRa is true for all $a \in A$.

or

R is reflexive if $(a, a) \in R \ \forall \ a \in A$.

Example 1. Let A be the set of all triangles (coplanar) and R stand for "equal in area to". Now any triangle $a \in A$ is equal in area to itself, *i.e.* $aRa \ \forall \ a \in R$. Hence R is reflexive.

Example 2. Let A be the set of all lines and R stand for "is parallel to". Now any line $a \in A$ is parallel to itself i.e. $aRa \ \forall \ a \in R$. Hence, R is reflexive.

Example 3. Let A be the set of all members of a family. Let R be defined by "is wife of". Clearly any wife $a \in A$ is not "the wife of" itself i.e. a is not R-related to a, $\forall \ a \in A$. Hence, R is not reflexive.

Example 4. (a) If $R_1 = \{(a, a), (a, b), (a, c), (b, b), (b, c)\}$ be a relation on $A = \{a, b, c\}$, then R_1 is reflexive relation since for every $a \in A$, $(a, a) \in R_1$.

(b) If $R_2 = \{(a, a), (a, b), (a, c), (b, c)\}$ be a relation on $A = \{a, b, c\}$ then R_2 is not a reflexive relation since for $b \in A$, $(b, b) \notin R_2$.

(ii) Symmetric

A relation R is called symmetric. If the second element is also related with first element in the same manner as the first element is relation with second element of each pair.

R is symmetric if $(a, b) \in R \Rightarrow (b, a) \in R$, $a, b \in A$.

or A relation R on a set A is defined as symmetric. If $aRb \Rightarrow bRa$, whenever $a, b \in A$

Example 1. Consider a set A of all students studying in a given college. Let R stand for "is a class-mate of" on the set A.

Clearly, if $a, b \in A$ and a "is a class-mate of" b, i.e. if aRb, then definitely b "is a class-mate of" a i.e. bRa.

So, $aRb \Rightarrow bRa$.

Hence, Relation R on set A is symmetric.

Example 2. Consider a set A of integers. Let R stand for "is equal to" on set A.

Definitely if $(a, b) \in R$ and $a = b \Rightarrow b = a$.

So, $aRb \Rightarrow bRa$.

So, Relation R on set A is symmetric.

Example 3. Let $R_1 = \{(a, a), (a, b), (a, c), (b, b), (b, a), (c, a)\}$ on $A = \{a, b, c\}$ is a symmetric relation.

Example 4. Let A be the set of lines in a plane. Then the relation "is parallel to" is a symmetric relation.

Let $a, b \in R$ because if $a \| b \Rightarrow b \| a$ on $aRb \Rightarrow bRa$.

Example 5. Let A be the set of lines in a plane. Then the relation "is perpendicular to" is a symmetric relation. Let $a, b \in A$

and of $a \perp b$ definitely $b \perp a$.

$aRb \Rightarrow bRa$.

So, Relation R on set A is symmetric.

Asymmetric Relation

A relation R on a set A is asymmetric if whenever $(a, b) \in R$ then $(b, a) \notin R$ for $a \neq b$ i.e. If $aRb \Rightarrow b\not{R}a$. It means that the presence of (a, b) in R excludes the possibility of presence of (b, a) in R. For example:

The Relation
$R_1 = \{(a, a), (a, b), (b, c), (c, a)\}$ on A = $\{a, b, c\}$ is a asymmetric relation.

Antisymmetric Relation

A relation R on a set A is antisymmetric. If for all. $a, b \in A$ $(aRb$ and $bRa) \Rightarrow a = b$.

Example 1. Let R standing for ">" be defined on the set N of all natural numbers. Clearly, if $n_1, n_2 \in N$ and of $n_1 R n_2$, then n_2 not related to n_1.

i.e. $n_1 > n_2$ then $n_2 \not> n_1$.

So Relation R on set N is antisymmetric.

Example 2. Let $R_1 = \{(x, y) \in R^2 \mid x \leq y\}$ is an antisymmetric relation on R since $x \leq y$ and $y \leq x$ is only possible. When $x = y$, then $(x, y) \in R$ and $(y, x) \in R$ implies $x = y$.

Example 3. Let $R = \{(x, y) \in N \mid x$ is a division of $y\}$ is an antisymmetric relation. Since x divides y and y divides x implies $x = y$.

Example 4. Let $R_1 = \{(a, b), (b, b), (b, c)\}$ on A = $\{1, 2, 3\}$ is an antisymmetric relation.

Transitive Relation

A relation R on a set A is called transitive if whenever aRb and bRc then aRc for $a, b, c \in A$.

i.e. aRb and $bRc \Rightarrow aRc$ for $a, b, c \in A$.

Example 1. The relation 'is parallel to' on the set of lines in a plane is transitive, because if a line x is parallel to the line y and if y is parallel to the line z then x is parallel to z.

Example 2. The relation 'is perpendicular to' on the set of lines in a plane is not transitive, because if a line x perpendicular to y and if y is perpendicular to the line z.

2.6 CONGRUENCE MODULO RELATION ON SET OF ALL INTEGERS

Let m be a positive integers and I be set of all integers.

The relation "congruence modulo m" is defined on all $a, b \in I$ by $a \equiv b \pmod{m}$ if and only if m divides $(a - b)$.

The symbol " $\equiv \pmod{m}$ " is read as "congruence modulo m".

"m divides $(a - b)$" or "m is a factor of $(a - b)$" is usually denoted by "$m \mid (a - b)$"

Thus,
(i) $69 \equiv 29 \pmod{4}$ i.e. 4 divides $(69 - 29)$
(ii) $64 \equiv 1 \pmod{7}$ i.e. 7 divides $(64 - 1)$

OR we can say that above relation is that $a \equiv b \pmod{m}$ iff b is the remainder when a is divided by m.

SOLVED EXAMPLES

Example 1: Prove that the relation R on the set N defined by xRy, if $x \neq y$, $x, y \in N$ is symmetric but neither reflexive nor transitive.

Solution: *Reflexivity* : Clearly xRx is not true,

∵ x is always equal to x.

Hence, R is not reflexive.

Symmetry : If $x \neq y$ then certainly $y \neq x$.

Hence, $xRy \Rightarrow yRx$.

Hence, R is symmetric.

Transitivity : If $x \neq y$ and $y \neq z$ then, $x \neq z$ is not always true.

Hence, xRy, yRz does not imply xRz for example when

1R2, 2R1 then 1R1 is not true.

Hence, R is not symmetric.

Example 2: Prove that the relation of divisibility on an integral domain is reflexive and transitive.

Solution: Let I stand for integral domain and let $x, y, z \in I$. Also let R relation is defined by "is divisible by".

Reflexivity : Since every element x is divisible by itself, hence xRx, $\forall x \in I$.

Thus R is reflexive.

Symmetry : If x is divisible by y then definitely y is not divisible by x. i.e.

xRy does not imply yRx.

Hence, R is not symmetric.

Transitivity: $xRy \Rightarrow$ y is a factor of x.
                   $yRz \Rightarrow$ z is a factor of y.
∴ $xRy, yRz \Rightarrow$ z is a factor of x.
                   $\Rightarrow$ x is divisible by z.
                   $\Rightarrow$ xRz.

Hence, R is transitive.

Example 3: Give an example of a relation which is:
(i) Reflexive and transitive but not symmetric.
(ii) Symmetric and transitive but not reflexive.
(iii) Reflexive and symmetric but not transitive.
(iv) Reflexive and transitive but neither symmetric nor antisymmetric.

Solution: Let A = {a, b, c}. Then let us take the relation :
(i) R = {(a, a), (b, b), (c, c), (a, c)}

It is clear that above relation is reflexive and transitive but not symmetric, since $(a, c) \in R$ but $(c, a) \notin R$.

(ii) R = {(a, a), (c, c), (a, c), (c, a)}

It is clear that above relation is symmetric and transitive but not reflexive, since $(c, c) \in R$.

(iii) R = {(a, a), (b, b), (3, 3), (a, b), (b, a), (b, c), (c, b)}

It is clear that above relation is reflexive and symmetric but not transitive, since $(a, b) \in R$ and $(b, c) \in R$ but $(a, c) \notin R$.

(iv) Let z^* be the set of all non-zero integers and R be the relation on z^* given by $(a, b) \in R$, if a is a factor b i.e. if a/b. Since a/a for all $a \in z^*$

a/b and $b/c \Rightarrow a/c$
Hence, R is reflexive and transitive.
For example 3/9 but 9/6 is not true.
Hence, R is not symmetric.

Irreflexive Relation

A relation R on a set A irreflexive if, for every $a \in A$, $(a, a) \notin R$. In other words, there is no $a \in A$ such that aRa. Then reflexive means that aRa is true for all a, and irreflexive means that aRa is true not for a.

Example 1. The relation R_1 = {(a, b), (a, c), (b, a), (b, c)} on A = {a, b, c} is irreflexive relation since $(x, x) \notin R_1$ for every $x \in R_1$.

Example 2. The relation $R_1 = \{(x, y) \in R^2 : x > y\}$ is an irreflexive relation since $x > x$ for any $x \in R$, where R is the set of real numbers.

Non-reflexive Relation

A relation R on a set A is non-reflexive if R is neither reflexive nor irreflexive, i.e. if aRa is true for some a and false for others. For example

R = {(a, b), (b, c), (b, b), (c, a)} on A = {a, b, c} is a non-reflexive relation because bRb is true but aRa and cRc are false.

EXERCISE 2.1

1. Show that on the set N, relation
   $$R = \{(x, y) : xy \text{ is odd}, x, y \in N\}$$
   is symmetric, transitive but not reflexive.

2. Show that on the set of all real numbers, the relation "less than" is transitive and anti-symmetric but not reflexive.

3. Prove that relation R on the set N defined by xRy if $x^2 - 4xy + 3y^2 = 0$
   is neither symmetric nor transitive but only reflexive.

4. For the following relations say whether each is reflexive, symmetric, transitive or antisymmetric.
   (i) The relation on {1, 2, 3, 4, 5} defined by $\{(a, b) : a - b \text{ is even}\}$
   (ii) The relation on {1, 2, 3, 4, 5} defined by $\{(a, b) : a + b \text{ is even}\}$
   (iii) The relation P, the set of all people, defined by $\{(a, b) : a \text{ and } b \text{ have an ancestor in common}\}$.

5. Prove that the relation R in the set I defined by aRb, if a and b are both odd, is symmetric and transitive but not reflexive.

6. Is there any set A in which every relation is symmetric.

7. Show that :
   (i) $741 \equiv 1 \pmod{10}$.
   (ii) $215 = 5 \pmod 7$.
   (iii) 2, 9, 10 are least positive integers to which 431, 109 × 431 and 109 respectively are congruent modulo 11.

8. Prove that :
   (i) $363 \equiv 99 \pmod{22}$ possesses a solution.
   (ii) $25x \equiv 12 \pmod{10}$ does not possess a solution.
   (iii) $35x \equiv 14 \pmod{21}$ has 7 incongruent solution (mod 21).

9. Find the incongruent solution of
   (i) $2x + 1 \equiv 4 \pmod 5$.
   (ii) $3x + 2 \equiv 0 \pmod 7$.
   (iii) $235x \equiv 54 \pmod 7$.

2.7 EQUIVALENCE RELATION

A relation R on a set A is called an equivalence relation on A, if it is reflexive, symmetric and transitive.

Existence of equivalence relation is denoted by the symbol $\cong$.

i.e. If $a, b, c \in A$ and if R be an equivalence relation on A then R is an equivalence relation. If

(i) $aRa, \forall a \in A$ (Reflexive)

(ii) $aRb \Rightarrow bRa$, where $a, b \in A$ (Symmetry)
(iii) $aRb, bRc \Rightarrow aRc$ (Transitivity)

Since a relation R is also regarded as a subset of A × A, alternative conditions in that order are as follows:

(i) $(a, a) \in R, \forall a \in A$
(ii) $(a, b), (b, a) \in R$ or $(a, b) \in R \Rightarrow (b, a) \in R$
(iii) $(a, b), (b, c), (c, a) \in R$ or $(a, b), (b, c) \in R \Rightarrow (a, c) \in R$

Smallest Equivalence Relation

An equivalence relation R on a set A is called the smallest equivalence relation. If R is smallest subset of A × A.

Clearly A × A contains n^2 elements, n elements are provided by reflexivity property in R. If no elements are provided in the set R by symmetric and transitive properties, then this is the smallest set.

Now 'is equal to' is an equivalence relation. For this R, reflexivity, symmetry and transitivity produce the same ordered pairs $(a, a) \; \forall a \in A$. Hence, it yields the minimum number of ordered pairs.

Largest Equivalence Relation

Since R on A is a subset of A × A, the largest subset of A × A is A × A itself. Hence A × A is the largest equivalence relation on A.

Theorem. Prove that 'congruence modulo m' is an equivalence relation on the set of all integers.

Proof. Let I be the set of all integers and let R defined on I stand for "congruence modulo m".

Thus aRb stands for

$$a \equiv b \pmod{m}, \; a, b \in I \text{ i.e. } m \mid (a - b)$$

(i) *Reflexivity:* Let $a \in I$

∵ $a - a = 0$ and 0 is divisible by m,

∴ $a \equiv a \pmod{m}, \forall a \in I$

Hence, R is reflexive.

(ii) *Symmetric:* Let $a, b \in I$

∴ $aRb \Rightarrow a \equiv b \pmod{m}$
$\Rightarrow m \mid (a - b)$
$\Rightarrow m \mid (b - a)$ because $a - b = -(b - a)$

$$\Rightarrow b \equiv a \pmod{m}$$
$$\Rightarrow bRa$$

Hence, R is symmetric.

(iii) *Transitivity* : Let $a, b, c \in I$. Also let aRb and bRc

$$aRb \Rightarrow m \mid (a-b) \text{ and } bRc \Rightarrow m \mid (b-c)$$
$$aRb, bRc \Rightarrow m \mid (a-b), m \mid (b-c)$$
$$\Rightarrow m \mid [(a-b) + (b-c)]$$
$$\Rightarrow m \mid (a-c)$$
$$\Rightarrow aRc$$

i.e. $a \equiv b \pmod{m}$, $b \equiv c \pmod{m} \Rightarrow a \equiv c \pmod{m}$

Hence, R is transitive.

Example 4: Show that "is similar to" on the set T of all coplanar triangles is an equivalence relation.

Solution: Let R stands for "is similar to".

(i) *Reflexivity* : Let $t \in T$. Hence, tRt, $\forall\, t \in T$

∴ Every triangle t is similar to itself.

Thus R is reflexive.

(ii) *Symmetry* : Let $t_1, t_2 \in T$, then $t_1 R t_2 \Rightarrow t_2 R t_1$ for

If triangle t_1 is similar to t_2 then triangle t_2 is also similar to t_1.

Hence, R is symmetric.

(iii) *Transitivity* : Let $t_1, t_2, t_3 \in T$, now if t_1 'is similar to' to t_2 and t_2 'is similar to' t_3 then we have t_1 'is similar to' t_3.

Hence, R is transitive.

Since R is reflexive, symmetric and transitive, hence R is an equivalence relation.

Example 5: If R and R' be equivalence relations on a set A, prove that $R \cap R'$ is an equivalence relation on A.

Solution: Let R and R' are defined on A.

∴ $R \subset A \times A$ and $R' \subset A \times A$

Hence, $R \cap R' \subset A \times A$

Now let $a, b, c \in A$

(i) *Reflexivity*: Since R and R' are equivalence relations,

∴ $aRa, aR'a\, \forall\, a \in A$

∴ $(a, a) \in R$ and $(a, a) \in R'$, $\forall\, a \in A$

Hence, $(a, a) \in R \cap R', \forall\ a \in A$

$\therefore\ R \cap R'$ is reflexive on A.

(ii) *Symmetry*: Let $(a, b) \in R \cap R'$ when $a, b \in A$

$\therefore\qquad (a, b) \in R \cap R' \Rightarrow (a, b) \in R$ and $(a, b) \in R'$

$\qquad\qquad\qquad\qquad \Rightarrow (b, a) \in R$ and $(b, a) \in R'$, ($\because$ R and R' are symmetric)

$\qquad\qquad\qquad\qquad \Rightarrow (b, a) \in R \cap R'$

$\therefore\ R \cap R'$ is symmetric.

(iii) *Transitivity*: Let $(a, b), (b, c) \in R \cap R'$

$\therefore\qquad (a, b), (b, c) \in R \cap R' \Rightarrow (a, b), (b, c) \in R$ and $(a, b), (b, c) \in R'$

$\qquad\qquad\qquad\qquad \Rightarrow (a, c) \in R$ and $(a, c) \in R'$

$\because$ R, R' are transitive.

$\qquad\qquad\qquad\qquad \Rightarrow (a, c) \in R \cap R'$

Hence, $R \cap R'$ is transitive.

Since $R \cap R'$ is reflexive, symmetric and transitive. Hence $R \cap R'$ is an equivalence relation.

Example 6: If R be a relation is the set of integers, z defined by

$R = \{(x, y) : x \in z, y \in z, (x - y) \text{ is divisible by } 6\}$

Then prove that R is an equivalence relation.

Solution: (i) *Reflexive*: Let $x \in z$. Then $x - x = 0$ and is divisible by 6.

i.e. xRx for $\forall\ x \in z$

Hence, R is reflexive.

(ii) *Symmetric*: $xRy \Rightarrow (x - y)$ is divisible by 6.

$\qquad\qquad\qquad \Rightarrow (x - y)$ is divisible by 6.

$\qquad\qquad\qquad \Rightarrow (y - x)$ is divisible by 6.

$\qquad\qquad\qquad \Rightarrow yRx$

Hence, R is symmetric.

(iii) *Transitive*: xRy and $yRz \Rightarrow (x - y)$ is divisible by 6 and $(y - z)$ is divisible by 6.

$\qquad\qquad\qquad \Rightarrow [(x - y) + (y - z)]$ is divisible by 6.

$\qquad\qquad\qquad \Rightarrow (x - z)$ is divisible by 6.

$\qquad\qquad\qquad \Rightarrow xRz$

Hence, R is transitive.

Since R is reflexive, symmetry and transitive, so R is an equivalence relation.

Theorem: Let R and S be relation from A to B, show that

(i) If $R \subseteq S$ then $R^{-1} \subseteq S^{-1}$

(ii) $(R \cap S)^{-1} = R^{-1} \cap S^{-1}$

(iii) $(R \cup S)^{-1} = R^{-1} \cup S^{-1}$

Proof: (i) Suppose $R \subseteq S$. If $(a, b) \in R^{-1}$ then $(b, a) \in R$ and also $(b, a) \in S$ since $R \subseteq S$. Again $(b, a) \in S$ implies $(a, b) \in S^{-1}$. Therefore $R^{-1} \subseteq S^{-1}$.

(ii) Let $(a, b) \in (R \cap S)^{-1}$. Then $(b, a) \in R \cap S$, so that $(b, a) \in R$ and $(b, a) \in S$. This implies $(a, b) \in R^{-1}$ and $(a, b) \in S^{-1}$. Hence $(a, b) \in R^{-1} \cap S^{-1}$.

Therefore $(R \cap S)^{-1} \subseteq R^{-1} \cap S^{-1}$ ...(1)

Conversely, let $(a, b) \in R^{-1} \cap S^{-1}$. Then $(a, b) \in R^{-1}$ and $(a, b) \in S^{-1}$. This implies $(b, a) \in R$ and $(b, a) \in S$. So $(b, a) \in R \cap S$.

Hence $(a, b) \in (R \cap S)^{-1}$

Therefore $R^{-1} \cap S^{-1} \subseteq (R \cap S)^{-1}$ ...(2)

From equations (1) and (2) we have

$$(R \cap S)^{-1} = R^{-1} \cap S^{-1}$$

(iii) Let $(a, b) \in (R \cup S)^{-1}$, then $(b, a) \in (R \cup S)$ so that $(b, a) \in R$ or $(b, a) \in S$. This implies $(a, b) \in R^{-1}$ or $(a, b) \in S^{-1}$.

Hence $(a, b) \in R^{-1} \cup S^{-1}$.

Therefore $(R \cup S)^{-1} \subseteq R^{-1} \cup S^{-1}$ ...(1)

Conversely, let $(a, b) \in R^{-1} \cup S^{-1}$. Then $(a, b) \in R^{-1}$ or $(a, b) \in S^{-1}$.

This implies $(b, a) \in R$ or $(b, a) \in S$

So, $(b, a) \in R \cup S$, hence $(a, b) \in (R \cup S)^{-1}$

$\therefore (R \cup S)^{-1} \subseteq R^{-1} \cup S^{-1}$ ...(2)

by using equations (1) and (2) we get

$$(R \cup S)^{-1} = R^{-1} \cup S^{-1}$$

Theorem: Let R be a relation on A, prove that
(i) If R is reflexive, so is R^{-1}
(ii) R is symmetric if and only if $R = R^{-1}$
(iii) R is antisymmetric if and only if $R \cap R^{-1} \subseteq I_A$

Proof: (i) Suppose R is reflexive. Then $(a, a) \in R$ for all $a \in A$.

So, $(a, a) \in R^{-1}$ for all $a \in A$.

Therefore $(R \cup S)^{-1} \subseteq R^{-1} \cup S^{-1}$ is reflexive.

(ii) Suppose R is symmetric. Let $(a, b) \in R^{-1}$. Then $(b, a) \in R$ and hence $(a, b) \in R$ since R is symmetric. Therefore $R^{-1} \subseteq R$.

Similarly $R \subseteq R^{-1}$

Hence, $R = R^{-1}$

Conversely suppose $R = R^{-1}$. Let $(a, b) \in R$. Then $(a, b) \in R^{-1}$ and so $(b, a) \in R$. Hence R is symmetric.

$\Rightarrow$ R is symmetric if and only if $R = R^{-1}$.

(iii) Suppose R is antisymmetric. Let $(a, b) \in R \cap R^{-1}$. Then $(a, b) \in R$ and $(a, b) \in R^{-1}$. Again $(a, b) \in R^{-1}$ implies $(b, a) \in R$.

Thus $(a, b) \in R$ and also $(b, a) \in R$.

Hence, $b = a$ because R is antisymmetric.

This is true for all $(a, b) \in R \cap R^{-1}$. Hence every element of $R \cap R^{-1}$ is of the form (a, a) where $a \in A$.

Therefore $R \cap R^{-1} \subseteq I_A$

Conversely suppose $R \cap R^{-1} \subseteq I_A$. Let $(a, b) \in A \times A$ such that $(a, b) \in R$ and $(b, a) \in R$, i.e. $(a, b) \in R$ and $(a, b) \in R^{-1}$.

Then $(a, b) \in R \cap R^{-1}$. Since $R \cap R^{-1} \subseteq I_A$. If follows that $b = a$. Hence R is antisymmetric.
Hence, proved.

Theorem: Suppose R and S are relations on a set A. Prove that :
(i) If R and S are reflexive, then $R \cup S$ and $R \cap S$ are reflexive.
(ii) If R and S are symmetric, then $R \cup S$ and $R \cap S$ are symmetric.
(iii) If R and S are transitive, then $R \cap S$ is transitive.

Proof: (i) Suppose R and S are reflexive. Then $(a, a) \in R$ and $(a, a) \in S$ for all $a \in A$.

Therefore $(a, a) \in R \cup S$ and $(a, a) \in R \cap S$. Hence $R \cup S$ and $R \cap S$ are reflexive.

(ii) Suppose R and S are symmetric then by theorem. $R = R^{-1}$ and $S = S^{-1}$.

Now by using $(R \cup S)^{-1} = R^{-1} \cup S^{-1}$ we can say $(R \cup S)^{-1} = R \cup S$ and using $(R \cap S)^{-1}$ = $R^{-1} \cap S^{-1}$, we can say $(R \cap S)^{-1} = R \cap S$. and using $(R \cap S)^{-1} = R^{-1} \cap S^{-1}$, we can say $(R \cap S)^{-1} = R \cap S$ and using $(R \cap S)^{-1} = R^{-1} \cap S^{-1}$, we can say $(R \cap S)^{-1} = R \cap S$. Hence, $R \cup S$ and $R \cap S$ are symmetric.

(iii) Suppose R and S are transitive. Let $(a, b) \in R \cap S$ and $(b, c) \in R \cap S$, then $(a, b) \in R$, $(a, b) \in S$, $(b, c) \in R$ and $(b, c) \in S$. Now $(a, b) \in R$ and $(b, c) \in R$ implies $(a, c) \in R$ and $(a, c) \in S$ and $(b, c) \in S$ implies $(a, c) \in S$. Therefore $(a, c) \in R \cap S$.

Hence, $(a, b) \in R \cap S$ and $(b, c) \in R \cap S$ implies $(a, c) \in R \cap S$. Therefore $R \cap S$ is transitive.

EXERCISE 2.2

1. Prove that the relation "=" on the set of all real numbers is an equivalence relation.
2. Let L be the set of all straight lines of the Eulerian plane, verify whether parallelism between two straight lines is an equivalence relation on L.
3. Let n be a fixed positive integer. Define a relation R on the set of all integers I as follows :

   aRb iff $n/(a - b)$ that is $(a - b)$ is divisible by n. Show that R is an equivalence relation on I.
4. m is said to be related to n if m and n are integers and $m - n$ is divisible by 13. Does this defines an equivalence relation?
5. A relation R on I (the set of integers) is defined as

   $R = \{(a - b) : a, b \in I$ and $a - b$ is divisible by $5\}$

   Show that R is an equivalence relation on I.
6. Let I be the set of integers. Let a relation aRb $(a, b \in I)$ be defined if $a - b$ is an even integer. Show that R is an equivalence relation.
7. N is the set of natural numbers. The relation R is defined on $N \times N$ as follows :

   $(a, b) \, R \, (c, d) \Leftrightarrow a + d = b + c$

   Prove that R is an equivalence relation.
8. N is the set of positive integers and $\sim$ be a relation on $N \times N$ defined by

   $(a, b) \sim (c, a)$ iff $ad = bc$

   check the relation for being an equivalence relation.
9. A relation R on the set of complex numbers is defined by $z_1 R z_2$ if and only if

$(z_1 - z_2)/(z_1 + z_2)$ is real. Show that R is an equivalence relation.

10. Which of the following are equivalence relations?
    (i) "Is the square of" for the set of natural numbers?
    (ii) "Has the same radius as" for the set of all circles in a plane?
    (iii) "$\subseteq$" for the set of sets {A, B, C, ...}.
    (iv) The set of real numbers : xRy, if $x = \pm y$.
    (v) The set of straight lines in the plane in which xRy if x is perpendicular to y.
    (vi) The set of straight line in the plane in which xRy if x is parallel to y.
    (vii) "$\leq$" for the set of real numbers.
    (viii) $X = \{(a, b) \mid a, b \in I\}$ on which $(a, b) R (c, d) \Rightarrow b - d = a - c$.
    (ix) 'a is less than or equal to b if there exist a non-negative c such that $a + c = b$' on the set R.

11. If R stands for "is at the same distance from the origin as" and is a relation on the set of all coplanar points, prove that R is an equivalence relation.

12. Show that the relation aRb defined by $|a| = |b|$, in the set of all real numbers, is an equivalence relation. Further show that the relation aRb defined by $|a| \geq |b|$ is an equivalence relation.

13. If R is a relation in the natural numbers N, defined by the open set "$x - y$ is divisible by 5" that is

    $R = \{(x, y) : x, y \in N, x - y \text{ is divisible by } 5\}$

    Prove that R is an equivalence relation.

14. Discuss the R-relation "$2x + 3y = 12$" defined on the set N of natural numbers, such that $x, y \in N$.

15. If R, R' are relations on a set A, test the truth of the following:
    (i) If R is an equivalence relation then:

    xRy, xRz $\Rightarrow$ yRz, x, y, z $\in$ A

    (ii) If R is reflexive $R \cap R^{-1} = \phi$

    (iii) If R, R' are transitive, then $R \cap R'$ is transitive.

16. Construct examples on the following R-relations.
    (i) R neither symmetric non-transitive but reflexive.
    (ii) R neither reflexive non-transitive but symmetric.
    (iii) R reflexive and symmetric but not transitive.
    (iv) R neither reflexive non-symmetric but transitive.
    (v) R reflexive and transitive but not symmetric.
    (vi) R symmetric and transitive but not reflexive.

17. Test the reflexivity, symmetry, transitivity of the following R-relations :

(i) When R stands for "is twice the are of" and is defined on all coplanar triangles.

(ii) When R on the set N is defined by xRy, if $x \neq y$, $x, y \in N$.

(iii) When R = {(1, 3), (3, 5), (5, 3), (5, 7)} on the set A = {1, 3, 5, 7}.

2.8 EQUIVALENCE CLASSES

Consider an equivalence relation R defined on a set S. If $x, y \in S$, so that yRx, then all such elements $y \in S$ constitute a subset of S. This subset of S is denoted by $[x]$ and is called equivalence set or equivalence class. Thus,

$$[x] = \{y : y \in S, yRx\}.$$

Here square brackets are used to denote an equivalence set.

Example 1. Let a set P of all coplanar straight lines, consider a particular straight line a_1 of P. Let R defined on P mean "is parallel to", then set of all straight lines a_2 "parallel to" a_1 is a subset of P. This subset of all a_2 straight lines is called equivalence set and is denoted by $[a_1]$. Also $a_1, a_2 \in P$ then $[a_2]$ represents a set of all straight lines in the given plane "parallel to" a given straight line a_2. Then

$a_1 \in [a_2]$ and $a_2 \in [a_2]$ and if $a_3 \in [a_1]$ and $a_3 \in [a_2]$ then $[a_1]$ and $[a_2]$ represent the same equivalence set as given by $[a_3]$.

Example 2. Let us consider an equivalence relation:

R = {(a, b), (b, a), (a, a), (b, b), (c, c), (d, d)} on P = {a, b, c, d} has the following equivalence classes.

$[a] = [b] = \{a, b\}, [c] = \{c\}, [d] = \{d\}$.

Properties of Equivalence Classes

(i) $a \in [a]$

(ii) If $b \in [a]$, then $[b] = [a]$

(iii) If $[a] \cap [b] \neq \phi$ then $[a] = [b]$

Proof. (i) Since relation R is an equivalence relation, by its reflexive property we have $aRa, \forall a \in S$. Hence by the definition of equivalence classes we have :

$x \in [a]$

(ii) Let $(a, b) \in S$. Now $\qquad b \in [a] \Rightarrow bRa.$

Also $x \in [b] \qquad\qquad \Rightarrow xRb \; \forall \; x \in [b]$

Hence, by transitivity xRb and $\quad bRa \Rightarrow xRa.$

$\Rightarrow x \in [a], \forall x \in [b]$.

$\therefore [b] \subseteq [a]$

Conversely, let $x \in [a]$ then :

$x \in [a] \Rightarrow xRa, \forall x \in [a]$

Also $b \in [a]$ $\Rightarrow bRa$

$\Rightarrow aRb$, for R is symmetric

$\Rightarrow a \in [b]$

but xRa and aRb $\Rightarrow xRb$ by transitivity of R

$\Rightarrow x \in [b], \forall x \in [a]$

$\therefore [a] \subseteq [b]$

Hence, $[a] = [b]$

(iii) Since $[a] \cap [b] \neq \phi$, let $[a] \cap [b] = [l, m, n]$

Clearly $l \in [a]$ and $l \in [b]$. Since $l \in [a]$, we have $[l] = [a]$ we have $[l] = [a]$ by (ii) property. Similarly $[l] = [b]$

Hence from transitivity of relation "=" we have $[a] = [b]$.

2.9 PARTITION OF A SET

A set {A, B, C, ...} of non-empty subset A, B, C, ... of a set S, is called a partition of S if

(i) $A \cup B \cup C \cup .... = S$

(ii) The intersection of every pair of distinct subsets is the empty set.

(iii) Clearly shows no two of A, B, C, ... have any element in common.

Example. Consider the set S = {5, 7, 9, 11, 13, 15, 17, 19} and its subsets X, Y and Z such that X = {5, 11, 13}, Y = {7, 9, 19}, Z = {15, 17}.

Clearly (i) X, Y, Z are non-empty.

(ii) $X \cup Y \cup Z = \{5, 7, 5, 11, 13, 15, 17, 19\} = S$

and (iii) $X \cap Y = \phi, Y \cap Z = \phi, Z \cap X = \phi$

Hence, the set {X, Y, Z} is a partition of the set S.

Theorem on a Partition of a Set

A relation R on S affects a partition of S, if and only if, R is an equivalence relation.

Proof. Consider an equivalence relation R on S defined by xRy, so that $Ay = [y] = \{x : x \in S, xRy\}$, for each $y \in S$ obviously $y \in [y]$. Hence S is the union of all the distinct A_a, A_b, A_c .... obtained by R.

Clearly, for any pair A_b, A_c of these subsets, $A_b \cap A_c = \phi$, since otherwise $A_b = A_c$ in which case A_b, A_c are not distinct. Hence, the set $\{A_a, A_b, A_c, ...A_y ....\}$ is the partition of S by R. (By definition).

Conversely. Let the set $A = \{A_a, A_b, A_c, ...\}$ be a partition of S. On S let us define a relation R given by xRy such that xRy, if and only if, $A_r \in A$ and $x, y \in A_r$, i.e.

$$xRy \Leftrightarrow x, y \in A_r \text{ and } A_r \in A \qquad ...(1)$$

Clearly $xRy \Leftrightarrow y, x \in A_r$ and $A_r \in A$ ...(2)

But by equation (1),

$y, x \in A_r$ and $A_r \in A \Leftrightarrow yRx$

From (2),

$xRy \Leftrightarrow yRx, \forall x, y \subset A_r$ and $A_r \in A$. Thus R is symmetric and reflexive.

Now assume xRz and yRz. Clearly, by the definition of R given above, there are subsets $A_r, A_k \in A$, such that $x, y \in A_r$ and $y, z \in A_k$ where $A_r = A_k$, not necessary distinct. Hence $A_r \cap A_k \neq \phi$ and therefore $A_r = A_k$.

Clearly, then $x, z \in A_r$ and hence xRz exists.

Thus, when xRy and yRz exist, we have established that xRz exists, i.e. $xRy, yRz \Leftrightarrow xRz$. Hence R is transitive and therefore R is an equivalence relation.

2.10 QUOTIENT SET

The set of all disjoint equivalence classes defined by an equivalence relation R on a set S, is called the quotient set of S, relative to R and is denoted by S/R. S/R is read as "S modulo R" or "S mod. R".

Forming a Quotient Set

Consider an element $a \in S$. From a set of all element $y \in S$, such that yRa. The set so obtained will be the equivalence Class $[a]$. If $[a] \neq S$, then there must be an element $b \in S$ and $b \notin [a]$. Now from an equivalence Class $[b]$ corresponding to the element b. Proceeding in this manner we shall get a set of equivalence classes $[a], [b], [c]$ .... where $[a], [b], [c]$ .... are distinct sets.

The set $\{[a], [b], [c], ...\}$ is the quotient set S/R. If T be the quotient set S/R, R may be regarded as mapping $f: S \to T$, such that

if yRx, then $f(y) = f(x) = t$, where $x, y \in S$ and $t \in T$.

It must be noted that, here all $y \in S$, so that yRx, are all mapped into a single element $t \in T$. Hence mapping is many one. Also every $t \in T$ has a pre image in S. Hence mapping is onto.

Example : If S = {3, 4} and its partition P = {{3}, {4}}, find an equivalence relation. Which induced this partition.

Solution. Consider the equivalence relation R given by "=".

The equivalence class corresponding to the element 3 of S is {3} and equivalence class corresponding to the element 4 of 5 is {4}. Hence, the partition {{3}, {4}}.

Product of Equivalence Relations

If R and S be equivalence relations in the sets X and Y respectively then the relation R × S in X × Y, defined by :

$(x, y) R \times S (z, u) \Leftrightarrow xRz, ySu$

is called product of relations R and S in that order. It can be verified that R × S is an equivalence relation in X × Y.

2.11 ORDERING RELATIONS AND ORDERED SET

Let there be a given set S. A binary relation R on S is called on ordering relation if, for $a, b, c \subset S$,

(i) aRb then bRa does not hold.
(ii) aRb and $bRc \Leftrightarrow aRc$.
(iii) $a \neq b$, then either aRb or bRa.

The relation R, satisfying above conditions, is said to establish ordering of set S.

The relations "is less than" and "is greater than" establish an ordering on any set of real numbers as well as on each of its subsets.

Similar Ordering

A relation R which orders the set S and another relation R' which orders the set S' are said to establish. Similar ordering if there exists a one-one mapping of the set S' on to the set S, satisfying the identity $xRy \equiv f(x) R' f(y)$, where $f(x)$ is image of x and $f(y)$ is image of y, $x, y \in S$ and $f(x), f(y) \in S$.

Ordering Sets

A set on which an ordering relation is defined is called an ordered set. Some sets are given in a natural order, in which for each two distinct elements, one precedes and one follows. Thus we can order a set of people according to their weight, their height, the alphabetical order of their names, etc. Thus a set S is ordered by assigning an order of precedence to any two distinct elements $x, y \in S$, i.e., by a rule which states, of two distinct elements x, y, that one precedes and other one follows, thus if a precedes b and b follows a, we write

$$a < b, b > a.$$

Thus, the definition of ordering relation given above incorporates this concept of ordering on a set. Thus, the ordered set {1, 2, 3, 4, 5} is not the same as the ordered set {5, 4, 3, 2, 1} though both sets are equal. Concept of equality of sets must be visualized as distinct from similarity of ordered sets.

Similar Sets : Two ordered sets are similar, written as $A \cong B$, if A and B are in one to one correspondence and this correspondence preserves the order i.e.

If $b = f(a)$, $b_1 = f(a_1)$ and $a < a_1 \Rightarrow b < b_1$.

Thus, the ordered set {1, 2, 3, 4, 5, 6, ...} is similar to ordered set {2, 3, 4, 5, 6, ...} but not similar to {3, 4, 5, 6, ...2, 1}. Order is usually determined by the way the elements appear in the set from left to right.

2.12 PARTIAL ORDER RELATION

Let P be a non-empty set. A partial order relation in P is a relation which is symbolized by $\leq$ and assumed to have the following properties :

(i) $x \leq x$ for every x (reflexivity).

(ii) $x \leq y$ and $y \leq x \Rightarrow x = y$ (anti-symmetry).

(iii) $x \leq y$ and $y \leq z \Rightarrow x \leq z$ (transitivity).

A non-empty set P in which there is defined a partial order relation is called a partially order relation.

Example 1. Let P be the set of all positive integers and let $m \leq n$ means m divides n.

Example 2. Let P be the set R of all real numbers and let $x \leq y$ have its usual meaning.

Example 3. Let P be the classes of all subsets of some universal set U and let $A \leq B$ means that A is a subset of B.

Example 4. Let P be the set of all real functions defined on a non-empty set X, and let $f \leq g$ means that $f(x) \leq g(x)$ for every x.

2.13 PARTIALLY ORDERED SET

A set S is said to be partially ordered by a binary relation R, if for arbitrary $a, b, c \in S$,

(i) R is reflexive i.e. $aRa \ \forall \ a \in S$.

(ii) R is antisymmetric, i.e. aRb and bRa iff $a = b$.

(iii) R is transitive, i.e. aRb and $bRc \Rightarrow aRc$.

Thus if S is partially ordered set with respect to R, then:

(i) every subset of S is also partially ordered, (Possibility of being totally ordered not excluded);

(ii) the elements $a \in S$, is called the first element of S, if $aRx \ \forall \ x \in S$; a is unique, if it exists;

(iii) the element $n \in S$ is called the last elements of S, if xRn, $\forall x \in S$; n is unique, if it exists;

(iv) the element $a \in S$ is called a minimum element of S if xRa implies $x = a$, $\forall x \in S$;

(v) the element $n \in S$ is called a maximal element of S if $nRx \Rightarrow n = x$, $\forall x \in S$.

2.14 TOTALLY ORDERED SET

An ordered set S is said to be totally ordered by an ordering relation R defined on S, if for every two arbitrary elements $a, b \in S$, either aRb or bRa.

Thus consider the set $S = \{1, 2, 3, 4, 9\}$. If R stands for "<" then for any two arbitrary elements say $3, 9 \in S$, we find either 3R9 or 9R3. In this case 3R9 holds. Hence for this R, S is totally ordered.

But if R stand for "is a factor of", then 1R2, 2R4, 3R9, 1R3, 1R4, 1R9 exist but neither 2R3, 3R4, 2R9, 4R9 exist. Hence for this R, S is only partially ordered.

Well Ordered Set

An ordered set S, each non-empty subset of which has a first element is called a well-ordered set.

Thus, the set of all natural numbers ordered by the relation "<" is well ordered but set Q of all rational numbers is not well ordered by the relation "<".

Theorem. Every well ordered set is totally ordered.

Proof. Consider a set S, well ordered by the relation R defined on it.

By definition, for arbitrary $a, b \in S$, the subset $\{a, b\}$ of S has a first element and either aRb or bRa. Hence S is totally ordered.

2.15 GEOMETRICAL REPRESENTATION OF ORDERING ON SET

Example 7: Consider the set

$S = \{1, 2, 3, 4, 9\}$, consider the following ordering R defined on it.

(i) R stands for "≤".

(ii) R stands for "is a factor of".

Solution: For this R, 1R2, 1R3, 1R4, 1R9, 2R3, 2R4, 2R9, 3R4, 3R9, 4R9 are true.

Hence, S is totally ordered set with respect to "<".

It has 1 as first element and 9 as last element.

We mark a point O representing the first element 1. Join it to the elements of set in the order given by $1 \le 2 \le 3 \le 4 \le 9$ by arrows.

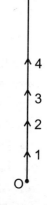

Figure 2.1

Thus totally ordered set gives a straight line as shown in the Fig. 1.

(*ii*) For this R, 1R2, 1R3, 1R4, 1R8, 2R4, 3R9 are true but neither 2R3, 3R4, 4R9, nor 3R2, 4R3, 9R4 are true. Thus this is partially ordered with respect to this R. Diagrammatically this can be represented as shown in the adjoining Figure 2.2.

Upper Bound: An element $a \in A$ is called an upper bound of B, if xRa, $\forall\, x \in B$ where B is a subset of A, partially ordered by R.

Lower Bound: An element $a \in A$ is called a lower bound of subset B of A if aRx, $\forall\, x \in B$ when B is partially ordered by R.

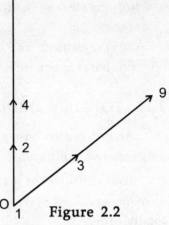

Figure 2.2

If a lower bound of B dominates every other lower bound of B then it is called the greatest lower bound or infimum of B and is represented by in $f(B)$.

If an upper bound of B precedes very other upper bound of B, then it is called the least upper bound or supremum of B and is represented by sup (B).

Example: Let Q be the set of rational numbers in natural order and A be a subset given by:
$$A = \{x : x \in Q,\, 8 < x^3 < 15\}$$

Hence A has both upper as well as lower bounds. In fact there are infinite number of lower bounds and infinitely many upper bounds of A in Q. But there is no sup (A) where as inf (A) = 2.

Zorn's Lemma: If X is a non-empty partially ordered set such that every totally ordered subset P of X has an upper bound, then P contains at least one maximal element.

Example 8: Draw a diagram showing partial ordering on the set

$= \{\{a\}, \{b\}, \{c\}, \{a, b\}, \{b, c\}, \{a, c\}, \{a, b, c\}, \phi\}$ with respect to binary relation "$\subseteq$".

Solution: Since $\phi \subseteq \{a\} \subseteq \{a, b\} \subseteq \{a, b, c\}$

$\phi \subseteq \{b\} \subseteq \{a, b\} \subseteq \{a, b, c\}$

$\phi \subseteq \{c\} \subseteq \{a, c\} \subseteq \{a, b, c\}$

therefore the adjoining Figure 3 gives the diagrammatic representation of partial ordering.

Example 9: If $S = \{2, 3, 4, 5, 6, 7, 8, 9, 10\}$ is ordered by "is a multiple of", find :

(*i*) First element

(*ii*) Last element

(*iii*) All minimal elements

(*iv*) All maximal elements.

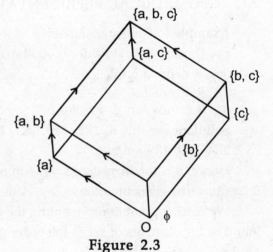

Figure 2.3

Solution: (i) There is no element in S "which is a multiple of" every other element. Hence there is no first element.

(ii) All element of S are not "a multiple of" the same element in S. Hence there is no last element.

(iii) Here $xR6 \Rightarrow x = 6$, $xR8 \Rightarrow x = 8$, $xR9 \Rightarrow x = 9$, $xR10 \Rightarrow x = 10$
$\therefore$ 6, 8, 9, 10 are minimal element.

(iv) Now $2Rx \Rightarrow 2 = x$, $3Rx \Rightarrow 3 = x$, $5Rx \Rightarrow 5 = x$.
Hence, 2, 3, 5 are minimal element.

Example 10: Show that there cannot be two minimal elements in a totally ordered set.

Solution: Let, if possible, there be two distinct minimal elements a, b in a totally ordered set.

As S is totally ordered, $\qquad \therefore$ either aRb or bRa.
But a is a minimal element $\qquad \therefore bRa \Rightarrow b = a$
Also b is a minimal element $\qquad \therefore aRa \Rightarrow b = a$

Hence in either case $a = b$. Thus there can be only one minimal element.

Example 11: Let $X = \{a, b, c, d, e\}$ and $R = \{(a, b), (a, a), (b, a), (b, b), (c, c), (d, d), (d, e), (e, e)\}$ be an equivalence relation on X. Determine the partitions corresponding to R^{-1}, if it is an equivalence relation.

Solution: We know that if R is an equivalence relation on X, then R^{-1} is also an equivalence relation on X. Thus, R^{-1} is

$R^{-1} = \{(b, a), (a, a), (a, b), (b, b), (c, c), (d, d), (e, d), (d, e), (e, e)\}$

The partition corresponding to R^{-1}

$$P = \{\{a, b\}, \{c\}, \{d, e\}\}$$

Example 12: Show that the relation $(x, y) R (a, b) \Leftrightarrow x^2 + y^2 = a^2 + b^2$ is an equivalence relation on the plane and describe the equivalence classes.

Solution: $(x, y) R (a, b) \qquad \Rightarrow \qquad x^2 + y^2 = a^2 + b^2$
$\Rightarrow \qquad a^2 + b^2 = x^2 + y^2$
$\Rightarrow \qquad (a, b) R (x, y)$

Hence, R is symmetric.

Now $(x, y) R (a, b)$ and $(a, b) R (c, d) \qquad \Rightarrow \qquad x^2 + y^2 = a^2 + b^2$ and $a^2 + b^2 = c^2 + d^2$
$\Rightarrow \qquad x^2 + y^2 = c^2 + d^2$

Hence, R is transitive $\qquad \Rightarrow \qquad (x, y) R (c, d)$

Again $(x, y) R (x, y) \Leftrightarrow x^2 + y^2 = x^2 + y^2$
Hence, R is reflexive.

$\Rightarrow$ Hence, R is an equivalence relation.

Now, for any point (x, y) the sum $x^2 + y^2$ is the square of its distance from the origin. The equivalence classes are therefore, the sets of points in the plane which have the same distance from the origin. Thus, the equivalence classes are concentric circles centred on the origin.

Example 13: If R be a relation in the set of integers z defined by :

R = {(x, y) : x∈ z, y∈ z, (x− y) is divisible by 3}.

Describe the distinct equivalence classes of R.

Solution: We can readily verify that R is an equivalence relation on z. we can determine the members of equivalent classes as follows.

For each integer a

[a]	= {x∈ z : xRa}
	= {x∈ z : (x− a) is divisible by 3}
	= {x∈ z : (x− a) = 3k, for some integer k}
	= {x∈ z : x = 3k+ a, for some integer k}

In particular

[0]	= {x∈ z : x = 3k+ 0, for some integer k}
	= {x∈ z : x = 3k, for some integer k}
	= {.... −9, −6, −3, 0, 3, 6, 9, ...}
[1]	= {x∈ z : x = 3k+ 1, for some integer k}
	= {.... −8, −5, −2, 1, 4, 7, ...}
[2]	= {x∈ z : x = 3k+ 2, for some integer k}
	= {.... −7, −4, −1, 2, 5, 8, 11, ...}

There are no equivalence classes, because every integers is already accounted for, in one of [0], [1], [2]. The three equivalence classes are (i) the set of all.

2.16 REPRESENTATION OF RELATIONS

There are many ways of representing relations on finite sets.

By Graphs

Let X and Y are two finite sets and R is a relation from X to Y. For graphical representation of a relation on a set, each element of the set is represented by a point. These points are called nodes or vertices. An arc is drawn from each point to its related point. If the pair $x \in X$, $y \in Y$ is in the relation, the corresponding nodes are connected by arcs called edges or arcs. The arcs start at the first element of the pairs and they go to the second element of the pair. The direction is indicated by an arrow. All arcs with an arrow are called directed arcs. The resulting pictorial representation of R is called a directed graph on digraph of R. An edge of the form (a, a) is represented using an arc from the vertex a back to itself. Such an edge is called loop. The actual location of the vertex is immaterial. The main idea is to place the vertices in such a way that the graph is easy to read.

For example, let A = {2, 4, 6}, B = {4, 6, 8} and R be the relation from the set A to the set

B given by : xRy mean "x is a factor of y", the R = {(2, 4), (2, 6), (2, 8), (4, 4), (6, 6), (4, 8)}. This relation R from A to B is given in the following Figure 2.4

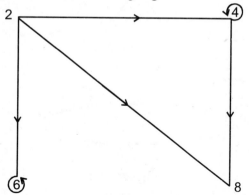

Figure 2.4.

Here this is to be remembered that the digraph of the inverse of a relation has exactly the same edges of the digraph of the original relation, but the directions of the edges are reversed.

Representation of Types of Relations by Graph

The directed graph representing a relation can be used to determine whether the relation has various properties.

  (*i*) A relation is reflexive if and only if there is a loop at very vertex of the directed graph, so that ordered pair of the form (*a*, *a*) occurs in the relation. It helps to identity a relation when it is presented in the graphical form. If no vertex has a loop, then the relation is irreflexive.

  (*ii*) A relation is symmetric if and only if for every edge between distinct vertices in its digraph, there is an edge in the opposite direction, so that (*b*, *a*) is in the relation whenever (*a*, *b*) is in the relation. A relation is antisymmetric if no two distinct points in the digraph have an edge going between them in both directions.

  (*iii*) A relation is transitive if and only if whenever there is a directed edge from vertex *a* to vertex *b* and from vertex *b* to vertex *c*, then there is also a directed edge from *a* to *c*.

Note that every digraph determines a relation, so that we may recover R from the graph. Domain and Range can be found easily if the relation is represented by a graph. Every node with an outgoing arc belongs to the domain, and every node with an incoming arc belongs to the range.

Example 14: Draw the directed graph that represents the relation

$$R = \{(a, a), (b, b), (a, b), (b, c), (c, b), (c, a), (c, c)\}$$
$$X = \{a, b, c\}$$

Solution: Each of there pairs corresponds to an edge of the directed graph with (*a*, *a*), (*b*, *b*), (*c*, *c*) corresponding to loop.

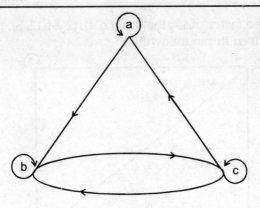

Figure 2.5

Example 15: Determine whether the relation for directed graph shown below are reflexive, symmetric, anti-symmetric and for transitive.

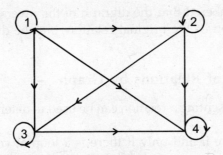

Figure 2.6

Solution. Reflexive: The relation is reflexive since the digraph of the relation has a self loop at every vertex.

Symmetric: The relation is not symmetric. The digraph of the relation has a direct edge from 3 to 2. This means that there are elements 2 and 3 such that 2R3 but 3R2.

Antisymmetric: The relation is antisymmetric. The digraph of the relation has at most one directed edge between each pair of vertices.

Transitive: The relation is transitive. The digraph of the relation has the property whenever there are directed edge from x to y and y to z, there is also a directed edge from x to z.

Matrix of a Relation: Let $A = \{a_1, a_2, a_3, ..., a_n\}$ and $B = \{b_1, b_2, b_3, ..., b_n\}$ are finite sets. Containing m and n elements respectively and let R be a relation from A to B. Then R can be represented by the mn matrix.

$$M_R = \{M_{ij}\} \text{ where}$$

$$M_{ij} = \begin{cases} 1 & \text{if } (a_i, b_j) \in R \\ 0 & \text{if } (a_i, b_j) \notin R \end{cases}$$

The matrix M_R is called the matrix of R.

Example 16: Let R be the relation from the set A = {1, 3, 4} on itself and defined by R = {(1, 1), (1, 3), (3, 3), (4, 4)} then the matrix of R can be found as follows.

Solution: Let M_R denote the matrix of R. Number of rows in M_R = Number of elements in A = 3.

Since the relation from the set A on itself, the number of columns in MR is also 3. So M_R is a 3 × 3 matrix.

We have $\quad a_1 = 1, a_2 = 3$ and $a_3 = 4$
And $\quad b_1 = 1, b_2 = 3$ and $b_3 = 4$
Since 1R1, we have $\quad m_{11} = 1$ as $(a_1, b_1) = (1, 1) = 1$
1R3, we have $\quad m_{12} = 1$ as $(a_1, b_2) = (1, 3) = 1$
3R3, we have $\quad m_{22} = 1$ as $(a_2, b_2) = (3, 3) = 1$
4R4, we have $\quad m_{33} = 1$ as $(a_3, b_3) = (4, 4) = 1$
and all other elements of M_R are zero.

Hence, $\quad M_R = \begin{bmatrix} m_{11} & m_{12} & m_{13} \\ m_{21} & m_{22} & m_{23} \\ m_{31} & m_{32} & m_{33} \end{bmatrix} = \begin{bmatrix} 1 & 1 & 0 \\ 0 & 1 & 0 \\ 0 & 0 & 1 \end{bmatrix}$

Conversely, given sets A and B with $n(A) = m$ and $n(B) = n$, an $m \times n$ matrix whose entries are zeros and ones determines a relation.

Example 17: Let A = {a_1, a_2, a_3} and B = {b_1, b_2, b_3, b_4}, which ordered pairs in the relation R represented by the matrix:

$$M_R = \begin{array}{c} \\ a_1 \\ a_2 \\ a_3 \end{array} \begin{array}{c} b_1 \; b_2 \; b_3 \; b_4 \\ \begin{bmatrix} 0 & 1 & 0 & 0 \\ 1 & 0 & 1 & 1 \\ 1 & 0 & 1 & 0 \end{bmatrix} \end{array}$$

Solution: Since R contains of those ordered pairs (a_i, b_j) with $m_{ij} = 1$. It follows that R = {$(a_1, b_2), (a_2, b_1), (a_2, b_3), (a_2, b_4), (a_3, b_1), (a_3, b_3)$}

If the relation is given by a matrix, then the domain is given by the rows that contains at least one non-zero entry, and the range is similarly, given by the columns with at least one non-zero entry.

If R and S are relation on a set A using operations on Boolean matrices, we can show:

$$M_{R \cup S} = M_R \vee M_S$$
$$M_{R \cap S} = M_R \vee M_S$$
$$M_{R^{-1}} = M_R^1$$

where = M_R^1 is the transpose of M_R.

The matrix representing a relation can be used to determine. Whether the relation has various properties. Let $M_R = [m_{ij}]$ represent the matrix of relation R.

(*i*) **Reflexive:** If all the elements in the main diagonal of the matrix representation of a relation are 1, then the relation is reflexive. The main diagonal elements of the matrix M_R is the set of all m_{ii}.

If all m_{ii} = then the relation is reflexive. Thus

$$M_R = \begin{bmatrix} 1 & 0 & 1 \\ 0 & 1 & 0 \\ 0 & 0 & 1 \end{bmatrix}$$ is the representation of a reflexive relation R.

If all $m_{ii} = 0$, the relation is irreflexive.

$$\begin{bmatrix} 0 & 0 & 1 \\ 0 & 0 & 0 \\ 1 & 1 & 0 \end{bmatrix}$$

Hence, reflexivity and irreflexivity depend only on the diagonal.

(*ii*) **Symmetric:** If the representative matrix of a relation is symmetric with respect to the main diagonal, i.e. $m_{ij} = m_{ji}$ for all values of i and j then the relation is symmetric (i.e. $M_R = M_R^1$).

A relation is antisymmetric if and only if $m_{ij} = 1$ necessitates that $m_{ji} = 0$. The following matrices illustrate the notations of symmetry and anti-symmetry.

Symmetric

$$\begin{bmatrix} 1 & 0 & 1 \\ 0 & 1 & 0 \\ 1 & 0 & 0 \end{bmatrix}$$

Antisymmetric

$$\begin{bmatrix} 0 & 0 & 1 \\ 0 & 1 & 0 \\ 0 & 1 & 0 \end{bmatrix}$$

(*iii*) **Transitive:** There is no simple way to test whether a relation R is a transitive by examining the matrix M_R. A relation R is transitive if and only if its matrix $M_R = [m_{ij}]$ has the property if $m_{ij} = 1$ and $m_{jk} = 1$, then $m_{ik} = 1$. This statement simply means R is transitive if $M_R.M_R$ has 1 in position i, k. Thus, the transitivity of R means that if $M_R^2 = M_R.M_R$ has a 1 in any position then M_R must have a 1 in the same position. Thus R is transitive if and only if $M_R^2 + M_R = M_R$.

Example 18: Let A = {1, 2, 3, 4} and let R be a relation on A whose matrix is

$$M_R = \begin{bmatrix} 1 & 1 & 1 & 1 \\ 0 & 0 & 0 & 0 \\ 1 & 1 & 1 & 1 \\ 0 & 1 & 0 & 0 \end{bmatrix}$$

show that R is transitive.

Solution:
$$M_R^2 = M_R \cdot M_R$$

$$= \begin{bmatrix} 1 & 1 & 1 & 1 \\ 0 & 0 & 0 & 0 \\ 1 & 1 & 1 & 1 \\ 0 & 1 & 0 & 0 \end{bmatrix} \cdot \begin{bmatrix} 1 & 1 & 1 & 1 \\ 0 & 0 & 0 & 0 \\ 1 & 1 & 1 & 1 \\ 0 & 1 & 0 & 0 \end{bmatrix} = \begin{bmatrix} 1 & 1 & 1 & 1 \\ 0 & 0 & 0 & 0 \\ 1 & 1 & 1 & 1 \\ 0 & 0 & 0 & 0 \end{bmatrix}$$

So,
$$M_R^2 + M_R = \begin{bmatrix} 1 & 1 & 1 & 1 \\ 0 & 0 & 0 & 0 \\ 1 & 1 & 1 & 1 \\ 0 & 0 & 0 & 0 \end{bmatrix} + \begin{bmatrix} 1 & 1 & 1 & 1 \\ 0 & 0 & 0 & 0 \\ 1 & 1 & 1 & 1 \\ 0 & 1 & 0 & 0 \end{bmatrix} = \begin{bmatrix} 1 & 1 & 1 & 1 \\ 0 & 0 & 0 & 0 \\ 1 & 1 & 1 & 1 \\ 0 & 1 & 0 & 0 \end{bmatrix} = M_R$$

2.17 MATRIX REPRESENTATION OF COMPOSITION

The matrix for the composite of relations can be found using the Boolean. Product of the matrices. Suppose, R is a relation from A to B and S be a relation from B to C. Suppose that A, B and C have elements m, n and p respectively. The matrices represented by R, S and S_oR are denoted by $M_R = [r_{ij}]_{m \times n}$
$M_S = [s_{ij}]_{n \times p}$ and $M_{RoS} = [t_{ij}]_{m \times p}$ respectively. It follows that $t_{ij} = 1$ if and only if $r_{ik} = S_{kj} = 1$ for some k. From the definition of Boolean product, this means that
$$M_{RoS} = M_R \cdot M_S$$

Theorem: Let A, B and C be finite sets. Let R be a relation from A to B and S be a relation from B to C. Show that
$$M_{ROS} = M_R \cdot M_S$$
where M_R and M_S represent relation matrices of R and S respectively.

Proof: Let $A = \{a_1, a_2, ... a_m\}$, $B = \{b_1, b_2, ..., b_n\}$, $C = \{c_1, c_2, ... c_n\}$. Suppose $M_R = [a_{ij}]$, $M_S = [b_{ij}]$ and $M_{RoS} = [d_{ij}]$. Then $d_{ij} = 1$ if and only if $(a_i, c_j) \in \text{RoS}$, which means that for some k, $(a_i, b_k) \in R$ and $(b, c_j) \in S$. In other words $a_{ik} = 1$ and $b_{kj} = 1$ for some k between 1 and n. If $d_{ij} = 0$, then either $(a_i, a_k) \notin R$ or $(a_k, a_j) \notin S$. This condition is identical to the condition needed for $M_R \cdot M_S$ to have a 1 or 0 in position i, j and thus $M_{RoS} = M_R \cdot M_S$.

By using above theorem, we can say:
- Since the Boolean matrix multiplication is not commutative in general so the composition on relations is not commutative in general.
  i.e. RoS ≠ SoR

- The distribution law is valid for Boolean matrix multiplication. So composition of relation is also satisfies the distributive law, that is :
  (R ∪ S)oT = RoT ∪ SoT

- The Boolean matrix is associative so composition on relations is also associative, that is
  Ro(SoT) = (RoS)oT

Let M_R and M_S denotes the matrices of relations R and S respectively. Then

$$M_R = \begin{bmatrix} 1 & 0 & 1 \\ 0 & 1 & 0 \\ 0 & 0 & 0 \end{bmatrix} \quad M_S = \begin{bmatrix} 0 & 1 & 0 \\ 1 & 1 & 0 \\ 0 & 0 & 1 \end{bmatrix}$$

The matrix of composite relations can be found by using the Boolean product of the matrices:

$$M_{RoS} = M_R \cdot M_S = \begin{bmatrix} 1 & 0 & 1 \\ 0 & 1 & 0 \\ 0 & 0 & 0 \end{bmatrix} \cdot \begin{bmatrix} 0 & 1 & 0 \\ 1 & 1 & 0 \\ 0 & 0 & 1 \end{bmatrix} = \begin{bmatrix} 0 & 1 & 1 \\ 1 & 1 & 0 \\ 0 & 0 & 0 \end{bmatrix}$$

The non-zero entries in the matrix tells us which elements are related to RoS.

Closure of Relations

Let R be a relation on a set A. R may or may not have some property P, such as reflexivity, symmetry or transitivity. If there is a relation S with property P containing R such that S is a subset of every relation with P containing R, then S is called the closure of R with respect to P.

Reflexive Closure

The reflexive closure $R^{(r)}$ of a relation R is the smallest reflexive relation that contains R as a subset. To find the reflexive closure, one therefore has to know what pairs have to be added to the relation to make it reflexive. Given a relation R on a set A, the reflexive closure of R can be formed by adding to R all pairs of the form (a, a) with $A \in R$, not already in R (since for reflexive closure xRx be true for all x). Thus

$$R^{(r)} = R \cup I_A$$

where $I_A = \{(a, a) : a \in A\}$ is the diagonal relation on A. For example, on the set S = {1, 2, 3, 4}, the relation R = {(1, 2), (2, 1), (1, 1), (2, 2)} is not reflexive since, for example 3R3. We can specify this relation with reflexivity by adding (3, 3) and (4, 4) to R, since these are the only pairs of the form (a, a) that are not in R.

The reflexive closure of the < (less than) relation on real numbers is obtained by adding the identity relation which is the equality relation on real numbers and the union of < and = is ≤. Thus, the ≤ relation is the reflexive closure of the relation <.

Symmetric Closure

The symmetric closure $R^{(r)}$ is the smallest symmetric relation that contains R as a subset. A symmetric relation contains (x, y) if it contains (y, x). Since the inverse relation R^{-1} contains (y, x) if (x, y) is in R, the symmetric closure of a relation can be constructed by taking the union of R and R^{-1}, that is $R \cup R^{-1}$ is the symmetric closure of R, where

$$R^{-1} = \{(y, x) : (x, y) \in R\}.$$

Example. If $R = \{(1, 2), (4, 3), (2, 2), (2, 1), (3, 1)\}$ be a relation on $S = \{1, 2, 3, 4\}$. Find the symmetric closure.

Solution. The symmetric closure can be found by taking the union of R and R^{-1}.

Now, $R^{-1} = \{(2, 1), (3, 4), (2, 2), (1, 2), (1, 3)\}$

So $R^{(s)} = R \cup R^{-1} = \{(1, 2), (2, 1), (4, 3), (3, 4), (3, 1), (1, 3)\}$

Transitive Closure

The relation obtained by adding the least number of ordered pairs to ensure transitivity is called the transitive closure of the relation. The transitive closure of R is denoted by R^+. Let a relation R is defined on A and A contains m elements, one never needs more than m steps. Consequently, to make a relation R transitive, one has to add all pairs of R^2, all pairs of R^3, ... all pairs of R^m, unless these pairs are already in R. Thus one can calculate R^+ as the union of the terms of the form R^k.

$$R^+ = R \cup R^2 \cup R^3 ... \cup R^m$$

This follows since there is a path in R^+ between two vertices if and only if there is path between these vertices in R^i, for some positive integer 1 with $1 \leq m$

Thus, if A be a set and R be a binary relation on A. The transitive closure of R on A, denoted by R^+, satisfies three properties:

(i) R^+ is transitive
(ii) $R \subseteq R^+$
(iii) If is any other transitive relation that contains R, then $R^+ \subseteq S$.

We can say, if M denotes the relational matrix of a relation R, then the symmetric closure of R, denoted by M_S, can be obtained from

$$M_S = M \vee M^1$$

The reflexive closure of R, denoted by M_T, can be obtained from :

$$M_1 = M \vee M^2 \vee M^3 \vee ... \vee M^n$$

In form of graphical representation of R we can say:

Reflexive closure: We add all the missing arrows from points to themselves in the digraph of R.

Symmetric closure: We add all the missing reverses of all the arrows in the digraph of R.

Transitive closure: We add an arrow connecting a point x to y whenever some sequence of arrows in the digraph of R connected x to y and there was not an arrow from x to y already.

Theorem Statement:

(i) If R is reflexive, so are $R^{(s)}$ and R^+

(ii) If R is symmetric so are $R^{(r)}$ and R^+
(iii) If R is transitive so is $R^{(r)}$.

Example 19: Let R = {(1, 2), (2, 3), (3, 1)} and A = {1, 2, 3}, find the reflexive symmetric and transitive closure of R using,
  (i) Composition of relation R.
  (ii) Composition of matrix relation R.
  (iii) Graphical representation of R.

Solution: (i) The symmetric closure of R, denoted by $R^{(r)}$, is given by :

$$R^{(r)} = R \cup I_A = \{(1, 2), (2, 3), (3, 1)\} \cup \{(1, 1), (2, 2), (3, 3)\}$$
$$= \{(1, 1), (1, 2), (2, 2), (2, 3), (3, 1), (3, 3)\}$$

The symmetric closure of R, denoted by $R^{(s)}$, is given by :

$$R^{(s)} = R \cup R^{-1} = \{(1, 2), (2, 3), (3, 1)\} \cup \{(2, 1), (3, 2), (1, 3)\}$$
$$= \{(1, 2), (1, 3), (2, 1), (2, 3), (3, 1), (3, 2)\}$$

Now   RoR = {(1, 2), (2, 3), (3, 1)} o {(1, 2), (2, 3), (3, 1)}

R^2 = {(1, 3), (2, 1), (3, 2)}

R^3 = R^2oR = {(1, 3), (2, 1), (3, 2)} o {(1, 2), (2, 3), (3, 1)}
  = {(1, 1), (2, 2), (3, 3)}

R^4 = R^3oR = {(1, 1), (2, 2), (3, 3)} o {(1, 2), (2, 3), (3, 1)}
  = {(1, 2), (2, 3), (3, 1)} = R

Thus, R^5 = R^4oR = RoR = R^2
  R^6 = R^5oR = R^2oR = R^3 and so on

Hence the transitive closure of R, denoted by R^+ is given by :

$R^+ = R \cup R^2 \cup R^3$ = {(1, 1), (1, 2), (1, 3), (2, 1), (2, 2), (2, 3), (3, 1), (3, 2), (3, 3)}

(ii) Let M be the relation matrix of R. Then

$$M = \begin{bmatrix} 0 & 1 & 0 \\ 0 & 0 & 1 \\ 1 & 0 & 0 \end{bmatrix}$$

The symmetric closure matrix of R, denoted by M_S is given by

$$M_S = M \vee M^1$$

$$= \begin{bmatrix} 0 & 1 & 0 \\ 0 & 0 & 1 \\ 1 & 0 & 0 \end{bmatrix} \vee \begin{bmatrix} 0 & 0 & 1 \\ 1 & 0 & 0 \\ 0 & 1 & 0 \end{bmatrix} = \begin{bmatrix} 0 & 1 & 1 \\ 1 & 0 & 1 \\ 1 & 1 & 0 \end{bmatrix}$$

So that $R^{(s)}$ = {(1, 2), (1, 3), (2, 1), (2, 3), (3, 1), (3, 2)}

The reflexive closure matrix of R, denoted by M_R is given by

$$M_R = M \vee I_3$$

$$= \begin{bmatrix} 0 & 1 & 0 \\ 0 & 0 & 1 \\ 1 & 0 & 0 \end{bmatrix} \vee \begin{bmatrix} 1 & 0 & 0 \\ 0 & 1 & 0 \\ 0 & 0 & 1 \end{bmatrix} = \begin{bmatrix} 1 & 1 & 0 \\ 0 & 1 & 1 \\ 1 & 0 & 1 \end{bmatrix}$$

So that $R^{(r)} = \{(1, 1), (1, 2), (2, 2), (2, 3), (3, 1), (3, 3)\}$

Now $M^2 = M \cdot M = \begin{bmatrix} 0 & 1 & 0 \\ 0 & 0 & 1 \\ 1 & 0 & 0 \end{bmatrix} \cdot \begin{bmatrix} 0 & 1 & 0 \\ 0 & 0 & 1 \\ 1 & 0 & 0 \end{bmatrix} = \begin{bmatrix} 0 & 0 & 1 \\ 1 & 0 & 0 \\ 0 & 1 & 0 \end{bmatrix}$

Now $M^3 = M^2 \cdot M = \begin{bmatrix} 0 & 0 & 1 \\ 1 & 0 & 0 \\ 0 & 1 & 0 \end{bmatrix} \cdot \begin{bmatrix} 0 & 1 & 0 \\ 0 & 0 & 1 \\ 1 & 0 & 0 \end{bmatrix} = \begin{bmatrix} 1 & 0 & 0 \\ 0 & 1 & 0 \\ 0 & 0 & 1 \end{bmatrix}$

The transitive closure relation matrix R, denoted by M_T i given by:

$$M_T = M \vee M^2 \vee M^3$$

$$= \begin{bmatrix} 0 & 1 & 0 \\ 0 & 0 & 1 \\ 0 & 0 & 0 \end{bmatrix} \vee \begin{bmatrix} 0 & 0 & 1 \\ 1 & 0 & 0 \\ 0 & 1 & 0 \end{bmatrix} \vee \begin{bmatrix} 1 & 0 & 0 \\ 0 & 1 & 0 \\ 0 & 0 & 1 \end{bmatrix} = \begin{bmatrix} 1 & 1 & 1 \\ 1 & 1 & 1 \\ 1 & 1 & 1 \end{bmatrix}$$

So that $R^+ = \{(1, 1), (1, 2), (1, 3), (2, 1), (2, 2), (2, 3), (3, 1), (3, 2), (3, 3)\}$

(*iii*) The graphical representation of R

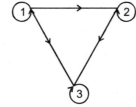

Figure 2.7

To find out the reflexive closure representation of R we add all the arrows from points to themselves.

To find the symmetric closure representation of R, we add missing reverses of all the arrow in graphical representation of R.

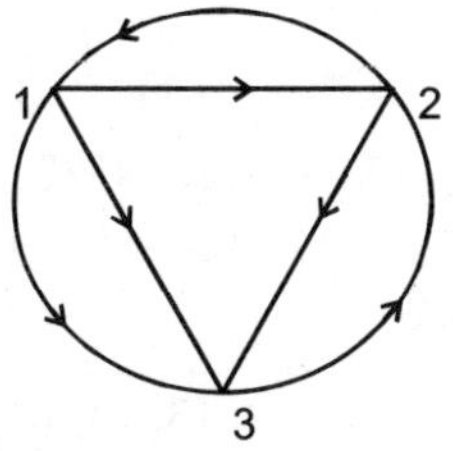

Figure 2.8

To find transitive arrow, we add arrow 1 to 1 since $1 \to 2 \to 3 \to 1$. Similarly 2 to 2 and 3 to 3. Again we add arrow 1 to 3, since $1 \to 2 \to 3$. Similarly, 2 to 1 and 3 to 2.

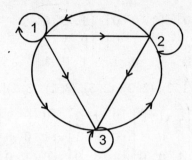

Figure 2.9

Example 20. Consider a relation R whose directed graph is shown in the figure below. Determine its inverse R^{-1} and complement R^1.

Solution. The elements of relation R from the given directed graph is

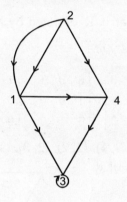

Figure 2.10

R = {(1, 2), (1, 3), (2, 1), (3, 3), (2, 4), (3, 4), (4, 1)}

(i) The inverse of relation R is:

R^{-1} = {(2, 1), (3, 1), (1, 2), (4, 2), (3, 3), (4, 3), (1, 4)}.

The directed graph of R^{-1} is shown below:

Thus, the directed graph of R^{-1} can be obtained from directed graph of R by reversing the directions of arrows on each arc.

The universal relation R × R on the set {1, 2, 3, 4} is :

R × R = {(1, 1), (1, 2), (1, 3), (1, 4), (2, 1), (2, 2), (2, 3), (2, 4), (3, 1), (3, 2), (3, 3), (3, 4), (4, 1), (4, 2), (4, 3), (4, 4)}

Thus the complement of R is :

R' = {(1, 1), (1, 4), (2, 2), (2, 3), (3, 2), (3, 1), (4, 4), (4, 3), (4, 2)}.

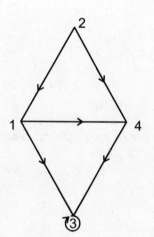

Figure 2.11

The directed graph of R' can be shown as in figure 2.12

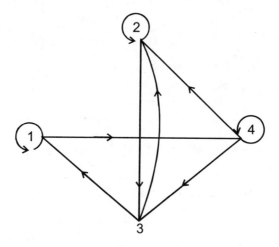

Figure 2.12

Example 21: Consider a relation R defined on A = {1, 2, 3} whose matrix representation is given below. Determine its inverse R^{-1} and complement R'.

$$M_R = \begin{bmatrix} 1 & 0 & 0 \\ 1 & 1 & 1 \\ 0 & 0 & 1 \end{bmatrix}$$

Solution: We have

$$M_{R^{-1}} = (M_R') = \text{transpose of } M_R$$

$$= \begin{bmatrix} 1 & 1 & 0 \\ 0 & 1 & 0 \\ 0 & 1 & 1 \end{bmatrix}$$

So, the inverse of R is

$$R^{-1} = \{(1, 1), (1, 2), (2, 2), (3, 2), (3, 3)\}$$

where R = {(1, 1), (2, 1), (2, 2), (2, 3), (3, 3)}

Again, we know M_R' can be obtained by changing 0 into 1 and 1 into 0 in M_R.

$$M_{R'} = \begin{bmatrix} 0 & 1 & 1 \\ 0 & 0 & 0 \\ 1 & 1 & 0 \end{bmatrix}$$

$$R' = \{(1, 2), (1, 3), (3, 1), (3, 2)\}$$

Example 22: Let A = {1, 2, 3} and B = {a, b, c, d}. Let R and S be the relations from A to B with the Boolean matrices.

$$M_R = \begin{bmatrix} 1 & 0 & 1 & 0 \\ 0 & 1 & 0 & 0 \\ 1 & 0 & 0 & 1 \end{bmatrix} \text{ and } M_S = \begin{bmatrix} 0 & 1 & 0 & 0 \\ 1 & 0 & 0 & 1 \\ 0 & 1 & 1 & 0 \end{bmatrix}$$

(i) Find Boolean matrices for R^{-1} and S^{-1}.

(ii) Find Boolean matrices for $(R \cap S)oR^{-1}$ and $RoR^{-1} \cap SoR^{-1}$.

(iii) Find Boolean matrices for $So(R^{-1} \cup S^{-1})$ and $SoR^{-1} \cup SoS^{-1}$.

(iv) Using (ii) and (iii) show that $(R \cap S)oR^{-1} \neq RoR^{-1} \cap SoR^{-1}$ and $So(R^{-1} \cup S^{-1}) = SoR^{-1} \cup SoS^{-1}$.

Solution: (i) Boolean matrices for R^{-1} and S^{-1} are :

$$M_{R^{-1}} = (M_R)' = \begin{bmatrix} 1 & 0 & 1 \\ 0 & 1 & 0 \\ 1 & 0 & 0 \\ 0 & 0 & 1 \end{bmatrix}$$

$$M_{S^{-1}} = (M_S)' = \begin{bmatrix} 0 & 1 & 0 \\ 1 & 0 & 1 \\ 0 & 0 & 1 \\ 0 & 1 & 0 \end{bmatrix}$$

(ii) Again, $M_{R \cap S} = M_R \wedge M_S$

$$= \begin{bmatrix} 1 & 0 & 1 & 0 \\ 0 & 1 & 0 & 0 \\ 1 & 0 & 0 & 1 \end{bmatrix} \wedge \begin{bmatrix} 0 & 1 & 0 & 0 \\ 1 & 0 & 0 & 1 \\ 0 & 1 & 1 & 0 \end{bmatrix} = \begin{bmatrix} 0 & 0 & 0 & 0 \\ 0 & 0 & 0 & 0 \\ 0 & 0 & 0 & 0 \end{bmatrix}$$

$M_{(R \cap S)oR^{-1}} = M_{R \cap S} \cdot M_{R^{-1}}$

$$= \begin{bmatrix} 0 & 0 & 0 & 0 \\ 0 & 0 & 0 & 0 \\ 0 & 0 & 0 & 0 \end{bmatrix} \cdot \begin{bmatrix} 1 & 0 & 1 \\ 0 & 1 & 0 \\ 1 & 0 & 0 \\ 0 & 0 & 1 \end{bmatrix} = \begin{bmatrix} 0 & 0 & 0 \\ 0 & 0 & 0 \\ 0 & 0 & 0 \end{bmatrix}$$

Now, $M_{RoR^{-1}} = M_R \cdot M_{R^{-1}}$

$$= \begin{bmatrix} 1 & 0 & 1 & 0 \\ 0 & 1 & 0 & 0 \\ 1 & 0 & 0 & 1 \end{bmatrix} \cdot \begin{bmatrix} 1 & 0 & 1 \\ 0 & 1 & 0 \\ 1 & 0 & 0 \\ 0 & 0 & 1 \end{bmatrix} = \begin{bmatrix} 1 & 0 & 1 \\ 0 & 1 & 0 \\ 1 & 0 & 1 \end{bmatrix}$$

$$M_{SoR^{-1}} = M_S \cdot M_{R^{-1}}$$

$$= \begin{bmatrix} 0 & 1 & 0 & 0 \\ 1 & 0 & 0 & 1 \\ 0 & 1 & 1 & 0 \end{bmatrix} \cdot \begin{bmatrix} 1 & 0 & 1 \\ 0 & 1 & 0 \\ 1 & 0 & 0 \\ 0 & 0 & 1 \end{bmatrix} = \begin{bmatrix} 0 & 1 & 0 \\ 1 & 0 & 1 \\ 1 & 1 & 0 \end{bmatrix}$$

$$M_{RoR^{-1}} \wedge M_{SoR^{-1}} = \begin{bmatrix} 1 & 0 & 1 \\ 0 & 1 & 0 \\ 1 & 0 & 1 \end{bmatrix} \wedge \begin{bmatrix} 0 & 1 & 0 \\ 0 & 0 & 1 \\ 1 & 1 & 0 \end{bmatrix} = \begin{bmatrix} 0 & 0 & 0 \\ 0 & 0 & 0 \\ 1 & 0 & 0 \end{bmatrix}$$

(c) $M_{R^{-1}} \vee M_{S^{-1}} = \begin{bmatrix} 0 & 1 & 0 \\ 1 & 0 & 0 \\ 0 & 0 & 1 \end{bmatrix} \vee \begin{bmatrix} 1 & 0 & 1 \\ 0 & 0 & 1 \\ 0 & 1 & 0 \end{bmatrix} = \begin{bmatrix} 1 & 1 & 1 \\ 1 & 0 & 1 \\ 0 & 1 & 1 \end{bmatrix}$

$$M_{So(R^{-1} \cup S^{-1})} = M \cdot M_{R^{-1} \cup S^{-1}}$$

$$= \begin{bmatrix} 0 & 1 & 0 & 0 \\ 1 & 0 & 0 & 1 \\ 0 & 1 & 1 & 0 \end{bmatrix} \cdot \begin{bmatrix} 1 & 1 & 1 \\ 1 & 1 & 1 \\ 1 & 0 & 1 \\ 0 & 0 & 1 \end{bmatrix} = \begin{bmatrix} 1 & 1 & 1 \\ 1 & 1 & 1 \\ 1 & 1 & 1 \end{bmatrix}$$

Again

$$M_{SoR^{-1}} = M_S \cdot M_{R^{-1}}$$

$$= \begin{bmatrix} 0 & 1 & 0 & 0 \\ 1 & 0 & 0 & 1 \\ 0 & 1 & 1 & 0 \end{bmatrix} \cdot \begin{bmatrix} 1 & 0 & 1 \\ 0 & 1 & 0 \\ 1 & 0 & 0 \\ 0 & 0 & 1 \end{bmatrix} = \begin{bmatrix} 0 & 1 & 0 \\ 1 & 0 & 1 \\ 1 & 1 & 0 \end{bmatrix}$$

and

$$M_{SoS^{-1}} = M_S \cdot M_{S^{-1}}$$

$$= \begin{bmatrix} 0 & 1 & 0 & 0 \\ 1 & 0 & 0 & 1 \\ 0 & 1 & 1 & 0 \end{bmatrix} \cdot \begin{bmatrix} 0 & 1 & 0 \\ 1 & 0 & 1 \\ 0 & 0 & 1 \\ 0 & 1 & 0 \end{bmatrix} = \begin{bmatrix} 1 & 0 & 1 \\ 0 & 1 & 0 \\ 1 & 0 & 1 \end{bmatrix}$$

Now, $M_{SoR^{-1}} \vee M_{SoS^{-1}} = \begin{bmatrix} 0 & 1 & 0 \\ 1 & 0 & 1 \\ 1 & 1 & 0 \end{bmatrix} \vee \begin{bmatrix} 1 & 0 & 1 \\ 0 & 1 & 0 \\ 1 & 0 & 1 \end{bmatrix} = \begin{bmatrix} 1 & 1 & 1 \\ 1 & 1 & 1 \\ 1 & 1 & 1 \end{bmatrix}$

(iv) from (ii) we get

$$M_{(R \cap S) o R^{-1}} \neq M_{R o R^{-1}} \wedge M_{S o S^{-1}}$$

$$\therefore (R \cap S) o R^{-1} \neq R o R^{-1} \cap S o R^{-1}$$

and from (c), we see

$$M_{S o (R^{-1} \cup S^{-1})} = M_{S o R^{-1}} \vee M_{S o S^{-1}}$$

$$\therefore \text{So,}\, (R^{-1} \cup S^{-1}) = S o R^{-1} \cup S o S^{-1}$$

Note : (a) $So(R \cup T) = SoR \cup SoT$

(b) $(R \cap S) o T \subseteq R o T \cap S o T$ and equality need not hold.

(c) $Ro(S \cap T) \subseteq RoS \cap RoT$ and equality need not hold.

For example :

$$M_R = \begin{bmatrix} 1 & 1 \\ 0 & 0 \end{bmatrix}, M_S = \begin{bmatrix} 0 & 0 \\ 0 & 1 \end{bmatrix} \text{ and } M_T = \begin{bmatrix} 0 & 1 \\ 0 & 0 \end{bmatrix}$$

$$M_{Ro(S \cap T)} = \begin{bmatrix} 1 & 1 \\ 0 & 0 \end{bmatrix} \cdot \begin{bmatrix} 0 & 0 \\ 0 & 0 \end{bmatrix} = \begin{bmatrix} 0 & 0 \\ 0 & 0 \end{bmatrix} \text{ and}$$

$$M_{RoS \cap RoT} = \begin{bmatrix} 1 & 1 \\ 0 & 0 \end{bmatrix} \cdot \begin{bmatrix} 0 & 0 \\ 0 & 1 \end{bmatrix} \wedge \begin{bmatrix} 1 & 1 \\ 0 & 0 \end{bmatrix} \cdot \begin{bmatrix} 0 & 1 \\ 0 & 0 \end{bmatrix}$$

$$= \begin{bmatrix} 0 & 1 \\ 0 & 0 \end{bmatrix} \wedge \begin{bmatrix} 0 & 1 \\ 0 & 0 \end{bmatrix} = \begin{bmatrix} 0 & 1 \\ 0 & 0 \end{bmatrix}$$

2.18 WARSHALL'S ALGORITHM

A more efficient method for computing the transitive closure of a relation is known as Warshall's algorithm named after Stephen Warshall.

The goal of this approach is to generate a sequence of Matrices $P^0, P^1, P^2, \ldots P^k, \ldots P^m$ for a graphs of vertices.

With $P^n = P$ (the path matrix). Initially, $P^0 = A$ (the adjacency matrix).

The first iteration consists of exploring the existence of paths from any vertex to any other either directly via an edge or indirectly through the intermediate or pivot vertex, say v_1. P^1 denotes the resulting matrix with its general element $P^{(1)}_{ij}$ obtained as follows:

$$P^{(1)}_{ij} \begin{cases} 1, & \text{If there exists an edge from } v_i \text{ to } v_j \text{ on there is a path (of length 2) from } v_i \\ & \text{to } v_i \text{ and } v_i \text{ to } v_j. \\ 0, & \text{otherwise.} \end{cases}$$

The second iteration is to explore any paths from any vertex to any other with v_1 or v_2 or both as pivots. We compute P^2 and define its general element $P_{ij}^{(2)}$ as follows:

$$P_{ij}^{(2)} = \begin{cases} 1, & \text{If there exists an edge from } v_i \text{ to } v_j \text{ or a path from } v_i \text{ to } v_j \text{ using only pivots} \\ & \text{(intermediate vertices) from } \{v_1, v_2\}. \\ 0, & \text{otherwise.} \end{cases}$$

In general, during the kth iteration is to explore any paths from the get $\{v_1, v_2, \ldots v_k\}$. The result of kth iteration is to compute P^k, where

$$P_{ij}^{(k)} = \begin{cases} 1, & \text{If there exists an edge from } v_i \text{ to } v_j \text{ or a path from } v_1 \text{ to } v_2 \text{ using only pivots} \\ & \text{from } \{v_1, v_2, \ldots v_k\}. \\ 0, & \text{otherwise.} \end{cases}$$

Now, we can compute $P_{ij}^{(k)}$ from the previous iteration as follows:

$$P_{ij}^{(k)} = P_{ij}^{(k-1)} \vee \left(P_{ik}^{(k-1)} \wedge P_{kj}^{(k-1)} \right)$$

That is:

$$P_{ij}^{(k)} = 1 \text{ if } P_{ij}^{(k-1)} = 1 \text{ or both } P_{ik}^{(k-1)} = 1 \text{ and } P_{kj}^{(k-1)} = 1$$

The only way that the value of $P_{ij}^{(k)}$ can change 0 is to find a path through v_k, that is, there is a path from v_j to v_k and a path from v_k to v_j.

Algorithm: *Warshall Algorithm* : Procedure Warshall (M_R : $n \times n$ zero-obe matrix)

$P^{(0)} = M_R$.

From $k = 1$ to n

begin for $i = 1$ to n

begin for $j = 1$ to n

$$P_{ij}^{(k)} = P_{ij}^{(k-1)} \vee \left(P_{ik}^{(k-1)} \wedge P_{kj}^{(k-1)} \right)$$

end

end $\{P_{ij}^{(n)}\}$

Example 23: Let R be a relation with given directed graph. Find the matrix of transitive closure of R using Warshall algorithm.

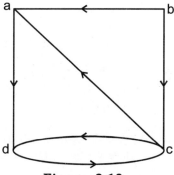

Figure 2.13

Solution:
$$M_R = \begin{bmatrix} 0 & 0 & 0 & 1 \\ 1 & 0 & 1 & 0 \\ 1 & 0 & 0 & 1 \\ 0 & 0 & 1 & 0 \end{bmatrix} = P^{(0)}$$

$P^{(1)}$ has 1 as its (i, j)th entry if there is a path (edge) from v_i to v_j that has only $v_i = a$ as an intermediate vertex. Note that $P^{(0)}$ has a 1 in location (1, 4), (2, 1), (2, 3), (3, 1), (3, 4) and (4, 3), $P^{(1)}$ will also have a 1 in the corresponding locations. This is because all paths of length one can still be used since they have no intermediate vertex. The changes in $P^{(1)}$ occurs for

$$P^{(1)}_{(2,4)} = 1 \text{ because } P^{(0)}_{(2,1)} = 1 \text{ and } P^{(0)}_{(1,4)} = 1$$

$$P^{(1)} = \begin{bmatrix} 0 & 0 & 0 & 1 \\ 1 & 0 & 1 & 1 \\ 1 & 0 & 0 & 1 \\ 0 & 0 & 1 & 0 \end{bmatrix}$$

$P^{(2)}$ has 1 as its (i, j)th entry if there is a path from v_i to v_j that along $v_1 = a$ on $v_2 = b$ as its intermediate vertices. $P^{(1)}$ has 1 in locations (1, 4), (2, 1), (2, 3), (3, 1), (3, 4), (4, 3) and (2, 4). Since there is no 1 in 2nd column of $P^{(1)}$, there is no edge that have b as terminal vertex. So no new path is obtained when we permit b as terminal vertex and, therefore, no new 1's arc inserted in $P^{(1)}$. Thus

$$P^{(2)} = P^{(1)} = \begin{bmatrix} 0 & 0 & 0 & 1 \\ 1 & 0 & 1 & 1 \\ 1 & 0 & 0 & 1 \\ 0 & 0 & 1 & 0 \end{bmatrix}$$

$P^{(3)}$ has 1 has its (i, j)th entry, if there is a path from v_i to v_j that has only $v_1 = a$, $v_2 = b$ and/or $v_3 = c$ as intermediate vertices. The changes in $P^{(3)}$ occurs for.

$$P^{(3)}_{(4,4)} = 1 \text{ because } P^{(2)}_{(4,3)} = 1 \text{ and } P^{(2)}_{(3,4)} = 1$$

$$P^{(3)}_{(4,1)} = 1 \text{ because } P^{(2)}_{(4,3)} = 1 \text{ and } P^{(2)}_{(3,1)} = 1$$

$$P^{(3)} = \begin{bmatrix} 0 & 0 & 0 & 1 \\ 1 & 0 & 1 & 1 \\ 1 & 0 & 0 & 1 \\ 1 & 0 & 1 & 1 \end{bmatrix}$$

Finally, $P^{(4)}$ has 1 as its (i, j)th entry, if there is a path from v_i to v_j that has only $v_1 = a$, $v_2 = b$, $v_3 = c$ and/or $v_4 = d$ as intermediate vertices. The changes in $P^{(4)}$ occurs.

For

$P^{(4)}_{(1,1)} = 1$ because $P^{(3)}_{(1,4)} = 1$ and $P^{(3)}_{(4,1)} = 1$

$P^{(4)}_{(1,3)} = 1$ because $P^{(3)}_{(1,4)} = 1$ and $P^{(3)}_{(4,3)} = 1$

$P^{(4)}_{(3,3)} = 1$ because $P^{(3)}_{(3,4)} = 1$ and $P^{(3)}_{(4,3)} = 1$

$$P^{(4)} = \begin{bmatrix} 1 & 0 & 1 & 1 \\ 1 & 0 & 1 & 1 \\ 1 & 0 & 1 & 1 \\ 1 & 0 & 1 & 1 \end{bmatrix}$$ is the matrix of transitive closure.

Example 24: Using Warshall algorithm find all transitive. Closure of the relation $R = \{(1, 2), (2, 3), (3, 3)\}$ on the set $A = \{1, 2, 3\}$.

Solution: The matrix of R is

$$M_R = \begin{bmatrix} 0 & 1 & 0 \\ 0 & 0 & 1 \\ 0 & 0 & 1 \end{bmatrix} = P^{(0)}$$

$P^{(1)}$ has 1 as its (i, j)th entry if there is a path from v_i to v_j that has only $v_1 = 1$ as an intermediate vertex. Since the Ist column of $P^{(0)}$ has no 1, there is no edge that have 1 as terminal vertex. So no new path is obtained and, therefore, no new 1's are inserted in $P^{(1)}$. Thus

$$P^{(1)} = P^{(0)} = \begin{bmatrix} 0 & 1 & 0 \\ 0 & 0 & 1 \\ 0 & 0 & 1 \end{bmatrix}$$

$P^{(2)}$ has 1 as its (i, j)th entry if there is a path from v_i to v_j that has only $v_1 = 1$ and/or $v_2 = 2$ as intermediate vertices. The changes in $P^{(2)}$ occurs for :

$P^{(2)}_{(1,3)} = 1$ because $P^{(1)}_{(1,2)} = 1$ and $P^{(1)}_{(2,3)} = 1$

$$P^{(2)} = \begin{bmatrix} 0 & 1 & 1 \\ 0 & 0 & 1 \\ 0 & 0 & 1 \end{bmatrix}$$

Finally, $P^{(3)}$ has 1 as its (i, j)th entry, if there is a path from v_i to v_j that has only $v_1 = 1$, $v_2 = 2$ and/or $v_3 = 3$ as its path. Observe that there is no edges that have 3 as intermediate vertex. So no new path is obtained and, therefore, no new 1's are inserted in $P^{(3)}$.

$$P^{(3)} = P^{(2)} = \begin{bmatrix} 0 & 1 & 1 \\ 0 & 0 & 1 \\ 0 & 0 & 1 \end{bmatrix}$$ is matrix of transitive closure of R.

Hence, the transitive closure of R,

$$R^+ = \{(1, 2), (1, 3), (2, 3), (3, 3)\}$$

EXERCISE 2.3

1. Partition the set $\left\{1, 2, 5, \sqrt{2}, \sqrt{3}, \pi, e, \dfrac{1}{2}, \dfrac{1}{3}, \dfrac{1}{4}, 3, 4\right\}$.

2. If R is the set of all real numbers, find on equivalence relation including the partitioning S = {0}, {x : x is a +ve real number}, {x : x is –ve real number}.

3. Draw diagrams to represent ordering 'divides' on the sets
   (i) {1, 3, 5, 6, 9, 10}   (ii) {2, 4, 5, 8, 15, 45, 60}

4. Draw diagrams to represent ordering 'less than' defined on the sets :
   (i) {3, 5, 7, 9, 11}   (ii) {3, 6, 8, 5, 7}
   (iii) {1, 3, 5, 15, 30, 45}

5. Show that the ordered set of all subsets A = {a, b, c} has ϕ as minimal element and A as maximal element with respect to the binary relation '$\subseteq$'.

6. Show that the ordered set {2, 4, 5, 15, 160} has a last element but no first element with respect to a binary relation 'is a factor of'.

7. Test the truth or otherwise of the following:
   (i) If partially ordered set A has only one maximal element a, then has also a last element.
   (ii) If a is the only maximal element in totally ordered set A, it is also a last celement.

8. Let A = {2, 3, 4, 7, 9, 12} be the set ordered by relation "divides". Find an upper bound, a lower bound, supremum and infimum, if any.

9. Let A = {0, 1, 2, 3, 4}. Show that the relation R = {(0, 0), (0, 4), (1, 1), (1, 3), (2, 2), (3, 1), (3, 3), (4, 0), (4, 4)} is an equivalence relation. Find the distinct equivalence classes of R.

10. Let R be the relation of congruence modulo 3. Which of the following equivalence classes are equal?
    [7], [– 4], [– 6], [7], [– 4], [27], [19]

CHAPTER 3

FUNCTION

3.1 CORRESPONDENCE

Let us consider the set $A = \{a_1, a_2, a_3 \ldots\ldots a_{n3}\}$ of all authors who have written the books which from the set $B = \{b_1, b_2, b_3 \ldots.. b_n\}$. Let us concern ourselves here, with the natural association of each book of the set B with authors of Set A.

This process of associating an element of B with an element of A may result in associating b_1 with a_2, b_2 with a_3, b_5 with a_2, b_7 with a_1, ..., b_3 with a_5. This association of the elementary one set B with the elements of another set A, is called correspondence.

3.2 TYPES OF CORRESPONDENCE

(i) One to one correspondence

If each element of A corresponds to one and only one element of another set B, and each element of B corresponds to one and only one element of A, we say there is one to one correspondence between the elements of A and the elements of B.

Let,
$$A = \{a_1, a_2, a_3, a_4, a_5\}$$
$$B = \{b_1, b_2, b_3, b_4, b_5\}$$

Let,
$a_1 \to b_2$, $a_2 \to b_3$, $a_3 \to b_4$, $a_4 \to b_5$, $a_5 \to b_1$ or graphically we can show.

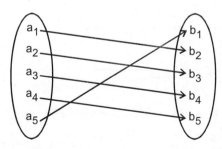

So there is a one to one correspondence between elements of set A and elements of set B.

(ii) Many to one correspondence

If at least two elements of set A correspond to only one element of set, then there is many to one correspond between set A and set B.

We can see it graphically as —

Let, $A = \{a_1, a_2, a_3, a_4, a_5\}$ and $B = \{b_1, b_2, b_3, b_4, b_5\}$

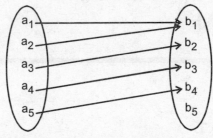

(iii) One to many correspondence

If some elements of A correspond to more than one element of another set B, then this type of correspondence is called one to many correspondence.

Let, $A = \{a_1, a_2, a_3, a_4, a_5\}$ $B = \{b_1, b_2, b_3, b_4, b_5\}$

Figure below shows one to many correspondence

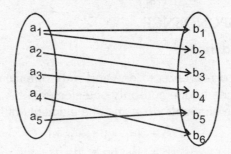

(iv) Many to many correspondence:
If one or more elements of A corresponds to one or more elements of B, then this type of correspondence is called many to many correspondence.

Figure below show many to many correspondence.

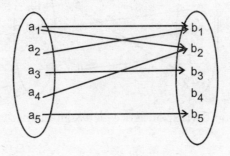

3.3 MAPPING (FUNCTION)

Let X and Y be two given sets. If y be any given rule or operation there corresponds to each element $x \in X$, a unique element $y \in Y$, then this correspondence denoted by f, is called mapping of X into Y or f is called function of X to Y.

f-image: The element $y \in Y$, which due to mapping f, corresponds to an element $x \in X$ is denoted by the symbol $f(x)$, i.e. $y = f(x)$.

$f(x)$ is called the f-image of x or the value of the function for x.

f-set: The set of all f-images of the element of X, is called image set and is denoted by $f(X)$ or $\{f(x)\}$.

The statement "the mapping f of X to Y is dented by $f : X \to Y$.

3.4 TYPES OF MAPPING (FUNCTIONS)

There are two types of mapping:

(1) Onto or Surjective Mapping: In this mapping, every element of Y is an f-image of some of $x \in X$, i.e., there is no element in Y which has no correspondence with any element of X such a correspondence is called the mapping of X onto Y.

Hence the mapping $f : X \to Y$ is said to be onto mapping if

f-set $\qquad \{f(x)\} = Y, \forall\ x \in x$

Let $\qquad X = \{x_1, x_2, x_3, x_4\}\ Y = \{y_1, y_2, y_3, y_4\}$

Then figure below shows onto mapping.

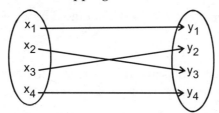

(2) Into or Injective Mapping: In this mapping at least one element Y is not an f-image of any $x \in X$. Such a correspondence is called mapping of X into Y.

Hence, the mapping $f : X \to Y$ is called mapping of X into Y if $\{f(x)\} \subset y, \forall\ x \in X$.

Figure below shows into mapping

Let, $\qquad X = \{x_1, x_2, x_3, x_4, x_5\}\ Y = \{y_1, y_2, y_3, y_4, y_5, y_6, y_7\}$

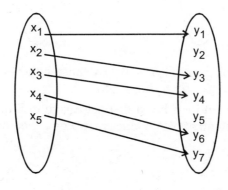

Equivalent sets: Two sets whose element can be placed in one to one correspondence are called equivalent set or cardinally equivalent sets.

Countable or denumerable set: If one to one correspondence exists between a set A and the set N of all natural numbers the set A is called countable or denumerable set

Domain and Co-domain: In any mapping $f:X \to Y$, the set X is called the domain and Y is called the co-domain of mapping f.

Range of mapping: The f-set is called the range of f.

One-one mapping: A mapping $f: X \to Y$ is called one-one mapping if every distinct element of X is mapped to a distinct element of Y, *i.e.* no element of Y is the image of more than one element in X.

$f: X \to Y$ is one-one if for $x_1, x_2, \in X$

$f(x_1) = f(x_2) \Rightarrow x_1 = x_2$

Many-One Mapping: A mapping $f:X \to Y$ is called many one mapping if at least one element of Y has two or more pre-images in X.

Symbolically

$f:X \to Y$ is many-one, if for $x_1, x_2, \in X$.

$f(x_1) = f(x_2)$ does not imply $x_1 = x_2$

$f(x_1) = f(x_2) \qquad \Rightarrow \quad x_1 \neq x_2$ when $x_1, x_2 \in X$

$x_1 \neq x_2 \qquad \Rightarrow f(x_1) = f(x_2)$ when $x_1, x_2 \in X$

Kinds of Mappings: Mappings are of four kinds—

(i) One-One into Mapping: Let $f: A \to B$ be defined by $a_1 \to b_1, a_2 \to b_2, a_3 \to b_3, a_4 \to b_4$ where $A = \{a_1, a_2, a_3, a_4\}$ and

$B = \{b_1, b_2, b_3, b_4, b_5, b_6\}$

Here no two elements have the same image. Hence the mapping is one-one. Also there are element in B, which has no pre-image in A. Hence mapping is into

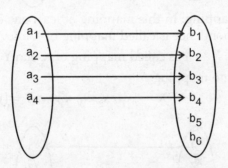

(ii) One-One on to Mapping: Let $f: A \to B$, be defined by $a_1 \to b_1, a_2 \to b_2, a_3 \to b_3, a_4 \to b_4$ where $A = \{a_1, a_2, a_3, a_4\}$, $B = \{b_1, b_2, b_3, b_4\}$.

Here each distinct element in A has a distinct image in B and there is no element in B which is not f-image of any element of A. Hence mapping is one-one onto also called bijective mapping.

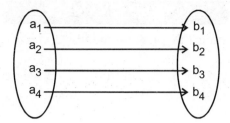

(*iii*) **Many-One into Mapping:** Let A = {$a_1, a_2, a_3, a_4, a_5, a_6$} and B = {$b_1, b_2, b_3, b_4$} and let $f: A \to B$ be defined by $a_1 \to b_1, a_2 \to b_2, a_3 \to b_3, a_4 \to b_3, a_5 \to b_3, a_6 \to b_4$

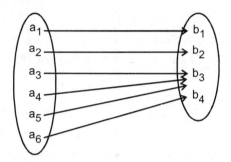

(*iv*) **Many-one-onto Mapping:** Here let the mapping $f: A \to B$ be defined by $a_1 \to b_1, a_2 \to b_2, a_3 \to b_2, a_4 \to b_3$ where
A = {a_1, a_2, a_3, a_4} and B = {b_1, b_2, b_3, b_4}

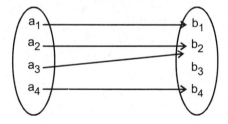

3.5 CLASSIFICATION OF FUNCTION

(*i*) **Constant mapping or constant function:** The function f or mapping f defined on a set x, such that $f(x) = a \ \forall \ x \in X$ is called a constant function on X.

i.e. every element of X is mapped onto the same single element a of another given set.

(*ii*) **Identity mapping:** If every element of a set X is mapped onto itself then the mapping is called identity mapping *i.e.* $x: X \to X$ is defined by $f(x) = x, \forall \ x \in X$.

(*iii*) **Transformation:** A mapping $f: x \to x$, *i.e.* mapping of a set into is called a transformation, *i.e.* mapping f is called transformation if domain of f - co-domain of f.

(*iv*) **Inclusion mapping:** A mapping $f: X \to Y$; defined by $f(x) = x$; if $X \subseteq Y$ is called inclusion map of X to Y.

SOLVED EXAMPLES

Example 1: For a mapping $f : X \to Y$; defined by $f(x) = x - 2$, $\forall\, x \in X$, find f-image of $x = 0, 3, 5, -1, -2 \in X$.

Solution: By putting $x = 0, 3, 5, -1, 2$ in $f(x) = x - 2$ we get
$f(0) = 0 - 2 = -2$; $f(3) = 3 - 2 = 1$; $f(5) = 5 - 2 = 3$; $f(-1) = -1 - 2 = -3$
$f(-2) = -2 - 2 = -4$

Hence Images are $-2, 1, 3, -3 -4$, respectively.

Example 2: For each mapping defined by the following set of ordered pairs, find out domain and the range.

(i) $\{(1, 1), (2, 4) (3, 9) (4, 16) \ldots$ and inf. $\}$

(ii) $f = \{(x, y) : x$ is the positive integer, $y = -x^2\}$.

Solution: Domain is found out by forming the set of first components of ordered pairs.

Hence domain = $\{1, 2, 3, 4 \ldots$ and inf$\}$.

Range is find out by forming the set of 2nd component of ordered pairs.

$\therefore$ Range = $\{1, 4, 9, 16 \ldots\}$.

(ii) Hence x is first component of ordered pairs and each x is positive integer. Hence domain is the set of all positive integers.

$\therefore$ Domain = $\{1, 2, 3, 4, \ldots$ and inf.$\}$

By putting $x = 1, 2, 3\ldots$ in $y = -x^2$ we get $y = -1, -4, -9, \ldots$

These are second component of orders pairs.

Hence range = $\{-1, -4, -9, \ldots$ an inf$\}$.

Example 3: Given $A = \{2, 3, 4\}$, $B = \{2, 5, 6, 7\}$ construct an example of each of the following.

(i) An injective mapping from A to B.

(ii) A mapping from A to B which is not injective

(iii) A mapping from B to A.

Solution: (i) An injective (i.e. one-one) mapping A to B may be defined as–
$$f = \{(2,5), (3,7), (4,6)\}.$$

(ii) A mapping from A to B which is not injective may be defined as–
$$g = \{(2, 2), (3, 5), (4, 2)\}$$

(iii) A mapping from B to A may be defined as–
$$h = \{(2, 2), (5, 3), (6\ 4), (7, 4)\}$$

Example 4: If $X = \{1, 2, 3, 4, 5\}$ and $Y = \{1, 3, 5, 7, 9\}$ Find $X \cap Y$ and $(X - Y) \cup (Y - X)$. Determine which of the following sets are (i) mappings (ii) relations (iii) neither; of X to Y.

(a) $F = \{\{x, y\} : y = x + 2, x \in X, y \in Y\}$

(b) $F = \{(1, 1)\} (2, 1), (3, 3) (4, 3) (5, 5)\}$

(c) $F = \{(1, 1)\} (1, 3), (3, 5) (3, 7) (5, 7)\}$

(d) $F = \{(1, 3)\} (2, 5), (4, 7) (5, 9) (3, 1)\}$

Solution: $X \cap Y = \{1, 3, 5\}$

$(X - Y) \cup (Y - X) = \{2, 4\} \cup (7, 9) = \{2, 4, 7, 9\}$

(a) We rewrite F as follows:

$\therefore 1 \in x$ and $y = x + 2 = 1 + 2 = 3 \in Y$

$\therefore (1, 3) \in F$

$\therefore 2 \in X$ but $y = 2 + 2 = 4 \notin Y$, we have $(2, 4) \notin F$

$\therefore 3 \in X$ and $y = 3 + 2 = 5 \in Y$, we have $(3, 5) \in F$

$\therefore 4 \in X$ and $y = 4 + 2 = 6 \notin$ we have $(4, 5) \notin F$.

Finally since $5 \in X$ and $y = 5 + 2 = 7 \in X$, we have $(5, 7) \in F$.

Hence, $F = \{1, 3\}, (3,5), (5, 7)\}$.

Hence, F is a relation from X to X, since $F \subset X * Y$ but it is not a mapping since the elements 2 and 4 of the domain X have no image in Y under F.

(b) Hence, F is a mapping from X to Y since each element of X has a unique image in Y under F. So F is a relation from X to Y.

Hence, F is a many one into mapping.

(c) Here, F is a relation from X to Y. Since $F \subset X$. Y but F is not a mapping since the element $1 \in X$ has two images 1 and 3 in X so that the image is not unique.

Example 5: Is $g = \{(1, 1) (2, 3), (3, 5) (4, 7)$ a function? If this is described by the formula $2(x) = \alpha x + \beta$, then what should be assigned to α and β?

Solution: Here g is a function since the image of every element of the domain $\{1, 2, 3, 4\}$ is unique. If x is any element of the domain, then clearly its image is, given by

$$g(x) = 2x - 1$$

Hence, $\alpha = 2$ and $\beta = -1$

Example 6: Suppose f is the collection of the ordered pairs of real numbers and $x = 6$ is the first element of some ordered pair in f. Suppose the vertical line through $x = 6$ intersects the graph of f twice. Is f a function? Why or why not?

Solution: f is not a function. The graph of the function f consists of points represented by the ordered pairs of the form $(x, f(x))$.

If the vertical line through $x = 6$ is cut by the graph of f twice, then it means that the element 6 of the domain of f has two images. Hence f is not a function as for t to be a function each element of the domain must have unique image.

Example 7: Given $A = \{x : \pi/6 \leq x \leq \pi/3\}$ and

$f(x) = \cos x - x(1 + x)$; find $f(A)$

Solution: We know that as x increases from 0 to $\pi/2$, cos x decreases from 1 to 0. Therefore

$\pi/6 \leq x \leq \pi/3 \Rightarrow \cos(\pi/6) \geq \cos x \geq \cos(\pi/3)$

$\Rightarrow 1/2 \leq \cos x \leq \sqrt{3}/2$ ...(1)

Again since $\pi/6 \leq x < \pi/3$ ...(2)

$\therefore 1 + \pi/6 \leq 1 + x \leq 1 + \pi/3$ ... (3)

Multiplying (2) and (3) we get ....

$\pi/6 * (1 + \pi/6) \leq x(1 + x) \leq \pi/3 (1 + \pi/3)$

or $-\pi/6 (1+\pi/6) \geq - x(1+x) \geq -\pi/3 (1+\pi/3)$

or $-\pi/3 (1 + \pi/3) \leq - x(1+x) \leq -\pi/6 (1+\pi/6)$ ... (4)

By Adding (1) and (4), we get

$1/2 - \pi/3 (1+\pi/3) \leq \cos x - (1 + x) \leq \sqrt{3}/2 \times \pi/6 (1 + \pi/6)$

i.e. $1/2 - \pi/3 (1+\pi/3) \leq f(x) \leq \sqrt{3}/2 - \pi/6 (1 + \pi/6)$

Hence,

$$f(A) = \{y : 1/2 - \pi/3 (1 + \pi/3) \leq y \leq \sqrt{3}/2 - \pi/6 (1+\pi/6)\}$$

Example 8: Let A and B be two sets each with a finite number of elements. Assume that there is injective mapping form A to B and that there is an injective mapping from B to A. Prove that there is a bijective mapping from A to B.

Solution: Let f be an injective mapping from A to B. Since f is one-one; number of elements in A is less than or equal to the number of elements in B, that is $n(A) \leq n(B)$.

Similarly sine there exists in injective mapping

$g:B \to A$, we have $n(B) \leq n(A)$

Hence, $n(A) = n(B)$

Since the number of elements in A and B is the same, we can define a bijective mapping from A to B

For if $A = \{a_1, a_2 .... a_n\}$

and $B = \{b_1, b_2...b_n\}$

Then one such bijective mapping is–

$$h = \{(a_1, b_1), (a_2, b_2)....(a_n, b_n)\}$$

In fact, we can define many such bijective mapping from A to B.

Example 9: Let f be a one-one function with domain (x, y, z) and range $\{1, 2, 3\}$. It is given that exactly one of the following statement is true and the remaining two are false. $f(x) = 1, f(g) \neq 1, f(z) \neq 2$. Determine $f^{-1}(1)$.

Solution : There are three posibilities:

(a) The statement $f(a) = 1$ is true the statement $f(y) \neq 1, f(z) \neq 2$ are both false.

(b) The statement $f(y) \neq 1$ is true and the statement $f(x) = 1, f(z) \neq 2$ are false.

(c) The statement $f(z) \neq 1, f(y) \neq 1$ are false.

In case (a), the true statement are :

$$f(x) = 1, f(y) = 1, f(z) = 2$$

If then f is not a one-one mapping which is a contradiction so this possibility is true out.

In case (b) the true statement must be:

$$f(x) = 2 \text{ or } 3, f(y) = 2 \text{ or } 3, f(z) = 2$$

So in the case also f cannot be a one-one mapping since in this case two elements x, y, z must have the same image.

In case (c), the true statement are
f(x) 2 or 3, f(y) = 1, f(z) = 1 or 3
Hence, $f^{-1}(1) = \{y\}$

Example 10: Let A = $\{x : -1 \leq x \leq 1\}$ = B. For each of the following functions from A to B, find whether it is surjective, injective or bijective.

(i) $f(x) = x/2$;
(ii) $g(x) = |x|$,
(iii) $h(x) = x|x|$,
(iv) $f(x) = x^2$
(v) $l(x) = \sin \pi x$

Solution: (i) f is one-one since $x, y \in A$, $x \neq y \Rightarrow \frac{1}{2}x \neq \frac{1}{2}y \Rightarrow f(x) \neq f(y)$.

Thus in this case f is injective

But f is not surjective, since 1 is not the image of any element in A. In fact, in this case we have $f(A) = \left\{x : \frac{-1}{2} \leq x \leq \frac{1}{2}\right\}$ which is a proper subset of BC = A.

(ii) Since $f(x) = 1$ and $f(-1) = 1$ etc.; f is many one mapping. Also in this case $f(A) = \{x : 0 \leq x \leq 1\}$ which is a proper subset of B (=A). Hence f is into mapping. Thus in this case f, is neither surjective non injective.

(iii) We may write the mapping h as follows:
$$h(x) = x^2 \text{ if } x \geq 0$$
and
$$h(x) = -x^2 \text{ if } x < 0$$

Now it can be easily seen that h is bijective.

(iv) $h(-1) = 1; h(1) = 1$

So it can be easily seen that f is a many one into mapping. So it is neither surjective non injective.

(v) The mapping l here is many one onto. It is many one since $l(-1) = \sin(-\pi) = 0, l(1) = \sin(\pi) = 0, l(0) = \sin 0 = 0$

So that $l(-1) = l(+1) = l(0) = 0$ etc.

It is onto since $\sin \pi x$ takes all value from -1 to 1 as x varies from -1 to 1.

Example 11: Let A = $\{x : 0 \leq x \leq 2\}$ and B = $\{1\}$. Give an example of a function. From A to. Can you define a function from B to A which is surjective? Give reasons for your Answer?

Solution: A function $f : A \to B$ is defined by–
$$f(x) = 1 \text{ for all } x \in A$$

Since A contains only one element whereas B contains infinite number of elements (*i.e.* all number from 0 to 2), it is not possible to define a function from B to A which is surjective.

Example 12: If $x = \{-1, 1\}$ and mapping $f : x \to x$ be defined by $f(x) = x^3$, prove that the mapping is one-one onto.

Solution: On putting $x = -1, 1$ in $f(x) = x^3$, we have–
$$f(-1) = (-1)^3 = -1, f(1) = 1^3 = 1$$
Here each distinct element has a distinct image

Hence, mapping is one-one.

Here image set = $\{-1, 1\}$ = Co-domain = $\{-1, 1\}$

Hence, mapping is onto

So, mapping is one-one onto.

Example 13: Prove that mapping $f : I \to R$, defined by $f(x) = x^2$, when I is the set of all integers and R is the set of all real number, is many one and onto

Solution: Many one. Let $x_1, x_2 \in I$
$$f(x_1) = x_1^2 \text{ and } f(x_1 = x_2^2)$$
$$f(x_1) = f(x^2)$$
$$\Rightarrow \quad x_1^2 = x_2^2$$
$$\Rightarrow \quad x_1 = \pm x_2$$

Thus $f(x_1) = f(x_2)$ also yields $x_1 \neq x_2$. Hence mapping is many one.

Into: Let $y = f(x) \in R$. Hence $y = x^2$ give $x = \pm \sqrt{y}$. Clearly if y is negative, x is imaginary and therefore does not belong to I. Hence, there are $y \in R$ which do not have any pre image $x \in I$. Hence it is an into mapping.

So the mapping is many- one into.

Example 14: If the set $X = \{x : x \in R, x \nsubseteq 0\}$, Prove that the map $f : X \to X$ given by $f(x) = 1/x$ is one-one and onto.

Solution: one-one : Let $x_1, x_2 \in X$

$\therefore \qquad f(x_1) = 1/x_1$ and $f(x_2) = 1/x_2$

clearly $\qquad f(x_1) = f(x_2) \Rightarrow 1/x_1 = 1/x_2 \Rightarrow x_1 = x_2$.

Hence mapping is one-one.

Onto: Let $y \in X$ so that $y = f(x) = Yx$

$\therefore \qquad x = 1/y$.

Thus if y is non-zero real then x is also non-zero and real and $x \in X$. Hence for every $y \in X$, there is a pre-image $\in x$. Hence mapping is onto.

So, it is one-one and onto mapping

Example 15: If X be the set of all triangles and Y that of positive number, prove that $f(X) \to Y$ defined $f(\Delta)$ = area of $\Delta, \Delta \subset x$ is many one onto mapping.

Solution: Many one: Since two different triangles may have the same area, *i.e.*, two distinct element of x may have the same f-image in Y. Hence mapping is many-one.

Onto: For every positive number $y \in Y$, we can construct a triangle $\Delta \in X$ such that $y = f(\Delta)$. Thus every y is image of some $\Delta \in X$. Hence mapping is onto.

So mapping is many one onto.

Example 16: Find the domain and the range of $f(x) = x^2/(1 + x^2)$ (x is real). Is the function one-one?

Solution: Since for every real x, $1+x^2 \neq 0$, therefore $x^2/(1 + x^2)$. is a real number for all real x. Hence the domain of $f(x)$ is the set R of all real number. The range of f consists of all real number y such that $f(x) = y$ for real x.

Now $$f(x) = y \Rightarrow x^2/(1+x^2) = y$$

$$\Rightarrow \quad x^2 = y + yx^2 \Rightarrow x = \sqrt{y/(1-y)}$$

Since x is real, we must have $y(1 - y) \geq 0$ ($y \neq 1$) which is satisfied of $0 \leq y < 1$.

Hence the range of $f = \{ y : y$ is real and $0 \leq g < 1\}$.

3.6 OPERATIONS ON FUNCTIONS

Just like you can add, subtract, multiply, or divide numbers, you can do those same operations with functions.

Suppose you have two functions $f(x) = 2x^2 - 3x + 4$ and $g(x) = x + 3$.

Addition

To add these two functions, we have two ways to write the notation. We can write addition as $(f + g)(x)$ which is how most math books indicate addition or $f(x) + g(x)$ which is what most students find most useful. To find we only need to add the two functions together.

$$f(x) + g(x) = (2x^2 - 3x + 4) + (x + 3)$$

The parentheses that were included in the previous step were not absolutely necessary. However, it can be helpful to use parentheses to separate the functions. It is also helpful to use this notation when subtracting and multiplying functions.

Now we need to combine like terms (link to basic operations-simplifying.doc)

$$(2x^2 - 3x + 4) + (x + 3) = 2x^2 - 2x + 7$$

So we have

$$(f + g)(x) = f(x) + g(x) = 2x^2 - 2x + 7$$

Subtraction

As with addition, we can write the notation for subtraction of two functions as $(f - g)(x)$ or $f(x) - g(x)$. However, with addition it did not matter if we wrote $(g + f)(x)$ or $(f + g)(x)$. The answer is the same in both cases. You might want to verify this for yourself. But with subtraction, the order definitely matters. Think about it this way. When you are adding two numbers, you can put them in any order, but the same is NOT true when subtracting two numbers. The same holds true when working with functions. So if we want to find $(f - g)(x)$, we must first write down $f(x)$ and then write down $g(x)$ and then perform the subtraction.

$$(f - g)(x) = f(x) - g(x) = (2x^2 - 3x + 4) - (x + 3)$$

You should be able to see how helpful the parentheses are in this problem, since we will have to change the sign of every term in the second set of parentheses. At this point, we need to combine like terms to get our answer.

$$(2x^2 - 3x + 4) - (x + 3) = 2x^2 - 3x + 4 - x - 3 = 2x^2 - 4x + 1$$

So we have

$$(f - g)(x) = f(x) - g(x) = 2x^2 - 4x + 1$$

Multiplication

As with addition, the order that multiplication is listed it will not affect the final answer. And as with subtraction, the use of parentheses will be important to us.

To find $(f \bullet g)(x) = f(x) \bullet g(x)$ we will simply multiply our two functions.

$$(f \bullet g)(x) = f(x) \bullet g(x) = (2x^2 - 3x + 4)(x + 3)$$

To simplify, we must make use of the distributive property. Remember the distributive property says that we have to multiply every term in the first set of parentheses by every term in the second set of parentheses.

$$(2x^2 - 3x + 4)(x + 3) = (2x^2)(x) + (2x^2)(3) + (-3x)(x) + (-3x)(3) + (4)(x) + (4)(3)$$

After using the distributive property, we will simplify and combine like terms.

$$(2x^2)(x) + (2x^2)(3) + (-3x)(x) + (-3x)(3) + (4)(x) + (4)(3) = 2x^3 + 6x^2 - 3x^2 - 9x + 4x + 12$$
$$= 2x^3 + 3x^2 - 5x + 12$$

So $(f \bullet g)(x) = f(x) \bullet g(x) = 2x^3 + 3x^2 - 5x + 12$

Division

In some ways, division of two functions can be the easiest of the four operations. To find

$$\left(\frac{f}{g}\right)(x) = \frac{f(x)}{g(x)}$$

we only have to create a rational function by putting $f(x)$ in the numerator and $g(x)$ in the denominator. This gives us

$$\left(\frac{f}{g}\right)(x) = \frac{f(x)}{g(x)} = \frac{2x^2 - 3x + 4}{x + 3}$$

Notice that when working with addition, subtraction, and multiplication, we did not worry about the domain of our newly created functions. However, with division when we create a rational function, we have to be concerned about having zero in the denominator.

Let's call our newly created function $h(x)$. So we have $h(x) = \dfrac{2x^2 - 3x + 4}{x + 3}$. The domain of $h(x)$ is all real numbers except $x = -3$.

You can see that combining functions is not quite as easy as combining two numbers, but the process is very similar. There is another way to combine two functions which is called composition of functions.

Evaluation

In each of the operations we looked at, we found a new function and that was the end of our problem. It is possible to be asked to evaluate this newly created function at a particular value.

Let us go back and revisit our addition problem. We had $f(x) = 2x^2 - 3x + 4$ and $g(x) = x + 3$ and $(f + g)(x) = 2x^2 - 2x + 7$ found. Suppose we were also asked about $(f + g)(3)$. It is the same process used for evaluating any other function. We need to substitute the value of 3 into the new function everywhere we see an x. Doing so will give us $(f + g)(3) = 2(3)^2 - 2(3) + 7 = 18 - 6 + 7 = 19$.

You should substitute the value of 3 into both the f function and the g function and then add those values to make sure that also gives you 19. This shows that it is possible to evaluate each function individually and then combine the two values. However, it is usually a more expedient method to combine the two functions and then do the evaluation. But it does provide a good way for you to check your work.

3.7 GROWTH OF FUNCTIONS

A function $f(n)$ is *asymptotically nonnegative* if $f(n)$ is nonnegative for all values sufficiently large values of n (*i.e.*, for all values of n above a certain value).

'Big-Oh'

$O(g(n)) = \{f(n) :$ there exist constants $c > 0$, $n_0 > 0$ such that $0 \leq f(n) \leq cg(n) \ \forall \ n \geq n_0\}$

Note. $O(g(n))$ refers to a *set of functions,* yet (by convention) we will use the equality symbol ('=') when dealing with big-Oh (the same holds for big-Omega and Theta), *i.e.*, we will say $f(n) = O(g(n))$ when we should actually say that $f(n) \in O(g(n))$. It is important to keep this distinction in mind.

'Big- Omega'

$\Omega(g(n)) = \{f(n) :$ there exist constants $c > 0$, $n_0 > 0$ such that $0 < cg(n) < f(n) \ \forall \ n \geq n_0\}$

'Theta'

$\theta(g(n)) = \{f/(n) :$ there exist constants $c_1 > 0$, $c_2 > 0$, $n_0 > 0$ such that $0 \leq c_1 g(n) \leq f(n) \leq c_2 g(n) \ \forall \ n \geq n_0\}$

Note, $f(n) = \theta(g(n))$ if and only if $f(n) = O(g(n))$ and $f(n) = \Omega(g(n))$

EXERCISE 3.1

1. Which of the following sets of ordered pairs define a mapping? Give reasons to support your answer.

    (i) {(1, 0), (2, 0) (0, 2) (1, 3)} (ii) {(a, b), (b, c), (c, d), (d, f)}

2. For each of the mapping defined by the following sets of ordered pairs, find out the domain and the range:
   (i) (1, 2)
   (ii) {(2, 1), (3, 3), (4, 9), (5, 9)}
   (iii) $f = \{(x, y): x$ is positive integer, $y^2 = x\}$
   (iv) $f = \{(x, y); y = x^2, x \in I, y \in R\}$
3. If A = { 1, 2, 3, 4} is mapped to B = {11, 17, 19, 20, 23}, such that $1 \to 11, 2 \to 17 \to 20$ $4 \to 23$, discuss the nature of mapping and express it as a set of ordered pair.
4. If the $X = \{x_1, x_2, x_3, x_4\}, y = \{1, 2, 3\}$ and $x_1 \to x_1, x_2 \to 2, x_3 \to 1 \; x_4 \to 3$. Prove that it is a many one onto mapping of X to Y.
5. If A = {1, 2, 3....} and $f(x) = 2x, \forall \; x \in A$. Prove that map $f: A \to A$ is one-one into.
6. If mapping $f = (x, y) : y = \{3x + 5, x \in R\}$ prove that this defines one-one onto mapping of real number on themselves.
7. Let Y = {1, 0} and map $f: N \to Y$ be defined by $f(zn) = 1; f(2n - 1) = 0 \; \forall \; n \in N$. Prove that mapping is many one and onto.
8. If c is the set of all complex number and R, the set of all real number, prove that– map $f : c \to R$ defined by $f(z) = |z|, \forall \; z \in c$ is neither one-one nor onto.
9. Discuss the following mapping $f : R \to R$ defined by
   (i) $f(x) = \log x, x \in R$
   (ii) $f(X) = \tan x, x \in R$
   (iii) $y = + \sqrt{x} \; x, x \in R$
10. If $f: N \to N$ be denoted by $f(n) = n - (-1)^n \; \forall \; n \in N$, prove that it is one-one onto mapping.
11. If R be the set of real number show that the mapping $f: R \to R$, defined by $f(x) = \cos x, x \in R$. is neither one-one nor onto. Modify the domain and the range so that it may be both one-one and onto.
12. Find the domain and range of the functions.
    (i) $g = |x|$,
    (ii) $y = 1 - |x|$
13. If S and T are two non-empty sets, prove that there exists a one to one correspondence between SXT and TXS.
14. If the set A is finite prove that :
    (i) If $f: A \to A$ is onto, then f is one-one
    (ii) if $f: A \to A$ is one-one then f is onto.
    (iii) statement (i) and (ii) are false, if A is finite.
15. Each of the following formula defines a function from R to R. Find the range of each function.
    (i) $g(x) = \sin x$
    (ii) $h(x) = x^2 + 1$

3.8 RECURSIVELY DEFINED FUNCTIONS

A function is said to be recursively defined of the function definition refers to itself. In order for the definition not to be circular, the function definition must have the following two properties:

(i) There must be certain arguments, called base values, for which the function does not refer to itself.

(ii) Each time the function does refers to itself, the argument of the function must be closer to a base value.

A recursive function with these two properties is said to be well defined but we take some examples to clarify recursion function.

(i) Factorial Function: The product of the positive integers from 1 to n, inclusive, is called "on factorial" and is usually denoted by $n!$ or $\lfloor n$.

i.e. $n!\ 1.\ 2.\ 3\ ...\ (n-1)\ .\ n$

The factorial function is defined for all non negative integers so–

$0! = 1,\ 1! = 1 \quad 2! = 1.2 \quad = 2,\ 3! \quad = 1.\ 2.\ 3 = 6$

$4! = \quad\quad\quad\quad 1.\ 2.\ 3.\ 4\ = 24,\ 5! \quad = 1.\ 2.\ 3.\ 4.\ 5 = 120$

and so on.

$n! = n\ (n–1)!$

Defining factorial function

(a) If $n = 0$ then $n! = 1$

(b) If $n > 0$ then $n! = n\ (n–1)!$

We observe that the above definition of $n!$ is recursive, since it refers to itself when it uses $(n–1)!$ However:

(i) The value of $n!$ is explicitly given when $n = 0$ (0 is the base value).

(ii) The value of $n!$ for arbitrary n is defined in terms of a smaller value of n which is closer to the base value 0.

Example: Let us calculate 5! using the reserve definitions. This calculation remains following eleven steps:

(1) $5! = 5.\ 4!$

(2) $4! = 4.\ 3!$

(3) $3! = 3.2!$

(4) $2! = 2.1!$

(5) $1! = 1.0!$

(6) $0! = 1$

(7) $1! = 1.\ 1 = 1$

(8) $2! = 2.1 = 2$

(9) $3! = 3.2 = 6$

(10) $4! = 4.\ 6 = 24$

(11) $5! = 5.24 = 120$

Step 1. This defines 5! in terms of 4!, so we must postpon evaluating 5! until we evaluate 4! this postponing is indicated by identing the next step.

Step 2. 4! is defined in terms of 3! so we must postpone evaluating 4! until we evaluate 3!

Step 3. This defines 3! in terms of 2!

Step 4. This defines 2! in terms of 1!

Step 5. This defines 1! in terms of 0!

Step 6. This step can explicitly evaluate 0! since 0 is the base value of the Recursive definition.

Steps 7 to 11. We backtracks, using 0! to find 1! using 1! to find. 2! and finally 4! to find 5! This backtracing is indicted by the "reverse" indention.

(ii) Fibonacci sequence

The celebrated Fibonacci sequence is as follows–

0, 1, 2, 3, 5, 8, 13, 21, 34, 55......

i.e., $F_0 - 0$ and $F_1 = 1$ and each succeeding terms is the sum of the two preceding terms.

Definition of Fibonacci sequence

(i) If $n = 0$ or $n = 1$ then $F_n = n$

(ii) If $n > 1$, then $F_n = F_{n-2} + F_{n-1}$

This is another example of a recursive function, since the definition refers to itself when it uses F_{n-2} and F_{n-1}. However:

(1) The base values are 0 and 1

(2) The value of F_n is defined in terms of smaller values of n which are closer to the base value. Accordingly, this function is well defined.

(iii) Ackermann Function

The ackermann function is a function with two arguments, each of which can be assigned any non-negative integers, *i.e.* 0, 1, 2 ... This function can be defined as follow:

Definition of Ackermann Function

(i) If $m = 0$ then $A(m, n) = n + 1$

(ii) If $m \neq 0$ but $n = 0$, then $A(m, n) = A(m-1, 1)$

(iii) If $m \neq 0$ and $n \neq 0$, then $A(m, n) = A(m - 1, A(m, n - 1))$ once more, we have a recursive definition, since the definition refers to itself in parts. (ii) and (iii).

Observe that $A(m, n)$ is explicitly given only when $m = 0$.

The base criteria are the pairs.

(0, 0) (0, 1) (0, 2) (0, 3) ..... (0, n)....

Although it is not obvious from the definition, the value of any $A(m, n)$ may eventually be expressed in terms of the value of the function on one or more of the base pairs.

Example 17: Use the definition of the Ackermann function to find $A(1, 3)$.

Solution: We have the following 15 steps:

(1) $A(1, 3) = A(0, A(1, 2))$

(2) $A(1, 2) = A(0, A(1, 1))$

(3) $A(1, 1) = A(0, A(1, 0)$

(4) $A(1, 0) = A(0, 1)$

(5) A (0, 1) = 1 + 1 = 2
(6) A (1, 0) = 2
(7) A (1, 1) = A (0, 2)
(8) A (0, 2) = 2 + 1 = 3
(9) A (1, 1) = 3
(10) A (1, 2) = A (0, 3)
(11) A (0, 3) = 3 + 1 = 4
(12) A (1, 2) = 4
(13) A (1, 3) = A (0, 4)
(14) A (0, 4) = 4 + 1 = 5
(15) A (1, 3) = 5

The forward indention indicates that we are postponing an evaluation and are recalling the definition and the backward indention indicates that we are backtracing. We can observe that (i) of the definition is used in steps 5, 8, 1 and 14 (ii) in step 4; and (iii) in steps 1, 2, and 3. In the other steps we are backtracking with substitutions.

Equal Mapping. Mapping $f : X \to Y$ and $g : X \to Y$ are called equal mappings if $f(x) = g(x)$, $\forall\ x \in X$. Equal mappings f and g then, are written as $f = g$.

Inverse Mapping. Let $f : X \to Y$ be a one-one mapping. Then the map
$f^{-1} : f[X] \to X$

Which associates at each element $y \in f[X]$ the element $x \in X$ such that $f(x) = y$ is called the inverse map of f. If f is one-one and onto than f^{-1} is a function from Y into X.

3.9 PRODUCT OR COMPOSITIONS OF MAPPINGS (FUNCTIONS)

Let X, Y, Z be three sets and let
$f : X \to X$ and $g ; Y \to Z$

Then under the map f an element $x \in X$ is mapped to an element $y = f(x) \in y$ which in turn is mapped by g to an element $z \in N$ such that $z = g(f(x)) \in z$. Hence we have a mapping of X onto Z. This new mapping is called composition of mappings found g is denoted by gof. It is defined by –

$$(gof)(x) = g(f(x))$$

It can be shown that if X, Y, Z and W are four sets and f, g, h three mappings.
$f : X \to Y, g : Y \to Z\ h : Z \to W$, then $(hog)\ of = ho\ (gof)$

Hence the composition of mappings is association.

Further if f, g are one-one onto then gof is also one-one a onto, and $(gof)^{-1} = f^{-1}\ og^{-1}$.

Example 18: Let a and b be positive integers, and suppose Q is defined recursively as follows:

$$Q(a, b) = \begin{cases} 0 & \text{if } a < b \\ Q(a-b, b)+1 & \text{if } b \le a \end{cases}$$

(a) Find (i) Q (2, 5) (ii) Q (12, 5)

(b) What does this function Q do? Find Q (5861, 7)

Solution: (a) (i) Q (2, 5) = 0 since 2 < 5.

(ii) Q (12, 5) = Q (7, 5) + 1 = [Q (2, 5) + 1] + 1 = Q (2,5) + 2.
= 0 + 2 = 2

(b) Each time b is subtracted from a, the value of Q is increased by 1 Hence Q (a, b) finds the quotient when a is divide by b.

Thus Q (5861, 7) = 837

Example 19: Let n denote a positive integer. Suppose a function L is defined recursively as follows:

$$L(n) = \begin{cases} 0 \text{ if } n=1 \\ L[n/2]+1 \text{ if } n>1 \end{cases}$$

Find (L (25)) and describe what this function does?

Solution:
$$L(25) = L(12) + 1 = L(6) + 1 + 1 = L(3) + 1 + 2$$
$$= L(1) + 1 + 3 = L(1) + 4 = 0 + 4 = 4$$

Each time n is divided by 2, then value of L is increased by 1. Hence L is the greatest integer such that

$$2^L < n$$

or

$$L(n) = [\log_2 n]$$

Example 20: Let the functions f and g be defined by $f(x) = 2x + 1$ and $g(x) = 2x^2 - 1$. Find the formula defining the function gof.

Solution:
$$gof = g\{f(x)\} = g(2x + 1) = 2(2x - 1)^2 - 1$$
$$= 2[4x^2 + 1 + 4x] - 1.$$
$$= 4x^2 + 4x + 1$$

Example 21: If $f : Q \to Q$ be defined by $f(x) = 3x + 4$, $x \in Q$ where Q is the set of all rational numbers, find the inverse mapping (f^{-1}).

Solution: Let $x_1, x_2 \in Q$. Hence $f(x_1) = 3x_1 + 4$, $f(x_2) = 3x_2 + 4$

∴

$$f(x_1) = f(x_2) \Rightarrow 3x_1 + 4 = 3x_2 + 4$$

⇒

$$x_1 = x_2$$

Hence mapping is one-one

Also let y be an arbitrary element of co-domain Q

$$y = 3x + 4 \Rightarrow x = (y - 4)/3.$$

This shows for any rational value of y, x is rational.

Thus every y has its pre-image in Q. Hence mapping is onto

∴ Inverse map exists

This inverse map $f^{-1} : Q \to Q$, therefore is defined by–

$$f^{-1}(y) = (y-4)/3, \forall y \in Q$$

Example 22: If $x = \{s : x \in R, -\pi/2 \le x \le \pi/2\}$
$$y = \{y : y \in R, -1 \le y \le 1\}$$
and

Prove that the mapping $f : X \to Y$ given $f(x) = \sin x$, $x \in X$ is one-one and onto. Find also the inverse map $f^{-1} Y \to X$.

Solution: $f(x) = \sin x, x \in R, -\pi/2 \le x \le -\pi/2$
$$f(x_1) = \sin x_1, f(x_2) = \sin(x_2)$$
$$f(x_1) = f(x_2) \Rightarrow \sin x_1 = \sin x_2$$
$$x_1 = n\pi \pm x_2 \text{ But here } -\pi/2 \le x < \pi/2$$

So,
$$x_1 = x_2.$$

Hence, mapping is one-one

Let,
$$y = \sin x \Rightarrow x = \sin^{-1}(y)$$

The value of $\sin^{-1}(y)$ is $[-1, 1]$ which is the given limit of y.
Thus every y has its pre-image. Hence mapping is onto so it is one-one onto.
Since $f : X \to Y$ is one-one, onto f^{-1} exists.

Let
$$y = f(x) = \sin x \text{ be any element of } y.$$
$\therefore x = \sin^{-1} y, \forall y \in Y$. Hence $f^{-1} : Y \to X$ is defined by–
$$x = \sin^{-1} y, \forall y \in Y$$

Example 23: If map $f : R \to R$ be defined by $f(x) = X^2$, where R is the set of all real number find $f^{-1}([-1, 1])$

Solution: Interval $[-1, 1]$ can be subdivided into two intervals $[-1, 0]$ and $[0, 1]$.
If $x \in [-1, 0]$ then x is negative but $f(x) = x^2$ gives positive values of $f(x)$.
Hence no element in the interval $[-1, 0]$ can be an f-image of any $x \in R$.
Also in the interval $[0, 1]$ each element of this interval is an f-image of some $x \in R$ domain.

Again $f(x) = y = x^2$ gives $x = \pm \sqrt{y}$

Hence for every f-image y lying in the interval $[0, 1]$ of $x \in R$ x lies in the interval $[-1, 1]$
Thus $f^{-1}[(-1, 1)] = [-1, 1] = \{x : x \in R, -1 \le x \le 1\}$

Example 24: Prove that $f^{-1}[A \cup B] = f^{-1}[A] \cup f^{-1}[B]$

Solution: Here $x \in f^{-1}[A \cup B] \Rightarrow f(x) \in A \cup B$
$\Rightarrow \qquad\qquad\qquad f(x) \in A \text{ or } f(x) \in B$
$\Rightarrow \qquad\qquad\qquad x \in f^{-1}(A) \text{ or } x \in f^{-1}(B)$
$\Rightarrow \qquad\qquad\qquad x \in f^{-1}[A] \cup f^{-1}[B]$
$\qquad\qquad\qquad f^{-1}[A \cup B] \subseteq f^{-1}[A] \cup f^{-1}[B]$ ...(1)

Conversely, $\qquad y \in f^{-1}[A] \cup f^{-1}[B] \Rightarrow y \in f^{-1}[A] \text{ or } y \in f^{-1}[B]$
$\Rightarrow \qquad\qquad\qquad f(y) \in A \text{ or } f(y) \in B$
$\Rightarrow \qquad\qquad\qquad f(y) \in A \cup B$

$$y \in f^{-1}[A \cup B]$$
$$\therefore \quad f^{-1}[A] \cup f^{-1}[B] \subseteq f^{-1}[A \cup B] \qquad \ldots(2)$$

By using equations (1) and (2), we get
$$f^{-1}[A \cup B] = f^{-1}[A] \cup f^{-1}[B]$$

Example 25: Given that $f: R \to R$ is defined by $f(x) = x + 3$ and map $: R \to R$ is given by $g(x) = x^2 + 2x - 9$; find $(gof)x$ and $(fog)x$.

Solution:
$$(gof)x = g[f(x)] = g[x+3] = (x+3)^2 + 2(x+3) - 9$$
$$= x^2 + 9 + 6x + 2x + 6 - 9 = x^2 + 8x + 6$$
$$(fog)x = f[g(x)] = f(x^2 + 2x - 9) = x^2 + 2x - 9 + 3$$
$$= x^2 + 2x - 6$$

Example 26: If the map $f: R \to R$ is defined by $f(x) = \sin^2 x$, and the map $g: R \to R$ is defined by $g(x) = x^3$, find mappings gof and fog of $R \to R$. also find $(gof)1$, $(fog)1$.

Solution: Here $(gof)x = g[f(x)]$, $\forall x \in R$ (domain)

Now $f(x) = \sin^2 x \in R$ (co-domain), $\forall x \in R$ (domain)

$$\therefore \qquad (gof)x = g(\sin^2 x), \text{ where } \sin^2 x \in R \qquad \ldots(i)$$

But when $x \in R$, $g(x) = x^3$

$$g(\sin^2 x) = (\sin^2 x)^3 = \sin^6 x \qquad \therefore \sin^2 x \in R$$
$$(gof)x = \sin^6 x, \forall x \in R$$

Hence map $gof: R \to R$ is given by $(gof)x = \sin^6 x$, $x \in R$

Now
$$(fog)x = f[g(x)], \forall x \in R$$
$$= f(x^3), \forall x \in R$$
$$= \sin^2 x^3, \forall x \in R.$$

Thus map $fog: R \to R$ is given by $(fog)x = \sin^2 x^3$, $\forall x \in R$

Hence,
$$(gof)1 = \sin^6 1$$
$$(fog)1 = \sin^2 1$$

Clearly $(gof)1 \neq (fog)1$

EXERCISE 3.2

1. Let a and b integers, and suppose $Q(a, b)$ is defined recursively by—
$$Q = (a, b) = \begin{cases} 5 & \text{if } a < b \\ Q(a-b, b+2) + a & \text{if } a \geq b \end{cases}$$

2. Let A and B be finite sets and $f: A \to B$. Then show that—
   (i) If f is one-one, then $|A| \leq |B|$ (ii) If f is onto then $|B| \leq A$
   (iii) If f is a bijection, then $|A| = |B|$

3. Show that the function $f: R \to R$ given by $f(x) = \cos x$ for all $x \in R$ is neither one to one nor onto.

4. If the function $f: R \to R$ defined by $f(x) = x^2$, then find $f^{-1}(4)$ and $f^{-1}(-4)$.

5. If the function f : R → R defined by
$$f(x) = \begin{cases} 3x-4 & x>0 \\ -3x+2 & x \leq 0 \end{cases} \text{ then find.}$$

   (i) $f(0), f(2/3), f(-2)$ (ii) $f^{-1}(0), f^{-1}(2), f^{-1}(-7)$

6. The function $f : R \to R$ is defined by–
$$f(x) = \begin{cases} 3x-12 & \text{for } x>3 \\ 2x^2+3 & \text{for } -2 < x \leq 3 \\ 3x^2-7 & \text{for } x \leq -2 \end{cases}$$

   then find out $f^{-1}(5)$

7. Prove the following for Ackermann function.
   (i) $A(1, y) = y + 2$ (ii) $A(2, y) = 2y + 3$

8. If the mapping f and g are given by $f = \{(1, 2) (3,5) (4, 1)\}$
   $g = \{(2, 3) (5, 1), (1, 3)\}$ then write down pairs in the mapping fog and gof.

9. Let the functions $f: R \to R$ and $g : R \to R$ be defined by $f(x) = 2x + 2$, $g(x) = x^2 + 1$. Find the formula for gof and fog.

CHAPTER 4

NATURAL NUMBERS

4.1 NATURAL NUMBERS

In mathematics, there are two conventions for the set **of natural numbers:** it is either the set of positive integers {1, 2, 3, ...} according to the traditional definition or the set of non-negative integers {0, 1, 2, ...} according to a definition first appearing in the nineteenth century.

Natural numbers have two main purposes: counting ("there are 6 coins on the table") and ordering ("this is the 3rd largest city in the country"). These purposes are related to the linguistic notions of cardinal and ordinal numbers, respectively. A more recent notion is that of a nominal number, which is used only for naming.

Properties of the natural numbers related to divisibility, such as the distribution of prime numbers, are studied in number theory. Problems concerning counting and ordering, such as partition enumeration, are studied in combinatorics.

History of Natural Numbers and the Status of Zero

The natural numbers had their origins in the words used to count things, beginning with the number 1. The first major advance in abstraction was the use of numerals to represent numbers. This allowed systems to be developed for recording large numbers. The Babylonians, for example, developed a powerful place-value system based essentially on the numerals for 1 and 10. The ancient Egyptians had a system of numerals with distinct hieroglyphs for 1, 10, and all the powers of 10 up to one million. A stone carving from Karnak, dating from around 1500 BC and now at the Louvre in Paris, depicts 276 as 2 hundreds, 7 tens, and 6 ones; and similarly for the number 4,622.

A much later advance in abstraction was the development of the idea of zero as a number with its own numeral. A zero digit had been used in place-value notation as early as 700 BC by the Babylonians but they omitted it when it would have been the last symbol in the number. The Olmec and Maya civilization used zero developed independently as a separate number as early as 1st century BC, but this usage did not spread beyond ,Mesoanierica. The concept as used in modern times originated with the Indian mathematician Brahmagupta in 628. Nevertheless, medieval computers (e.g. people who calculated the date of Easter), beginning with Dionysius Exiguus in 525, used zero as a number without using a Roman numeral to write it. Instead *nullus,* the Latin word for "nothing", was employed.

The first systematic study of numbers as abstractions (that is, as abstract entities) is usually credited to the Greek philosophers Pythagoras and Archimedes. Note that many Greek mathematicians did not consider 1 to be "a number", so to them 2 was the smallest number.

Independent studies also occurred at around the same time in India, China, and Mesoamerica.

Several set-theoretical definitions of natural numbers were developed in the 19th century. With these definitions it was convenient to include 0 (corresponding to the empty set) as a natural number. Including 0 is now the common convention among set theorists, logicians, and computer scientists. Many other mathematicians also include 0, although some have kept the older tradition and take 1 to be the first natural number. Sometimes the set of natural numbers with 0 included is called the set of whole numbers or **counting numbers.**

4.2 NOTATION

Mathematicians use N or $\mathbb{N}$ (an N in blackboard bold, displayed as ℕ in Unicode) to refer to the set of all natural numbers. This set is countably infinite: it is infinite but countable by definition. This is also expressed by saying that the cardinal number of the set is aleph-null (χ_0).

To be unambiguous about whether zero is included or not, sometimes an index "0" is added in the former case, and a superscript "*" or subscript "1" is added in the latter case:

$$\mathbb{N}_0 = \{0,1,2,...\}; \quad \mathbb{N}^* = \mathbb{N}_1 = \{1,2,...\}.$$

(Sometimes, an index or superscript "+" is added to signify "positive". However, this is often used for "nonnegative" in other cases, as $\mathbf{R}^+ = [0, \infty)$ and $\mathbf{Z}^+ = \{0, 1, 2,... \}$, at least in European literature. The notation "*", however, is standard for nonzero, or rather, invertible elements.)

Some authors who exclude zero from the naturals use the terms *natural numbers with zero, whole numbers,* or *counting numbers,* denoted W, for the set of nonnegative integers. Others use the notation **P** for the positive integers if there is no danger of confusing this with the prime numbers.

Set theorists often denote the set of all natural numbers including zero by a lower-case Greek letter omega ω. This stems from the identification of an ordinal number with the set of ordinals that are smaller.

Algebraic Properties

The addition and multiplication operations on natural numbers have several algebraic properties:

- **Closure** under addition and multiplication: for all natural numbers a and b, both $a + b$ and $a \times b$ are natural numbers.
- **Associativity:** for all natural numbers $a, b,$ and $c, a + (b + c) = (a + b) + c$ and $a \times (b \times c) = (a \times b) \times c.$
- **Commutativity:** for all natural numbers a and $b, a + b = b + a$ and $a \times b = b \times a.$

- **Existence of identity elements:** for every natural number a, $a + 0 = a$ and $a \times 1 = a$.
- **Distributivity** for all natural numbers a, b, and c, $a \times (b + c) = (a \times b) + (a \times c)$
- **No zero divisors:** if a and b are natural numbers such that $a \times b = 0$ then $a = 0$ or $b = 0$

Properties

One can recursively define an addition on the natural numbers by setting $a + 0 = a$ and $a + S(b) = S(a + b)$ for all a, b. Here S should be read as "successor". This turns the natural numbers (**N**, +) into a commutative monoid with identity element 0, the so-called free monoid with one generator. This monoid satisfies the cancellation property and can be embedded in a group. The smallest group containing the natural numbers is the integers.

If we define $1 := S(0)$, then $b + 1 = b + S(0) = S(b + 0) = S(b)$. That is, $b + 1$ is simply the successor of b.

Analogously, given that addition has been defined, a multiplication $\times$ can be defined via $a \times 0 = 0$ and $a \times S(6) = (a \times b) + a$. This turns (**N***, ×) into a free commutative monoid with identity element 1; a generator set for this monoid is the set of prime numbers. 'Addition and multiplication are compatible, which is expressed in the distribution law: $a \times (b + c) = (a \times b) + (a \times c)$. These properties of addition and multiplication make the natural numbers an instance of a commutative semiring. Semirings are an algebraic generalization of the natural numbers where multiplication is not necessarily commutative. The lack of additive inverses, which is equivalent to the fact that N is not closed under subtraction, means that N is *not* a ring.

If we interpret the natural numbers* as "excluding 0", and "starting at 1", the definitions of + and × are as above, except that we start with $a + 1 = S(a)$ and $a \times 1 = a$.

For the remainder of the article, we write ab to indicate the product $a \times b$, and we also assume the standard order of operations.

Furthermore, one defines a total order on the natural numbers by writing $a < b$ if and only if there exists another natural number c with $a + c = b$. This order is compatible with the arithmetical operations in the following sense: if a, b and c are natural numbers and $a < b$, then $a + c < b + c$ and $ac < bc$. An important property of the natural numbers is that they are well ordered: every non-empty set of natural numbers has a least element. The rank among well ordered sets is expressed by an ordinal number; for the natural numbers this is expressed as "ω".

While it is in general, not possible to divide one natural number by another and get a natural number as result, the procedure of *division with remainder*—is available as a substitute: for any two natural numbers a and b with $b \neq 0$ we can find natural numbers q and r such that

$a = bq + r$ and $r < b$.

The number q is called the *quotient* and r is called the *remainder* of division of a by b. The numbers q and r are uniquely determined by a and b. This, the Division algorithm, is key to several other properties (divisibility), algorithms (such as the Euclidean algorithm), and ideas in number theory.

Generalizations

Two generalizations of natural numbers arise from the two uses:

(a) A natural number can be used to express the size of a finite set; more generally a cardinal number is a measure for the size of a set also suitable for infinite sets; this refers to a concept of "size" such that if there is a bijection between two sets they have the same size. The set of natural numbers itself and any other countably infinite set has cardinality aleph-null (χ_0).

(b) Ordinal numbers "first", "second", "third" can be assigned to the elements of a totally ordered finite set, and also to the elements of well-ordered countably infinite sets like the set of natural numbers itself. This can be generalized to ordinal numbers which describe the position of an element in a well-order set in general. An ordinal number is also used to describe the "size" of a well-ordered set, in a sense different from cardinality: if there is an order isomorphism between two well-ordered sets they have the same ordinal number. The first ordinal number that is not a natural number is expressed as co; this is also the ordinal number of the set of natural numbers itself.

Many well-ordered sets with cardinal number χ_0 have an ordinal number greater than ω. For example, $\omega^{\omega^{\omega^{6+42}}1729+\omega^9+88} \cdot 3 + \omega^{\omega^{\omega}} \cdot 5 + 65537$ has cardinality χ_0. The least ordinal of cardinality χ_0 (*i.e.*, the initial ordinal) is ω.

For finite well-ordered sets, there is one-to-one correspondence between ordinal and cardinal numbers; therefore they can both be expressed by the same natural number, the number of elements of the set. This number can also be used to describe the position of an element in a larger finite, or an infinite, sequence.

Formal Definitions

Historically, the precise mathematical definition of the natural numbers developed with some difficulty. The Peano axioms state conditions that any successful definition must satisfy. Certain constructions show that, given set theory, models of the Peano postulates must exist.

Peano Axioms

The Peano axioms give a formal theory of the natural numbers. The axioms are:

(i) There is a natural number 0.
(ii) Every natural number a has a natural number successor, denoted by $S(a)$. Intuitively, $S(a)$ is $a + 1$.
(iii) There is no natural number whose successor is 0.
(iv) Distinct natural numbers have distinct successors: if $a \neq b$, then $S(a) \neq S(b)$.
(v) If a property is possessed by 0 and also by the successor of every natural number which possesses it, then it is possessed by all natural numbers. (This postulate ensures that the proof technique of mathematical induction is valid.)

It should be noted that the "0" in the above definition need not correspond to what we normally consider to be the number zero. "0" simply means some object that when combined with an appropriate successor function, satisfies the Peano axioms. All systems that satisfy these axioms are isomorphic, the name "0" is used here for the first element, which is the only element that is not a successor. For example, the natural numbers starting with one also satisfy the axioms, if the symbol 0 is interpreted as the natural number 1, the symbol S(0) as the number 2, etc. In fact, in Peano's original formulation, the first natural number *was* 1.

Constructions Based on Set Theory

A Standard Construction

A standard construction in set theory, a special case of the von Neumann ordinal, construction, is to define the natural numbers as follows:

We set $0 := \{\ \}$, the empty set,

and define $S(a) = a \cup \{a\}$ for every set a. $S(a)$ is the successor of a, and S is called the successor function.

If the axiom of infinity holds, then the set of all natural numbers exists and is the intersection of all sets containing 0 which are closed under this successor function.

If the set of all natural numbers exists, then it satisfies the Peano axioms.

Each natural number is then equal to the set of natural numbers less than it, so that

- $0 = \{\}$
- $1 = \{0\} = \{\{\}\}$
- $2 = \{0, 1\} = \{0,\{0\}\} = \{\{\},\{\{\}\}\}$
- $3 = \{0, 1, 2\} = \{0, \{0\}, \{0, \{0\}\}\} = \{\{\ \}, \{\{\ \}\}, \{\{\ \}, \{\{\ \}\}\}\}$
- $n = \{0, 1, 2, ..., n-2, n-1\} = \{0, 1, 2, ..., n-2\} \cap \{n-1\} = \{n-1\} \cup \{n-1\}$

and so on. When a natural number is used as a set, this is typically what is meant. Under this definition, there are exactly n elements (in the naive sense) in the set n and $n < m$ (in the naive sense) if and only if n is a subset of m.

Also, with this definition, different possible interpretations of notations like $\mathbf{R}^n$ (n-tuples versus mappings of n into $\mathbf{R}$) coincide.

Even if the axiom of infinity fails and the set of all natural numbers does not exist, it is possible to define what it means to be one of these sets. A set n is a natural number means that it is either 0 (empty) or a successor, and each of its elements is either 0 or the successor of another of its elements.

Other Constructions

Although the standard construction is useful, it is not the only possible construction. For example:

one could define $0 = \{\ \}$

and $S(a) = \{a\}$,

producing

0 = {}
1 = {0} = {{}}
2 = {1} = {{{}}}, etc.

Or we could even define 0 = { { } }

and $S(a) = a \cup \{a\}$

producing

0 = {{}}
1 = {{}, 0} = {{}, {{}}}
2 = {{}, 0, 1}, etc.

Arguably the oldest set-theoretic definition of the natural numbers is the definition commonly ascribed to Frege and Russell under which each concrete natural number n is defined as the set of all sets with n elements. This may appear circular, but can be made rigorous with care. Define 0 as { { } } (clearly the set of all sets with 0 elements) and define S(A) (for any set A) as $\{x \cup \{y\} \mid x \in A \wedge y \notin x\}$. Then 0 will be the set of all sets with 0 elements, 1 = S(0) will be the set of all sets with 1 element, 2 = S(1) will be the set of all sets with 2 elements, and so forth. The set of all natural numbers can be defined as the intersection of all sets containing 0 as an element and closed under S (that is, if the set contains an element n, it also contains S(n)). This definition does not work in the usual systems of axiomatic set theory because the collections involved are too large (it will not work in any set theory with the axiom of separation; but it does work in new foundations (and in related systems known to be relatively consistent) and in some systems of type theory.

4.3 NOTION OF PROOF

The proofs used in every day working mathematics are based on the same logical framework as the propositional calculus but their structure is not usually displayed in the format.

Direct Proof

We are typically faced with a set of hypothesis $H_1, H_2 ... H_n$ from which we want to infer a conclusion C. One of the most natural sorts of proof is the direct proof in which we show

$$H_1 \wedge H_2 \wedge .... H_n \Rightarrow C$$

Let us take some examples of direct proof.

SOLVED EXAMPLES

Example 1: Prove that if x and y are rational numbers then $x + y$ is rational.

Solution: Since x and y are rational number we can find integers p, q, m, n such that $x = p/q$ and $y = m/n$ then– $x + y = P/q + m/n = (Pn + mq)/qn$

Since put mq and qn are both integers, we conclude that $x + y$ is a rational number.

Example 2: Prove that product of two odd integers is an odd integers.

Solution: Assume m and n are two odd integers. Then there exists two integers r and t, so that $m = 2r + 1$ and $n = 2t + 1$ then—
$$mn = (2r + 1)(2t + 1) = 4rt + 2r + 2t + 1$$
$$= 2(rt + r + t) + 1$$

Which is odd interger.
Hence, proved.

Indirect Proof

Proofs that are not direct are called indirect. The two main types of indirect proof both use the negation of the conclusion, so they are often suitable when that negation is easy to state. The first type of proof is contrapositive proof. We give a proof by contrapositive of the statement in the following example.

Example : Prove that if $m + n \geq 73$ then $m \geq 37$ or $n \geq 37$, m and n being positive integers.

Solution : There seems to be no way to proof the given fact directly. Instead, one can prove by taking the contrapositive not "$m \geq 37$ or $n \geq 37$" implies not $m + n \geq 73$". By De Morgan's Law, the negation of "$m \geq 37$ or $n \geq 37$" is not $m \geq 37$ and $n \geq 37$" "$m \leq 37$ and $n \leq 36$". So the contrapositive proposition is if $m \leq 36$ and $n \leq 36$ then $m + n \leq 72$. This follows immediately from property of inequalitites : $a \leq c$ and $b \leq d$ imply that $a + b < c + d$ for all real number a, b, c, d.

Proof by Contradiction

The second type of indirect proof is known as proof by contradiction. In this type of proof, we assume the opposite of what we trying to prove and get a logical contradiction. Hence our assumption must have been false, and therefore what are originally required to prove must be true.

Example: Prove that $\sqrt{2}$ is irrational, by giving a proof by contradiction.

Solution: Suppose that $\sqrt{2}$ is irrational. We will show that this leads to a contradiction.

Under assumption that $\sqrt{2}$ is rational there exists interns a and b such that $\sqrt{2} = a/b$, where a and b have no common factor. Squaring both sides, we get
$$2 = a^2/b^2 \Rightarrow 2b^2 = a^2$$

Hence a^2 is a multiple of 2, and therefore even. This implies a is even. Hence $a = 2k$ for same integer k. Then $2b^2 = (2x)^2$ and therefore $b^2 = 4x^2$. Thus b^2 is even, b is even. But now a and b have a common factor of 2, which is a contradiction to the statement that a and b have no common factors.

Hence, our initial assumption that $\sqrt{2}$ is rational is false. Thus, $\sqrt{2}$ is irrational.

Proof by cases

An implication of the form
$H_1 V H_2 .... V H n \Rightarrow C$
is equivalent to $(H_1 \Rightarrow C)(H_2 \Rightarrow C)$ and .... and $(H_n \Rightarrow C)$.
and hence can be proved by cases, *i.e.*, by proving each of $H_1 \Rightarrow C, ... H_n \Rightarrow C$ separately.

Example: Prove that for every positive integer n, $n^3 + n$ is even.

Proof: Case (*i*) suppose n is even, then $n = 2k$ for some positive integer k. Now $n^3 + n = 8k^3 + 2k = 2(4k^3 + k)$ which is even.

Case (*ii*) Suppose n is odd. Then $n = 2k + 1$ for some positive integer k. Now
$$n^3 + n = (2k + 1)^3 + 2\ 2k + 1 = (8k^3 + 12x^2 + 6k + 1) + 2k + 1$$
$$= 2(4k^3 + 6k^2 + 4k + 1) \text{ which is even}$$

Hence, the sum $n^3 + n$ is even.

4.4 FALLACIES

A fallacy is an error in reasoning that results in an invalid argument. Several common fallacies arise in incorrect arguments. They superficially resemble those that valid by rules of inference, but are not infact valid. We discuss two types of fallacies.

(1) The fallacy of affirming the consequent (or affirming the converser).

(2) The fallacy of denying the hypothesis (or assuming the inverse). The fallacy of affirming the consequent has the following form.

$$p \Rightarrow q$$
$$\therefore \quad \frac{q}{p}$$

Example: Show that the following argument is invalid: if Ramesh solved this problem, then he obtained the answer 7.

Ramesh obtained the answer 7.

Therefore, Ramesh solved this problem correctly.

Solution: Let p and q be the propositions as

p : Ramesh solved this problem

q : Ramesh obtained the answer 7

Then this answer is of the form if $p \Rightarrow q$ and q, then p. This argument is faulty because the conclusion can be false even through $p \Rightarrow q$ and q are true. That is, in this implication $[(p \Rightarrow q) \wedge q] \Rightarrow p$ is not a tautology. It is possible, Ramesh obtained the correct answer 7 by luck, guessing or prior knowledge but the arguments and intermediate steps are wrong. Hence the argument is invalid.

The fallacy under lying this invalid argument form is called the fallacy of affirming the consequent because the conclusion of the argument valid follow from the premises $p \Rightarrow q$ were replaced by its coverse. Such a replacement is not allowed however, because a conditional statement is not logically equivalent to its converse.

Example 3: Test the validity of the following argument if two sides of a triangle are equal, then the opposite angles are equal.

Two sides of a triangle are not equal.

Therefore, the opposite angles are not equal.

Solution: Let p and q be the propositions as.

p: Two sides of a triangle are equal.
q: The opposite angles are equal.
Hence p: Two side of a triangle are not equal. Then this argument is of the form.

$$p \Rightarrow q$$
$$\frac{\sim p}{\sim p}$$

∴

The proposition $[Cp \Rightarrow q) \wedge \sim p] \Rightarrow \sim q$ is not a tautology, the argument is invalid.

The fallacy under lying this invalid argument form is called the fallacy of denying the hypothesis assuming the inverse because the incorrect argument is of the form $p \Rightarrow q$ and $\sim p$ imply $\sim q$ which is not allowed.

4.5 MATHEMATICAL INDUCTION

This is a method of proof by which the truth of a general theorem is deduced from those of particular, cases. The principle of mathematical induction states below essentially assets that the positive integer N begin with the number 1 and the rest are obtained by successively adding, *i.e.*, we begin with 1, then $2 = 1 + 1$, then $3 = 2 + 1$, then $4 = 3 + 1$, and so on.

Principle of Mathematical Induction: Let S be a set of positive integers with the following two properties:

(*i*) 1 belongs to S.

(*ii*) If k belongs to S, then $k+1$ belongs to S.

Then S is the set of all positive integers.

OR

Let p be a proposition defined on the integers $n \geq 1$ such that:

(*i*) $p(1)$ is true.

(*ii*) $p(n + 1)$ is true whenever $P(n)$ is true

Then p is true for every integer $n \geq 1$.

4.6 PROOF BY COUNTEREXAMPLES

A counterexample is an example that disproves a universal ("for all") statement. Obtaining counterexamples is a very important part of mathematics, because doing mathematics requires that you develop a critical attitude towards claims. When you have an idea or when someone tells you something, test the idea by trying examples. If you find a counterexample which show that the idea is false, that's good: Progress comes not only though doing the right thing, but also by correcting your mistakes.

Suppose you have a quantified statement

"All x's satisfy property p" : $\forall x P(x)$.

What is its negation?

$\sim \forall x P(x) \leftrightarrow \exists x \sim P(x)$

In words, the second quantified statement says: "There is an x which does not satisfy property p". In other words, to prove that "All x's satisfy property P" is false you must find and x which does not satisfy property P.

Example: To disprove the statement

"All professors like pizza"

you must find a professor who does not like pizza.

To disprove the statement

"For every real number x, $(x + 1)^2 = x^2 + 1$"

you must find a real number x for which $(x + 1)^2 \neq x^2 + 1$

On the other hand, to disprove the statement

"$x^3 + 5x + 3$ has a root between $x = 0$ and $x = 1$"

It is not enough to say "$x = 0.5$ is between $x = 0$ and $x = 1$, but $(0.5)^3 + 5(0.5) + 3 = 5.625 \neq 0$" The statement to be disproved is an existence statement :

'There is an x such that $0 < x < 1$ and $x^3 + 5x + 3 = 0$."

You can check that the negation is

"For all x, it is not the case that both $0 < x < 1$ and $x^3 + 5x + 3 = 0$."

To disprove the original statement is to prove its negation, but a single example will not prove this "for all" statement.

The point made in the last example illustrates the difference between "proof by example" — which is usually invalid — and giving a counterexample.

- A single example can't prove a universal statement (unless the universe consists of only one case!)
- A single counterexample can disprove a universal statement:

In many cases where you need a counterexample, the statement under consideration is and if then statement. So how do you give a counterexample to a conditional statement $P \rightarrow Q$? By basic logic, $P \rightarrow Q$ false when P is true and Q is false. Therefore:

- To give a counterexample to a conditional statement $P \rightarrow Q$, find a case where P is true but Q is false. example mix

Example: Consider real-valued functions defined on the interval $0 \leq x \leq 1$ Give a counterexample to the following statement:

"If the product of two functions is the zero function, then one of the functions is the zero function."

(The zero function is the function which produces 0 for all inputs—i.e. the constant function $f = 0$.)

Here are two functions whose product is the zero function, neither of which is the zero function:

$$f(x) = \begin{cases} \frac{1}{2} - x & \text{if } 0 \leq x \leq \frac{1}{2} \\ 0 & \text{if } \frac{1}{2} < x \leq 1 \end{cases} \quad g(x) = \begin{cases} 0 & \text{if } 0 \leq x \leq \frac{1}{2} \\ x - \frac{1}{2} & \text{if } \frac{1}{2} < x \leq 1 \end{cases}$$

Example 4: Give a counterexample to the following statement:

"If $\lim_{n \to \infty} a_n = 0$, then $\sum_{n=1}^{\infty} \frac{1}{n}$ converges".

Solution: You may recall this mistake from studying infinite series. The Harmonic series

$$\sum_{n=1}^{\infty} \frac{1}{n} = 1 + \frac{1}{2} + \frac{1}{3} + \frac{1}{4} + \cdots \text{ diverges, even though } \lim_{n \to \infty} \frac{1}{n} = 0.$$

The converse of the given statement — the Zero Limit Test — is true: If $\sum_{n=1}^{\infty} \frac{1}{n}$ converges, then $\lim_{n \to \infty} a_n = 0$. Or to put it another way (taking the contrapositive), if, $\sum_{n=1}^{\infty} \frac{1}{n}$ diverges.

For example, the series $\sum_{n=1}^{\infty} \frac{3n^2 + 1}{4n^2 - 3}$ diverges because

$$\lim_{n \to \infty} \frac{3n^2 + 1}{4n^2 - 3} = \frac{3}{4} \neq 0$$

Example 5: An algebraic identity is an equation which is true for all values of the variables for which both sides of the equation are defined.

Solution: For example, $\frac{1}{x} + 1 = \frac{x+1}{x}$ is an algebraic identity for real numbers: It is true for all $x \neq 0$.

Since an algebraic identity is a statement about all numbers in a certain set, you can prove that a statement is not an identity by producing a counterexample.

For example, I'll prove that "$(a + b)^2 = a^2 + b^2$" is not an algebraic identity, where $a, b \in \mathbb{R}$. I need to find specific real numbers a and b for which the equation is false.

If an equation is not an identity, you can usually find a counterexample by trial and error. In this case, if $a = 1$ and $b = 2$, then

$$(a + b)^2 = (1 + 2)^2 = 3^2 = 9, \text{ while } a^2 + b^2 = 1^2 + 2^2 = 5$$

So if $a = 1$ and $b = 2$, then $(a + b)^2 \neq a^2 + b^2$, and hence the statement is not an identity.

A common mistake is to say:

"$(a+b)^2 = a^2 + 2ab + b^2$, which is not the same as $a^2 + b^2$"

In the first place, how do you know $a^2 + 2ab + b^2$ is not the same as $a^2 + b^2$? It is no answer to say that they look different — after all, $(\sin \theta)^2 + (\cos \theta)^2$ looks very different than 1, but $(\sin \theta)^2 + (\cos \theta)^2 = 1$ is an identity.

In the second place, $a^2 + 2ab + b^2$ is the same as $a^2 + b^2$ if (for instance) $a = 17$ and $b = 0$ — and they're equal for many other values of and b.

To disprove an identity, you should always give a specific numerical counterexample.

Example 6: Give a counterexample which shows that "$\frac{1}{x+2} = \frac{1}{x} + \frac{1}{2}$" is not an identity.

Solution: An identity is only asserted for values of the variables for which both sides are defined. So the assertion here is actually.

$$\text{"}\frac{1}{x+2} = \frac{1}{x} + \frac{1}{2} \text{ for } x \neq 0 \text{ and } x = -2.\text{"}$$

Thus, $x = 1$ is a counterexample, since

$$\frac{1}{x+2} = \frac{1}{3}, \text{ while } \frac{1}{x} + \frac{1}{2} = \frac{1}{1} + \frac{1}{2} = \frac{3}{2}$$

You should not give $x = 0$ or $x = -2$ as a counterexample. For these values of x, one side of the purported identity is undefined. Therefore, these cases are not part of what is claimed, so they can't be counterexample.

Finally, do not confuse giving a counterexample with proof by contradiction. A counterexample disproves a statement by giving a situation where the statements is false; in proof by contradiction, you prove a statement by assuming its negation and obtaining a contradiction.

Second Form of Induction

Let p be a proposition which is defined on the integers $n \geq 1$ such that:

(i) $p(1)$ is true

(ii) $p(n)$ is true whenever $p(x)$ is true all $1 \leq k < n$.

Then P is true for every integer $n \geq 1$

Example 7: Prove the proposition that the sum of the first n positive integers is $n(n+1)/2$

i.e. $\qquad p(n) = 1 + 2 + ... + n = n(n+1)/2$

Solution: For $n = 1$

$$p(1) = \frac{1}{2}(1)(1+1) = 1$$

so proposition holds for $n = 1$.

Let proposition is true for n. i.e.,

$$P_{(n)} = 1 + 2 + 3 + ... + n = \frac{1}{2}n(n+1) \qquad ...(1)$$

Then we have to prove that proposition holds for $(n + 1)$.

i.e., $\qquad p(n+1) = 1 + 2 + ... + n + (n+1) = \frac{1}{2}(n+1)(n+2)$

By Adding $(n+1)$ in equation (1) both sides, we get

$$1 + 2 + 3 + \ldots + n + (n+1) = \frac{1}{2}(n+1)n + n + 1$$

$$= \frac{1}{2}\left[n^2 + n + 2n + 2\right]$$

$$= \frac{1}{2}\left[n(n+2) + (n+2)\right]$$

$$= \frac{1}{2}(n+1)(n+2)$$

Hence,

$$1 + 2 + 3 + \ldots (n+1) = \frac{1}{2}(n+1)(n+2)$$

Hence, proposition $p(n + 1)$ holds

So p is true for every $n \in N$

Example 8: Prove the proposition p that the sum of the square of first n positive integers is $n(n + 1)(2n + 1)/6$

i.e., $\quad p_{(n)} : 1^2 + 2^2 + \ldots + n^2 = \dfrac{n(n+1)(2n+1)}{6}$

Solution: $p_{(1)}$ is true since

$$1^2 = \frac{1(1+1)(2.1+1)}{6} = 6/6 = 1$$

Assuming $p(n)$ is true, we add $(n + 1)^2$ to both sides of $p(n)$ obtaining

$$1^2 + 2^2 + \ldots + n^2 + (n+1)^2 = \frac{n(n+1)(2n+1)}{6} + (n+1)^2$$

$$= \frac{n(n+1)(2n+1) + 6(n+1)^2}{6}$$

$$= \frac{(n+1)(n+2)(2n+3)}{6}$$

$$= \frac{(n+1)[\{(n+1)+1\}2(n+1)+1\}]}{6}$$

which is $p(n + 1)$. Thus $p(n + 1)$ is true whenever $p(n)$ is true. By the principle of induction, p is true for all $n \in N$.

Example 9: Suppose $a \neq 1$. Let p be the proposition on $n \geq 1$ defined by

$$p(n) : 1 + a + a^2 + \ldots + a^n = \frac{a^{n+1} - 1}{a - 1}$$

show p is true for all n.

Solution: $p(1)$ is true since

$$1 + a = \frac{a^2 - 1}{a-1} = \frac{(a+1)(a-1)}{(a-1)} = 1 + a$$

Assuming $p(n)$ is true, we add $(a^{n+1})^2$ to both sides of $p(n)$ obtaining

$$1 + a + a^2 + \ldots + a^n + a^{n+1} = \frac{a^{n+1} - 1}{a-1} + a^{n+1}$$

$$= \frac{a^{n+1} - 1 + a^{n+2} - a^{n+1}}{a-1}$$

$$= \frac{a^{n+2} - 1}{a-1}$$

which is $p(n+1)$. Thus $p(n+1)$ is true whenever $p(n)$ is true. By principle of induction p is true for all $n \in N$

Example 10: Prove by mathematical induction that $6^{n+2} + 7^{2n+1}$ is divisible by 43 for each positive integer n.

Solution: P(1) true since

$$6^{1+2} + 7^{2+1} = 6^3 + 7^3 = 216 + 343 = 559 = 43.13$$

Let $p(k)$ is true, *i.e.* $6^{k+2} + 7^{2k+1} = 43.m$ for some integers m.

We have to prove this is true P($k+1$)

i.e,. $6^{(k+1)+2} + 7^{2(k+1)+1} = 6^{k+3} + 7^{2k+3}$ is divisible by 43.

Now,

$$= 6^{k+3} + 7^{2k+1+2} = 6^{k+3} + 7^{2k+1} \cdot 7^2$$
$$= 6(6^{k+2} + 7^{2k+1}) + 43(7^{2k+1})$$
$$= 6.43m + 43(7^{2k+1})$$
$$= 43(6m + 7^{2k+1})$$

which is divisible by 43. Hence $p(k+1)$ is true. So by the principle of mathematical Induction $p(n)$ is true for all positive integers.

Example 11: If A is a set with n elements, show that A has 2^n element.

Solution: Let P(n) be the statement that a set with n elements has exactly 2^n subsets.

$p(1)$ is true because–

For $n = 1$, the set contains only one element say A = $\{a\}$, and it has only two (2^1) subjects, *i.e.* ϕ and $\{a_1\}$.

Let $p(k)$ is true, *i.e.* the set A = $\{a_1, a_2, a_3 \ldots a_k\}$ has 2^k subsets.

We have to prove it is true for $p(k+1)$

i.e., the set A = $\{a_1, a_2, a_3 \ldots a_k, a_k+1\}$ has 2^{k+1} subsets.

Now the subset of A containing ($k+1$) elements can be divided into two parts.

Part I: Subsets not containing ($2k+1$) and this number is 2^k.

Part II: By inductive hypothesis subsets containing, a_{k+1} which can be represented as $\{a_{k+1}\}, \{a_1, a_{k+1}\}, \{a_2, a_{k+1}\}, a_1, a_2, a_{k+1} \ldots \{a_1, a_2, a_{k+1}\}$ and this number is 2^k.

Total number of subsets of A = The number of subsets not contain a_{k+1} + The number of subsets containing a_{k+1}

$$= 2^k + 2^k = 2^{k+1}, \text{ which is } p(k+1)$$

i.e., By the principle of mathematical induction $p(n)$ is true for all positive n.

Example 12: If $a_n = a_{n-1} + a_{n-2}$ for all $n > 3$ and $a_1 = a_2 = 1$ prove that $2^{n-1} a_n = n \pmod 5$ for all $n > 1$

Solution: $p(n)$ be the statement $2^{n-1} a_n = n \pmod 5$ for all $n > 1$.

$p(1)$ is true because–

For $n = 1$, $2^0 a_1 = 1 \pmod 5 \Rightarrow a_1 = 1 \pmod 5 = 1$ which is true

Since $\qquad a_1 = 1$

Let $p(k)$ is true

i.e. $\qquad 2^{k-1} a_k = k \pmod 5$

we have to prove that— $P(k + 1)$ is true, i.e.,

$$2^{k+1-1} a_{k+1} = k + 1 \pmod 5$$

But $\qquad a_{k+1} = a_k + a_{k-1}$

or $\qquad 2^k a_{k+1} = 2^k a_n + 2^k a_{k-1}$

$$= 2(2^{k-1} a_k) + 2^2 (2^{k-2} a_{k-1})$$

By hypothesis, we have $2^{k-1} a_k = k \pmod 5$ and

$$2^{k-2} a_{k-1} = k - 1 \pmod 5$$

Therefore, $\qquad 2^{k+1-1} a_{k+1} = 2^k a_{k+1}$

$$= 2^k a_k + 2^k a_{k-1}$$
$$= 2^k + 2^2 (k-1) \bmod 5$$
$$= k + 1 \bmod 5$$

$[6 = 1 \bmod 5 \text{ and } -1 = 4 \pmod 5]$

Thus $p(k+1)$ is true.

Hence By mathematical induction $p(n)$ is true sfor all $n > 1$.

EXERCISE 4.1

1. Prove or disprove
   (i) The sum of two even integers is an even integers.
   (ii) The sum of three odd integers is on odd integer.
   (iii) The sum of two primes is never a prime integers.

2. Prove that $\sqrt{3}$ is irrational.
3. Prove that $|xy| = |x| \, |y|$
4. Show that

(i) $\left[\dfrac{n}{3}\right] + \left[\dfrac{2n}{3}\right] = n$ for all positive integers $n \geq 1$

(ii) $\left[\dfrac{m}{2}\right] + \left[\dfrac{m}{2}\right] = m$ for all integers m.

5. Prove by contradiction that the difference of any rational number and any irrational number is irrational.

6. Prove by contradictions that $6 - 7\sqrt{2}$ is irrational.

7. Prove that there is at most one real number a with the property $a + r = r$ for all real number a. Give direct and also proof by contradiction.

8. Prove that the structure [even integer, + x] is closed with respect to multiplication.

9. Show that each of the following inferences is fallacy.
    (i) If today is Rita's birthday, then today is June 21. Today is June 21, Hence today is Rita's birthday.
    (ii) If the client is guilty, then he has at the scene of the crime. The client was at the scence of the crime. Hence, the client is not guilty.

EXERCISE 4.2

Prove the following by the method of induction for all integers

1. $1 + 3 + 5 + \ldots + (2n - 1) = n^2$

2. $1^3 + 2^3 + 3^3 + \ldots + n^3 = \left[\dfrac{n(n+\psi)}{2}\right]^2$

3. $2 + 2^2 + 2^3 + \ldots + 2^n = 2(2^n - 1)$

4. $\dfrac{1}{1.2} + \dfrac{1}{2.3} + \dfrac{1}{3.4} + \ldots$ to n terms $= \dfrac{n}{n+1}$

5. $x^n - y^n$ is divisible by $(x + y)$ for all positive integral value of n.

6. $x^n - y^n$ is divisible by $(x+y)$ when n is an even integers.

7. $a + an + an^2 + \ldots + an^{n-1} = \dfrac{ar^n - a}{r - 1}$

8. $\dfrac{1}{3.5} + \dfrac{1}{5.7} + \dfrac{1}{7.9} + \ldots$ to n terms $= \dfrac{n}{3(2n+3)}$

9. $1 + 5 + 12 + 22 + 35 + \ldots$ to n terms $= \dfrac{n^2(n+1)}{2}$

10. $1^2 + 3^2 + 5^2 + \ldots + (2n-1)^2 = (4n^3 - n)/3$

11. $n^3 - 4n + 6$ is divisible by 3.

12. $1^n - 4^n$ is divisible by 7.
13. $10^{2n-1} + 1$ is divisible by 71.
14. $3.5^{2n} + 2^{3n+1}$ is divisible by 17
15. $7^{2n} + 16n - 1$ is divisible by 64.
16. $n(n+1)(n+2)$ is divisible by 6.
17. $n^3 + 2n$ is divisible by 3.
18. $4^{2n+1} + 3^{n+2}$ is an integer multiple of 13.
19. $7^n - 2^n$ is divisible by 5
20. Show that $(1+1/2)^n \geq 1 + n/2$
21. Show that $3n > n^3$ for each integer $n \geq 4$.
22. Show that $(1+ 5/_2)^n \geq 1 + 5n/2$
23. Show that $n^3 + 1 > n^2$ for each integer $n \geq 2$
24. Show that $4n > 3n$ for all $n \geq 0$
25. Show that $2n > n^2$, for all $n \geq 5$
26. Show that $n^2 > 2n+1$ for all $n \geq 3$

CHAPTER 5

ALGEBRAIC STRUCTURE

5.1 BINARY OPERATION

We are aware with arithmetic operations like addition, subtraction, multiplication and division. Wherein these operations on any two real numbers yield another real number.

Arithmetic operations combine two elements of the set of real numbers to give another element of the same set. Such operations are called 'binary operations or binary compositions'.

Let $x, y \in R$ then $x + y \in R$, $x - y \in R$, $x * y \in R$, and $\frac{x}{y} \in R$. So a binary operation is a rule, defined on a given set S, which assigns to any elements $a, b \in S$, a unique third element in S. It is denoted by the symbol '∘'.

'∘' assigns to any two elements $a, b \in S$, a unique third element $a \circ b \in S$.

An operation '∘' on a non-empty set S which is a mapping that associates with each ordered pair (a, b). When $a, b \in S$, a uniquely defined element $a \circ b$ of S, is called a binary operation.

In other words, a binary operation '∘' on a set S, is a mapping of $S \times S$ into S.

Example 1. Let the operation '∘' be ordinary addition '+' defined on the set N. Let $a, b \in N$. Hence $a \circ b = a + b = +$ ve (integer)

i.e. $(a + b) \in N$. Hence '+' is a binary operation.

Example 2. Consider a set S of all odd integers, on which the operation. '+' is defined. Clearly if $a, b \in S$, then $a + b$ is an even number. Thus $(a + b) \notin S$. Hence '+' on S is not a binary operation.

Example 3. Let '∘' stand for ordinary division '÷' defined on the set S of all non-zero integers. Let $a, b \in S$.

Clearly $a \circ b = a \div b$. But $a \div b$ may not be an integer. Hence $a \div b$ may not belong to S. Hence '÷' on S is not a binary operation.

5.2 TYPES OF BINARY OPERATIONS

(i) Commutative: If binary operation '∘' on a set S is such that $a \circ b = b \circ a$, $\forall\, a, b \in S$, then operation '∘' is called Commutative.

Example. Let R be the set of all real numbers and '∘' stand for ordinary '+'. If $a, b \in R$ then
$$a + b = b + a, \forall\, a, b \in R.$$
Hence '+' on R is commutative.

But if ∘ stands for ÷ than $a \circ b = a \div b$ and $b \circ a = b \div a$ definitely $a/b \neq b/a$ except when $a = \pm b$.

Hence '÷' on R is not a commutative operation.

(ii) Associative: An operation '∘' on a set S is said to be associative if, $\forall\, a, b, c \in S$, $a \circ (b \circ c) = (a \circ b) \circ c$.

Example 1. Ordinary '+' and '·' defined on the set of all real numbers or complex numbers or all integers on all rational numbers are associative, i.e.
$$a + (b + c) = (a + b) + c \quad \forall\, a, b, c \in S$$
$$a \cdot (b \cdot c) = (a \cdot b) \cdot c \quad \forall\, a, b, c \in S$$

But 'subtraction' and 'division' are not associative on the sets mentioned above, e.g.
$$a - (b - c) = a - b + c, \text{ while } (a - b) - c = a - b - c$$
$$a - (b - c) \neq (a - b) - c$$

Example 2: Let binary operation '∘' on the set R of all real numbers be taken by $a \circ b = a + 3b$, $\forall\, a, b \in R$.

Now let $a, b, c \in R$

∴ $\qquad a \circ (b \circ c) = a \circ (b + 3c) = a + 3(b + 3c) = a + 3b + 9c$.

and $\qquad (a \circ b) \circ c = (a + 3b) \circ c = (a + 3b) + 3c = a + 3b + 3c$

Hence, $\qquad a \circ (b \circ c) \neq (a \circ b) \circ c$

Hence, '∘' on R is not associative.

(iii) Distributive: Let '∘' and ⊕ be two binary operations defined on a given set S. Let $a, b, c \in S$.

If $\qquad a \circ (b \oplus c) = (a \circ b) \oplus (a \circ c)$

Operation '∘' is said to be left distribution with respect to ⊕.

If $(b \oplus c) \circ a = (b \circ a) \oplus (c \circ a)$, then '∘' is said to be right distributive with respect to '⊕'.

If an operation is left as well as right distributive we simply say it is distributive.

Example: Let I be the set of all integers. If $x, y, z \in I$ then let $x \circ y + x + 2y$ and $x \oplus y = 2xy$.

Clearly $\qquad x \circ (y \oplus z) = x \circ (2yz) = x + 2(yz) = x + 2yz;$
$$x \circ y = x + 2y; \ x \circ z = x + 2z$$
$\therefore \qquad (x \circ y) \oplus (x \circ z) = (x + 2y) \oplus (x + 2z)$
$$= 2(x + 2y)(x + 2z)$$
$$x \circ (y \oplus z) \neq (x \circ y) \oplus (x \circ z)$$

Hence '∘' is not left distributive with respect to ⊕.

Again $\qquad (y \oplus z) \circ x = 2yz \circ x = 2yz + 2x$

Now $\qquad (y \circ x) \oplus (z \circ x) = 2(y + 2x)(z + 2x)$

Clearly $\qquad (y \oplus z) \circ x \neq (y \circ x) \oplus (z \circ x)$

Hence '∘' is not right distributive with respect to ⊕.

But $\qquad x \oplus (y \circ z) = x \oplus (y + 2z) = 2x(y + 2z) = 2xy + 4xz$
$$= (x \oplus y) \circ (x \oplus z).$$

Hence ⊕ is left distributive with respect to ∘.

Also $\qquad (y \circ z) \oplus x = (y + 2z) \oplus x = 2(y + 2z)x$
$$= 2yx + 4zx = (y \oplus x) \circ (z \oplus x).$$

Hence ⊕ is also right distributive over ∘.

Identity of the operation: Let '∘' be the operation defined. On a set A. If there exists the element $e \in A$, such that
$$e \circ a = a, \ \forall \ a \in A$$
then e is called left identity with respect to operation '∘'.

But if $a \circ e = a, \ \forall \ a \in A$

then e is called right identity w.r.t operation '∘'.

Increase with respect to the operation: Let '∘' be the operation defined on the set A. If there exists an element $b \in A$, corresponding to $a \in A$. Such that
$$b \circ a = e$$
then b is left inverse of a in A with respect to operation '∘'.

Similarly, if $a \circ b = e$, then b is called right inverse of a with respect to operation ∘ in A. Inverse of a is denoted by the symbol a^{-1}.

SOLVED EXAMPLES

Example 1: Let '+' be the binary composition on the set I, of all integers the find a^{-1}.

Solution: Here zero is the identity element,

for $\qquad 0 + a = a, \ \forall \ a \in I$

Also $a \in I \ \Rightarrow \ -a \in I$ and $a + (-a) = 0$

i.e., $-a$ is inverse of a in I with respect to operation '+'.

Hence, a^{-1} is $(-a)$.

Example 2: Let '·' be ordinary multiplication defined on the set of non-zero real numbers, then find its inverse w.r.t. operation.

Solution: If R_0 denotes set of non-zero real numbers, then 1 is the identity for

$$a \cdot 1 = a \ \forall \ a \in R_0 \text{ and } 1 \in R_0.$$

Also $a \in R_0 \ \Rightarrow \ \dfrac{1}{a} \in R_0$ and $a \cdot \dfrac{1}{a} = 1$

Hence $1/a$ is inverse of a in R_0 with respect to the operation '·'.

So $\qquad a^{-1} = \dfrac{1}{a}.$

5.3 COMPOSITION TABLE

When number of elements in the given set is small, a very compact method of deciding whether the composition defined on it,

(i) is binary,

(ii) has identity in the set,

(iii) yields inverse in the set, and (iv) is commutative, is that of forming the composition table.

5.4 FORMATION OF COMPOSITION TABLE

Let $A = \{a, b, c, d\}$ on which is defined the composition '∘'. Let us draw four vertical lines and four horizontal lines intersecting each other. Let each column be marked by an element on the set. Now mark each row by an element of the set, order of elements in making columns being identical with the order of making rows. Let "∘" be marked in the table as shown.

"∘"	a	b	c	d
a	a∘a	a∘b	a∘c	a∘d
b	b∘a	b∘b	b∘c	b∘d
c	c∘a	c∘b	c∘c	c∘d
d	d∘a	d∘b	d∘c	d∘d

Let us complete the blanks in the first row : first element of the first row will be $a \circ a$, second will be $a \circ b$, third $a \circ c$ and fourth $a \circ d$.

Second row will have element $b \circ a$, $b \circ b$, $b \circ c$, $b \circ d$; third row will have $c \circ a$, $c \circ b$, $c \circ c$, $c \circ d$; and fourth row will have $d \circ a$, $d \circ b$, $d \circ c$, $d \circ d$

Now let $a \circ a = a$, $a \circ b = b$, ... etc., as shown in the below given table

"$\circ$"	a	b	c	d
a	a	b	c	d
b	b	c	d	a
c	c	d	a	b
d	d	a	b	c

The table gives the following information:
1. Since each element of the table belongs to the given set A.
    i.e., '$\circ$' is a binary operation.
2. Order of elements in 1st column is the same as order of markings of the columns and order of elements in the 1st row is same as order of markings of the rows of the table. It implies that a is the identity element.
3. Order of elements in each row, exactly corresponding to the order of elements in corresponding column, i.e., orders of elements in 1st, 2nd, 3rd and 4th columns are same as in 1st, 2nd, 3rd, 4th rows respectively. This will mean '$\circ$' is commutative.
4. Mark out identity element appearing whenever in the table. Corresponding elements in row-column meanings will be inverses of each other. We get a is the identity element. Thus a is inverse of a, b is inverse of d, c the inverse of c and d is the inverse of b.

Note : Whenever an operation in a set A is denoted additively by '+', the corresponding identity is called zero and is accordingly denoted by $\circ$. Also then inverse of an element $a \in A$ is denoted by $-a$. Similarly if operation is denoted multiplicatively by '$\cdot$', the corresponding identity is called unity and is denoted by 1.

5.5 ALGEBRAIC STRUCTURES

A non-empty set together with one or more binary operations defined on the set is called a Algebraic Structure or Mathematical Structure.

Or we can say an algebraic structure is a set together with closed operation defined over the set.

It G is a non-empty set and '$\circ$' is a binary operation on it then Algebraic structure formed by them is denoted by $(G, \circ)$.

Isomorphic

A homomorphic which is injective (one-to-one) is called a monomorphism, a homomorphism which is surjective (on to) is called an epimorphism and when the homomorphism is a bijection then it is called an isomorphism. If there exists an isomorphism between two structures then they are said to be isomorphic.

The word isomorphic means 'same shape' and so it seems reasonable to expect that isomorphisms should be able to partition the set of all algebraic structure into equivalence classes.

Example 3: The two structures $(\{\phi, S\}, \cap, \cup)$ and $(\{0, 1\}, \wedge, \vee)$ (as defined in the below given table) are Isomorphic.

$\vee$	0	1	2	3	4
0	0	1	2	3	4
1	1	1	2	3	4
2	2	2	2	3	4
3	3	3	3	3	4
4	4	4	4	4	4

$\wedge$	0	1	2	3	4
0	0	0	0	0	0
1	0	1	1	1	1
2	0	1	2	2	2
3	0	1	2	3	3
4	0	1	2	3	4

$a \vee b$ = the maximum of a and b.

$a \wedge b$ = the minimum of a and b.

Solution: Let $\phi(\phi) = 0$ and $\phi(S) = 1$. Clearly ϕ is a bijection. Also $\phi(\phi \cap \phi) = \phi(\phi) = 0 = 0 \wedge 0 = \phi(\phi) \wedge \phi(\phi)$.

Group: Consider a mathematical system consisting of a non-empty set S and an operation $*$ defined on set S. The system is a group under the following group axioms :

(G$_1$) Closure axiom: G is closed under the binary operation $*$.

Thus if $a, b \in G$, we have

$$a * b \in G, \quad \forall\ a, b \in G;$$

i.e., $a * b$ is an element of G. This means '$*$' is a binary operation.

(G$_2$) Associative axiom: The operation '$*$' on G is associative.

Thus if $a, b, c \in G$ we have

$$a * (b * c) = (a * b) * c \quad \forall\ a, b, c \in G$$

(G$_3$) Identity axiom: These exists the element $e \in G$ such that :

$$a * e = e * a = a \quad \forall\ a \in G$$

e is called the identity element of G, for the operation '$*$'.

(G₄) Inverse axiom: Every element belonging to G has its inverse.

Thus $\forall\ a \in G$, there exists a^{-1}, such that

$$a * a^{-1} = a^{-1} * a = e,\ a^{-1} \in G$$

5.6 SOME DEFINITIONS

(*i*) **Finite Group:** If a group consists of a finite number of elements, it is called a finite group.

(*ii*) **Infinite Group:** If a group contains an infinite number of elements, it is called an infinite group.

(*iii*) **Order of a Group:** The number of elements in a finite group is called the order of the group.

Let a group consists of *m* elements, it is called a group of *m*th order.

(*iv*) **Abelian Group or Commutative Group:** It is addition to group axioms, operation is also commutative, it is called Abelian Group or Commutative Group.

Thus for an abelian group with composition '*' defined in it, we must have

$$a * b = b * a\ \forall\ a, b, \in G.$$

or we can say

A mathematical system consisting of a set S and a binary operation * forms a commutative group. If it exhibits the following properties :

— Closure Property
— Associative Property
— Identity Property
— Inverse Property
— Commutative Property

Example 4: Prove that the set of cube roots of unity is abelian finite group with respect to multiplication.

Solution: Let 1, ω, ω^2 be cube roots of unity so that $\omega^3 = 1$. Let us form the composition table for the set G = {1, ω, ω^2} with respect to multiplication.

'*'	1	ω	ω^2
1	1	ω	ω^2
ω	ω	ω^2	1
ω^2	ω^2	1	ω

Here $\omega^3 = 1$ and $\omega^4 = \omega^3 . \omega = \omega$.

1. Since all elements in the composition table belong to G, hence G is closed under binary operation *.

2. Since multiplication of complex number is associative therefore multiplication is associative in G.
3. It is clear from first row of the table that 1 is the identity element in G.
4. Inverse of 1, ω, ω^2 are 1, ω^2, ω respectively and they all belong to G.

Hence G is a multiplication group and it contains. Finite number of elements $\Rightarrow$ Hence it is a finite group.

Example 5: The table given below defines a certain binary operation $\oplus$ on the Set A = {a, b, c, d, e}. Show that the set A forms a group with respect to the binary operation $\oplus$ and find the inverse of each element.

$\oplus$	a	b	c	d	e
a	a	b	c	d	e
b	b	c	d	e	a
c	c	d	e	a	b
d	d	e	a	b	c
e	e	a	b	c	d

Solution:
1. Since all elements in the composition table belong to A. Hence A is closed under binary operation $\oplus$.
2. Since $(a \oplus b) \oplus c = b \oplus c = d$
   $a \oplus (b \oplus c) = a \oplus d = d$
   So $\oplus$ is associative in A.
3. Since
   $a \oplus a = a,\ b \oplus a = a \oplus b = b,\ c \oplus a = a \oplus c = c,$
   $d \oplus a = a \oplus d = d,\ e \oplus a = a \oplus e = e.$
   So, 'a' is the identity element in A.
4. Inverse:
   Here $d \oplus c = a$ (left identity) and $c \oplus d = a$ (Right identity)
   $\Rightarrow$ c is the inverse of d.
   Similarly d is the inverse of c

and since $e \oplus b = a$ and $b \oplus e = a$ so e is the inverse of b and b is the inverse of e. Inverse of a is a.

Since (A, $\oplus$) has all four properties of group. Hence the system (A, $\oplus$) forms a group.

Example 6: Does {1, – 1} forms a group under multiplication.
Solution: First we make the composition table
Let A = {1, – 1}

*	1	−1
1	1	−1
−1	−1	1

1. Since all element of the composition table belong to A. So A is closed under the binary operation $*$.
2. Since multiplication of integers is associative, so A is associative under the operation $*$.
3. The identity element for multiplication is 1 which is in the Group A.
4. Since $1*1 = 1$ so inverse of 1 is 1

   $-1*-1 = 1$ inverse of -1 is -1

Hence, (A, $*$) has all four properties of a Group.

Hence the system (A, $*$) forms a Group.

Example 7: Prove that the set of all integers (including zero) with additive, binary operation is an infinite abelian group.

Solution: Let I be the set of all integers and $x, y, z \in I$.

(i) Closure: $x, y \in I \Rightarrow x, y$ are integers.

$\Rightarrow x + y$ is an integer.

$\Rightarrow (x + y) \in I$

Hence the group I is closed under the operation +.

(ii) Associative Property: We know that the addition of integers x, y, z is associative.

i.e., if $x, y, z \in I$ then $x + (y + z) = (x + y) + z$.

Hence, addition in I is associative.

(iii) Existence of Intensity: Let $x \in I$, also $0 \in I$.

Here $\qquad 0 + x = x + 0, \forall x \in I$

Hence, 0 is additive identity in I.

(iv) Existence of Inverse: Here $x \in I \Rightarrow -x \in I$.

But $\qquad (-x) + x = 0 \ \forall x \in I$

Hence, $-x$ is inverse of $x, \forall x \in I$.

Hence inverse of each element in I belongs to I.

It is clear that number of elements in I is infinite.

$\Rightarrow$ I is a infinite group.

(v) Commutative Property: Since addition of integers is commutative.

i.e., $\qquad x + y = y + x, \forall x, y \in I.$

Since I contains all five properties of an abelian group under the operation +.

So I is an infinite abelian group.

Example 8: Prove that the set of all non-singular square matrices of order n with real elements is a group with respect to matrix multiplication. Is it an abelian group?

Solution: Set G be the set of all $n \times n$ non-singular matrices over the set of real numbers. Let matrices A, B, C ∈ G.

(i) Closure: Since produce of two square matrices of the same order is a square matrices of the same order.

∴ AB ∈ G, ∀ A, B ∈ G

Hence G is closed under the operation multiplication.

(ii) Associative Property: We know that product of three matrices A, B, C ∈ G is associative, *i.e.* A(BC) = (AB)C by the theory of matrices.

Hence G is associative under this operation.

(iii) Identity Element: Let I denote the unit matrix of order n. Hence I ∈ G.

Also IA = A ∀ A ∈ G.

Hence, I is multiplicative identity.

(iv) Inverse Elements: By theory of matrices every non-singular square matrix A possesses a non-singular square matrix A^{-1} as its inverse. Such that

$$A^{-1}A = I \ \forall \ A \in G.$$

Hence, G is a group with this composition.

(v) Commutative Property: Since the product of two matrices is not commutative.

i.e., AB ≠ BA

Hence G is a non-abelian group.

Example: Determine whether the following sets form a group for the operation defined in the set.

(i) K = {...., – 4, – 2, 0, 2, 4, ...} with the operation of addition.

(ii) Z = set of integers including zero with the operation ∗ defined as :

$a \ast b = a - b$; $a, b \in Z$.

EXERCISE 5.1

1. Do {a, b} and the binary operation denoted by ∗ form a group if the operation is defined on {a, b} by the table given below.

∗	a	b
a	b	b
b	a	a

2. Show that the set A = $\{-3, -2, -1, 0, 1, 2, 3\}$ is not a group with respect to addition.

3. Prove that the set of matrices $A_2 = \begin{bmatrix} \cos\alpha & -\sin\alpha \\ \sin\alpha & \cos\alpha \end{bmatrix}$ where α is real, forms a group under multiplication.

4. For the binary operation denoted by $\oplus$ and defined on {0, 1} in below given table, determine whether the system forms a group.

$\oplus$	0	1
0	0	1
1	1	0

Example 9: A set of 2 × 2 real matrices $\begin{bmatrix} a & b \\ c & d \end{bmatrix}_{2\times 2}$ with binary multiplication is a group. When $ad - bc \neq 0$.

Solution: Let G be the group of 2 × 2 real matrices.

(i) Closure: Let $A, B \in G$

and
$$A = \begin{bmatrix} a & b \\ c & d \end{bmatrix}, B = \begin{bmatrix} p & q \\ r & s \end{bmatrix}$$

$$A * B = \begin{bmatrix} a & b \\ c & d \end{bmatrix} * \begin{bmatrix} p & q \\ r & s \end{bmatrix}$$

$$= \begin{bmatrix} ab+br & aq+rs \\ cp+dr & cq+ds \end{bmatrix}$$

$= (ab + br)(cq + ds) - (cp + dr)(aq + rs)$
$= abcq + apds + prcq + brds - cpaq - cpbs - draq - drbs.$
$= ps(ad - bc) + rq(cb - ad)$
$= (ad - bc)(ps - rq)$
$\neq 0$

So $A * B \in G$

Hence, Group G is closed under the operation multiplication.

(ii) Associative Property: We know that in matrix operation
$$(AB)C = A(BC)$$
Hence multiplication is associative in G.

(iii) Identity : Let e be the identity.

So $$\begin{bmatrix} a & b \\ c & d \end{bmatrix} e = \begin{bmatrix} a & b \\ c & d \end{bmatrix}$$

$$\Rightarrow \quad e = \begin{bmatrix} 1 & 0 \\ 0 & 1 \end{bmatrix}$$

Hence, identity exists in G.

(iv) Inverse : $\begin{bmatrix} a & b \\ c & d \end{bmatrix} \begin{bmatrix} a & b \\ c & d \end{bmatrix}^{-1} = e = \begin{bmatrix} 1 & 0 \\ 0 & 1 \end{bmatrix}$

Example 10: Show that the set of all odd integers with addition is not a group.

Solution: Let Set A = $\{1, 3, 5, 7, 9, ...\}$

1. Closure: Here $1, 3 \in A$

But $\qquad 1 + 3 = 4 \notin A$

So Set A is not closed under the operation '+'.

Hence, it is not a group.

Example: Prove that set of zero and even integers with addition is an abelian group.

Solution: Let the Set A = $\{0, 2, 4, 6, 8, ...\}$

Let $x, y, z \in A$ and A be the set of zero and even integers.

(i) Closure: $x, y \in A$, since x, y are even integers.

$\Rightarrow x + y$ is an even integer.

$\Rightarrow x + y \in A$

Hence closure property Holds.

(ii) Associative Property: We know that the addition of integers is associative.

If $x, y, z \in A$ then

$$x + (y + z) = (x + y) + z$$

Hence, addition in A is associative.

(iii) Existence of Identity: Let $x \in A$, also $0 \in A$

Hence, $0 + x = x + 0 \; \forall \; x \in A$

Hence, 0 is additive identity in A.

(iv) Existence of Inverse: Here $x \in A \Rightarrow -x \in A$

But $\qquad (-x) + x = 0 \; \forall \; x \in A$

So $-x$ is inverse of $x \;\forall\; x \in A$.

Hence inverse of each element in A belongs to A.

(v) Existence of Abelian: Here $x, y \in A$

and $\qquad\qquad x \circ y = y \circ x \;\forall\; x, y$

Also addition of integers is commutative.

i.e. $\qquad\qquad x + y = y + x$

$\forall\; x, y \in A$

Hence, it is an infinite abelian group.

Example 11: Show that the set $\{0 \pm n, \pm 2n, ... \pm kn....\}$ is an additive group, n being a fixed integer.

Solution: Set $A = \{0 \pm n, \pm 2n, ... \pm kn....\}$ and also let $x, y, z \in A$.

(i) Closure : $x, y \in A \; x, y$ are integers.

$\Rightarrow x + y$ is an integer. (because $0 \pm n = \pm kn \in A$)

$\Rightarrow (x + y) \in A$

Hence closure property holds.

(ii) Associative Property: We know that addition of integers x, y, z is associative i.e. if $x, y, z \in A$ then

$$x + (y + z) = (x + y) + z$$

(iii) Existence of Identity: Let $x \in A$. Also $0 \in A$

Hence $\qquad\qquad 0 + x = x + 0 \;\forall\; x \in A$

Hence 0 is additive identity in A.

(iv) Existence of Inverse: Here $x \in A, -x \in A$. But

$$(-x) + (x) = 0 \qquad \forall\; x \in A.$$

Hence, $-x$ is inverse of $x \;\forall\; x \in A$.

Thus, A forms an additive group in which number of element is infinite. Hence it is an additive group.

Example 12: Prove that the set $\{1, -1, i, -i\}$ is an abelian multiplication finite group of order 4.

Solution:

(i) Closure: $1, -1 \in A$ and also $x, y, z \in A$

$\therefore \qquad\qquad -1 * -1 = -1 \in A$

$\qquad\qquad i \times i = i^2 = -1 \in A$

$$i \times i = 1 \in A$$
$$-i \times i = -1 \in A$$

Hence closure property holds.

(ii) Associative Property: Let $x, y, z \in A$.

Now $\quad\quad\quad x \times (y \times z) = z \times (x \times y)$

Since $\quad\quad\quad 1 \times (-1 \times i) = -i$

$\quad\quad\quad\quad\quad i \times (1 \times -i) = -i$

Hence associative property holds.

(iii) Existence of Identity: Let $1 \in A$. Also $-1 \in A$.

But $\quad\quad\quad 1 \times -1 = -1 \; \forall \, x \in A$.

Hence, -1 is the multiplication identity.

(iv) Existence of Inverse: Here $x \in A \Rightarrow -x \in A$.

But $\quad\quad\quad 1 \times -1 = -1 \; \forall \, x \in A$.

Hence, -1 is inverse of $1 \; \forall \, 1 \in A$.

Hence, inverse of each element in A belongs to A.

Thus A forms an multiplication group in which numbers is 4. Hence it is finite group of order 4.

(v) Existence of Abelian: Also multiplication of integers is commutative.

i.e. $\quad\quad\quad x \times y = y \times x \; \forall \, x, y \in A.$

Hence it is an finite abelian group of order 4.

Example 13: Show that the set {1} forms a group with respect to multiplication.

Solution: Let Set A = {1} and $x, y, z \in A$

where $x = 1, y = 1, z = 1$.

(i) Closure: $1, 1 \in A$

$\Rightarrow 1 \times 1$ is an integer.

$\Rightarrow 1 \in A$ Hence closure property holds.

(ii) Associative Property: We know that multiplication of integers 1, 1, 1 is associative i.e. if $1, 1, 1 \in A$ then

$$x \times (y \times x) = (x \times y) \times x$$

Hence multiplication in A is associative.

(iii) Existence of Identity: Let $x \in A$ and $1 \in I$.

Hence $\quad\quad\quad 1 \times x = x \times 1 \; \forall \, x \in 1.$

Hence, 1 is additive identity in A.

(iv) Existence of Inverse: Here $x \in A \Rightarrow \dfrac{1}{x}$

But $\qquad\qquad x \times \dfrac{1}{x} = 1 \qquad \forall\, x \in A$

Hence $\dfrac{1}{x}$ is inverse of $x\ \forall\, x \in A$

Hence, inverse of each element in A belongs to A.

Thus, A forms an multiplication group.

Example 14: Show that the set of integers with respect to multiplication is not a group.

Solution: Let A = Set of integers and $x, y, z \in A$.

Where x, y, z are integers.

(i) Closure: Let $x, y \in A$

$\Rightarrow x \times y$ is an integer.

$x \times y \in A$

Hence, closure property holds.

(ii) Associative Property: We know that multiplication of integers x, y, z is associative i.e., if $x, y, z \in A$ then

$$x \times (y \times z) = (x \times y) \times z$$

Hence, addition in A is associative.

(iii) Existence of Identity: Let $x \in A$ also $1 \in A$

Hence $\qquad\qquad 1 \times x = x \times 1 = x \qquad \forall\, x \in A$

Hence 1 is the multiplication identity.

(iv) Existence of Inverse: Here $x \in A$

But $\dfrac{1}{x}$ is not a integer

So $\dfrac{1}{x} \notin A$ For example Let $x = 2$

$$\dfrac{1}{2} \notin A$$

Hence, inverse does not exist.

Hence, A is not a group with respect to multiplication.

Example 15: Show that the set of vectors with vector multiplication as composition is not a group.

Solution: Let V be the set of all vectors and let '*' denote vector multiplication. Also let $v_1, v_2, v_3 \in V$.

(*i*) **Closure Property :** $v_1, v_2 \in V$, then $v_1 \times v_2$ is a vector.

∴ $v_1 \times v_2 \in V$, $\forall v_1, v_2 \in V$

Hence, closure property holds.

(*ii*) **Associative Property:** Now $a \times (b \times c) = (a \cdot c)b - (a \cdot b) \cdot c$

and $\qquad (a \times b) \times c = (a \cdot c)b - (b \cdot c) \cdot a$

Clearly $\qquad (a \times b) \times c \neq (a \times b) \times c$

Hence * is not associative.

Hence V is not a group for the operation '*'.

Example 16: Prove that n, nth roots of unity from a multiplicative abelian group.

Solution: Let $x^n = 1 = \cos(2r\pi) + i\sin(2r\pi)$

∴ $\qquad x = \cos\dfrac{2r\pi}{n} + i\sin\dfrac{2r\pi}{n}$, $r = 0, 1, 2, \ldots n-1$

$\qquad\qquad = e^{2r\pi i/n}$, $r = 0, 1, 2, \ldots n-1$

Hence the set of n, nth root of unity is given by

$$A = \left\{ e^{2r\pi i/n} : r = 0, 1, 2, \ldots n-1 \right\}$$

(*i*) **Closure Property:** Let $a, b \in A$ and $a = e^{2\pi r_1 i/n}$, $b = e^{2\pi r_2 i/n}$
where $0 < r_1, r_2 < n - 1$.

∴ $\qquad ab = e^{2\pi r_1 i/n} \cdot e^{2\pi r_2 i/n} = e^{2\pi i(r_1 + r_2)/n} \in A$ if $(r_1 + r_2) < n - 1$

But if $r_1 + r_2 > n - 1$, let $r_1 + r_2 = n + P$ where $0 \leq P \leq n - 2$.

∴ $\qquad ab = e^{2\pi r(n+P)/n} = e^{2\pi i} \cdot e^{2\pi P i/n} = e^{2\pi P i/n} \in A$

For $0 \leq P \leq n - 2$.

Hence $a, b \in A \Rightarrow ab \in A$, $\forall a, b \in A$.

Hence closure property holds.

(*ii*) **Associative Properties:** Since n, nth roots of unity are complex numbers, and multiplication of complex numbers is associative, hence product of roots is associative.

or $\qquad e^{2\pi r_1 i/n} (e^{2\pi r_2 i/n} \cdot e^{2\pi r_3 i/n}) = e^{2\pi i(r_1 + r_2 + r_3)/n}$

$\qquad\qquad\qquad = (e^{2\pi r_1 i/n} \cdot e^{2\pi r_2 i/n}) e^{2\pi r_3 i/n}$

Hence Associative property holds.

(*iii*) **Existence of Identity:** Putting $r = 0$ in $e^{2r\pi i/n}$

We get $\qquad e^{2r\pi i/n} = e^0 = 1$

$\therefore$ $1 \in A$. Also $\quad 1 \cdot e^{2\pi r i/n} = e^{2\pi r i/n}, \quad \forall\, e^{2\pi r i/n} \in A$.

Hence, 1 is identity and $1 \in A$.

(iv) Existence of Inverse: Let $a = e^{2\pi i r/n}$ and $b = e^{2\pi (n-r)i/n}$.

Hence, $\quad ab = e^{2\pi r i/n} \cdot e^{2\pi (n-r)i/n} = e^{2\pi i} = 1, \quad \forall\, a \in A$.

Hence, $e^{2\pi(n-r)i/n}$ is inverse of $e^{2\pi r i/n}$.

Also $e^{2\pi(n-r)i/n} \in A$.

Hence, inverses of elements of A belong to A.

Also $a \cdot b = b \cdot a$.

Hence, A is abelian group with respect to multiplication.

Example 17: Verify that the totality of all positive rationals form as group under the composition defined by

$$a * b = ab/2$$

Solution: Let Q^+ be the set of all positive rationals and let $a, b, c \in Q^+$.

(i) Closure Property: Here $a * b = \dfrac{ab}{2}$ a rational.

$\therefore$ $ab \in Q^+$, $\forall\, a, b \in Q^+$.

Hence, closure property holds.

(ii) Associative Property: Now $a * (b * c) = a * \left(\dfrac{bc}{2}\right)$

by definition

$$= \dfrac{abc}{4} = \dfrac{ab}{2} \cdot \dfrac{c}{2}$$

$$= (a * b) * \dfrac{c}{2} = (a * b) * c$$

Hence associative property holds.

(iii) Existence of Identity: Let e be the left identity.

$\therefore \quad\quad\quad e * a = a \Rightarrow \dfrac{ea}{2} = a$

$\Rightarrow \quad\quad\quad e = 2.\quad\quad$ Also $2 \in Q^+$

Hence 2 is identity and $2 \in Q^+$.

(iv) Existence of Inverse: Let b be left inverse of a.

$$b * a = a = 2$$

$\Rightarrow \quad \dfrac{ba}{2} = 2$

$\Rightarrow \quad b = \dfrac{4}{a} \in Q^+ \qquad \because a \in Q^+$

Hence inverse of $\forall\, a \in Q^+, \in Q^+$.

Thus $(Q^+, *)$ is a group.

(v) Commutative Law: Clearly $(a * b) = \dfrac{ab}{2} = \dfrac{ba}{2} = b * a$

Hence $*$ is commutative and $(Q^+, *)$ is an abelian group.

EXERCISE 5.2

1. Define a group and state group axioms.
2. Show that $G = \{0\}$ forms a group with additive composition.
3. Prove that the set $\{1, -1, i, -i\}$ is an abelian multiplication finite group of order 4.
4. Show that four matrices
$\begin{bmatrix} 1 & 0 \\ 0 & 1 \end{bmatrix}, \begin{bmatrix} -1 & 0 \\ 0 & 1 \end{bmatrix}, \begin{bmatrix} 1 & 0 \\ 0 & -1 \end{bmatrix}, \begin{bmatrix} -1 & 0 \\ 0 & -1 \end{bmatrix}$ form a multiplicative group.
5. Show that the set of all $m \times n$ matrices having their elements integers (rational, real or complex numbers) is an infinite Abelian group with addition of matrices as composition.
6. Show that the set $\{0, \pm n, \pm 2n, \ldots \pm kn \ldots\}$ is an additive group, n being a fixed integer.
7. Show that the set of all odd integers with addition is not a group.
8. Prove that the set of zero and even integers with addition is an abelian group.
9. Show that the set $\{1\}$ forms a group with respect to multiplication.
10. Show that the set of integers with respect to multiplication is not a group.
11. Show that the set of all non-zero rational number with multiplication composition is a group.
12. Show that the set I of integers with binary composition $\circ$ defined by $a \circ b = a - b$, when $a, b \in I$ is not a group.
13. Prove that the set I of all integers is a group for the operation $\circ$ defined by $a \circ b = a + b + 1$, $a, b \in I$.

14. Show that the set of all vectors with vector addition as composition is an Abelian group.
15. Show that the set of all integral powers of 2 forms a group with respect to multiplication.
16. Prime that the set of numbers $\cos\theta + i\sin\theta$, where θ runs over all rational numbers, forms an infinite multiplicative Abelian group.
17. Show that the following composition table for the system $(\{a, b, c, d\}, \oplus)$ defines an abelian group.

$\oplus$	a	b	c	d
a	a	b	c	d
b	b	a	d	c
c	c	d	b	a
d	d	c	a	b

5.7 SOME PROPERTIES OF GROUPS

Theorem 5.1. Left Cancellation Law:

If $a, b, c \in G$ then

$$a * b = a * c \Rightarrow b = c$$

where $*$ is any operation.

Proof. Let a^{-1} is left inverse of a.

Hence $\quad a * b = a * c \Rightarrow a^{-1} * (a * b) = a^{-1} * (a * c)$

$\Rightarrow \quad (a^{-1} * a) * b = (a^{-1} * a) * c \quad$ (By Associative Property)

$\Rightarrow \quad e * b = e * c \quad$ (By Inverse Property)

$\Rightarrow \quad \boxed{b = c} \quad$ (By Identity Property)

Theorem 5.2. The left identity e is also the right identity.

$(e * a) = a * e = a \ \forall \ a \in$ Group G

Proof. Let a^{-1} be left inverse of a.

Hence $\quad a^{-1} * (a * e) = (a^{-1} * a) * e \quad$ (By Associative Property)

$\qquad \qquad \qquad = e * e \quad$ (By Inverse Property)

$\qquad \qquad \qquad = e \quad$ (By Identity Property)

$\qquad \qquad \qquad = a^{-1} * a \quad$ (By Inverse Property)

∴ $a * e = a$ By left cancellation law.

But $\quad\quad\quad\quad\quad\quad\quad e * a = a$

Hence $\boxed{e * a = a = a * e}$

Theorem 5.3. The left inverse a^{-1} of an element a in a group is also the right inverse of a,

i.e. $a^{-1} * a = e = a * a^{-1}$

Proof. Here $\quad a^{-1} * (a * a^{-1}) = (a^{-1} * a) * a^{-1}$ (By Associative Property)

$\quad\quad\quad\quad\quad\quad\quad\quad\quad\quad = e * a^{-1}$ (By Inverse Property)

$\quad\quad\quad\quad\quad\quad\quad\quad\quad\quad = a^{-1}$ (By Identity Property)

$\quad\quad\quad\quad\quad\quad\quad\quad\quad\quad = a^{-1} * e$ (By 2nd Theorem)

∴ $\quad\quad\quad\quad\quad a * a^{-1} = e$ (By 1st Theorem)

Hence $\boxed{a^{-1} * a = e = a * a^{-1}}$

Corollary : To prove $b * a = c * a \Rightarrow b = c$. (Right Cancellation Law).

Proof. $b * a = c * a \Rightarrow (b * a) * a^{-1} = (c * a) * a^{-1}$

$\Rightarrow \quad\quad\quad\quad\quad b * (a * a^{-1}) = c * (a * a^{-1})$ (By Associative Property)

$\Rightarrow \quad\quad\quad\quad\quad\quad\quad b * e = c * e$ (By 3rd Theorem)

$\Rightarrow \quad\quad\quad\quad\quad\quad\quad\quad\quad b = c$ (By 2nd Theorem)

Theorem 5.4. Identity is its own inverse i.e. $e = e^{-1}$.

Proof. Since $e * e = e$.

$\Rightarrow e$ is inverse of e.(By Inverse Property)

Hence $\boxed{e = e^{-1}}$

Theorem 5.5. For $a, b \in G$, each of the equations $ax = b$ and $ya = b$ has a unique solution in G, where binary operation is being denoted multiplicatively for convenience.

Proof. To prove the above statement, first we prove that these equations. Possess a solution in G the we have that this solution is unique.

Now $ax = b \Rightarrow a^{-1}(ax) = a^{-1}b$

$\Rightarrow \quad\quad\quad\quad\quad\quad (a^{-1}a)x = a^{-1}b$ (By Associative Property)

$\Rightarrow \quad\quad\quad\quad\quad\quad\quad ex = a^{-1}b$ (By Inverse Property)

$\Rightarrow \quad\quad\quad\quad\quad\quad\quad\quad x = a^{-1}b$ (By Identity Property)

Hence, $ax = b$ has a solution $a^{-1}b$ and by closure property $a^{-1}b$ belongs to G.
Similarly we can say that :

$ya = b$ has a solution given by $y = ba^{-1}$ and $ba^{-1} \in G$.

Thus given equations have solutions in G.

Now to prove that they have unique solution. Let us assume that, if possible, $ax = b$. Possesses two distinct solutions x_1, x_2 in G.

$$\therefore \quad ax_1 = b \text{ and } ax_2 = b$$
$$\Rightarrow \quad ax_1 = ax_2 \text{ by transitivity of equality}$$
$$\Rightarrow \quad x_1 = x_2 \text{ (by left cancellation law)}$$

Hence solutions are not distinct. Hence $ax = b$ has unique solution $a^{-1}b$ in G.
Similarly $ya = b$ has a unique solution $b^{-1}a$ in G.

Corollary 1. Identity element e a group is unique.

Proof. By theorem V equation $ax = a$ or $ya = a$ has unique solution $a^{-1}a$ or aa^{-1} in G.
But $a^{-1}a = aa^{-1} = e$

$\Rightarrow$ e is unique element in G.

Hence, identity element e in a group is unique.

The second way to prove the above statement :

Let if possible e_1, e_2 be two identities in G.

$$\therefore \quad e_1 e_2 = e_1 \quad \text{(since } e_2 \text{ is identity in G)}$$

and $e_1 e_2 = e_2$ (since e_1 is identity in G)

$\Rightarrow$ $\boxed{e_1 = e_2}$

Hence, Identity element e in a group is unique.

Corollary 2. Inverse of an element a in a group is unique.

1st Way: By theorem V, the equation $ax = e$ or $ya = e$ has a unique solution.
But $ax = e$ has $x = a^{-1}e$ as solution and $a^{-1}e = a^{-1}$
Hence a^{-1} is unique.

2nd Way: If possible, let b, c be inverse of a. Since b is inverse of a.

$$\therefore \quad b * a = e \qquad \qquad \qquad ...(1)$$

where $*$ represents any operation.
Since e is inverse of a.

$$\therefore \quad a * c = e \qquad \qquad \qquad ...(2)$$

Now $a * c = e \Rightarrow b * (a * c) = b * e$

$$\Rightarrow \quad (b * a) * c = b$$
$$\Rightarrow \quad e * c = b \qquad \qquad \text{(By eqn. (1))}$$
$$\Rightarrow \quad \boxed{c = b}$$

Thus b is not distinct from c.

Hence, Inverse of an element a in group is unique.

Theorem 5.6. $\forall\ a \in G$, the inverse of inverse of a is a, i.e. $(a^{-1})^{-1} = a$.

Proof. We know that

$$a^{-1} * a = e$$

∴ also $\qquad (a^{-1}) * (a^{-1}) = e$

∵ $(a^{-1})^{-1}$ is inverse of a^{-1}.

Hence $\qquad a^{-1} * a = e,$

$$a^{-1} * (a^{-1})^{-1} = e$$

$\Rightarrow \qquad a^{-1} * a = a^{-1} * (a^{-1})^{-1}$

$$a = (a^{-1})^{-1} \qquad \text{(By left cancellation law)}$$

Hence proved.

Theorem 5.7. $\forall\ a, b \in G$, $(a * b)^{-1} = b^{-1} * a^{-1}$.

Proof. Here $(b^{-1} * a^{-1}) * (a * b) = b^{-1} * (a^{-1} * a) * b$

(By Associative Property)

$\qquad\qquad = b^{-1} * e * b \qquad$ (By Inverse Law)

$\qquad\qquad = b^{-1} * (e * b) \qquad$ (By Associative Law)

$\qquad\qquad = b^{-1} * b \qquad$ (By Identity Law)

$\qquad\qquad = e \qquad$ (By Inverse Law)

Hence by definition $b^{-1} * a^{-1}$ is inverse of $a * b$.

i.e., $b^{-1} * a^{-1} = (a * b)^{-1}$

Hence, proved.

This is also called Reversal Law of Inverse.

Generalization of the Theorem : Now we can say that :

If $a, b, c .... k \in G$ then

$$\left(a * b * c * .... * l * k\right)^{-1} = k^{-1} * e^{-1} * .... * c^{-1} * b^{-1} * a^{-1}$$

Theorem 5.8. For any $a \in$ Group G, prove that

$a^m * a^n = a^{m+n}$. When $m, n \in I$.

Proof. 1st Case : m, n both + ve integers.

By definition

$a^m * a^n \Rightarrow (a*a*a... \text{ to } m \text{ terms}) * (a*a*a... \text{ to } n \text{ terms})$

$\Rightarrow a*a*a.... \text{ to } (m+n) \text{ terms.}$ (By Associative Property)

$\Rightarrow a^{m+n}$ (By definition)

2nd Case : If m is $-$ ve integer and n is $+$ive integer.

Let $m = -r$, where r is + ve integer.

Hence $\quad a^m * a^n = a^{-r} * a^n$

$= (a^{-1} * a^{-1} * a^{-1} .... \text{ up to } r \text{ terms}) *$

$(a * a * a .... \text{ up to } n \text{ terms})$ ...(1)

If $r < n$, we have by (1)

$a^m * a^n = (a^{-1} * a^{-1} * a^{-1} .... * a * a * a * a) *$

$(a * a * .... \text{ up to } n-r \text{ terms})$

$= e * a^{n-r}$ (By Associative Property and by Inverse Property)

$= a^{n-r}$ (By Identity Property)

$= a^{-r+n}$

$= a^{m+n} \quad \because -r = m$

If $r > n$, we have by (1)

$a^m * a^n = (a^{-1} * a^{-1} * a^{-1} .... \text{ up to } n-r \text{ terms}) *$

$(a^{-1} * a^{-1} * a^{-1} * .... * a * a * a * a)$

$= (a^{-1})^{r-n} * e$ (By Associative and Inverse Property)

$= (a^{-1})^{r-n}$ (By Identity Property)

$= a^{-r+n}$

$= a^{m+n}$

(Since $-r = m$)   Hence proved.

3rd Case : m, n both negative.

Let $m = -r, n = -s$ where r, s are both positive.

$\therefore \quad a^m * a^n = a^{-r} * a^{-s} = (a^{-1})^r * (a^{-1})^s$

$= (a^{-1})^{r+s}$ by Ist case ($\because r, s$ are both positive)

$= (a)^{-r-s}$ by definition.

$= a^{m+n}$ Hence proved.

4th Case : When $m = 0$ or $n = 0$.

Let $m = 0$. Now $a^0 = a^{1+(-1)} = a * a^{-1}$ (By 2nd Case)

$\qquad\qquad\qquad\qquad\qquad\quad = e$ (By Inverse Property)

Hence $\qquad a^m * a^n = a^0 * a^n = e * a^n = a^n = a^{0+n} = a^{m+n}$

Hence proved.

Theorem 5.9. Prove that $a^{-m} = (a^m)^{-1}$, $m \in I$ where $a \in G$, G being the group.

Proof. 1st Case : m is $+$ ve integer.

$a^m = a \cdot a \cdot a \ldots$ to m factors.

$\therefore \qquad (a^m)^{-1} = (a \cdot a \cdot a \ldots a)^{-1} = a^{-1} \cdot a^{-1} \ldots a^{-1}$ (By Reversal Law)

$\qquad\qquad\quad = (a^{-1})^m$ \hfill (By definition)

$\qquad\qquad\quad = a^{-m}$ \hfill (By definition)

Hence the result.

2nd Case : m is $-$ve integer.

Let $m = -p$, where p is positive integer.

$\therefore a^m = a^{-p} = (a^{-1})^p = a^{-1} \cdot a^{-1} \ldots$ to p terms.

$\qquad (a^m)^{-1} = (a^{-1} \cdot a^{-1} \ldots a^{-1})^{-1} = (a^{-1})^{-1} \cdot (a^{-1})^{-1} \ldots$ to p terms.

$\qquad\qquad\quad = a \cdot a \cdot a \ldots$ to p terms. $\therefore (a^{-1})^{-1} = a$.

$\qquad\qquad\quad = a^p = a^{-m}$.   Hence proved.

3rd Case : $m = 0$

Here $a^m = a^0 = e$

$\because \qquad\qquad (a^{-m})^{-1} = e^{-1} = e = a^0 = a^{-m}$.   Hence proved.

5.8 ALTERNATIVE DEFINITION OF A GROUP

Theorem 5.10. A set G with a binary composition denoted multiplicatively is a group iff

(i) the composition is associative and (ii) $\forall\ a, b \in G$, the equations $ax = b$ and $ya = b$ have unique solution in G.

Proof. Necessary Conditions : Let $(G, *)$ be the group.

Hence from group axioms, composition is binary as well as associative. Also let a^{-1} be the inverse of a, where $a \in G$.

Then $ax = b \Rightarrow a^{-1}(ax) = a^{-1}b$

$\Rightarrow (a^{-1}a)x \Rightarrow a^{-1}b$ (By Associative Property)

$\Rightarrow \qquad\qquad\qquad ex = a^{-1}b$ (By existence of Inverse)

$\Rightarrow \qquad\qquad\qquad x = a^{-1}b$ (By existence of Identity)

But by closure axiom :

$$a^{-1}b \in G \therefore x \in G$$

$\therefore$ $ax = b$ has a solution in G.

Let if possible $ax = b$ have two solutions x_1, x_2.

i.e. $\qquad ax_1 = b$ and $ax_2 = b$.

$\therefore \qquad ax_1 = ax_2$.

$\Rightarrow$ $\boxed{x_1 = x_2}$ (By left cancellation law)

Thus, solution is unique.

Similarly,

$$ya = b$$
$\Rightarrow \qquad a^{-1}(ay) = a^{-1}b$
$\Rightarrow \qquad (a^{-1}a)y = a^{-1}b$ (By Associative Property)
$\Rightarrow \qquad ex = a^{-1}b$ (By existence of Inverse)

Now let a be an arbitrary element in G and $b = e$, so that the equation $ya = b$ becomes $ya = e$.

But solution of this equation is unique in G. Let C $\Re$ G be the solution.

$\therefore$ $Ca = e$. Hence C is left inverse of a in G.

Thus left identity exists and each element has its left inverse. Hence G is a graph.

Theories II. Prove that a finite set G with an associative multiplicatively denoted binary composition is a group if right and left cancellation laws hold good in G, *i.e.,*

$$ax = bx \Rightarrow a = b \text{ and } xa = xb \Rightarrow a = b.$$

Proof : Let $G = \{a_1, a_2, a_3, ..., a_h, ..., a_n\}$, where all the n elements in G are distinct. Let n be any one element of G. Consider the n 'Products'.

$$a_1 a_r, a_2 a_{r2}, ..., a_h a_{rh}$$

All these 'Products' are distinct, for it possible, let

$$a_p a_r = a_q a_r \text{ where } a_p a_q \in G$$

But, by right cancellation $a_p a_r = a_q a_r \Rightarrow a_p = a_q$.

But by hypothesis $a_p \neq a_q$. Hence all products are distinct.

Also by closure product $a_1 a_r, a_2 a_{rw} .... a_n n_{rh} \in G$. Hence these are the n elements of G, order of elements have may be different from that given in G above.

Thus, the equation $xa_n = a_1$ has a unique solution in G, *i.e.,* the equation $xa = b$, where $a, b \in G$ has a unique solution in G.

Similarly, by applying left cancellation law, we can prove that $ay = b$, where $ab \in G$ has a unique solution in G. Maximum operation is associative. Hence by 1st theorem, G is a group.

Example 18: Prove that if $a, b \in$ graph G, then $(ab)^2 = a^2 b^2$ iff G is abelian.

Solution: Let G be abelian and $a, b \in G$

$$\therefore \quad (ab)^2 = (ab)(ab)$$
$$= a(ba)b \quad \text{by } G_2.$$
$$= a(ab)b \quad \therefore G \text{ is abelian, i.e. } ab = ba.$$
$$= (ab)(bb) \quad \text{by } G_2.$$
$$= a^2b^2 \quad \text{Hence proved.}$$

Converse : Let $a, b \in G$ and $(ab)^2 = a^2b^2$

Now $\quad (ab)^2 = a^2b^2 \Rightarrow (ab)(ab) = (aa)b^2$

$\Rightarrow \quad a(ba)b = a(ab^2)$

$\Rightarrow \quad bab = ab^2 \quad$ by left cancellation law

$\Rightarrow \quad (ba)b = (ab)b$

$\Rightarrow \quad ba = ab.\quad$ by right cancellation law

$\Rightarrow$ G is abelian

Example 19: If G a group such that $(ab)^3 = a^3 b^3$ for three consecutive integers M, $\forall$ $a, b \in G$, prove that G is abelian.

Solution: Let M, M + 1, M + 2 be three consecutive integers :

Hence we are given that

$$(ab)^M = a^M b^M \qquad \ldots(1)$$
$$(ab)^{M+1} = a^{M+1} b^{M+1} \qquad \ldots(2)$$
$$(ab)^{M+2} = a^{M+2} b^{M+2} \qquad \ldots(3)$$

Now $\quad (ab)^{M+2} = a^{M+2} b^{M+2} \Rightarrow (ab)^{M+1}(ab) = (a^{M+1} a)(b^{M+1} b)$

$\Rightarrow \quad (a^{M+1} b^{M+1})(ab) = a^{M+1}(ab^{M+1} b)\quad$ by (2) and G_2

$\Rightarrow \quad a^{M+1}(b^{M+1} ab) = a^{M+1}(ab^{M+1} b)\quad$ by G_2

$\Rightarrow \quad b^{M+1} ab = ab^{M+1} b\quad$ by left cancellation law

$\Rightarrow \quad (b^{M+1} a)b = (ab^{M+1})b\quad$ by G_2

$\Rightarrow \quad b^{M+1} a = ab^{M+1}\quad$ by right cancellation law

$\Rightarrow \quad a^M(b^{M+1} a) = a^M(ab^{M+1})\quad$ by operating a^M on both sides.

$\Rightarrow \quad (a^M b^M)(ba) = a^{M+1} b^{M+1}\quad$ by G_2

$\Rightarrow \quad (a^M b^M)(ba) = (ab)^{M+1}\quad$ by (2)

$\Rightarrow \quad (ab)^M (ba) = (ab)^M (ab)\quad$ by G_2 and (2)

$\Rightarrow \quad ba = ab\quad$ by left cancellation law

$\Rightarrow$ G is abelian   Hence proved

EXERCISE 5.3

1. If in group G, $a, b \in G$ and operation is denoted multiplicatively, then prove that—

(i) $aa = a \Rightarrow a = e$,

(ii) $a^{-1} b^{-1} = b^{-1} a^{-1}$ if $ab = ba$

(*iii*) $a^{-1} b = b^{a-1}$ if $ab = ba$

2. Prove that a group G with composition denoted. Multiplicatively is an abelian group, if.
   (*i*) each element is its own inverse.
   (*ii*) $b^{-1} a^{-1} b a = e$, $\forall\ a, b \in G$
   (*iii*) it has only form elements.

3. When G is abelian group, prove that $\forall\ a, b \in G$,
   $(ab)^n = a^n b^n$, $n \in G$.

4. Let S be the totality of the pairs (a, b) such that $a, b \in R$ and $a \neq 0$. If composition $*$ in S is defined by $(a, b) * (c, d) = (ac, bc, + d)$ verify that $(S, *)$ is a group.

5. Prove that a non-compensative group has at least six elements.

6. If corresponding to any element $a \in$ group G, there is an element O_a which satisfies a condition. $a + O_a = a$ and $O_a + a = a$, then show that it is necessary that $O_a = 0$ where O is additive identity of G.

7. Is the mathematical system $(Q, \div)$ a group when Q is the set of all rational numbers and $\div$ is ordinary division?

8. Show that in a group with even number of elements there is at least one element besides identity which is its own inverse.

◻◻

CHAPTER 6

SUBGROUPS

6.1 SUBGROUPS AND SUBGROUP TESTS

A *subgroup* of a group G is a subset of G which is a subgroup in its own right (with the same group operation).

There are two subgroup tests, resembling the two subring tests.

Proposition (First Subgroup Test): *A non-empty subset H of a group G is a subgroup of G if, for any $h, k \in H$, we have $hk \in H$ and $h^{-1} \in H$.*

Proof: We have to show that H satisfies the group axioms. The conditions of the test show that it is closed under composition (G0) and inverses (G3). The associative law (G1) holds in H because it holds for all elements of G. We have only to prove (G2), the identity axiom.

We are given that H is non-empty, so choose $h \in H$. Then by assumption, $h^{-1} \in H$, and then (choosing $k = h^{-1}$) $I = hh^{-1} \in H$.

We can reduce the number of things to be checked from two to one proposition **(Second Subgroup Test)** : A non-empty subset H of a group G is a subgroup of G if, for any $h, k \in H$, we have $hk^{-1} \in H$.

Proof: Choosing $k = h$, we see that $I = hh^{-1} \in H$.. Now using I and h in place of h and k, we see that $h^{-1} = 1h^{-1} \in H$. Finally, given $h, k \in H$, we know that $k^{-1} \in H$, so $hk = h(k^{-1})^{-1} \in H$. So the conditions of the First Subgroup test hold.

Example: Look back to the Cayley tables in the last chapter, in the first case, $\{e, y\}$ is a subgroup. In the second case. $\{e, a\}$, $\{e, b\}$ and $\{e, e\}$ are all subgroups.

6.2 CYCLIC GROUPS

If g is an element of a group G, we define the powers g^n of G (for $n \in \mathbb{Z}$) as follows: if n is positive, then g^n is the product of n factors g; $g^0 = -1$; and $g^{-n} = (g^{-1})^n$. The usual laws of exponents hold: $g^{m+n} = g^m . g^n$ and $g^{mn} = (g^m)^n$.

A *cyclic group* is a group C which consists of all the powers (positive and negative) of a single element. If C consists of all the powers of g, then we write $C = (g)$, and say that C is *generated by g*.

Proposition: *A cyclic group is Abelian.*

Proof: Let $C = (g)$, Take two elements of C, say g^m and g^n. Then,
$$g^m \cdot g^n = g^{m+n} = g^n \cdot g^m$$

Let $C = (g)$. Recall the *order* of g the smallest positive integer n such that $g^n = 1$ (if such n exists – otherwise the order is infinite).

Proposition: *Let g be an element of the group G. Then the set of all powers (positive and negative) of g forms a cyclic subgroup of G. Its order is equal to the order of g.*

Proof: Let $C = \{g^n : n \in \mathbb{Z}\}$. We apply the Second Subgroup test: if $g^m, g^n \in C$, then, $(g^m)(g^n)^{-1} = g^{m-n} \in C$. So C is a subgroup.

If g has infinite order, then no positive power of g is equal to 1. It follows that all the powers g^n for $n \in \mathbb{Z}$ are different elements. (For if $g^m = g^n$, with $m > n$, then $g^{n-m} = 1$.) So C is infinite.

Suppose that g has finite order n. We claim that any power of g is equal to one of the elements $g^0 = 1, g^1 = g \ldots g^{n-1}$. Take any power g^m. Using the division algorithm in $\mathbb{Z}$, write $m = nq + r$, where $0 \le r \le n - 1$. Then
$$g^m = g^{nq+r} = (g^n)^q \cdot g^r = 1 \cdot g^r = g^r$$

Furthermore, the elements $1, g \ldots g^{n-1}$ are all different: for if $g^r = g^s$, with $0 \le r < s \le n-1$, then $g^{s-r} = 1$, and $0 < s - r < n$, contradicting the fact that n is the order of g (the smallest exponent i such that $g^i = 1$).

Example: The additive group of the ring $\mathbb{Z}/n\mathbb{Z}$ is a cyclic group of order n, generated by $\overline{1} = n\mathbb{Z} + 1$. Remember that the group operation is addition here, and the identity element is zero, so in place of $g^n = 1$ we have $n = \overline{1} = \overline{0}$, which is true in the integers mod n; moreover it is true that no smaller positive multiple of $\overline{1}$ can be zero.

Proposition: *Let G be a cyclic group of finite order n. Then g has a cyclic subgroup of order m for every m which divides n; and these are all the subgroups of G.*

Proof: Let $G = (g) = \{1, g, g^2 \ldots g^{n-1}\}$. If m divides n, let $n = mk$, and put $h = g^k$. Then $h = g^k$. Then $h^m = (g^k)^m = g^n = 1$, and clearly no smaller power of h is equal to 1; so h has order m, and generates a cyclic group of order m.

Now let H be any subgroup of G. If $H = \{1\}$, then H is the unique cyclic subgroup of order 1 in G, so suppose not. Let g^m be the smallest positive power of g which belongs to H. We claim that, if $g^k \in H$, then m divides k. For let $k = mq + r$, where $0 \le r \le m - 1$. Then
$$g^r = g^{mq+r} g^{-mq} = g^k (g^m)^{-q} \in H$$

So, $r = 0$ (since m was the smallest positive exponent of an element of H. So H is generated by g^m. Now $g^n = 1 \in H$, so m divides n, and we are done.

Cosets

Given any subgroup H of a group G we can construct a partition of G into "cosets" of H, just as we did for rings. But for groups, things are a bit more complicated.

Because the group operation may not be commutative, we have to define two different sorts of cosets.

Let H be a subgroup of a group G. Define a relation $\sim_r$ on G by the rule
$$x \sim_r y \text{ if and only if } yx^{-1} \in H$$
We claim that $\sim_r$ is an equivalence relation:

Reflexive : For any $x \in G$, we have $xx^{-1} = 1 \in H$, so $x \sim_r x$,

Symmetric : Suppose that $x \sim_r y$, so that $h = yx^{-1} \in H$. Then $h^{-1} = (yx^{-1})^{-1} = xy^{-1} \in H$, so $y \sim_r x$.

Transitive: Suppose that $x \sim_r y$ and $y \sim_r z$, so that $h = yx^{-1} \in H$ and $k = zy^{-1} \in H$.

Then, $kh = (zy^{-1})(yx^{-1}) = zx^{-1} \in H$, so $x \sim_r z$.

The equivalence classes of this equivalence relation are called the *right cosets* of H in G.

A right coset is a set of elements of the form $Hx = \{hx : h \in H\}$, for some fixed element $x \in G$ called the "coset representative". For,
$$y \in Hx \Leftrightarrow y = hx \text{ for some } h \in H \leftrightarrow yx^{-1} \in H \Leftrightarrow x \sim_r y.$$
We summarise all this as follows:

Proposition: *If H is a subgroup of the of the group G, then G is partitioned into right cosets of H in G, sets of the form $Hx = \{hx : h \in H\}$.*

In a similar way, the relation $\sim_r$ defined on G by the rule,
$$x \sim_r y \text{ if and only } x^{-1}y \in H$$
is an equivalence relation on G, and its equivalence classes are the *left cosets of* H in G, the sets of the form $xH = \{xh : h \in H\}$.

If G is an abelian group, the left and right cosets of any subgroup coincide, since
$$Hx = \{hx : h \in H\} = \{xh : h \in H\} = xH.$$
This is not true in general:

Example: Let G be the symmetric group S_3, and let H be the subgroup $\{1, (1,2)\}$ consisting of all permutations fixing the point 3. The right cosets of H in G are

$H1 = \{1, (1, 2)\}$,

$H(1, 3) = \{(1, 3), (1, 2, 3)\}$

$H(2, 3) = \{(2, 3), (1, 3, 2)\}$

while the left cosets are:

$1H = \{1, (1, 2)\}$,

$(1, 3) H = \{(1, 3), (1, 3, 2)\}$

$(2, 3) H = \{(2, 3), (1, 3, 2)\}$

We see that, as expected, both right and left cosets partition G, but the two partitions are not the same. But each partition divides G into three sets of size 2.

Lagrange's Theorem

Lagrange's Theorem states a very important relation between the orders of a finite group and any subgroup.

Theorem (Lagrange's Theorem): *Let H be a subgroup of a finite group G. Then the order of H divides the order of G.*

Proof: We already know from the last section that the group G is partitioned into the right cosets of H. We show that every right coset Hg contains the same number of elements as H.

To prove this, we construct a bijection ϕ from H to Hg. The bijection is defined in the obvious way: ϕ maps h to hg.

- ϕ is one-to-one: suppose that $\phi(h_1) = (\phi) h_2$, that is, $h_{1}g = h_{2}g$. Cancelling the g (by the cancellation law, or by multiplying by g^{-1}, we get $h_1 = h_2$.
- ϕ is onto: by definition, every element in the coset Hg has the form hg for some $h \in$ H, that is, it is $\phi(h)$.

So, ϕ is a bijection, and $\{H_g\} = |H|$.

Now, if m is the number of right cosets of H in G, then $m|H| = |G|$, so $|H|$ divides $|G|$.

Remarks: We see that $/G//|H|$ is the number of right cosets of H in G. This number is called the *index,* of H in G.

We could have used left cosets instead, and we see that $/G|/|H|$ is also the number of left cosets. So these numbers are the same. In fact, there is another reason for this.

Exercise: Show that the set of all inverses of the elements in the right coset Hg form the left coset $g^{-1}H$. So there is a bijection between the set of right cosets and the set of left cosets of H.

In the example of preceding section, we had a group S_3 with a subgroup having three right cosets and three left cosets: that is, a subgroup with index 3.

Corollary: *Let g be an element of the finite group G. Then the order of g divides the order of G.*

Proof: Remember, first, that the word "order" here has two quite different meanings: the order of a group is the number of elements it has: while the order of an element is the smallest n such thai $g^n = 1$.

However, we also saw that if the element g has order m, then the set $\{1, g, g^2, ..., g^{m-1}\}$ is a cyclic subgroup of G having order m. So, by Lagrange's Theorem m divides the order of G.

Example: Let $G = S_3$. Then the order of G is 6. The element (1) (2, 3) has order 2, while the element (1, 3, 2) has order 3.

6.3 HOMOMORPHISMS AND NORMAL SUBGROUPS

This section is similar to the corresponding one for rings. Homomorphisms are maps preserving the structure, while normal subgroups do the same job for groups as ideals do for rings: that is, they are kernels of homomorphisms.

Isomorphism

Just as for rings, we say that groups are isomorphism if there is a bijection between them which preserves the algebraic structure.

Formally, let G_1 and G_2 be groups. The map $\theta : G_1 \to G_2$ is an *isomorphism* if it is a bijection from G_1 to G_2 and satisfies
$$(gh)\theta = (g\theta)(h\theta) \text{ for all } g, h \in G_1.$$
Note that, as before, we write the map θ to the right of its argument: that is, $g\theta$ is the image of g under the map θ. If there is an isomorphism from G_1 to G_2, we say that the groups G_1 and G_2 are *isomorphic*.

Example: Let G_1 be the additive group of $\mathbb{Z}/2\mathbb{Z}$, and let G_2 be the symmetric group S_2. Their Cayley tables are:

+	$\bar{0}$	$\bar{1}$
$\bar{0}$	$\bar{0}$	$\bar{1}$
$\bar{1}$	$\bar{1}$	$\bar{0}$

·	1	(1,2)
1	1	(1,2)
(1,2)	(1,2)	1

The map θ that takes $\bar{0}$ to 1, and $\bar{1}$ to $(1, 2)$, is clearly an isomorphism from G_1 to G_2.

Homomorphisms

An isomorphism between groups has two properties: it is a bijection; and it preserves the group operation. If we relax the first property but keep the second, we obtain a homomorphism. Just as for rings, we say that a function $\theta : G_1 \to G_2$ is

- a *homomorphism* if it satisfies
$$(gh)\theta = (g\theta)(h\theta); \qquad \ldots(3.1)$$
- a *monomorphism* if it satisfies (3.1) and is one-to-one;
- an *epimorphism* if it satisfies (3.1) and is onto;
- an *isomorphism* if it satisfies (3.1) and is one-to-one and onto.

We have the following lemma, proved in much the same way as for rings:

Lemma: Let $\theta : G_1 \to G_2$ be a homomorphism. Then $1\theta = 1$; $(g^{-1})\theta = (g\theta)^{-1}$; and $(gh^{-1})\theta = (g\theta)(h\theta)^{-1}$, for all $g, h \in G_1$.

Now, if $\theta : G_1 \to G_2$ is a homomorphism, we define the *image* of θ to be the subset
$$\{x \in G_2 : x = g\theta \text{ for some } g \in G_1\}$$
of G_2, and the *kernel* of θ to be the subset
$$\{g \in G_1 : g\theta = 1\}$$
of G_1.

Proposition: Let $\theta : G_1 \to G_2$ be a homomorphism.
(a) $\mathrm{Im}(\theta)$ *is a subgroup of* G_2.
(b) $\mathrm{Ker}(\theta)$ *is a subgroup of* G_1.

Proof: We use the Second Subgroup Test in each case.
(a) Take $x, y \in \mathrm{Im}(\theta)$, say $x = g\theta$ and $y = h\theta$ for $g, h \in G_1$. Then $xy^{-1} = (gh^{-1})\theta \in \mathrm{Im}(\theta)$, by the Lemma.
(b) Take $g, h \in \mathrm{Ker}(\theta)$. Then $g\theta = h\theta = 1$, so $(gh^{-1})\theta = 1^{-1}1 = 1$, so $gh^{-1} \in \mathrm{Ker}(\theta)$.

Example: Colour the elements 1, (1, 2, 3) and (1, 3, 2) red, and the elements (1, 2), (2, 3) and (1, 3) blue. We see that the Cayley table has the "simplified form"

.	red	blue
red	red	blue
blue	blue	red

This is a group of order 2, and the map θ taking 1, (1, 2, 3) and (1, 3, 2) to red and (1, 2), (2, 3) and (1, 3) to blue is a homomorphism. Its kernel is the subgroup {1, (1, 2, 3), (1, 3, 2)}.

6.4 NORMAL SUBGROUPS

A normal subgroup is a special kind of subgroup of a group. Recall from the last chapter that any subgroup H has right and left cosets, which may not be the same. We say that H is a *normal subgroup* of G if the right and left cosets of H in G are the same; that is, if $Hx = xH$ for any $x \in G$.

There are several equivalent ways of saying the same thing. We define
$$x^{-1} Hx = \{x^{-1} hx : h \in H\}$$
for any element $x \in G$.

Proposition: *Let H be a subgroup of G. Then be following are equivalent:*
(a) H is a normal subgroup, that is, $Hx = xH$ for all $x \in G$;
(b) $x^{-1} Hx = H$ for all $x \in G$:
(c) $x^{-1} hx \in H$, for all $x \in G$ and $h \in H$.

Proof: If $Hx = xH$, then $x^{-1} Hx = x^{-1} xH = H$ and conversely. So (a) and (b) are equivalent.

If (b) holds then every element $x^{-1} hx$ belongs to $x^{-1} Hx$ and so to H, so (c) holds. Conversely, suppose that (c) holds. Then every element, of $x^{-1} Hx$ belongs to H, and we have to prove the reverse inclusion. So take $h \in H$. Putting $y = x^{-1}$, we have $k = y^{-1} hy = xhx^{-1} \in H$, so $h \in x^{-1} Hx$, finishing the proof.

Now the important thing about normal subgroups is that, like ideals, they are kernels of homomorphisms.

Proposition: Let $\theta: G_1 \to G_2$ be a homomorphism. Then Ker (θ) is a normal subgroup of G_1.

Proof: Let $H = \text{Ker}(\theta)$. Suppose that $h \in H$ and $x \in G$. Then
$$(x^{-1} hx)\theta = (x^{-1})\theta \cdot h\theta \cdot x\theta = (x\theta)^{-1} \cdot 1 \cdot x\theta = 1,$$
so $x^{-1} hx \in \ker(\theta) = H$. By part (c) of the preceding; proposition H is a normal subgroup of G.

There are a couple of situations in which we can guarantee that a subgroup is normal.

Proposition: *(a) If G is Abelian then every subgroup H of G is normal.*
(b) If H has index 2 in G, then H is normal in G.

Proof: (a) If G is Abelian. then $xH = Hx$ for all $x \in G$.
(b) Recall that this means that H has exactly two cosets (left or right) in G. One of these

cosets is H itself; the other must consist of all the other elements of G that is, G/H. This is the case whether we are looking at left or right cosets. So the left and right cosets are the same.

Remark: We saw in the last chapter an example of a group S_3 with a non-normal subgroup having index 3 (that is, just three cosets). So we cannot improve this theorem from 2 to 3.

In our example in the last section, the subgroup $\{1, (1, 2, 3), (1, 3, 2)\}$ of S_3 has index 2, and so is normal, in S_3 this also follows from the fact that it is the kernel of a homomorphism.

For the record, here is a normal Subgroup test:

Proposition (Normal subgroup test): *A non-empty subset H of a group G is a normal subgroup of G if the following hold:*

(a) for any $h, k \in H$, we have $hk^{-1} \in H$;

(b) for any $h \in H$ and $x \in G$, we have $x^{-1} hx \in H$.

Proof: (a) is the condition of the second subgroup test, and we saw that (b) is a condition for a subgroup to be normal.

6.5 QUOTIENT GROUPS

Let H be a normal subgroup of a group G. We define the *quotient group* G/H as follows :

- The elements of G/H are the cosets of H in G (left or right does not matter, since H is normal);
- The group operation is defined by $(Hx)(hy) = Hxy$ for all $x, y \in G$: in other words, to multiply cosets, we multiply their representatives.

Proposition: *If H is normal subgroup of G, then the quoteint group G/H as defined above is a group. Moreover, the map q from G to G/H defined by $xq = Hx$ is a homomorphism whose kernel is H and whose image is G/H.*

Proof: First we have to show that the definition of the group operation is a good one. In other words, suppose that we choose different representatives x' and y' for the cosets Hx and Hy; is it true that $Hxy = Hx'y'$? We have $x' = hx$ and $y' = ky$, for some $h, k \in H$. Now xk belongs to the left coset xH. Since H is normal, this is equal to the right coset Hx, so that $xk = lx$ for some $l \in H$. Then $x'y' = hxky = (hl)(xy) \in Hxy$, since $hl \in H$. Thus the operation is indeed well defined.

Now we have to verify the group axioms:

(G0) Closure is clear since the product of two cosets is a coset.

(G1) Given three cosets Hx, Hy,Hz, we have

$((Hx)(Hy))(Hz) = (Hxy)(Hz) = H(xy)z = Hx(yz) = (Hx)(Hyz) = (Hx)((Hy)(Hz))$,

using the associative law in G;

(G2) The identity is $H1 = H$, since $(H1)(Hx) = H(1x) = Hx$ for all $x \in G$.

(G3) The inverse of Hx is clearly Hx^{-1}.

Finally, for the map θ, we have

$$(xy)\theta = Hxy = (Hx)(Hy) = (x\theta)(y\theta).$$

So θ is a homomorphims. Its image consists of all cosets Hx, that is, Im(θ) = G/H. The identity element of G/H is (as we saw in in the proof of (G2)) the coset H; and Hx = H if and only if x ∈ H, so that Ker(θ) = H'.

The map θ in the above proof is called the natural homomorphism from G to G/H. We see that, if H is a normal subgroup of G, then it is the kernel of the natural homomorphism from G to G/H. So normal subgroups are the same thing as kernels of homomorphisms.

Example: Let $G = S_3$, and let H be the subgroup {1, (1, 2, 3), (1, 3, 2)}. We have observed that H is a normal subgroup. It has two cosets namely HI = H and H(1, 2) = {(1, 2), (2, 3), (1, 3). The rules for multiplication of these cosets will be the same as the rules for multiplying the elements 1 and (1, 2). So G/H is isomorphic to the group {1, (1, 2)} of order 2.

The Isomorphism Theorems

The Isomorphism Theorems for groups look just like the versions for rings.

Theorem (First Isomorphism Theorem): *Let G_1 and G_2 be groups, and let $\theta: G_1 \to G_2$ be a homomorphism. Then:*

(a) Im(θ) is a subgroup of G_2;

(b) Ker(θ) is a normal subgroup of G_1;

(c) $G_1/\text{Ker}(\theta) \cong \text{Im}(\theta)$

Proof: We already proved the first two parts of this theorem. We have to prove (c). That is, we have to construct a bijection ϕ from G/N to Im(θ), where N = Ker(θ), and prove that it preserves the group operation.

The map ϕ is defined by (Nx)ϕ = xθ. We have

$$(Nx)\phi = (Ny)\phi \Leftrightarrow x\theta = y\theta \Leftrightarrow (xy^{-1})\theta = 1 \Leftrightarrow xy^{-1} \in \text{Ker}(\theta) = N \Leftrightarrow Nx = Ny$$

so ϕ is well defined and one-to-one. It is clearly onto, Finally,

$$(Nx)\phi \cdot (Ny)\phi = (x\theta)(y\theta) = (xy)\theta (Nxy)\theta = ((Nx)(Ny))\phi$$

so ϕ preserves the group operation as required.

The same picture as for rings may be useful:

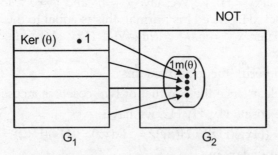

The parts on the left are the cosets of N = Ker(θ), where N itself is the topmost part. Each coset of N maps to a single element of lm(θ), and the correspondence between cosets and elements of lm(θ) is the bijection of the last part of the theorem.

The other two theorems will be stated without proof. You are encouraged to try the proofs for yourself: they are very similar to the proofs for rings.

Theorem (Second Isomorphism Theorem): Let N be a normal subgroup of the group G. Then there is a one-to-one correspondence between the subgroups of G/N and the subgroup of G containing N, given as follows : to a subgroup H of G containing N corresponds the subgroup H/N of G/N. Under this correspondence, normal subgroups of G/N correspond to normal subgroups of G containing N; and, if M is a normal subgroup of G containing N, then,

$$(G/N)/(M/N) \cong G/M$$

For the next theorem we need to define, for any two subsets A, B of a group G, the set

$$AB = \{ab : a \in A, b \in B\}$$

of all products of an element of A by an element of B.

Theorem (Third Isomorphism Theorem): Let G be a group, H a subgroup of G, and N a normal subgroup of G. Then

(a) HN is a subgroup of G containing N;

(b) H ∩ N is an normal subgroup of H;

(c) $H/(H \cap N) \cong (HN)/N$.

We end this section with one fact about groups which doesn't have an obvious analog for rings.

Proposition: Let H and K be subgroups of the group G.

(a) *HK is a subgroup of G if and only if HK = KH.*

(b) *If K is a normal subgroup of G, then HK = KH.*

Proof: (a) Suppose that HK = KH. Then every element of the form kh (for $k \in K$ and $h \in$ H) can also be expressed in the form $h'k'$ (for $h' \in$ H and $k' \in$ K). Now we apply the subgroup test to HK. Take $h_1k_1, h_2k_2 \in$ HK. We want to know if $h_1k_1 (h_2k_2)^{-1} \in$ HK. This expression is $h_1k_1k_2^{-1}h_2^{-1}$. Now $k_1k_2^{-1} \in$ K, so $(k_1k_2^{-1})h_2^{-1} \in$ KH; so we can write this element as $h'k'$, for some $h' \in$ H and $k' \in$ K. Then,

$$h_1k_1k_2^{-1}h_2^{-1} = (h_1h')k' \in \text{HK}.$$

as required.

Conversely, suppose that HK is a subgroup. We have to show that HK = KH, that is, every element of one is in the other. Take any element $x \in$ HK. Then $x^{-1} \in$ HK, so $x^{-1} = hk$, for some $h \in$ H and $k \in$ K. Then $x = k^{-1} h^{-1} \in$ KH. The reverse inclusion is proved similarly.

(b) If K is a normal subgroup, then the Third Isomorphism Theorem shows that HK is a subgroup, so that HK = KH by Part (a).

Exercise: If H and K are subgroups of G, show that,

$$[HK] = \frac{[H].[K]}{[H \cap K]}$$

whether or not HK is a subgroup.

[**Hint:** there are [H].[K] choices of an expression hk. Show that every element in HK can be expressed as such a product in [H ∩ K] different ways.]

Example: Let $G = S_3$, $H = \{1, (1, 2)\}$ and $K = \{1,(2, 3)\}$. Then H and K are two subgroups of G, each of order 2, and $H \cap K = \{1\}$, so [HK] = 4. Since 4 doesn't divide 6, Lagrange's

Theorem shows that HK cannot be a subgroup of G. This shows, once again, that H and K are not normal subgroups of G.

Conjugacy

Conjugacy is another equivalence relation on a group which is related to the idea of normal subgroups.

Let G be a group, we say that two elements $g, h \in G$ are conjugate if $h = x^{-1} gx$ for some element $x \in G$.

Proposition: (a) *Conjugacy is an equivalence relation on G.*

(b) *A subgroup H of G is normal if and only if it is a union of (some of the) conjugacy classes in G.*

Proof: (a) Write $g \sim h$ to mean that $h = x^{-1} gx$ for some $x \in G$. Then
- $g = 1^{-1} g$, so $g \sim g$: $\sim$ is reflexive.
- If $h = x^{-1} gx$, then $g = (x^{-1})^{-1} h(x^{-1})$: so $\sim$ is symmetric.
- Suppose that $g \sim h$ and $h \sim k$. Then $h = x^{-1} gx$ and $k = y^{-1} hy$ for some x, y. Then $k = y^{-1} x^{-1} gxy = (xy)^{-1} g(xy)$. so $g \sim k$: $\sim$ is transitive.

(b) The condition that H is a union of conjugacy classes means that, if $h \in H$, then any element conjugate to h is also in H. We saw in Proposition 3.19 that this condition is equivalent to normality of H.

Exercise: Let $G = S_3$. Show that the conjugacy classes in G are $\{1\}$. $\{(1, 2, 3), (1, 3, 2)\}$, and $\{(1, 2). (2, 3), (1, 3)\}$. (We will look at conjugacy in symmetric groups in the next section.)

6.6 PERMUTATION GROUP

A permutation group is a group G whose elements are permutations of a given set M, and whose group operation is the composition of permutations in G (which are thought of as bijective functions from the set M to itself); the relationship is often written as (G, M). Note that the group of *all* permutations of a set is the symmetric group; the term *permutation group* is usually restricted to mean a subgroup of the symmetric group. The symmetric group of n elements is denoted by S_n; if M is any finite or infinite set, then the group of all permutations of M is often written as Sym(M). The application of a permutation group to the elements being permuted is called its group action; it has applications in both the study of symmetries, combinatorics and many other branches of Mathematics, Physics and Chemistry.

Closure properties: As a subgroup of a symmetric group, all that is necessary for a permutation group to satisfy the group axioms is that it contain the identity permutation, the inverse permutation of each permutation it contains, and be closed under composition of its permutations. A general property of finite groups implies that a finite subset of a symmetric group is again a group if and only if it is closed under the group operation.

Examples: Permutations are often written in *cyclic form, e.g.* during cycle index computations, so that given the set M= {1, 2, 3, 4}, a permutation g of M with $g(1) = 2$, $g(2) = 4$, $g(4) = 1$ and $g(3) = 3$ will be written as (1, 2, 4)(3), or more commonly, (1, 2, 4) since 3 is left unchanged; if the objects are denoted by a single letter or digit, commas are also dispensed with, and we have a notation such as (1 2 4).

Consider the following set G of permutations of the set M = {1, 2, 3, 4}:
- $e = (1)(2)(3)(4)$
  This is the identity, the trivial permutation which fixes each element.
- $a = (1\ 2)(3)(4) = (1\ 2)$
  This permutation interchanges 1 and 2, and fixes 3 and 4.
- $b = (1)(2)(3\ 4) = (3\ 4)$
  Like the previous one, but exchanging 3 and 4, and fixing the others.
- $ab = (1\ 2)(3\ 4)$
  This permutation, which is the composition of the previous two exchanges simultaneously 1 with 2, and 3 with 4.

G forms a group, since $aa = bb = e$, $ba = ab$, and $baba = e$. So (G,M) forms a permutation group.

The Rubik's Cube puzzle is another example of a permutation group. The underlying set being permuted is the coloured subcubes of the whole cube. Each of the rotations of the faces of the cube is a permutation of the positions and orientations of the subcubes. Taken together, the rotations form a generating set, which in turn generates a group by composition of these rotations. The axioms of a group are easily seen to be satisfied; to invert any sequence of rotations, simply perform their opposites, in reverse order.

The group of permutations on the Rubik's Cube does not form a complete symmetric group of the 20 corner and face cubelets; there are some final cube positions which cannot be achieved through the legal manipulations of the cube. More generally, *every* group G is isomorphic to a permutation group by virtue of its regular action on G as a set; this is the content of Cayley's theorem.

6.7 SYMMETRIC GROUPS AND CAYLEY'S THEOREM

Cayley's Theorem is one of the reasons why the symmetric groups form such an important class of groups: in a sense, if we understand the symmetric groups completely, then we understand all groups!

Theorem: *Every group is isomorphic to a subgroup of some symmetric group.*

Before we give the proof, here is a small digression on the background. Group theory began in the late 18th century: Lagrange, for example, proved his theorem in 1770. Probably the first person to write down the axioms for a group in anything like the present form was Dyck in 1882. So what exactly were group theorists doing for a hundred years.

The answer is that Lagrange, Galois, etc. regarded a group as a set G of permutations with the properties :
- G is closed under composition;
- G contains the identity permutation;
- G contains the inverse of each of its elements.

In other words, G is a subgroup of the symmetric group.

Thus, Cayley's contribution was to show that every group (in the modern sense) could be regarded as a group of permutations; that is, every structure which satisfies the group axioms can indeed be thought of as a group in the sense that Lagrange and others would have understood.

In general, systems of axiom in mathematics are usually not invented out of the blue, but are an attempt to capture some theory which already exists.

Proof of Cayley's Theorem

Cayley

We begin with an example. Here is the Cayley table of a group we have met earlier: it is $C_2 \times C_2$, or the additive group of the field of four elements. When we saw it in Chapter 7, its elements were called e, a, b, c, now I will call them g_1, g_2, g_3, g_4.

·	g_1	g_2	g_3	g_4
g_1	g_1	g_2	g_3	g_4
g_2	g_2	g_1	g_4	g_3
g_3	g_3	g_4	g_1	g_2
g_4	g_4	g_3	g_2	g_1

Now consider the four columns of this table. In each column, we see the four group elements $g_1, \ldots, g_4$, each occurring once; so their subscripts form a permutation of $\{1, 2, 3, 4\}$. Let π_i be the permutation which is given by the ith column.

For example, for $i = 3$, the elements of the column are (g_3, g_4, g_1, g_2), and so π_3 is the permutation which is $\begin{pmatrix} 1 & 2 & 3 & 4 \\ 3 & 4 & 1 & 2 \end{pmatrix}$ in two-line notation, or $(1, 3)(2, 4)$ in cycle notation.

The four permutations which arise in this way are:

$$\pi_1 = 1$$
$$\pi_2 = (1, 2)(3, 4)$$
$$\pi_3 = (1, 3)(2, 4)$$
$$\pi_4 = (1, 4)(2, 3)$$

Now we claim that $\{\pi_1, \pi_2, \pi_3, \pi_4\}$ is a subgroup H of the symmetric group S_4, and that the map θ defined by $g_i\theta = \pi_i$ is an isomorphism from G to H. (This means that, if $g_ig_j = g_k$, then $\pi_i\pi_j = \pi_k$, where permutations are composed in the usual way.) This can be verified with a small amount of checking.

You might think that it would be easier to use rows instead of columns in this argument. In the case of an Abelian group, like the one in this example, of course it makes no difference since the Cayley table is symmetric; but for non-Abelian groups, the statement would not be correct for rows.

So now we come to a more precise statement of Cayley's Theorem. We assume here that the group is finite; but the argument works just as well for infinite groups too.

Theorem (Cayley's Theorem): Let $G = \{g_1, ..., g_n\}$ be a finite group. For $j \in \{1, ..., n\}$, let π_j be the function from $\{1, ..., n\}$ to itself defined by the rule

$$i\pi_j = k \text{ if and only if } g_i g_j = g_k$$

Then,

(a) π_j is a permutation of $\{1, ..., n\}$:
(b) the set $H = \{\pi_1, ..., \pi_n\}$ is a subgroup of S_n;
(c) the map $\theta : G \to S_n$ given by $g_j \theta = \pi_j$ is a homomorphism with kernel $\{1\}$
(d) G is isomorphic to H.

Proof: (a) To show that π_j is a permutation, it is enough to show that it is one-to-one, since a one-to-one function on a finite set is onto. (For infinite groups, we would also have to prove that it is onto). So suppose that $i_1 \pi_j = i_2 \pi_j = k$. This means, by definition, that $g_{i_1} g_j = g_{i_2} g_j = g_k$. Then by the cancellation law, $g_{i_1} = g_{i_2} = g_k g_j^{-1}$ and so $i_1 = i_2$.

(c) Clearly the image of θ is H. So if we can show that θ is a homomorphism, the fact that H is a subgroup of S_n will follow.

Suppose that $g_j g_k = g_l$. We have to show that $\pi_j \pi_k = \pi_l$, in other words (since these are functions) that $(i\pi_j)\pi_k = i\pi_l$ for any $i \in \{1, ..., n\}$. Define numbers p, q, r by $g_i g_j = g_p$, $g_p g_k = g_q$, and $g_i g_l = g_r$. Then $i\pi_j = p$, $p\pi_k = q$, (so $i\pi_j \pi_k = q$), and $i\pi_l = r$. So we have to prove that $q = r$. But

$$g_q = g_p g_k = (g_i g_j) g_k = g_i (g_j g_k) = g_i g_l = g_r,$$

so $q = r$.

Now the kernel of θ is the set of elements g_j for which π_j is the identity permutation. Suppose that $g_j \in \text{Ker}(\theta)$. Then $i\pi_j = i$ for any $i \in \{1 ....., n\}$. This means that $g_i g_j = g_i$, so (by the cancellation law) if is the identity. So $\text{Ker}(\theta) = \{1\}$.

Now the First Isomorphism Theorem shows that $H = \text{Im}(\theta)$ is a subgroup of S_n (that is, (b) holds), and that

$$G \cong G/\{1\} = G/\text{Ker}(\theta) \cong H.$$

that is, (d) holds. So we are done.

Remark: There may be other ways to find a subgroup of S_n isomorphic to the given group G. For example, consider the group $G = 1S_3$, a non-Abelian group of order 6, whose Cayley table we wrote down in Chapter 7. From this table, you could find a set of six permutations in the symmetric group S_6 which form a subgroup of S_6, isomorphic to G. But G is already a subgroup of S_3!

6.8 CONJUGACY IN SYMMETRIC GROUPS

We finish this chapter with two more topics involving symmetric groups. First, how do we tell whether two elements of S_n are conjugate?

We define the *cycle structure* of a permutation $g \in S_n$ to be the list of cycle lengths of g when it is expressed in cycle notation. (We include all cycles including those of length 1.) The order of the terms in the list is not important. Thus, for example, the permutation (1, 7) (2, 6, 5)(3, 8, 4) has cycle structure [2, 3, 3].

Proposition: Two *permutations in S_n are conjugate in S_n if and only if they have the same cycle structure.*

Proof: Suppose that $(a_1, a_2, \ldots, a_r)$ is a cycle of a permutation g. This means that g maps
$$a_1 \to a_2 \to \ldots \to a_r \to a_1$$
We claim that $h = x^{-1} gx$ maps
$$a_1 x \to a_2 x \to \ldots \to a_r x \to a_1 x$$
so that it has a cycle $(a_1, a_2, \ldots, a_r)$. This is true because
$$(a_i x)h = (a_i x)(x^{-1} gx) = a_i gx = a_{i+1} x$$
for $i = 1, \ldots, r-1$, and $(a_r x)h = a_1 x$.

This shows that conjugate elements have the same cycle structure. The recipe is given g in cycle notation, replace each element a_i in each cycle by its image under x to obtain $x^{-1} gx$ in cycle notation.

Conversely, suppose that g and h have the same cycle structure. We can write h under g so that cycles correspond. Then the permutation x which takes each point in a cycle of g to the corresponding point in a cycle of h is the one we require, satisfying $h = x^{-1} gx$, to show that g and h are conjugate.

Example: The permutations $g = (1, 7)(2, 6, 5)(3, 8, 4)$ and $h = (2, 3)(1, 5, 4)(6, 8, 7)$ are conjugate. We can take x to be the permutation given by $\begin{pmatrix} 1 & 7 & 2 & 6 & 5 & 3 & 8 & 4 \\ 2 & 3 & 1 & 5 & 4 & 6 & 8 & 7 \end{pmatrix}$ in two-line notation, or $(1, 2)(3, 6, 5, 4, 7)(8)$ in cycle notation.

6.9 THE ALTERNATING GROUPS

You have probably met the sign of a permutation in linear algebra; it is used in the formula for the determinant. It is important in group theory too.

Let g be an element of S_n, which has k cycles when written in cycle notation (including cycles of length 1). We defined its *sign* to be $(-1)^{n-k}$.

Note that the sign depends only on the cycle structure; so *conjugate permutations have the same sign.*

Theorem: *The function sign is a homomorphism from the symmetric group S_n to the multiplicative group $\{+1, -1\}$. For $n \geq 2$, it is onto; so its kernel (the set of permutations of $\{1, \ldots, n\}$ with sign $+$) is a normal subgroup of index 2 in S_n, called the alternating group A_n.*

Example: For $n = 3$, the permutations with sign $+1$ are $1, (1, 2, 3)$ and $(1, 3, 2)$, while those with sign -1 are $(1, 2), (2, 3)$ and $(1, 3)$. We have seen that the first three form a normal subgroup.

Proof: We define a *transposition* to be a permutation of $\{1, \ldots, n\}$ which inter-changes two points i and j and fixes all the rest. Now a transposition has cycle structure $[2, 1, 1, \ldots, 1]$, and so has $n - 1$ cycles; so its sign is $(-1)^1 = -1$.

We show the following two facts:

(a) Every permutation can be written as the product of transpositions.

(b) If t is a transposition and g any permutation, then $\mathrm{sign}\,(gt) = -\,\mathrm{sign}\,(g)$.

Now the homomorphism property follows. For take any $g, h \in S_n$. Write $h = t_1 t_2 \ldots t_r$, where $t_1, \ldots, t_r$ are transpositions. Then applying (b) r times, we see that $\mathrm{sign}\,(gh) = \mathrm{sign}(g)(-1)^r$. But also $\mathrm{sign}\,(h) = (-1)^r$ (using the identity instead of g), so $\mathrm{sign}\,(gh) = \mathrm{sign}(g)\,\mathrm{sign}(h)$. Thus sign is a homomorphism. Since $\mathrm{sign}\,(1,2) = -1$, we see that $\mathrm{Im}(\mathrm{sign}) = \{+1, -1\}$. So, if A_n denotes $\mathrm{Ker}(\mathrm{sign})$, the First Isomorphism Theorem shows that $S_n/A_n \cong \{\pm 1\}$, so that A_n has two cosets in S_n (that is, index 2).

Proof: Of (a): Take any permutation g and write it in cycle notation, as it product of disjoint cycles. It is enough to show that each cycle can be written as a product of transpositions. Check that

$$(a_1, a_2, \ldots, a_r) = (a_1, a_2)(a_1, a_3) \ldots (a_1, a_r)$$

Proof: Of (b): Again, write g in cycle notation. Now check that, if t interchanges points in different cycles of g, then in the product gt these two cycles are "stitched together" into a single cycle while, if t interchanges points in the same g-cycle, then this cycle splits into two in, gt. For example,

$$(1, 3, 5, 7)(2, 4, 6) \cdot (3, 4) = (1, 4, 6, 2, 3, 5, 7)$$
$$(1, 2, 5, 3, 8, 6, 4, 7) \cdot (3, 4) = (1, 2, 5, 4, 7)(3, 8, 6).$$

So multiplying g by a transposition changes the number of cycles by one (either increases or decreases), and so multiplies the sign by -1.

The proof shows another interesting fact about permutations. As we saw, every permutation can be written as a product of transpositions.

Corollary: *Given any two expressions of $g \in S_n$ as a product of transpositions, the numbers of transpositions used have the same parity, which is even if $\mathrm{sign}\,(g) = +1$ and odd if $\mathrm{sign}\,(g) = -1$.*

Proof: We saw that if, g is the product of r transpositions, then $\mathrm{sign}(g) = (-1)^r$. This must be the same for any other expression for g as a product of transpositions.

Example: $(1, 2) = (1, 3)(2, 3)$; $(1, 3)$; one expression uses one transposition, the other uses three.

6.10 SOME SPECIAL GROUPS

In the final section we make the acquaintance of some further types of groups, and investigate more closely the groups S_4 and S_5.

Normal Subgroups of S_4 and S_5

In this section, we find all the normal subgroups of the groups S_4 and S_5. There are two possible approaches we could take. We could find all the subgroups and check which ones on our list are normal. But, for example, S_5 has 156 subgroups, so this would be quite a big job! The approach we will take is based on the fact that a subgroup of G is normal if and only if it is a union of conjugacy classes. So we find the conjugacy classes in each of these groups, and then figure out how to get some of them together to form a subgroup (which will automatically be normal).

Recall from the last chapter that two permutations in S_n, are conjugate if and only if they have, the same cycle structure. So first we list the possible cycle structures and count the permutations with each structure. For S_4 we get the following table. (We list the sign in the last column: we know that all the permutations with sign $+1$ must form a normal subgroup. The sign is of course $(-1)^{n-k}$, where k is the number of cycles.

Cycle structure	Number	Sign
[1, 1, 1, 1]	1	+
[2, 1, 1]	6	−
[2, 2]	3	+
[3, 1]	8	+
[4]	6	−
Total	24	

How do we compute these numbers? There is a general formula for the number of permutations with given cycle structure. If you want to use it to check, here it is. Suppose that the cycle structure is $[a_1, a_2, ..., a_r]$, and suppose that in this list the number 1 occurs m_1 times, the numuer 2 occurs m_2 times, and so on. Then the number of permutations with this cycle structure is

$$\frac{n!}{1^{m_1} m_1! 2^{m_2} m_2! ...}$$

So for example, for the cycle structure [2, 2], we have two 2s and nothing else, so $m_2 = 2$, and the number of permutations with cycle structure [2, 2] is $4!/(2^2 2!) = 3$.

In small cases we can argue directly. There is only one permutation with cycle structure [1, 1, 1, 1], namely the identity. Cycle structure [2, 1, 1] describes transpositions, and there are six of these (the number of choices of the two points to be transposed). For cycle structure [2, 2] we observe that the six transpositions fall into three complementary pairs, so there are three such elements. For [3, 1], there are four choices of which point is fixed, and two choices of a 3-cycle on the remaining points. Finally, for [4] a 4-cycle can start at any point, so we might as well assume that the first point is 1. Then there are $3! = 6$ ways to put the remaining points into a bracket $(1, , ,)$ to make a cycle.

Having produced this table, how can we pick some of the conjugacy classes to form a subgroup? We know that a subgroup must contain the identity, so the first class must be included. Also, by Lagrange's Theorem, the order of any subgroup must divide the order of the group. So, unless we take all live classes, we cannot include a class of size 6. (For them the order of the subgroup would be at least 7, so necessarily 8 or 12, and we cannot make up 1 or 5 elements out of the remaining classes.) So the only possibilities are :

- Take just the class {1}. This gives the trivial subgroup, which is certainly normal.
- Take {1} together with the class [2, 2], giving four elements. We have to look further at this.
- Take {1} together with the classes [2, 2] and [1, 3]. There are all the even permutations, so do form a normal subgroup, namely the alternating group A_4.

- Take all five classes. This gives the whole of S_4, which is a normal subgroup of itself.

The one case still in doubt is the possibility that the set
$$V_4 = \{1, (1, 2)(3, 4), (1, 3)(2, 4), (1, 4)(2, 3)\}$$
is a normal subgroup. Of course, if it is a subgroup, then it is normal, since it consists of two conjugacy classes. And it is a subgroup; our example of Cayley's Theorem produced precisely this subgroup! It is called V from the German word vier, meaning "four"; it is sometimes referred to as the "four-group".

We have proved:

Proposition: *The group S_4 has four normal subgroups. These are the identity, the four-group V_4, the alternating group A_4 and the symmetric group S_4.*

What about the factor groups? Clearly $S_4/\{1\} \cong S_4$, while $S_4/S_4 \cong \{1\}$. We know that $S_4//A_4$ is isomorphic to the multiplicative group $\{\pm 1\}$, which is a cyclic group of order 2. One case remains:

Proposition: $S_4/V_4 \cong S_3$.

Proof: There are many ways to prove this, here is the simplest.

Consider the subgroup S_3 of S_4 consisting of all permutation fixing the point 4. We have $|S_3| = 6$, $|V_6| = 4$, and $S_3 \cap V_4 = \{1\}$ (by inspection of the elements of (V_4). so $[S_3V_4] = 24$; that is, $S_3V_4 = S_4$. Now, by the Third Isomorphism Theorem,
$$S_4/V_4 = S_3V_4/V_4 \cong S_3/(S_3 \cap V_4) = S_3/\{1\} = S_3.$$

We now look at S_5 and show :

Proposition: *The group S_5 has three normal subgroups: the identity, the alternating group A_5, and the symmetric group S_5.*

We will be more brief here. The table of conjugacy classes looks like this :

Cycle structure	Number	Sign
[1, 1, 1, 1, 1]	1	+
[2, 1, 1, 1]	10	−
[2, 2, 1]	15	+
[3, 1, 1]	20	+
[3, 2]	20	−
[4, 1]	30	−
[5]	24	+
Total	120	

Number these classes as $C_1 \ldots C_7$ in order. We have to choose some of them including C_1 such that the sum of the numbers is a divisor of 120. All classes except C_1 and C_7 have size divisible by 5: so if we do not include C_7 then the total divides 24, which is easily seen to be impossible. So we must have C_7. Now, since we are trying to build a subgroup, it must be closed under composition; so any cycle type which can be obtained by multiplying together two 5-cycles has to be included. Since
$$(1, 2, 3, 4, 5)(1, 5, 4, 2, 3) = (1, 3, 2),$$

$$(1, 2, 3, 4, 5)(1, 2, 3, 5, 4) = (1, 3)(2, 5)$$

both classes C_3 and C_4 must be included. Now $C_1 \cup C_3 \cup C_4 \cup C_7$ is the alternating group A_5, and if there is anything else, we must have die entire symmetric group.

6.11 DIHEDRAL GROUPS

One important source of groups is as symmetries of geometric figures. Here is an example. Consider a square, as shown in the figure. (We have marked various axes of symmetry as dotted lines.)

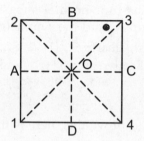

Now the square has eight symmetries, four rotations and four reflections. They are given in the table, together with their effect as permutations of the four vertices of the square. The rotations are taken clockwise.

Symmetry	Permutation
Identity	1
Rotation through 90° about O	(1, 2, 3, 4)
Rotation through 180° about O	(1, 3)(2, 4)
Rotation through 270° about O	(1, 4, 3, 2)
Reflection about AC	(1, 2) (3, 4)
Reflection about BD	(1, 4) (2, 3)
Reflection about 13	(2, 4)
Reflection about 24	(1, 3)

The eight symmetries form a group (with the operation of composition). The corresponding permutations form a group, a subgroup of the symmetric group S_4, which is isomorphic to the group of symmetries. This group is non-Abelian. (This can be seen by composing symmetries, or by composing permutations. For example, $(1, 2, 3, 4)(1, 3) = (1, 2)(3, 4)$, while $(1, 3)(1, 2, 3, 4) = (1, 4)(2, 3)$).

More generally, a regular n-gon has $2n$ symmetries, n rotations and n reflections, forming a group which is Known as the *dihedral group* D_{2n}. (Thus, the group in the table above is D_8. You should be wanted that some people refer to what I have called D_{2a} as simply D_n).

Here are some properties of dihedral groups, which should be clear from the figure.
- The n rotations form a cyclic subgroup C_n of D_{2n}. (This subgroup has index 2 in D_{2n} so it is a normal subgroup.)
- If n is odd, then every reflection axis joins a vertex to the midpoint of the opposite side; while if n is even, then $n/2$ axes join the midpoints of opposite sides and $n/2$ join opposite vertices.
- Any reflection has order 2.
- If a is a rotation and b is a reflection, then $bab = a^{-1}$.

The last condition says: reflect the figure, rotate it clockwise, and then reflect again; the result is an anticlockwise rotation. This gives another proof that the rotations form a normal subgroup. For let a be a rotation, and x any element. If x is a rotation, then $ax = xa$, so $x^{-1}ax = a$. If x is a reflection, then $x^{-1} = x$, and so $x^{-1}ax = a^{-1}$. So any conjugate of a rotation is a rotation.

The definition of D_4 is not clear, since there is no regular 2-gon. But if we take the pattern in D_{2n}, and apply it for $n = 2$, we would expect a group with a cyclic subgroup of order 2, and all elements outside this subgroup having order 2. This is a description of the four-group.

Moreover, D_6 is the group of symmetries of an equilateral triangle, and the six permutations of the vertices comprise all possible permutations.

So the following is true.

Proposition: (a) *The dihedral group D_4 is isomorphic to the four-group V_4.*
(b) *The dihedral group D_6 is isomorphic to the symmetric group S_3*

6.12 SMALL GROUPS

How many different groups of order n are there? (Here. "different" means non-isomorphic".) This is a hard question, and the answer is not known in general: it was only six years ago that the number of groups of order 1024 was computed: the number is 49, 487, 365, 422. (The result of this computation was announced at Queen Mary.)

We will not be so ambitious. For small n the number of groups of order n is given in this table. We will verify the table up to $n = 7$.

Order	1	2	3	4	5	6	7	8
Number of groups	1	1	1	2	1	7	1	5

Clearly there is only one group of order 1. The result for $n = 2, 3, 5, 7$ follows from the next Proposition.

Proposition: A group of prime order p is isomorphic to the cyclic group C_p.

Proof: Take an element g of such a group G, other than the identity. By Lagrange's Theorem, the order of g must divide p, and so the order must be 1 or p (since p is prime). But $g \neq 1$, so its order is not 1; thus it is p. So G is a cyclic group generated by g.

Next we show that there are just two groups of order 4. We know two such groups already: the cyclic group C_4, and the four-group V_4.

Let G be a group of order 4. Then the order of any element of G divides 4, and so is 1, 2 or 4 by Lagrange's Theorem. If there is an element of order 4, then G is cyclic', so suppose not. Then we can take $G = \{1, a, b, c\}$. where $a^2 = b^2 = c^2 = 1$. What is ab? It cannot be 1. since $ab = 1 = a^2$ would imply $a = b$ by the Cancellation Law. Similarly $ab \neq a$ and $ab \neq b$, also by the Cancellation Law (these would imply $b = 1$ or $a = 1$ respectively). So $ab = c$. Similarly, all the other products are determined: the product of any two non-identity elements is the third. In other words, the Cayley table is

.	1	a	b	c
1	1	a	b	c
a	a	1	c	b
b	b	c	1	a
c	c	b	a	1

We recognise the Klein four-group. So there is just one such group (up to isomorphism), giving two altogether.

To deal with order 6, we need a couple of preliminary results.

Proposition: *A group of even order must contain an element of order 2.*

Proof: Take any group G of order n, and attempt to pair up the elements of G with their inverses. Suppose that we can form m pairs accounting for $2m$ elements. The elements we have failed to pair up are the ones which satisfy $g = g^{-1}$, or $g^2 = 1$; these include the identity (one element) and all the elements of order 2. So there must be $n - 2m - 1$ elements of order 2. If n is even, then $n - 2m - 1$ is odd, and so cannot be zero: so there is an element of order 2 in G.

Proposition: *A finite group in which every element has order 2 (apart from the identity) is Abelian.*

Proof Take any $g, h \in G$. We have

$$(gh)^2 = ghgh = 1$$
$$g^2h^2 = gghh = 1$$

so by cancellation. $hg = gh$. Thus G is Abelian.

Now let G be a group of order 6. If G contains an element of order 6, then it is cyclic: so suppose not. Now all its elements except the identity have order 2 or 3. The first proposition above shows that G contains an element a of order 2. The second shows that it must also have an element of order 3. For, suppose not. Then all non-identity elements of G have order 2. If g and g are two such elements, then it is easy to see that $\{1, g, h, gh = hg\}$ is a subgroup of order 4. contradicting Lagrange's Theorem.

So let a be an element of order 3 and an element b of order 2. The cyclic subgroup $(a) = \{1, a, a^{-1} = a^2\}$ of G has order 3 and index 2, so if is normal. So $b^{-1} ab \in (a)$, whence $b^{-1} ab = a$ or $b^{-1} ab = a^{-1}$.

If $b^{-1}ab = a$, then $ab = ba$, so $(ab)^i = a^i b^i$ for all i. Then the powers of ab are
$(ab)^2 = a^2 b^2 = a^2$, $(ab)^3 = a^3 b^3 = b$, $(ab)^4 = a^4 b^4 = a$,
$(ab)^5 = a^5 b^5 = a^2 b$, $(ab)^6 = a^6 b^6 = 1$,
so the order of ab is 6, contradicting our case assumption. So we must have $b^{-1}ab = a^{-1}$, or $ba = a^{-1}b = a^2 b$.

Now using this, the Cayley table G is completely determined: all the elements have the form $a^i b^j$, where $i = 0, 1, 2$ and $j = 0, 1$; to multiply $a^i b^j$ by $a^k b^l$, we use the condition $ba = a^{-1}b$ to jump the first b over the as to its right if necessary and the conditions $a^3 = b^2 = 1$ to reduce the exponents. For example,

$$a^2 b.ab = a^2(ba)b = a^2(a^2 b)b = a^4 b^2 = a.$$

So there is only one possible group of this type. Its Cayley table is :

	1	a	a^2	b	ab	$a^2 b$
1	1	a	$a^2 b$	b	ab	$a^2 b$
a	a	a^2	1	ab	$a^2 b$	b
a^2	a^2	1	a	$a^2 b$	b	ab
b	b	$a^2 b$	ab	1	a^2	a
ab	ab	b	$a^2 b$	a	1	a^2
$a^2 b$	$a^2 b$	ab	b	a^2	a	1

These relations are satisfied in the group S_3, if we take $a = (1, 2, 3)$ and $b = (1, 2)$. So there is such a group; and so there are two groups of order 6 altogether (the other being the cyclic group;. Alternatively, we could observe that the above relations characterise the dihedral group D_6, so the two groups are C_6 and D_6.

We see, incidentally, that $S_3 \cong D_6$, and that this is the smallest non-Abelian group.

6.13 POLYHEDRAL GROUPS

You have probably seen models of the live famous regular polyhedron: the tetrahedron, the cube (or hexahedron), the octahedron, the dodecahedron, and the icosahedron. These beautiful figures have been know since antiquity.

What are their symmetry groups?

Here I will just consider the groups of rotations; the extra symmetries realised by reflections in three-dimensional space make the situation a bit more complicated. As in the case of the dihedral groups, these groups can be realised as permutation groups, by numbering the varies and reading off the permutation induced by any symmetry.

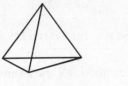 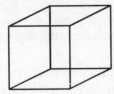 

Figure 6.1: Tetrahedron cube and octahedron

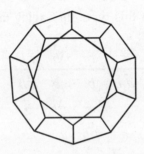 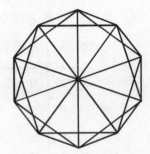

Figure 6.2: Dodecahedron and Icosahedron

Moreover, there are five figures, but only three groups. Apart from the tetrahedron, the figures fall into "dual pairs": the figure whose vertices are the face centres of a cube is an octahedron and *vice versa,* and a similar relation holds between the dodecahedron and the icosahedron. Dual pairs have the same symmetry group. (The face centres of the tetrahedron are the vertices of another tetrahedron, so this figure is "self-dual".) The following result describes the three symmetry groups.

Proposition: (*a*) The tetrahedral group is isomorphic to A_4.

(*b*) The octahedral group is isomorphic to S_4.

(*c*) The icosahedral group is isomorphic to A_6.

Proof: I will outline the proof. First we Compute the orders of these groups. If a figure has m faces, each a regular polygon with n sides, then the number of rotational symmetries is mn. For imagine the figure with one face on the table. I can pick it up and rotate it so that any of the m faces is on the table, in any of the n possible orientations. Now we have

- Tetrahedron: $m = 4$, $n = 3$, group order 12.
- Cube: $m = 6$, $n = 4$, group order 24.
- Octahedron: $m = 8$, $n = 3$ group order 24.
- Dodecahedron: $m = 12$, $n = 5$, group order 60.
- Icosahedron: $m = 20$, $n = 3$, group order 60.

We see that the symmetry groups of dual polyhedra have the same order, as they should.

(a) Any symmetry of the tetrahedron permutes the four vertices. So the symmetry group is a subgroup of S_4 of order 12. To see that it is A_4, we simply have to observe that every symmetry is an even permutation of the vertices. A rotation about the line joining a vertex to the midpoint of the opposite face has cycle structure [3, 1] while a rotation about the line joining the midpoints of opposite edges has cycle structure [2, 2]. (Alternatively, a subgroup of S_4 of order 12 has index 2 and is therefore normal; and we have worked out the normal subgroups of S_4. The only one of order 12 is A_4).

(b) Consider the cube. It has four diagonals joining opposite vertices. Any symmetry induces a permutation of the four diagonals. It is not hard to see that the map from symmetries to permutations is one-to-one. So the group is isomorphic to a subgroup of S_4 of order 24, necessarily the whole of S_4.

(c) This is the hardest to prove. But in fact it is possible to embed a cube so that its vertices are eight of the 20 vertices of the dodecahedron in five different ways. These five inscribed cubes are permuted by any rotation, so we have a subgroup of S_5 of order 60. This subgroup has index 2 and so is normal, so it must be A_5.

EXERCISE 6.1

1. Show that the additive group of integers is a subrgoup of the additive group of national numbers.
2. Show that with respect to addition the set of all even integers is a sub-group of set of all integers.
3. Prove that the integral multiples of 5 form subgroup of additive group of integers.
4. Show that four rotation of a savare lamina about an axis through its centre, perpendicular to its. Plane form a cyclic additive group.
5. Show that if a commutative group of order 6 contains an element of orders 3, then it is cyclic.
6. Shows that additive groups.
   $G = \{..., 3, -2, -1, 0, 1, 2, 3, ...\}$
   and $G' = \{..., -6, -4, -2, 0, 2, 4, 6, ...\}$ are isomorphic.
7. Prove that multiplicative group $G = \{1, -1, i, -i\}$ is isomorphic to the group G' of residue classes modulo 4 under additive composition of residue classes.
8. If G be the additive group of all integers and $G' = \{2m : m \in I, I \text{ is the set of integers}\}$ be a multiplicative group, show that G is isomorphic to G'.
9. If R is the additive group of real numbers and R^+, the multiplicative group of all positive real numbers, prove that the map $f : R \to R^+$ defined by $f(x) = e^x$ is an isomorphism.
10. Prove that the multiplicative group $G = \{1, \omega, \omega^2\}$ is isomorphic to the additive group G' of residue classes (mod 3), where $\omega^3 = 1$.
11. Show that any two cyclic group of the same order are isomorphic.

EXERCISE 6.2

1. Find fg, gf if $f = \begin{pmatrix} 1 & 2 & 3 \\ 2 & 1 & 3 \end{pmatrix}$, $g = \begin{pmatrix} 1 & 2 & 3 \\ 1 & 3 & 2 \end{pmatrix}$.

2. (i) If $A = \begin{pmatrix} 1 & 2 & 3 \\ 3 & 1 & 2 \end{pmatrix}$, $B = \begin{pmatrix} 1 & 2 & 3 \\ 2 & 3 & 1 \end{pmatrix}$, find A^{-1}, B^{-1} and verify that $(AB)^{-1} = B^{-1}A^{-1}$.

   (ii) Find the inverse of $\begin{pmatrix} 1 & 2 & 3 & 4 \\ 1 & 3 & 4 & 2 \end{pmatrix}$.

3. Prove that permutation $\begin{pmatrix} 1 & 2 & 3 & 4 \\ 3 & 4 & 1 & 2 \end{pmatrix}$ is inverse of itself.

4. Verify the following
   (i) $(1\ 2\ 4\ 5\ 3)(3\ 2\ 1\ 5\ 4) = (2\ 3\ 5)$
   (ii) $\begin{pmatrix} 1 & 2 & 3 & 4 & 5 & 6 & 7 & 8 \\ 2 & 4 & 3 & 8 & 6 & 5 & 7 & 1 \end{pmatrix} = (1\ 2\ 4\ 8)(5\ 6)$

5. Show that the set P_3 of permutations on three symbols. 1, 2, 3, is a finite non-abelian group of order 6 with respect to permutation multiplication as composition.

6. Show that the multiplicative group $G = \{1, -1\}$ is isomorphic to the permutation group $G' = \{I, (a, b)\}$.

7. Show that the multiplicative group $G' = \{1, W, W^2\}$ is isomorphic to the permutation group $G = \{I, (a, b, c), (a, c, b)\}$ on three symbols, where $W^3 = 1$.

8. Prove Langrange's theorem on order of a finite subgroup by using concept of cosets.

9. If G is the symmetric group of order three and H is a subgroup of G given by $H = \{I, (a, c, b), (abc)\}$, then prove that H is a normal subgroup of G.

10. If G is a group of order h and H a subgroup of order m, then prove that $n = mj$, where j is index of H in G.

11. If G is group and H is a subgroup of index 2 in G then H is a normal subgroup of G. Hence or otherwise show that $H = \{(1), (12)\}$ is not a normal subgroup of P_3.

12. Let H be the additive group of integers and H be a subgroup of I such that
    $$H = \{mx : x \subseteq I\}$$
    where m is a fixed integer. Write the elements of quotient group I/H, Also form a composition table for I/H when $m = s$.

13. Prove that every subgroup of an abelian group is normal.

14. If H is a subgroup of G and N is a normal subgroup of G then prove that
    (i) HN is a subgroup of G.
    (ii) N is a normal subgroup of HN.

15. If H is a subgroup of G and N is a normal subgroup of G, prove that H∩N is a normal subgroup of H.

16. If H_1 and H_2 are normal subgroup of G, prove that H_1H_2 is normal subgroup of G.

CHAPTER 7

RINGS AND FIELDS

7.1 INTRODUCTION

A ring can be thought of as a generalisation of the integers, $\mathbb{Z}$. We can add and multiply elements of a ring, and we are interested in such questions as factorisation into primes, construction of "modular arithmetic", and so on'.

Definition of a Ring

Our first class of structures are *rings*. A ring has two operations: the first is called *addition* and is denoted by + (with infix notation); the second is called *multiplication,* and is usually denoted by juxtaposition (but sometimes by with infix notation).

In order to be a ring, the structure must satisfy certain rules called *axioms*. We divide these into three part. The name of the ring is R.

We define a *ring* to be a set R with two binary operations satisfying the following axioms:

(i) Axioms for Addition

- (A0) *(Closure law)* For any $a, b \in R$, we have $a + b \in R$.
- (A1) *(Associative law)* For any $a, b, c \in R$, we have $(a + b) + c = a + (b + c)$.
- (A2) *(Identity law)* There is an element $0 \in R$ with the property that $a + 0 = 0 + a = a$ for all $a \in R$. (The element 0 is called the *zero element* of R.)
- (A3) *(Inverse law)* For any element $a \in R$, there is an element $b \in R$ satisfying $a + b = b + a = 0$. (We denote this element b by $-a$, and call it the *additive inverse* or *negative* of a.)
- (A4) *(Commutative law)* For any $a, b \in R$, we have $a + b = b + a$.

(ii) Axioms for Multiplication

- (M0) *(Closure law)* For any $a, b \in R$, we have $ab \in R$.
- (Ml) *(Associative law)* For any $a, b, c \in R$, we have $(ab)c = a(bc)$.

(iii) Mixed Axiom

- (D) *(Distributive laws)* For any $a, b, c \in R$, we have $(a + b)c = ac + bc$ and $c(a + b) = ca + cb$.

Remark 1. The closure laws (A0) and (M0) are not strictly necessary. If + is a binary operation, then it is a function from $R \times R$ to R, and so certainly $a + b$ is an element of R for all $a, b \in R$. We keep these laws in our list as a reminder.

Remark 2. The zero element 0 denoted by (A2) and the negative $-a$ denoted by (A3) are not claimed to be unique by the axioms. We will see later on that there is only one zero element in a ring, and that each element has only one negative.

Axioms (M0) and (M1) parallel (A0) and (A1). Notice that we do not require multiplicative analogues of the other additive axioms. But there will obviously be some rings in which they hold. We state them here for reference.

Further multiplicative properties :

(M2) (*Identity law*) There is an element $1 \in R$ such that $a1 = 1a = a$ for all $a \in R$. (The element 1 is called the *identity element* of R).

(M3) (*Inverse law*) For any $a \in R$, if $a \neq 0$, then there exists an element $b \in R$ such that $ab = ba = 1$. (We denote this element b by a^{-1}, and call it the *multiplicative inverse* of a.)

(M4) (*Commutative law*) For all $a, b \in R$, we have $ab = ba$.

A ring which satisfies (M2) is called a *ring with identity;* a ring which satisfies (M2) and (M3) is called a *division ring;* and a ring which satisfies (M4) is called a *commutative ring*. (Note that the term "commutative ring" refers to the fact that the multiplication is commutative; the addition in a ring is always commutative!) A ring which satisfies all three further properties (that is, a commutative division ring) is called *a field*.

Examples of Rings

1. The integers

The most important example of a ring is the set $\mathbb{Z}$ of integers, with the usual addition and multiplication. The various properties should be familiar to you; we will simply accept that they hold. $\mathbb{Z}$ is a commutative ring with identity. It is not a division ring because there is no *integer b* satisfying $2b = 1$. This ring will be our prototype for several things in the course.

Note that the set $\mathbb{N}$ of natural numbers, or non-negative integers, is not a ring, since it fails the inverse law for addition. (There is no non-negative integer b such that $2 + b = 0$).

2. Other number systems

Several other familiar number systems, namely the rational numbers $\mathbb{Q}$, the real numbers $\mathbb{R}$, and the complex numbers $\mathbb{C}$, are fields. Again, these properties are assumed to be familiar to you.

3. The quaternions

There do exist division rings in which the multiplication is not commutative, that is, which are not fields, but they are not so easy to find. The simplest example is the ring *of quaternions,* discovered by Hamilton in 1843.

On 16 October 1843 (a Monday), Hamilton was walking in along the Royal Canal with his wife to preside at a Council meeting of the Royal Irish Academy. Although his wife talked to him now and again Hamilton hardly heard, for the discovery of the quaternions, the first noncommutative [ring] to be studied, was taking shape in his mind. He could not resist the impulse to carve the formulae for the quaternions in the stone of Broom'e Bridge (or Brougham Bridge as he called it) as he and his wife passed it.

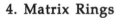

Instead of adding just one element i to the real numbers, Hamilton added three. That is, a *quaternion* is an object of the form $a + bi + cj + dk$, where

$$i^2 = j^2 = k^2 = -1, \quad ij = -ji = k, \quad jk = -kj = i, \quad ki = -ik = j.$$

It can be shown that all the axioms (A0)-(A4), (M0)-(M3) and (D) are satisfied.

For example, if a, b, c, d are not all zero, then we have

$$(a+bi+cj+dk)\left(\frac{a-bi-cj-dk}{a^2+b^2+c^2+d^2}\right) = 1$$

The ring of quaternions is denoted by $\mathbb{H}$, to commemorate Hamilton.

4. Matrix Rings

We briefly denned addition and multiplication for matrices in the last chapter. The formulae for addition and multiplication of $n \times n$ matrices, namely

$$(A + B)_{ij} = A_{ij} + B_{ij}, \quad (AB)_{ij} = \sum_{k=1}^{n} A_{ik}B_{kj}$$

just depend on the fact that we can add and multiply the entries. In principle these can be extended to any system in which addition and multiplication are possible. However, there is a problem with multiplication, because of the $\sum_{k=1}^{n}$, which tells us to add up n terms. In general we can only add two things at a time, since addition is a binary operation, so we have to make the convention that, for example, $a + b + c$ means $(a + b) + c$, $a + b + c + d$ means $(a + b + c) + d$, and so on. We will return to this point in the next subsection. Now we have the following result:

Proposition 7.1: Let R be a ring. Then the set $M_n(R)$ of $n \times n$ matrices over R, with addition and multiplication defined in the usual way, is a ring. If R has an identity, then $M_n(R)$ has an identity; but it is not in general a commutative ring or a division ring.

We will look at the proof later, once we have considered addition of n terms.

5. Polynomial Rings

In much the same way, the usual rules for addition of polynomials,

$$\left(\Sigma a_i x^i\right) + \left(\Sigma b_i x^i\right) = \Sigma (a_i + b_i) x^i, \quad \left(\Sigma a_i x^i\right) + \left(\Sigma b_i x^i\right) = \Sigma d_i x^i$$

where

$$d_i = \sum_{k=0}^{i} a_k b_{i-k}$$

can be extended to polynomials with coefficients in any algebraic structure in which addition and multiplication are defined. As for matrices, we have to be able to add an arbitrary number of terms to make sense of the definition of multiplication. We have the result:

Proposition 7.2: Let R be a ring, then the set $R[x]$ of polynomials over R, with addition and multiplication defined in the usual way, is a ring. If R is commutative, then so is $R[x]$; if R has an identity, then so does $R[x]$; but it is not a division ring.

Again we defer looking at the proof.

6. Rings of Sets

The idea of forming a ring from operations on sets is due to George Boole, who published in 1854 *An investigation into the Laws of Thought, on Which are founded the Mathematical Theories of Logic and Probabilities.* Boole approached logic in a new way reducing it to algebra, in much the same way as Descartes had reduced geometry to algebra.

George Boole

The familiar set operations of union and intersection satisfy some but not all of the ring axioms. They are both commutative and associative, and satisfy the distributive laws both ways round; but they do not satisfy the identity and inverse laws for addition.

Boole's algebra of sets works as follows. Let P(A), the *power set* of A, be the set of all subsets of the set A. Now we define addition and multiplication on to P(A) be the operations of symmetric difference and intersection respectively:

$$x + y = x \Delta y, \quad xy = x \cap y$$

Proposition 7.3: The set P(A), with the above operations, is a ring; it is commutative, has an identity element, but is not a field if $|A| > 1$. It satisfies the further conditions $x + x = 0$ and $xx = x$ for all x.

We will not give a complete proof, but note that the empty set is the zero element (since $x \Delta \emptyset = x$ for any set x), while the additive inverse $-x$ of x is equal to x itself (since $x \Delta x = \emptyset$ for any x). Check the other axioms for yourself with Venn diagrams.

A ring satisfying the further condition that $xx = x$ for all x is called a *Boolean ring*.

7. Zero Rings

Suppose that we have any set R with a binary operation + satisfying the additive axioms (A0)-(A4). (We will see later in the course that such a structure is called an abelian group.) Then we can make R into a ring by defining $xy = 0$ for all $x, y \in R$. This is not a very exciting rule for multiplication, but it is easy to check that all remaining axioms are satisfied.

A ring in which all products are zero is called a zero ring. It is commutative, but doesn't have an identity (if $|R| > 1$).

8. Direct Sum

Let R and S be any two rings. Then we define the direct sum $R \oplus S$ as follows. As a set, $R \oplus S$ is just the cartesian product $R \times S$. The operations are given by the rules

$$(r_1, s_1) + (r_2, s_2) = (r_1 + r_2, s_1 + s_2), \qquad (r_1, s_1)(r_2, s_2) = (r_1 r_2, s_1 s_2).$$

(Note that in the ordered pair $(r_1 + r_2, s_1 + s_2)$, the first + denotes addition in R, and the second + is addition in S.)

Proposition 7.4: If R and S are rings, then R ⊕ S is a ring. If R and S are commutative, then so is R ⊕ S; if R and S have identities, then so does R ⊕ S; but R ⊕ S is not a division ring if both R and S have more than one element.

The proof is straightforward checking.

9. Modular Arithmetic

Let $\mathbb{Z}_n$ denote the set of all congruence classes modulo n, where n is a positive integer. We saw in the first chapter that there are n congruence classes; so $\mathbb{Z}_n$ is a set with n elements:

$$\mathbb{Z}_n = \{[0]_n, [1]_n, ..., [n-1]_n\}$$

Define addition and multiplication on $\mathbb{Z}_n$ by the rules

$$[a]_n + [b]_n = [a+b]_n, \quad [a]_n[b]_n = [ab]_n$$

There is an important job to do here: we have to show that these definitions do not depend on our choice of representatives of the equivalence classes.

Proposition 7.5: For any positive integer n, $\mathbb{Z}_n$, is a commutative ring with identity. It is a field if and only if n is a prime number.

Here, for example, are the addition and multiplication tables of the ring $\mathbb{Z}_5$. We simplify the notation by writing x instead of $[x]_5$.

+	0	1	2	3	4
0	0	1	2	3	4
1	1	2	3	4	0
2	2	3	4	0	1
3	3	4	0	1	2
4	4	0	1	2	3

·	0	1	2	3	4
0	0	0	0	0	0
1	0	1	2	3	4
2	0	2	4	1	3
3	0	3	1	4	2
4	0	4	3	2	1

Note, for example, that $2^{-1} = 3$ in this ring.

10. Rings of Functions

The sum and product of continuous real functions are continuous. So there is a ring $C(\mathbb{R})$ of continuous functions from $\mathbb{R}$ to $\mathbb{R}$, with

$$(f+g)(x) = f(x) + g(x), \quad (fg)(x) = f(x)g(x)$$

There are several related rings, such as $C^1(\mathbb{R})$ (the ring of differentiable functions), $C_0(\mathbb{R})$ (the ring of continuous functions satisfying $f(x) \to 0$ as $x \to \pm\infty$), and $C([a, b])$ (the ring of continuous functions on the interval $[a, b]$. All these rings are commutative, and all except $C_0(\mathbb{R})$ have an identity (the constant function with value 1).

These rings are the subject-matter of Functional Analysis.

Properties of Rings

We have some business deferred from earlier to deal with. After that, we prove some basic properties of rings, starting from the axioms.

Uniqueness of zero element

The zero element of a ring is unique. For suppose that there are two zero elements, say z_1 and z_2. (This means that $a + z_1 = z_1 + a = a$ for all a and also $a + z_2 = z_2 + a$ for all a.) Then
$$z_1 = z_1 + z_2 = z_2$$

Exercise: Show that the identity element of a ring, if it exists is unique.

Uniqueness of Additive Inverse

The additive inverse of an element a is unique. For suppose that b and c are both additive inverses of a. (This means that $a + b = b + a = 0$ and $a + c = c + a = 0$ we know now that there is a unique zero element, and we call it 0.) Then
$$b = b + 0 = b(a + c) = (b + c) + c = 0 + c = c,$$
where we use the associative law in the third step.

Exercise: Show that the multiplicative inverse of an element of a ring, if it exists, is unique.

Adding more than two elements

The associative law tells us that if we have to add three elements, then the two possible ways of doing it, namely $(a + b) + c$ and $a + (b + c)$, give us the same result. For more than three elements, there are many different ways of adding them: we have to put in brackets so that the sum can be worked out by adding two elements at a time. For example, there are five ways of adding four elements:
$$((a + b) + c) + d, (a + (b + c)) + d, (a + b) + (c + d), a + ((b + c) + d), a + (b + (c + d)).$$

These are all equal. For the associative law $(a + b) + c = a + (b + c)$ shows that the first and second are equal, while the associative law for b, c, d shows that the fourth and fifth are equal. Also, putting $x = a + b$, we have
$$((a + b) + c) + d = (x + c) + d = x + (c + d) = (a + b) + (c + d)$$
so the first and third are equal; and similarly the third and fifth are equal.

In general we have the following. The proof works for any associative binary operation.

Proposition 7.6: Let * be an associative binary operation on a set A, and $a_1, \ldots, a_n \in A$. Then the result of evaluating $a_1 * a_2 * \ldots * a_n$, by adding brackets in any way to make the expression well defined, is the same, independent of bracketing.

Proof: The proof is by induction on the number of terms. For $n = 2$ there is nothing to prove; for $n = 3$, the statement is just the associative law; and for $n = 4$, we showed it above. Suppose that the result is true for fewer than n terms. Suppose now that we have two different bracketings of the expression $a_1 * a_2 * \ldots * a_n$. The first will have the form $(a_1 * \ldots * a_i) * (a_{i+1} * \ldots * a_n)$, with the terms inside the two sets of brackets themselves bracketed in some way. By induction, the result is independent of the bracketing of $a_1, \ldots, a_i$ and of $a_{i+1}, \ldots, a_n$.

Similarly, the second expression will have the form $(a_1 * \ldots * a_j) * (a_{j+1} * \ldots * a_n)$, and is independent of the bracketing of $a_1, \ldots, a_j$ and of $a_{j+1}, \ldots, a_n$.

Case 1: $i = j$. Then the two expressions are obviously equal.

Case 2: $i \neq j$; suppose, without loss, that $i < j$. Then the first expression can be written as
$$(a_1 * \ldots * a_i) * ((a_{i+1} * \ldots *a_j) * (a_{j+1} * \ldots * a_n)),$$
and the second as
$$((a_1 * \ldots * a_i) * (a_{i+1} * \ldots *a_j)) * (a_{j+1} * \ldots * a_n),$$
where each expression is independent of any further bracketing. By the associative law, these two expressions are equal: they are $x * (y * z)$ and $(x * y) * z$, where $x = a_1 * \ldots *a_i$, $y = a_{i+1} * \ldots *a_j$, and $z = a_{j+1} * \ldots *a_n$.

Note that this result applies to both addition and multiplication in a ring. As usual, we denote $a_1 + a_2 + \ldots + a_n$ by $\sum_{i=1}^{n} a_i$

Cancellation laws

Proposition 7.7: In a ring R, if $a + x = b + x$, then $a = b$. Similarly, if $x + a = x + b$, then $a = b$.

Proof: Suppose that $a + x = b + x$, and let $y = -x$. Then
$$a = a + 0 = a + (x + y) = (a + x) + y = (b + x) + y = b + (x + y) = b + 0 = b.$$

The other law is proved similarly, or by using the commutativity of addition. These facts are the *cancellation laws*.

A Property of Zero

One familiar property of the integers is that $0a = 0$ for any integer a. We do not have to include this as an axiom, since it follows from the other axioms. Here is the proof. We have $0 + 0 = 0$, so $0a + 0 = 0a = (0 + 0)a = 0a + 0a$, by the distributive law; so the cancellation law gives $0 = 0a$. Similarly $a0 = 0$.

It follows that if R has an identity 1, and $|R| > 1$, then $1 \neq 0$. For choose any element $a \neq 0$; then $1 a = a$ and $0a = 0$. It also explains why we have to exclude 0 in condition (M3): 0 cannot have a multiplicative inverse.

Commutativity of Addition

It turns out that, in a ring with identity, it is not necessary to assume that addition is commutative: axiom (A4) follows from the other ring axioms together with (M2).

For suppose that (A0)–(A3), (M0)–(M2) and (D) all hold. We have to show that $a + b = b + a$. Consider the expression $(1 + 1)(a + b)$. We can expand this in two different ways by the two distributive laws:

$$(1 + 1)(a + b) = 1(a + b) + 1(a + b) = a + b + a + b$$
$$(1 + 1)(a + b) = (1 + 1)a + (1 + 1)b = a + a + b + b$$

Hence $a + b + a + b = a + a + b + b$, and using the two cancellation laws we conclude that $b + a = a + b$.

This argument depends on the existence of a multiplicative identity. If we take a structure with an operation + satisfying (A0)-(A3) (we'll see later that such a structure is known as a group), and apply the "zero ring" construction to it (that is, $ab = 0$ for all a, b), we obtain a structure satisfying all the ring axioms except (A4).

Boolean Rings

We saw that a *Boolean ring* is a ring R in which $xx = x$ for all $x \in R$.

Proposition 7.8: A Boolean ring is commutative and satisfies $x + x = 0$ for all $x \in R$.

Proof: We have $(x + y)(x + y) = x + y$. Expanding the left using the distributive laws, we find that
$$xx + xy + yx + yy = x + y.$$
Now $xx = x$ and $yy = y$. So we can apply the cancellation laws to get

In particular, putting $y = x$ in this equation, we have $xx + xx = 0$, or $x + x = 0$, one of the things we had to prove.

Taking this equation and putting xy in place of x, we have
$$xy + xy = 0 = xy + yx,$$
and then the cancellation law gives us $xy = yx$, as required.

We saw that the power set of any set, with the operations of symmetric difference and intersection, is a Boolean ring. Another example is the ring $\mathbb{Z}_2$ (the integers mod 2).

Matrix Rings

In view of Proposition 7.6, the definition of the product of two $n \times n$ matrices now makes sense: $AB = D$, where
$$D_{ij} = \sum_{k=1}^{n} A_{ik} B_{kj}$$
So we are in the position to prove Proposition 7.1.

A complete proof of this proposition involves verifying all the ring axioms. The arguments are somewhat repetitive; I will give proofs of two of the axioms.

Axiom (A2): Let 0 be the zero element of the ring R, and let O be the zero matrix in $M_n(R)$, satisfying $O_{ij} = 0$ for all i, j. Then O is the zero element of $M_n(R)$: for, given any matrix A,
$$(O + A)_{ij} = O_{ij} + A_{ij} = 0 + A_{ij} = A_{ij}, \quad (A + O)_{ij} = A_{ij} + O_{ij} = A_{ij} + 0 = A_{ij}$$
using the properties of $0 \in R$. So $O + A = A + O = A$.

Axiom (D): the (i, j) entry of $A(B + C)$ is
$$\sum_{k=1}^{n} A_{ik}(B+C)_{kj} = \sum_{k=1}^{n} A_{ik} B_{kj} + A_{ik} C_{kj}$$
by the distributive law in R; and the (i, j) entry of $AB + AC$ is

Rings and Fields   181

$$\sum_{k=1}^{n} A_{ik}B_{kj} + \sum_{k=1}^{n} A_{ik}C_{kj}$$

Why are these two expressions the same? Let us consider the case $n = 2$. The first expression is

$$A_{i1}B_{1j} + A_{i1}C_{1j} + A_{i2}B_{2j} + A_{i2}C_{2j}$$

while the second is

$$A_{i1}B_{1j} + A_{i2}B_{2j} + A_{i1}C_{1j} + A_{i2}C_{2j}$$

(By Proposition 7.6, the bracketing is not significant.) Now the commutative law for addition allows us to swap the second and third terms of the sum; so the two expressions are equal. Hence $A(B + C) = AB + AC$ for any matrices A, B, C. For $n > 2$, things are similar, but the rearrangement required is a bit more complicated.

The proof of the other distributive law is similar.

Observe what happens in this proof: we use properties of the ring R to deduce properties of $M_n(R)$. To prove the distributive law for $M_n(R)$, we needed the distributive law and the associative and commutative laws for addition in R. Similar things happen for the other axioms.

Polynomial Rings

What exactly is a polynomial? We deferred this question before, but now is the time to face it.

A polynomial $\Sigma a_i x^i$ is completely determined by the sequence of its coefficients $a_0, a_1, \ldots$. These have the property that only a finite number of terms in the sequence are non-zero, but we cannot say in advance how many. So we make the following definition:

A *polynomial* over a ring R is an infinite sequence

$$(a_i)_{i \geq 0} = (a_0, a_1, \ldots)$$

of elements of R, having the property that only finitely many terms are non-zero; that is, there exists an n such that $a_i = 0$ for all $i > n$. If a_n is the last non-zero term, we say that the *degree* of the polynomial is n. (Note that, according to this definition, the all-zero sequence does not have a degree.) Now the rules for addition and multiplication are

$$(a_i) + (b_i) = (c_i) \quad \text{where} \quad c_i = a_i + b_i,$$

$$(a_i) + (b_i) = (d_i) \quad \text{where} \quad d_i = \sum_{j=0}^{i} a_j b_{i-j}$$

Again, the sum in the definition of multiplication is justified by Proposition 2.6. We think of the polynomial $(a_i)_{i \geq 0}$ of degree n as what we usually write as $\sum_{i=0}^{n} a_i x^i$; the rules we gave agree with the usual ones.

Now we can prove Proposition 7.2, asserting that the set of polynomials over a ring R is a ring. As for matrices, we have to check all the axioms, which involves a certain amount of tedium. The zero polynomial required by (A2) is the all-zero sequence. Here is a proof of

(M1). You will see that it involves careful work with dummy subscripts!

We have to prove the associative law for multiplication. So suppose that $f = (a_i)$, $g = (b_i)$ and $h = (c_i)$. Then the ith term of fg is $\sum_{j=0}^{i} a_j b_{i-j}$ and so tne ith term of $(fg)h$ is

$$\sum_{k=0}^{i} \left(\sum_{j=0}^{k} a_j b_{k-j} \right) c_{i-k}$$

Similarly the ith term of $f(gh)$ is

$$\sum_{s=0}^{i} a_s \left(\sum_{t=0}^{i-s} b_t c_{i-s-t} \right)$$

Each term on both sides has the form $a_p b_q c_r$, where $p, q, r \geq 0$ and $p + q + r = i$. (In the first expression, $p = j, q = k - j, r = i - k$; in the second, $p = s, q = t, r = i - s - t$.) So the two expressions contain the same terms in a different order. By the associative and commutative laws for addition, they are equal.

7.2 SUBRINGS

Definition and Test

Suppose that we are given a set S with operations of addition and multiplication, and we are asked to prove that it is a ring. In general, we have to check all the axioms. But there is a situation in which things are much simpler: this is when S is a subset of a set R which we already know to be a ring, and the addition and multiplication in S are just the restrictions of the operations in R (that is, to add two elements of S, we regard them as elements of R and use the addition in R).

Definition Let R be a ring. A *subring* of R is a subset S of R which is a ring in its own right with respect to the restrictions of the operations in R.

What do we have to do to show that S is a subring?

- The associative law (A1) holds in S. For, if $a, b, c \in S$, then we have $a, b, c \in R$ (since $S \subseteq R$), and so

$$(a + b) + c = a + (b + c)$$

since R satisfies (A1) (as we are given that it is a ring).
- Exactly the same argument shows that the commutative law for addition (A4), the associative law for multiplication (M1), and the distributive laws (D), all hold in S.
- This leaves only (A0), (A2), (A3) and (M0) to check.

Even here we can make a simplification, if $S \neq \emptyset$. For suppose that (A0) and (A3) hold in S. Given $a \in S$, the additive inverse $-a$ belongs to S (since we are assuming (A3)), and so $0 = a + (-a)$ belongs to S (since we are assuming (A0)). Thus (A2) follows from (A0) and (A3).

Rings and Fields

We state this as a theorem :

Theorem 7.1 (First Subring Test) *Let R be a ring, and let S be a non-empty sub set of R. Then S is a subring of R if the following condition holds:*

for all $a, b \in S$, we have $a + b, ab, -a \in S$.

Example. We show that the set S of even integers is a ring. Clearly it is a non-empty subset of the ring $\mathbb{Z}$ of integers. Now, if $a, b \in S$, say $a = 2c$ and $b = 2d$, we have
$$a + b = 2(c + d) \in S, \quad ab = 2(2cd) \in S, \quad -a = 2(-c) \in S,$$
and so S is a subring of $\mathbb{Z}$, and hence is a ring.

The theorem gives us three things to check. But we can reduce the number from three to two. We use $a - b$ as shorthand for $a + (-b)$. In the next proof we need to know that $-(-b) = b$. This holds for the following reason. We have, by (A3),
$$b(-b) = (-b) + b = 0$$
so that b is an additive inverse of $-b$. Also, of course, $-(-b)$ is an additive inverse of $-b$. By the uniqueness of additive inverse, $-(-b) = b$, as required. In particular, $a - (-b) = a + (-(-b)) = a + b$.

Theorem 7.2 (Second Subring Test) *Let R be a ring, and let S be a non-empty Subset of R. Then S is a subring of R if the following condition holds:*

for all $a, b \in S$, we have $a - b, ab \in S$.

Proof: Let S satisfy this condition: that is, S is closed under subtraction and multiplication. We have to verify that it satisfies the conditions of the First Subring Test. Choose any element $a \in S$ (this is possible since S is non-empty). Then the hypothesis of the theorem shows that $0 = a - a \in S$. Applying the hypothesis again shows that $- a = 0 - a \in S$. Finally, if $a, b \in S$, then $- b \in S$ (by what has just been proved), and so $a + b = a - (-b) \in S$. So we are done.

Cosets

Suppose that S is a subring of R. We now define a partition of R, one of whose parts is S. Remember that, by the Equivalence Relation Theorem, in order to specify a partition of R, we must give an equivalence relation on R.

Let $\equiv S$ be the relation on R defined by the rule

$a \equiv S\, b$ if and only if $b - a \in S$.

We claim that $\equiv S$ is an equivalence relation.

Reflexive: for any $a \in R$, $a - a = 0 \in S$, so $a \equiv S a$.

Symmetric: take $a, b \in R$ with $a \equiv S\, b$, so that $b - a \in S$. Then $a - b = -(b - a) \in S$, so $b \equiv S\, a$.

Transitive: take $a, b, c \in R$ with $a \equiv S\, b$ and $b \equiv S\, c$. Then $b - a, c - b \in S$. So
$$c - a = (c - b) + (b - a) \in S, \text{ so } a \equiv S\, c.$$

So $\equiv S$ is an equivalence relation. Its equivalence classes are called the *cosets* of S in R.

Example. Let n be a positive intger. Let $R = \mathbb{Z}$ and $S = n\mathbb{Z}$, the set of all multiples of n. Then S is a subring of R. (By the Second Subring Test, if $a, b \in S$, say $a = nc$ and $b = nd$, then $a - b = n(c - d) \in S$ and $ab = n(ncd) \in S$). In this case, the relation $\equiv S$ is just congruence mod n, since $a \equiv_S b$ if and only if $b - a$ is a multiple of n. The cosets of S are thus precisely the congruence classes mod n.

An element of a coset is called a *coset representative*. As we saw in the first chapter, it is a general property of equivalence, relations that any element can be used as the coset representative: if b is in the same equivalence class as a, then a and b define the same equivalence classes. We now give a description of cosets.

If S is a subset of R, and $a \in R$, we define $S + a$ to be the set
$$S + a = \{S + a : S \in S\}$$
consisting of all elements that we can get by adding a to an element of S.

Proposition 7.9. Let S be a subring of R, and $a \in R$. Then the coset of R containing a is $S + a$.

Proof: Let $[a]$ denote the coset containing a, that is,
$$[a] = \{b \in R : a \equiv_S b\} = \{b \in R : b - a \in S\}.$$
We have to show that $[a] = S + a$.

First take $b \in [a]$, so that $b - a \in S$. Let $s = b - a$. Then $b = s + a \in S + a$.

In the other direction, take $b \in S + a$, so that $b = s + a$ for some $S \in S$. Then $b - a = (s + a) - a = s \in S$, so $b \equiv_S a$, that is, $b \in [a]$.

So $[a] = S + a$, as required.

Any element of a coset can be used as its representative. That is, if $b \in S + a$, then $S + a = S + b$.

Here is a picture.

		R		
• 0		• a		
		$S + a$		
S		=		
		$S + b$		
		• b		

Note that $S + 0 = S$, so the subring S is a coset of itself, namely the coset containing 0.

In particular, the congruence class $[a]_n$ in $\mathbb{Z}$ is the coset $n\mathbb{Z} + a$, consisting of all elements obtained by adding a multiple of n to a. So the ring $\mathbb{Z}$ is partitioned into n cosets of $n\mathbb{Z}$.

7.3 HOMOMORPHISMS AND QUOTIENT RINGS

Isomorphism

Here are the addition and multiplication tables of a ring with two elements, which for now we will call o and i.

+	o	i
o	o	i
i	i	o

·	o	i
o	o	o
i	o	i

You may recognise this ring in various guises: it is the Boolean ring $P(X)$, where $X = \{x\}$ is a set with just one element x; we have $o = \emptyset$ and $i = \{x\}$. Alternatively it is the ring of integers mod 2, with $o = [0]_2$ and $i = [1]_2$.

The fact that these two rings have the same addition and multiplication tables shows that, from an algebraic point of view, we cannot distinguish between them.

We formalise this as follows. Let R_1 and R_2 be rings. Let $\theta : R_1 \to R_2$ be a function which is one-to-one and onto, that is, a bijection between R_1 and R_2. Now we denote the result of applying the function θ to an element $r \in R_1$ by $r\theta$ or $(r)\theta$ rather than by $\theta(r)$; that is, we write the function on the right of its argument.

Now we say that θ is an *isomorphism* from R_1 to R_2 if it is a bijection which satisfies

$$(r_1 + r_2)\theta = r_1\theta + r_2\theta, \quad (r_1 r_2)\theta = (r_1\theta)(r_2\theta) \quad \ldots(1)$$

This means that we "match up" elements in R_1 with elements in R_2 so that addition and multiplication work in the same way in both rings.

Example. To return to our earlier example, let $R_1 = P(\{x\})$ and let R_2 be the ring of integers mod 2, and define a function $\theta : R_1 \to R_2$ by

$$\emptyset\theta = [0]_2, \quad \{x\} = \theta = [1]2$$

Then θ is an isomorphism.

We say that the rings R_1 and R_2 are "isomorphic" if there is an isomorphism from R_1 to R_2. The word "isomorphic" means, roughly speaking, "the same shape": if two rings are isomorphic then they can be regarded as identical from the point of view of Ring Theory, even if their actual elements are quite different (as in our example). We could say that Ring Theory is the study of properties of rings which are the same in isomorphic rings.

So, for example, if R_1 and R_2 are isomorphic then:

- If R_1 is commutative, then so is R_2, and *vice versa*; and the same holds for the property of being a ring with identity, a division ring, a Boolean ring, a zero ring, etc.
- However, the property of being a ring of matrices, or a ring of polynomials, etc., are not necessarily shared by isomorphic rings.

We use the notation $R_1 \cong R_2$ to mean "R_1 is isomorphic to R_2". Remember that isomorphism is a relation between two rings. If you are given two rings R_1 and R_2 and asked whether they are isomorphic, **do not** say "R_1 is isomorphic but R_2 is not".

Homomorphisms

An isomorphism is a function between rings with two properties: it is a bijection (one-to-one and onto), and it preserves addition and multiplication (as expressed by equation (1)). A function which preserves addition and multiplication but is not necessarily a bijection is called a homomorphism. Thus, a *homomorphism* from R_1 to R_2 is a function $\theta : R_1 \to R_2$ satisfying

$$(r_1 + r_2)\theta = r_1\theta + r_2\theta, \quad (r_1 r_2)\theta = (r_1\theta)(r_2\theta).$$

You should get used to these two long words, and two others. A function $\theta : R_1 \to R_2$ is

- a *homomorphism* if it satisfies (1); (homo=similar)
- a *monomorphism* if it satisfies (1) and is one-to-one; (mono=one)

- an *epimorphism* if it satisfies (1) and is onto; (epi=onto)
- an *isomorphism* if it satisfies (1) and is one-to-one and onto (iso=equal)

For example, the function from the ring $\mathbb{Z}$ to the ring of integers mod 2, which takes the integer n to its congruence class $[n]_2$ mod 2, is a homomorphism. Basically this says that, if we only care about the parity of an integer, its congruence mod 2, then the addition and multiplication tables are

+	even	odd
even	even	odd
odd	odd	even

·	even	odd
even	even	even
odd	even	odd

and this ring is the same as the one at the start of this section.

Let $\theta : R_1 \to R_2$ be a homomorphism. The image of θ is, as usual, the set
$$\mathrm{Im}(\theta) = \{s \in R_2 : s = r\theta \text{ for some } r \in R_1\}.$$

We define the *kernel* of θ to be the set
$$\mathrm{Ker}(\theta) = \{r \in R_1 : r\theta = 0\},$$

the set of elements of R_1 which are mapped to the zero element of R_2 by θ. You will have seen a definition very similar to this in Linear Algebra.

The image and kernel of a homomorphism have an extra property. This is not the final version of this theorem: we will strengthen it in two ways in the next two sections. First, a lemma:

Lemma 7.1 Let $\theta : R_1 \to R_2$ be a homomorphism. Then

(a) $0\theta = 0$;

(b) $(-a)\theta = -(a\theta)$ for all $a \in R_1$;

(c) $(a-b)\theta = a\theta - b\theta$ for all $a, b \in R_1$.

Proof: We have
$$0 + 0\theta = 0\theta = (0+0)\theta = 0\theta + 0\theta,$$
and the cancellation law gives $0\theta = 0$.

Then
$$a\theta + (-a)\theta = (a-a)\theta = 0\theta = 0,$$
so $(-a)\theta$ is the additive inverse of $a\theta$, that is, $(-1)\theta = -(a\theta)$.

Finally, $(a-b)\theta = a\theta + (-b)\theta = a\theta - b\theta$.

Proposition 7.10 Let $\theta : R_1 \to R_2$ be a homomorphism. Then

(a) $\mathrm{Im}(\theta)$ is a subring of R_2;

(b) $\mathrm{Ker}(\theta)$ is a subring of R_1.

Proof: We use the Second Subring Test.

(a) First notice that $\mathrm{Im}(\theta) \neq 0$, since $\mathrm{Im}(\theta)$ contains 0, by the Lemma.

Take $a, b \in \mathrm{Im}(\theta)$, say, $a = x\theta$ and $b = y\theta$. Then $-b = (-y)\theta$, so
$$a - b = x\theta + (-y)\theta = (x-y)\theta \in \mathrm{Im}(\theta)$$
Also $ab = (x\theta)(y\theta) = (xy)\theta \in \mathrm{Im}(\theta)$. So $\mathrm{Im}(\theta)$ is a subring of R_2.

(b) First notice that Ker(θ) $\neq \emptyset$, since Ker(θ) contains 0, by the Lemma. Take $a, b \in$ Ker(θ), so that $a\theta = b\theta = 0$. Then
$$(a - b)\theta = a\theta - b\theta = 0 - 0 = 0,$$
$$(ab)\theta = (a\theta)(b\theta) = 0 \cdot 0 = 0,$$
so Ker(θ) is a subring.

Ideals

An ideal in a ring is a special kind of subring.

Let S be a subring of R. We say that S is an *ideal* if, for any $a \in S$ and $r \in R$, we have $ar \in S$ and $ra \in S$.

For example, let $R = \mathbb{Z}$ and $S = n\mathbb{Z}$ for some positive integer n. We know that S is a subring of R. Choose $a \in S$, say $a = nc$ for some $c \in \mathbb{Z}$. Then $ar = ra = n(cr) \in S$. So S is an ideal.

Any ring R has two trivial ideals: the whole ring R is an ideal; and the set $\{0\}$ consisting only of the zero element is an ideal.

There is an ideal test similar to the subring tests. We give just one form.

Theorem 7.3 (Ideal Test) Let R be a ring, and S a non-empty subset of R. Then S is an ideal if the following conditions hold:

(a) for all $a, b \in S$, we have $a - b \in S$;

(b) for all $a \in S$ and $r \in R$, we have $ar, ra \in S$.

Proof: Take $a, b \in S$. Then $ab \in S$ (this is a special case of (b), with $r = b$). So by the Second Subring Test, S is a subring. Then by (b), it is an ideal.

Now we can strengthen the statement that the kernel of a homomorphism is a subring.

Proposition 7.11 Let $\theta : R_1 \to R_2$ be a homomorphism. Then Ker(θ) is an ideal in R_1.

Proof: We already know that it is a subring, so we only have to check the last part of the definition. So take $a \in$ Ker(θ) (so that $a\theta = 0$), and $r \in R_1$. Then
$$(ar)\theta = (a\theta)(r\theta) = 0(r\theta) = 0,$$
and similarly $(ra)\theta = 0$. So $ar, ra \in$ Ker(θ).

We will see in the next section that it goes the other way too: every ideal is the kernel of a homomorphism. So "ideals" are the same thing as "kernels of homomorphisms".

Quotient Rings

Let I be an ideal of a ring R. We will define a ring, which we call the *quotient ring* or *factor ring*, of R by I, and denote by R/I.

The elements of R/I are the cosets of I in R. Thus each element of R/I is a set of elements (an equivalence class) of R. Remember that each coset can be written as $I + a$ for some $a \in R$. Now we have to define addition and multiplication. We do this by the rules
$$(I + a)(I + b) = I + (a + b)$$
$$(I + a)(I + b) = I + ab.$$

There is one important job that we have to do to prove that this is a good definition. Remember that any element of a coset can be used as a representative. So you might use the representatives a and b, while I use the representatives a' and b' for the same cosets. We need to show that the definitions do not depend on these choices; that is, we have to show that

$I + a = I + a'$ and $I + b = I + b'$ imply $I + (a + b) = I + (a' + b')$ and $I + ab = I + a'b'$

So suppose that $I + a = I + a'$ and $I + b = I + b'$. Then $a' \in I + a$, so $a' = s + a$ for some $s \in I$. Similarly, $b' = t + b$ for some $t \in I$. Now

$$a' + b' = (s + a)(t + b) = (s + t) + (a + b) \in I + (a + b)$$
$$a'b' = (s + a)(t + b) = st + sb + ta + ab \in I + ab$$

by using the associative and commutative laws for addition and the distributive laws. So the result is proved, once we justify the last step by showing that $s + t \in I$ and $st + sb + at \in I$. Remember that $s, t \in I$, so that $s + t \in I$ (as I is a subring); also $st \in I$ (since I is a subring) and $sb \in I$ and $at \in I$ (since I is an ideal), so the sum of these three expressions is in I.

Proposition 7.12. If I is an ideal of the ring R, then the set R/I, with operations of addition and multiplication defined as above, is a ring, and the map $\theta : R \to R/I$ defined by $r\theta = I + r$ is a homomorphism whose kernel is I.

Proof: We have well-defined operations of addition and multiplication, so (A0) and (M0) hold. The proofs of the other axioms are all very similar. Here is a proof of the first distributive law. Take three elements of R/I (that is, three cosets!), say $I + a, I + b, I + c$. Then

$$((I + a) + (I + b))(I + c) = (I + (a + b))(I + c)$$
$$= I + (a + b)c$$
$$= I + (ac + bc)$$
$$= (I + ac) + (I + bc)$$
$$= (I + a)(I + c) + (I + b)(I + c).$$

Here we use the distributive law in R to get from the second line to the third, while the other steps just use the definitions of addition and multiplication in R/I.

Next we show that θ is a homomorphism. This is true by definition:

$$(a + b)\theta = (I + a) + (I + b) = I + (a + b) = (a + b)\theta$$
$$(ab)\theta = (I + a)(I + b) = I + (ab) = (ab)\theta.$$

Finally we calculate $\text{Ker}(\theta)$. There is one important thing to note. The zero element of R/I is the coset $I + 0$. This is just the ideal I itself! So

$$\text{Ker}(\theta) = \{a \in R : a\theta = 0\} = \{a \in R : I + a = I\} = I,$$

since $I + a = I$ means that a is a representative for the coset I, that is, $a \in I$.

The map θ in this result is called the *natural homomorphism* from R to R/I. We see that, if I is any ideal of R, then I is the kernel of the natural homomorphism from R to R/I.

The Isomorphism Theorems

The Isomorphism Theorems are a number of results which look more closely at a homomorphism. The first one makes more precise the results we saw earlier about the image and kernel of a homomorphism.

Theorem 7.4 (First Isomorphism Theorem) Let R_1 and R_2 be rings, and let $\theta : R_1 \to R_2$ be a homomorphism. Then

(a) $\text{Im}(\theta)$ is a subring of R_2;

(b) $\text{Ker}(\theta)$ is an ideal of R_1;

(c) $R_1/\text{Ker}(\theta) \cong \text{Im}(\theta)$.

Proof: We already proved the first two parts of this theorem, in theorem 7.3 and proposition 7.11. We have to prove (c). Remember that this means that the rings $R_1/\text{Ker}(\theta)$ (the quotient ring, which is defined because $\text{Ker}(\theta)$ is an ideal in R_1) and $\text{Im}(\theta)$ (a subring of R_2) are isomorphic. We have to construct a map ϕ between these two rings which is one-to-one and onto, and is a homomorphism.

Put $I = \text{Ker}(\theta)$, and define ϕ by the rule

$$(I + r)\phi = r\theta$$

for $r \phi R_1$. On the face of it, this might depend on the choice of the coset representative r. So first we have to prove that, if $I + r = I + r'$, then $r\theta = r'\theta$. We have

$$I + r = I + r' \Rightarrow r' = s + r \text{ for some } s \in I = \text{Ker}(\theta)$$
$$\Rightarrow r'\theta = s\theta + r\theta = 0 + r\theta = r\theta$$

as required. So indeed ϕ is well defined.

In fact this argument also reverses. If $r\theta = r'\theta$, then $(r' - r)\theta = r'\theta - r\theta = 0$, so $r' - r \in \text{Ker}(\theta)$. This means, by definition, that r and r' lie in the same coset of $\text{Ker}(\theta) = I$, so that $I + r = I + r'$. This shows that ϕ is one-to-one.

To show that ϕ is onto, take $s \in \text{Im}(\theta)$. Then $s = r\theta$ for some $r \in R$, and we have $s = r\theta = (I + r)\phi$. So $\text{Im}(\theta) = \text{Im}(\theta)$ as required.

Finally,

$$((I + r_1) + (I + r_2))\phi = (r_1 + r_2)\theta = (r_1\theta) + (r_2\theta) = (I + r_1)\phi + (I + r_2)\phi,$$
$$((I + r_1)(I + r_2))\phi = (r_1 r_2)\theta = (r_1\theta)(r_2\theta) = (I + r_1)(I + r_2)\phi,$$

so ϕ is a homomorphism, and hence an isomorphism, as required.

Homomorphisms and Quotient Rings

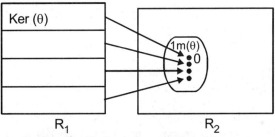

Figure 7.1: A homomorphism

We illustrate this theorem with a picture.

In the picture, the parts into which R_1 is divided are the cosets of the ideal $\text{ker}(\theta)$ (the set $\text{Ker}(\theta)$ itself has been taken to be the top part of the partition). The oval region inside R_2 is the subring $\text{Im}(\theta)$. Each coset of $\text{Ker}(\theta)$ maps to a single element of $\text{Im}(\theta)$.

The second Isomorphism Theorem is sometimes called the "Correspondence Theorem", since it says that subrings of R/I correspond in a one-to-one manner with subrings of R containing I.

Theorem 7.5 (Second Isomorphism Theorem) Let I be an ideal of the ring R. Then there is a one-to-one correspondence between the subrings of R/I and the subrings of R containing I, given as follows: to a subring S of R containing I corresponds the subring S/I of R/I. Under this correspondence, ideals of R/I correspond to ideals of R containing I; and, if J is an ideal of R containing I, then

$$(R/I)/(J/I) \cong R/J$$

Proof: If S is a subring of R containing I, then I is an ideal of S. (For applying the ideal test inside S means we have to check that I is closed under subtraction and under multiplication by elements of S; these are just some of the checks that would be required to show that it is an ideal of R. Now if $s \in S$, then the entire coset I + s lies in S, since S is closed under addition. So S/I is well defined: it consists of all the cosets of I which are contained in S. Clearly it is a subring of R/I. Thus, we have a mapping from subrings of R containing I to subrings of R/I.

In the other direction, let T be a subring of R/I. This means that T is a set of cosets of I which form a ring. Let S be the *union* of all the cosets in T. We will show that S is a subring of R. It obviously contains I (since I is the zero coset) and S/I = T follows.

Take $a, b \in S$. Then $I + a, I + b \in T$. Since T is a subring, we have $(I + a) - (I + b) = I + (a - b) \in T$ and $(I + a)(I + b) = I + ab \in T$, so $a - b \in S$ and $ab \in S$. By the Second Subring Test, S is a subring.

Next we show that ideals correspond to ideals. Let J be an ideal of R containing I. Then J/I is a subring of R/I, and we have to show that it is an ideal. Take $I + a \in J/I$ and $/ + r \in R/I$. Then $a \in J$ and $r \in R$, so $ar, ra \in J$, whence $(I + a)(I + r), (I + r)(I + a) \in J/I$. Thus J/I is an ideal of R/I. The converse is similar.

I will not give the proof that $(R/I)/(J/I) \cong R/J$: this will not be used in the course.

The Third Isomorphism Theorem needs a little more notation. Let A and B be two subsets of a ring R. Then we define A + B to consist of all sums of an element of A and an element of B:

$$A + B = \{a + b; a \in A, b \in B\}$$

Theorem 7.6 (Third Isomorphism Theorem) Let R be a ring, S a subring of R, and I an ideal of R. Then

(a) S + I is a subring of R containing I;

(b) S ∩ I is an ideal of S;

(c) $S/(S \cap I) \cong (S + I)/I$.

Proof: We could prove the three parts in order, but it is actually easier to start at the end! Remember the natural homomorphism θ from R to R/I with kernel θ. What happens when we restrict θ to S, that is, we only put elements of S into the function θ? Let θ denote this restriction. Then φ maps S to R/I. We find its image and kernel, and apply the First Isomorphism Theorem to them.

(a) The image of φ consists of all cosets I + s containing a coset representative in S. The union of all these cosets is I + S, so the image of φ is (I + S)/I. This is a subring of R/I (since it is the image of a homomorphism). By the Correspondence Theorem, S + I is a subring of R containing I.

(b) The kernel of φ consists of all elements of S mapped to zero by φ, that is, all elements $s \in S$ such that $s \in \text{Ker}(\theta) = I$. Thus, $\text{Ker}(\phi) = S \cap I$, and so $S \cap I$ is an ideal of S.

(c) Now the first isomorphism theorem shows that

$$S/(I + S) \cong \text{Im}(\phi) = (I + S)/I,$$

and we are done.

7.4 FACTORISATION

One of the most important properties of the "integers is that any number can be factorised into prime factors in a unique way. But we have to be a bit careful. It would be silly to try to factorise 0 or 1; and the factorisation is not quite unique, since (– 2) (–3) = 2 . 3, for example. Once we have the definitions straight, we will see that "unique factorisation" holds in a large class of rings.

Zero Divisors and Units

In this section, we will assume that our rings are always commutative.

Let R be a ring. We know that $0a = 0$ holds for all $a \in R$. It is also possible for the product of two non-zero elements of R to be zero. We say that a is a *zero-divisor* if

- $a \neq 0$, and
- there exists $b \in R$, with $b \neq 0$, such that $ab = 0$.

In other words, if the product of two non-zero elements is zero, then we call each of them a zero-divisor.

The ring $\mathbb{Z}$ has no zero-divisors, since if a and b are non-zero integers then obviously $ab \neq 0$. Also, a field has no zero divisors. For suppose that R is a field, and let a be a zero-divisor. Thus, $a \neq 0$, and there exists $b \neq 0$ such that $ab = 0$. Since R is a field, a has a multiplicative inverse a^{-1} satisfying $a^{-1} a = 1$. Then

$$0 = a^{-1} 0 = a^{-1}(ab) = (a^{-1} a)b = 1b = b,$$

contradicting our assumption that $b \neq 0$.

In the next example, we use the greatest common divisor function for integers: d is a greatest common divisor of a and b if it divides both of them, and if any other divisor of a and b also divides d. That is, 6 is a greatest common divisor of 12 and 18; but – 6 is also a greatest common divisor. We will live with this slight Awkwardness for a while, choosing gcd(a, b) to be the positive rather than the negative value.

Example 1: Let $R = \mathbb{Z}/n\mathbb{Z}$, the ring of integers mod n. Then the element $a \in R$ is a zero-divisor if and only if $1 < \gcd(a, n) < n$.

Proof: Suppose that a is a zero-divisor in R. This means that $a \neq 0$ in R (that is, a is not divisible by n, which shows that $\gcd(a, n) < n$), and there exists $b \in R$ with $b \neq 0$ and $ab = 0$. So, regarding a, b, n as integers, we have $n | ab$ but n does not divide either a or b. We are trying to prove that $\gcd(a, n) > 1$, so suppose (for a contradiction) that the greatest common divisor is 1. Since n and a are coprime, the fact that n divides ab means that n must divide b, which contradicts our assumption that $b \neq 0$ in R.

Conversely, suppose that $1 < d = \gcd(a, n) < n$. Then $a \neq 0$ as an element of R. Let $a = dx$ and $n = db$. Then n divides $nx = (db)x = (dx)b = ab$, but clearly n does not divide y. So, in the ring R, we have $ab = 0$ and $b \neq 0$. Thus a is a zero-divisor.

From now on we make another assumption about our rings: as well as being commutative, they will always have an identity element. We make a definition:

An *integral domain* is a commutative ring with identity which has no zero-divisors.

Example 2: $\mathbb{Z}$ is an integral domain. (This example is the "prototype" of an integral domain, and gives us the name for this class of rings.) Any field is an integral domain. The ring $\mathbb{Z}/n\mathbb{Z}$ is an integral domain if and only if n is a prime number. The last statement is true because a positive integer n has the property that every smaller positive integer a satisfies $\gcd(a, n) = 1$ if and only if n is prime.

Example 3: If R is an integral domain, then so is the ring R[x] of polynomials over R.

For suppose that f and g are non-zero polynomials, with degrees m and n respectively: that is,

$$f(x) = \sum_{i=0}^{n} a_i x^i, \quad g(x) = \sum_{i=0}^{m} b_i x^i$$

where $a_n \neq 0$ and $b_m \neq 0$. The coefficient of x^{m+n} in $f(x)g(x)$ is $a_n b_m \neq 0$ (because R is an integral domain). So $f(x)g(x) \neq 0$.

Let R be a ring with identity element 1; we assume that $1 \neq 0$. Let $a \in R$, with $a \neq 0$. An inverse of a is an element $b \in R$ such that $ab = ba = 1$. We say that a is a unit if it has an inverse. (We exclude zero because obviously 0 has no inverse: $0b = 0$ for any element b.)

An element a has at most one inverse. For suppose that b and c are inverses of a. Then
$$b = b1 = b(ac) = (bd)c = ac = c.$$

We write the inverse of the unit a as a^{-1}. Furthermore, a zero-divisor cannot be a unit. For, if $ba = 1$ and $ac = 0$, then
$$0 = b0 = b(ac) = (bc)c = 1c = c.$$

Lemma 7.2 Let R be a ring with identity. Then

(a) 1 is a unit;

(b) if u is a unit then so is u^{-1};

(c) if u and v are units then so is uv.

Proof: (a) $1.1 = 1$.

(b) The equations $uu^{-1} = u^{-1}u = 1$ show that the inverse of u^{-1} is u.

(c) Let u and v be units. We claim that the inverse of uv is $v^{-1}u^{-1}$. (Note the reverse order!) For we have
$$(uv)(v^{-1}u^{-1}) = u(vv^{-1})u^{-1} = u1u^{-1} = uu^{-1} = 1.$$
$$(v^{-1}u^{-1})(uv) = v^{-1}(u^{-1}u)v = v^{-1}1v = v^{-1}v = 1.$$

To help you remember that you have to reverse the order when you find the inverse of a product, this example may help. Suppose that u is the operation of putting on your socks, and v the operation of putting on your shoes, so that uv means "put on your socks and then your shoes". What is the inverse of uv?

Example 4: In the integral domain $\mathbb{Z}$, the only units are $+1$ and -1. For if $ab = 1$, then $a = 1$ or $a = -1$.

Example 5: Consider the ring $\mathbb{Z}/n\mathbb{Z}$, where $n > 1$. We already saw that a is a zero-divisor if and only if $1 < \gcd(a, n) < n$. We claim that a is a unit if and only if $\gcd(a, n) = 1$.

Suppose first that a is a unit, and that $d = \gcd(a, n)$. Then $d \mid a$ and $d \mid n$. Let b be the inverse of a, so that $ab = 1$ in R, which means that $ab = 1 \pmod{n}$, or $ab = xn + 1$. But then d divides ab and d divides xn, so d divides 1, whence $d = 1$.

To prove the converse, we use the Euclidean algorithm (more about this shortly), which shows that, given any two integers a and n, there are integers x and y such that $xa + yn = d$, where $d = \gcd(a, n)$. If $d = 1$, then this equation shows that $xa \equiv 1 \pmod{n}$, so that $xa = 1$ in $\mathbb{Z}/n\mathbb{Z}$, so that a is a unit.

This shows that every non-zero element of $\mathbb{Z}/n\mathbb{Z}$ is either a zero-divisor or a unit.

For example, for $n = 12$, we have:

1	unit	$1 \cdot 1 = 1$
2	zero-divisor	$2 \cdot 6 = 0$
3	zero-divisor	$3 \cdot 4 = 0$
4	zero-divisor	$4 \cdot 3 = 0$
5	unit	$5 \cdot 5 = 1$
6	zero-divisor	$6 \cdot 2 = 0$
7	unit	$7 \cdot 7 = 1$
8	zero-divisor	$8 \cdot 3 = 0$
9	zero-divisor	$9 \cdot 4 = 0$
10	zero-divisor	$10 \cdot 6 = 0$
11	unit	$11 \cdot 11 = 1$

We call two elements $a, b \in R$ *associates* if there is a unit $u \in R$ such that $b = ua$. Write $a \sim b$ to mean that a and b are associates. Thus, any unit is an associate of 1, while 0 is associate only to itself.

Being associates is an equivalence relation: it is
- reflexive since $a = a1$ and 1 is a unit;
- symmetric since, if $b = au$, then $a = bu^{-1}$, and u^{-1} is a unit;
- transitive since, if $b = au$ and $c = bv$ where u and v are units, then $c = a(uv)$, and uv is a unit.

Here we have invoked the three parts of the lemma above about units.

For example, in the ring $\mathbb{Z}/12\mathbb{Z}$, the associate classes are

$\{0\}$, $\{1, 5, 7, 11\}$, $\{2, 10\}$, $\{3, 9\}$ $\{4, 8\}$ $\{6\}$.

For example, the associate class containing 2 consists of 2, 2 . 5 = 10, 2 . 7 = 2, and 2 . 11 = 10.

Now we can define greatest common divisors properly.

Let R be an integral domain. (Remember: this means that R is a commutative ring with identity and has no divisors of zero.) We say that *a divides b in* R (written as usual as $a \mid b$) if there exists $x \in R$ with $b = ax$. Notice that every element divides 0, whereas 0 does not divide anything else except 0. Also, 1 divides any element of R, but the only elements which divide 1 are the units of R. [Check all these claims!]

Proposition 7.13 In an integral domain R, two elements a and b are associates if and only if $a \mid b$ and $b \mid a$.

Proof: Suppose that a and b are associates. Then $b = au$ for some unit u, so $a \mid b$. Also $a = bu^{-1}$, so $b \mid a$.

Conversely, suppose that $a \mid b$ and $b \mid a$. If $a = 0$, then also $b = 0$ and a, b are associates. So suppose that $a \neq 0$. Then there are elements x and y such that $b = ax$ and $a = by$. We have $axy = a$, so $a(1 - xy) = 0$. Since R is an integral domain and $a \neq 0$, we must have $1 - xy = 0$, or $xy = 1$. So x and y are units, and a and b are associates.

Now we say that d is a *greatest common divisor* of a and b if

- $d \mid a$ and $d \mid b$;
- if e is any element such that $e \mid a$ and $e \mid b$, then $e \mid b$.

We abbreviate "greatest common divisor" to *gcd*.

Notice that, in general, "greatest" does not mean "largest" in any obvious way. Both 6 and –6 are greatest common divisors of 12 and 18 in $\mathbb{Z}$, for example.

Proposition 7.14 If d is a *gcd* of two elements a, b in an integral domain R, then another element d' is a *gcd* of a and b if and only if it is an associate of d.

Proof: Suppose first that d and d' are both *gcds* of a and b. Then $d' \mid d$ and $d \mid d'$ (using the second part of the definition), so that d and d' are associates.

Conversely, suppose that d is a *gcd* of a and b (say $a = dx$ and $b = dy$), and d' an associate of d, say $d' = du$ for some unit u. Then

- $a = d'u^{-1}x$ and $b = d'u^{-1}y$, so $d' \mid a$ and $d' \mid b$;
- suppose that $e \mid a$ and $e \mid b$. Then $e \mid d$, say $d = ez$; so we have $d' = eu^{-1}z$ and $e \mid d'$.

Thus d' is a *gcd* of a and b.

Thus we can say: the greatest common divisor of a and b, if it exists, is "unique up to associate", that is, any two *gcds* are associates. We use the notation *gcd* (a, b) to denote some (unspecified) greatest common divisor. In the integers, we can make the convention that we choose the non-negative element of the associate pair as the *gcd*.

7.4.2 Unique Factorisation Domains

We are interested in the property of "unique factorisation" of integers, that is, any integer other than 0, +1, –1 can be uniquely factorised into primes. Of course, the factorisation is not quite unique, for two reasons:

(a) the multiplication is commutative, so we can change the order: $6 = 2 \cdot 3 = 3 \cdot 2$.

(b) we will see that -2 and -3 also count as "primes", and $6 = 2.3 = (-2) \cdot (-3)$.

By convention, 1 is not a prime, since it divides everything. The same holds for -1 (and only these two integers, since they are the only units in $\mathbb{Z}$.) Accordingly, we will specify that irreducible elements (the analogue of primes in a general domain) should not be zero or units, and that we only try to factorise elements which are not zero or a unit.

So we make the following definitions. Let R be an integral domain.

(a) An element $p \in R$ is *irreducible* if p is not zero or a unit, but whenever $p = ab$, then one of a and b is a unit (and the other therefore an associate of p).

(b) R is a *unique factorisation domain* if it has the following properties:
- every element $a \in R$ which is not zero or a unit can be written as a product of irreducibles;
- if $p_1, \ldots, p_m, q_1, \ldots, q_n$ are irreducibles and
$$p_1 p_2 \cdots p_m = q_1 q_2 \cdots q_n$$
then $m = n$ and, after possibly permuting the factors in one product, p_i and q_i are associates for $i = 1, \ldots, m$.

Note that, if an element p is irreducible, then so is every associate of p. If the second condition in the definition of a unique factorisation holds, we say that "factorisation is unique up to order and associates". As we saw, this is the best we can expect in terms of unique factorisation!

The ring $\mathbb{Z}$ is a unique factorisation domain; so is the ring F[x] of polynomials over any field F. We will prove these things later on; we will see that it is the Euclidean algorithm which is crucial to the proof, and the integers and polynomials over a field both have a Euclidean algorithm.

Note that, to decide whether a ring is a unique factorisation domain, we have first to check that it really is an integral domain, and second to find all the units (so that we know when two elements are associates).

Example: Here is an example of a ring which is an integral domain but not a unique factorisation domain. Let

$$\{a + b\sqrt{-5} : a, b \in \mathbb{Z}\}\}$$

We show first that R is a subring of $\mathbb{C}$. Take two elements of R, say $r = a + b\sqrt{-5}$ and $s = c + d\sqrt{-5}$, with $a, b, c, d \in \mathbb{Z}$. Then

$$r - s = (a - c) + (b - d)\sqrt{-5} \in R,$$
$$rs = (ac - 5bd) + (ad + bc)\sqrt{-5} \in R,$$

since $a - c, b - d, ac - 5bc, ad + bc \in \mathbb{Z}$. So the Subring Test applies.

R is clearly an integral domain: there do not exist two nonzero complex numbers whose product is zero.

What are the units of R? To answer this, we use the fact that $\left|a+b\sqrt{-5}\right|^2 = a^2 + 5b^2$.
Now suppose that $a+b\sqrt{-5}$ is a unit, say
$$\left(a+b\sqrt{-5}\right)\left(c+d\sqrt{-5}\right) = 1$$
Taking the modulus and squaring gives $(a^2 + 5b^2)(c^2 + 5d^2) = 1$

So $a^2 + 5b^2 = 1$ (it can not be -1 since it is positive). The only solution is $a = \pm 1$, $b = 0$. So the only units are ± 1, and so r is associate only to r and $-r$.

Now we show that 2 is irreducible. Suppose that
$$2 = \left(a+b\sqrt{-5}\right)\left(c+d\sqrt{-5}\right)$$
Taking the modulus squared again gives
$$4 = (a^2 + 5b^2)(c^2 + 5d^2)$$
So $a^2 + 5b^2 = 1$, 2 or 4. But the equation $a^2 + 5b^2 = 2$ has no solution, while $a^2 + 5b^2 = 1$ implies $a = \pm 1$, $b = 0$, and $a^2 + 5b^2 = 4$ implies $c^2 + 5d^2 = 1$, so that $c = \pm 1$, $d = 0$. So the only factorisations are
$$2 = 2 \cdot 1 = 1 \cdot 2 = (-2) \cdot (-1) = (-2) \cdot (-1) :$$
in each case, one factor is a unit and the other is an associate of 2.

In a similar way we can show that 3, $1 + \sqrt{-5}$ and $1 - \sqrt{-5}$ are irreducible. Now consider the factorisations

These are two factorisations into irreducibles, which are not equivalent up to order and associates. So R is not a unique factorisation domain!

Principal Ideal Domains

Let R be a commutative ring with identity. We denote by aR, or by (a), the set $\{ar : r \in R\}$ of all elements divisible by a.

Lemma 7.3 (a) is an ideal of R containing a, and if I is any ideal of R containing a then $(a) \subseteq I$.

Proof: We apply the Ideal Test. If $ar_1, ar_2 \in (a)$, then
$ar_1 - ar_2 = a(r_1 - r_2) \in (a)$.
Also, if $ar \in (a)$ and $x \in R$, then
$(ar)x = a(rx) \in (a)$.
So (I) is an ideal.

Since R has an identity element 1, we have $a = a1 \in (a)$.

Finally, if I is any ideal containing a, then (by definition of an ideal) we have $ar \in 1$ for any $r \in R$; that is, $(a) \subseteq I$.

Lemma 7.4 Let R be an integral domain. Then $(a) = (b)$ if and only if a and b are associates.

Proof: $(a) = (b)$ means, by definition, that each of a and b is a multiple of the other, that is, they are associates.

We call (a) the *ideal generated by* a and say that it is a *principal ideal*.

More generally, if $a_1, ..., a_n \in R$ (where R is a commutative ring with identity, then we let
$$(a_1,...,a_n) = \{r_1 a_1 + ... + r_n a_n : r_1, ..., r_n \in R\}.$$
Then it can be shown, just as above, that $(a_1, ..., a_n)$ is an ideal of R containing $a_1, ..., a_n$, and that any ideal which contains these elements must contain $(a_1, ... a_n)$. We call this the *ideal generated by* $a_1, ..., a_n$.

A ring R is a *principal ideal domain* if *every* ideal is principal. We will see later that $\mathbb{Z}$ is a principal ideal domain.

Proposition 7.15 Let R be a principal ideal domain. Then any two elements of R have a greatest common divisor; in fact, $d = \gcd(a, b)$ if and only if $(a, b) = (d)$.

Proof: Suppose that R is a principal ideal domain. Then (a, b), the ideal generated by a and b, is a principal ideal, so it is equal to (d), for some $d \in R$. Now we claim that $d = \gcd(a, b)$.

- $a \in (d)$, so d/a. Similarly d/b.
- Also, $d \in (a, b)$, so $d = ua + vb$ for some $u, v \in R$?. Now suppose that e/a and e/b, say $a = ep$ and $b = eq$. Then $d = ua + vb = e(up + vq)$, so that e/d.

The claim is proved.

Since any two *gcds* of a and b are associates, and any two generators of (a, b) are associates, the result is proved.

Example. The ring $\mathbb{Z}$ is a principal ideal domain. That means that the only ideals in $\mathbb{Z}$ are the sets $(n) = n\mathbb{Z}$, for $n \in \mathbb{Z}$. We will deduce this from a more general result in the next section.

Now it is the case that any principal ideal domain is a unique factorisation domain. We will not prove all of this. The complete proof involves showing two things: any element which is not zero or a unit can be factorised into irreducibles; and any two factorisations of the same element differ only by order and associates. We will prove the second of these two assertions. See the appendix to this chapter for comments on the first.

Lemma 7.5 Let R be a principal ideal domain; let p be irreducible in R, and $a, b \in R$. If p/ab, then p/a or p/b.

Proof: Suppose that $p \mid ab$ but that p does not divide a. Then we have $\gcd(a, p) = 1$, and so there exist $u, v \in R$ with $1 = ua + vp$. So $b = uab + vpb$. But $p \mid uab$ by assumption, and obviously $p \mid vpb$; so p/b, as required.

This lemma clearly extends. If p is irreducible and divides a product $a_1 a_2 ... a_n$, then p must divide one of the factors. For either p/a_1 or $p/a_2 .... a_n$; in the latter case, proceed by induction.

Theorem 7.7 Let R be a principal ideal domain, and suppose that
$$a = p_1 p_2 .... p_m = q_1 q_2 ... q_n$$
where $p_1, ..., p_m, q_1, ..., q_n$ are irreducible. Then $m = n$ and, after possibly permuting the factors, p_i and q_i are associates for $i = 1, ..., m$.

Proof: Obviously p_1 divides $q_1 ... q_n$, so p_1 must divide one of the factors, say $p_1 \mid q_i$. Since p_1 and q_i are irreducible, they must be associates. By permuting the order of the qs

and adjusting them by unit factors, we can assume that $p_1 = q_1$. Then $p_2 \ldots p_m = q_2 \ldots q_n$, and we proceed by induction.

Example. Here is an example of an integral domain which is not a principal ideal domain. Consider the ring $R = \mathbb{Z}[x]$ of polynomials over the integers. Let I be the set of all such polynomials whose constant term is even. Then I is an ideal in R: if f and g are polynomials with even constant term, then so is $f - g$, and so is fh for any polynomial h. But I is not a principal ideal. For I contains both the constant polynomial 2 and the polynomial x of degree 1. If $I = (a)$, then a must divide both 2 and x, so $a = \pm 1$. But $\pm 1 \not\subset I$.

The polynomials 2 and x are both irreducible in R, and so their *gcd* is 1. But 1 cannot be written in the form $2u + xv$ for any polynomials u and v.

The ring $\mathbb{Z}[x]$ is a unique factorisation domain (see the Appendix to this chapter).

Euclidean Domains

Any two integers have a greatest common divisor, and we can use the Euclidean algorithm to find it. You may also have seen that the Euclidean algorithm works for polynomials we now give the algorithm in a very general form.

Let R be an integral domain. A *Euclidean function* on R is a function d from the set $R \setminus \{0\}$ (the set of non-zero elements of R) to the set $\mathbb{N}$ of non-negative integers satisfying the two conditions

(a) for any $a, b \in R$ with $a, b \neq 0$, we have $d(ab) \geq d(a)$;

(b) for any $a, b \in R$ with $b \neq 0$, there exist $q, r \in R$ such that
- $a = bq + r$;
- either $r = 0$, or $d(r) < d(b)$.

We say that an integral domain is a *Euclidean domain* if it has a Euclidean function.

Example. Let $R = \mathbb{Z}$, and let $d(a) = |a|$ for any integer a.

Example. Let $R = F[x]$; the ring of polynomials over F, where F is a field. For any non-zero polynomial $f(x)$, let $d(f(x))$ be the degree of the polynomial $f(x)$ (the index of the largest non-zero coefficient).

Both of these examples are Euclidean functions.

(a) In the integers, we have $d(ab) = |ab| = |a|.|b| \geq |a| = d(a)$, since $b \neq 0$. In the polynomial ring $F[x]$, we have
$$d(ab) = \deg(ab) = \deg(a) + \deg(b) \geq \deg(a),$$
since if the leading terms of a and b are $a_n x^n$ and $b_m x^m$ respectively then the leading term of ab is $a_n b_m x^{n+m}$.

(b) In each case this is the "division algorithm": we can divide a by b to obtain a quotient q and remainder r, where r is smaller than the divisor b as measured by the appropriate function d.

You will have seen how to use the Euclidean algorithm to find the greatest common divisor of two integers or two polynomials. The same method works in any Euclidean domain. It goes like this. Suppose that R is a Euclidean domain, with Euclidean function d. Let a and

b be any two elements of R. If $b = 0$, then $gcd(a, b) = a$. Otherwise, proceed as follows. Put $a = a_0$ and $b = a_1$. If a_{i-1} and a_i have been constructed, then
- if $a_i = 0$ then $gcd(a, b) = a_{i-1}$;
- otherwise, write $a_{i-1} = a_i q + r$, with $r = 0$ or $d(r) < d(a_i)$, and set $a_{i+1} = r$; repeat the procedure for a_i and a_{i+1}.

The algorithm terminates because, as long as $a_i \neq 0$, we have
$$d(a_i) < d(a_{i-1}) < \ldots < d(a_1)$$
Since the values of d are non-negative integers, this chain must stop after a finite number of steps.

To see that the result is correct, note that, if $a = bq + r$, then
$$gcd(a, b) = gcd(b, r)$$
(as an easy calculation shows: the common divisors of a and b are the same as the common divisors of b and r. So we have $gcd(a_{i-1}, a_i) = gcd(a, b)$ as long as a_i is defined. At the last step, $a_i = 0$ and so $gcd(a, b) = gcd(a_{i-1}, 0) = a_{i-1}$.

The algorithm can also be used to express $gcd(a,b)$ in the form $ua + vb$ for some $u, v \in R$. For a and b themselves are both expressible in this form; and, if $a_{i-1} = u_{i-1} a + v_{i-1} b$ and $a_i = u_i a + v_i b$ then with $a_{i-1} = qa_i + a_{i+1}$, we have
$$a_{i+1} = a_{i-1} - qa_i = (u_{i-1} - qu_i) a + (v_{i-1} - qv_i) b.$$

Example: Find $gcd(204, 135)$. We have
$$204 = 135.1 + 69,$$
$$135 = 69.1 + 66,$$
$$69 = 66.1 + 3,$$
$$66 = 3.22,$$
so $gcd(204,135) = 3$. To express $3 = 204u + 135v$, we have
$$69 = 204.1 - 135.1,$$
$$66 = 135.69 = 135.2 - 204.1,$$
$$3 = 69.66 = 204.2 - 135.3.$$

We will show that a Euclidean domain is a unique factorisation domain. First we need one lemma. Note that, if a and b are associates, then $b = au$, so $d(b) \geq d(a)$, and also $a = bu^{-1}$, so $d(a) \geq d(b)$, so we have $d(a) = d(b)$.

Lemma 7.6 Let R be a Euclidean domain. Suppose that a and b are non-zero elements of R such that $a \mid b$ and $d(a) = d(b)$. Then a and b are associates.

Proof: Let $a = bq + r$ for some q, r, as in the second part of the definition. Suppose that $r \neq 0$. Now $b = ac$ for some element c; so $a = acq + r$. Thus, $r = a(1 - cq)$, and since $r \neq 0$ we have $d(r) > d(a)$, contrary to assumption. So $r = 0$. Then $b \mid a$; since we are given that $a \mid b$, it follows that a and b are associates.

Theorem 7.8 (a) A Euclidean domain is a principal ideal domain,

(b) A Euclidean domain is a unique factorisation domain.

Proof: (a) Let R be a Euclidean domain, and let I be an ideal in R. If I = {0}, then certainly

$I = (0)$ and I is principal. So suppose that I is not $\{0\}$. Since the values of $d(x)$ for $x \in I$ are non-negative integers, there must be a smallest value, say $d(a)$. We will claim that $I = (a)$.

First, take $b \in (a)$, say $b = ax$. Then $b \in I$, by definition of an ideal.

Next, take $b \in I$. Use the second part of the definition of a Euclidean function to find elements q and r such that $b = aq + r$, with either $r = 0$ or $d(r) < d(a)$. Suppose that $r \neq 0$. Then $b \in I$ and $aq \in I$, so $r = b - aq \in I$; but $d(r) < d(a)$ contradicts the fact that $d(a)$ was the smallest value of the function d on the non-zero elements of I. So the supposition is impossible; that is, $r = 0$, and $b = aq \in (a)$.

So $I = (a)$ is a principal ideal.

(b) Again let R be a Euclidean domain. We show that any nonzero non-unit of R can be factorised into irreducibles. We showed in the last section that the factorisation is unique (because R is a principal ideal domain)

Choose any element $a \in R$ such that $a \neq 0$ and a is not a unit. We have to show that a can be factorised into irreducibles. The proof is by induction on $d(a)$; so we can assume that any element b with $d(b) < d(a)$ has a factorisation into irreducibles.

If a is irreducible, then we have the required factorisation with just one term. So suppose that $a = bc$ where b and c are not units. If $d(b) < d(a)$ and $d(c) < d(a)$ then, by induction, each of b and c has a factorisation into irreducibles; putting these together we get a factorisation of a. So suppose that $d(a) \geq d(b)$. We also have $d(b) \geq d(a)$, by the first property of a Euclidean function; so $d(a) = d(b)$. We also have $b \mid a$; by the Lemma before the Theorem, we conclude that a and b are associates, so that c is a unit, contrary to assumption.

Corollary 7.1 (a) *$\mathbb{Z}$ is a principal ideal domain and a unique factorisation domain.*

(b) *For any field F, the ring $F[x]$ of polynomials over F is a principal ideal domain and a unique factorisation domain.*

Proof: This follows from the theorem since we have seen that these rings are integral domains and have Euclidean function, and so are Euclidean domains.

Theorem 7.2 Euclidean domain $\Rightarrow$ principal ideal domain $\Rightarrow$ unique factorisation domain.

Proof: We proved most of this: we showed that a Euclidean domain is a principal ideal domain, and that in a principal ideal domain factorisations are unique if they exist. The proof that factorisations into irreducibles always exist in a principal ideal domain is a little harder.

Neither implication reverses. We saw that $\mathbb{Z}[x]$ is not a principal ideal domain, though it is a unique factorisation domain (see below). It is harder to construct a ring which is a principal ideal domain but not a Euclidean domain, though such rings do exist.

Another way to see the increasing strength of the conditions from right to left is to look at greatest common divisors.

- In a unique factorisation domain, any two elements a and b have a greatest common divisor d (which is unique up to associates).
- In a principal ideal domain, any two elements a and b have a greatest common divisor d (which is unique up to associates), and d can be written in the form $d = xa + yb$.

- In a Euclidean domain, any two elements a and b have a greatest common divisor d (which is unique up to associates), and d can be written in the form $d = xa + yb$; moreover, the gcd, and the elements x and y, can be found by the Euclidean algorithm.

You will also meet the theorem known as *Gauss's Lemma:*

Theorem 7.10 If R is a unique factorisation domain, then so is R[x].

This result shows that $\mathbb{Z}[x]$ is a unique factorisation domain, as we claimed above.

7.5 FIELDS

As you know from linear algebra, fields form a particularly important class of rings, since in linear algebra the scalars are always taken to form a field.

Although the ring with a single element 0 would technically qualify as a field according to our definition, we always rule out this case. Thus,

A field must have more than one element.

Another way of saying the same thing is that, in a field, we must have $1 \neq 0$. (If there is any element $x \neq 0$ in a ring with identity, then $1 \cdot x = x \neq 0 = 0 \cdot x$, and so $1 \neq 0$).

The "standard" examples of fields are the rational, real and complex numbers, and the integers mod p for a prime number p.

In this chapter, we will see how new fields can be constructed. The most important method of construction is *adjoining a root of a polynomial.* The standard example of this is the construction of $\mathbb{C}$ by adjoining the square root of -1 (a root of the polynomial $x^2 + 1 = 0$) to $\mathbb{R}$. We will also see that finite fields can be constructed in this way.

Also we can build fields as *fields of fractions;* the standard example is the construction of the rationals from the integers.

Maximal Ideals

In this chapter, R always denotes a commutative ring with identity. As above, we assume that the identity element 1 is different from the zero element 0: that is, $0 \neq 1$.

An ideal I of R is said to be proper if $I \neq R$. An ideal I is maximal if $I \neq R$ and there does not exist an ideal J with $I \subset J \subset R$; that is, any ideal J with $I \subseteq J \subseteq R$ must satisfy $J = I$ or $J = R$.

Lemma 7.7 Let R be a commutative ring with identity. Then R is a field if and only if it has no ideals except $\{0\}$ and R.

Proof: If $u \in R$ is a unit, then the only ideal containing u is the whole ring R. (For, given any ideal I with $u \in I$, and any $r \in R$, we have $r = u(u^{-1} r) \in I$, so $I = R$.) If R is a field, then every non-zero element is a unit, and so any ideal other than $\{0\}$ is R.

Conversely, suppose that the only ideals are 0 and R. We have to prove that multiplicative inverses exist (axiom (M3)). Take any element $a \in R$ with $a \neq 0$. Then $(a) = R$, so $1 \in (a)$. This means that there exists $b \in R$ with $ab = 1$, so $b = a^{-1}$ as required.

Proposition 7.16 Let F be a commutative ring with identity, and I a proper ideal of R. Then R/I is a field if and only if I is a maximal ideal.

Proof: By the Second Isomorphism Theorem, ideals of R/I correspond to ideals of R containing I. Thus, I is a maximal ideal if and only if the only ideals of R/I are zero and the whole ring, that is, R/I is a field (by the Lemma).

Proposition 7.17 Let R be a principal ideal domain., and I = (a) an ideal of R. Then

(a) I = R if and only if a is a unit;

(b) I is a maximal ideal if and only if a is irreducible.

Proof: (a) If a is a unit, then for any $r \in R$ we have $r = a(a^{-1} r) \in (a)$, so $(a) = R$. Conversely, if $(a) = R$, then $1 = ab$ for some $b \in R$, and a is a unit.

(b) Since R is a PID, any ideal containing (a) has the form (b) for some $b \in R$. Moreover, $(a) \subseteq (b)$ if and only if $b|a$. So (a) is maximal if and only if, whenever $b|a$, we have either b is a unit (so $(b) = R$) or b is an associate of a (so $(b) = (a)$).

Corollary 7.2 $\mathbb{Z}/n\mathbb{Z}$ is a field if and only if n is prime.

Proof: $\mathbb{Z}$ is a principal ideal domain, and irreducible's are just the prime integers.

The field $\mathbb{Z}/p\mathbb{Z}$, for a prime number p, is often denoted by $\mathbb{F}_p$.

Adding the Root of a Polynomial

The other important class of principal ideal domains consists of the polynomial rings over fields. For these, Propositions 7.16 and 7.17 give the first part of the following result.

Proposition 7.18 Let F be a field and $f(x)$ an irreducible polynomial over F. Then K = F[x]/ \{f(x)\} is a field. Moreover, there is an isomorphism from F to a subfield of K; and, if α denotes the coset $\{f(x)\} + x$, then we have the following, where n is the degree of $f(x)$, and we identify an element of F with its image under the isomorphism:

(a) every element of k can be uniquely written in the form

$$c_0 + c_1\alpha + c_2\alpha^2 + \ldots + c_{n-1}\alpha^{n-1};$$

(b) $f(\alpha) = 0$.

Before proving this, we notice that this gives us a construction of the complex numbers; Let $F = \mathbb{R}$, and let $f(x) = x^2 + 1$ (this polynomial is irreducible over $\mathbb{R}$). Use the notation i instead of a for the coset $(f(x)) + x$. Then we have $n = 2$, and the two parts of the proposition tell us that

(a) every element of K can be written uniquely as $a + bi$, where $a, b \in R$;

(b) $i^2 = -1$.

Thus, $K = \mathbb{R}[x]/(x^2 + 1)$ is the field $\mathbb{C}$. The general theory tells us that this construction of $\mathbb{C}$ does produce a field; it is not necessary to check all the axioms.

Proof: (a) Let I denote the ideal $(f(x))$. Remember that the elements of the quotient ring F[x]/I are the cosets of I in F[x]. The isomorphism θ from F to K = F[x]/I is given by

$$a\theta = I + a \quad \text{for} \quad a \in F.$$

Clearly θ is one-to-one; for if $a\theta = b\theta$, then $b - a \in I$, but I consists of all multiples of the irreducible polynomial $f(x)$, and cannot contain any constant polynomial except 0, so $a = b$. It is routine to check that θ preserves addition and multiplication. From now on, we identify a with the coset $I + a$, and regard F as a subfield of F[x]/I.

Let $g(x) \in F[x]$. Then by the Euclidean algorithm we can write
$$g(x) = f(x)q(x) + r(x),$$
where $r(x) = 0$ or $r(x)$ has degree less than n. Also, since $g(x) - r(x)$ is a multiple of $f(x)$, it belongs to I, and so the cosets $I + g(x)$ and $I + r(x)$ are equal. In other words, every coset of I in F[x] has a coset representative with degree less than n (possibly zero). This coset representative is unique, since the difference between any two coset representatives is a multiple of $f(x)$.

Now let $r(x) = c_0 + c_1 x + c_2 x^2 + \ldots + c_{n-1} x^{n-1}$. We have
$$\begin{aligned}1 + r(x) &= I + (c_0 + c_1 x + c_2 x_2 + \ldots + c_{n-1} x_{n-1}) \\ &= (I + c_0) + (I + c_1)(I + x) + (I + c_2)(I + x)^2 + \ldots + (I + c_{n-1})(I + x)^{n-1} \\ &= c_0 + c_1 \alpha + c_2 \alpha^2 + \ldots + c_{n-1} \alpha^{n-1}.\end{aligned}$$

Here, in the second line, we use the definition of addition and multiplication of cosets, and in the third line we put $I + x = \alpha$ and use our identification of $I + c = c\theta$ with c for $c \in F$.

So we have the required representation. Clearly it is unique.

(b) As before, if $f(x) = a_0 + a_1 x + \ldots + a_n x^n$, we have $I + f(x) = I$ (since $f(x) \in I$), and so
$$\begin{aligned}0 &= I + 0 \\ &= I + (a_0 + a_1 x + \ldots + a_n x^n) = (I + a_0) + (I + a_1)(I + x) + \ldots + (I + a^n)(I + x)^n \\ &= a_0 + a_1 \alpha + \ldots + a_n \alpha^n \\ &= f(\alpha)\end{aligned}$$

Finite Fields

Suppose that $f(x)$ is an irreducible polynomial of degree n over the field $\mathbb{F}_p$ of integers mod p. Then $K = F_p[x]/(f(x))$ is a field, by Proposition 7.17. According to that proposition, its elements can be written uniquely in the form
$$c_0 + c_1 \alpha + \ldots + c_{n-1} \alpha^{n-1}$$
for $c_0, \ldots, c_{n-1} \in \mathbb{F}_p$. There are p choices for each of the n coefficients $c_0, c_1, \ldots, c_{n-1}$ giving a total of p^n elements altogether. Thus:

Proposition 7.19 Let $f(x)$ be an irreducible polynomial of degree n over $\mathbb{F}_p$. Then $K = F_p \mathbb{F}_p[x]/((f(x))$ is a field containing p^n elements.

Example. Let $p = 2$ and $n = 2$. The coefficients of a polynomial over $\mathbb{F}_2$ must be 0 or 1, and so there are just four polynomials of degree 2, namely x^2, $x^2 + 1$, $x^2 + x$ and $x^2 + x + 1$. We have

$$x^2 = x \cdot x, \qquad x^2 + x = x \cdot (x + 1), \qquad x^2 + 1 = (x + 1) \cdot (x + 1)$$

(remember that $1 + 1 = 0$ in $\mathbb{F}_2$!), and so the only irreducible polynomial is $x^2 + x + 1$. Thus, there is a field consisting of the four elements $0, 1, \alpha, 1 + \alpha$, in which $\alpha^2 + \alpha + 1 = 0$, that is, $\alpha^2 = 1 + \alpha$ (since $-1 = +1$ in $\mathbb{F}_2$!) The addition and multiplication tables are easily found (with $\beta = 1 + \alpha$) to be

+	0	1	α	β
0	0	1	α	β
1	1	0	β	α
α	α	β	0	1
β	β	α	1	0

·	0	1	α	β
0	0	0	0	0
1	0	1	α	β
α	0	α	β	1
β	0	β	1	α

We have, for example,
$$\alpha + \beta = \alpha + 1 + \alpha = 1,$$
$$\alpha\beta = \alpha(1 + \alpha) = \alpha + \alpha^2 = 1$$
$$\beta^2 = (1 + \alpha)^2 = 1 + \alpha = \beta.$$

The basic facts about finite fields were one of the discoveries of Évariste Galois, the French mathematician who was killed in a duel in 1832 at the age of 19. Most of his mathematical work, which is fundamental for modern algebra, was not published until fifteen years after his death, but the result on finite fields was one of the few papers published during his lifetime.

Evariste Galois

Galois proved the following theorem:

Theorem 7.11 The number of elements in a finite field is a power of a prime. For any prime power p^n, there is a field with p^n elements, and any two finite fields with the same number of elements are isomorphic.

We commemorate Galois by using the term *Galois field* for finite field. If $q = p^n$, then we often denote the field with q elements by GF(q). Thus the field on the preceding page is GF(4). (Note that GF(4) is *not* the same as $\mathbb{Z}/4\mathbb{Z}$, the integers mod 4, which is not a field!)

Field of Fractions

In this section we generalise the construction of the rational numbers from the integers. [This section and the two following were not covered in the lectures, but you are encouraged to read them for interest.]

Theorem 7.12 Let R be an integral domain. Then there is a field F such that

(a) R is a subring of F;

(b) every element of F has the form ab^{-1}, for $a, b \in R$ and $b \ne 0$.

The field F is called *the, field of fractions* of R, since every element of F can be expressed as a fraction a/b.

We will build F as the set of all fractions of this form. But we have to answer two questions?

• When are two fractions equal?

• How do we add and multiply fractions?

Thus, we start with the set X consisting of all ordered pairs (a, b), with $a, b \in R$ and $b \ne 0$. (That is, $X = R \times (R\setminus\{0\})$.) The ordered pair (a, b) will "represent" the fraction a/b. So at this point we have to answer the first question above: when does $a/b = c/d$? Multiplying up by bd, we see that this holds if and only if $ad = bc$. Thus, we define a relation ~ on X by the rule

$(a, b) \sim (c, d)$ if and only if $ad = bc$.

We have to show that this is an equivalence relation.

reflexive: $ab = ba$, so $(a, b) \sim (a, b)$.

symmetric: If $(a, b) \sim (c, d)$, then $ad = bc$, so $cb = da$, whence $(c, d) \sim (a, b)$.

transitive: Suppose that $(a, b) \sim (c, d)$ and $(c, d) \sim (e, f)$. Then $ad = bc$ and $cf = de$. So $adf = bcf = bde$. This means that $d(af - be) = 0$. But $d \neq 0$ and R is an integral domain, so we conclude that $af = be$, so that $(a, b) \sim (e, f)$.

Now we let F be the set of equivalence classes of the relation $\sim$. We write the equivalence class containing (a, b) as a/b. Thus we do indeed have that $a/b = c/d$ if and only if $ad = bc$.

Now we define addition and multiplication by the "usual rules":

- $(a/b) + (c/d) = (ad + bc)/(bd)$;
- $(a/b)(c/d) = (ac)/(bd)$.

(To see where these rules come from, just calculate these fractions in the usual way!) Again, since $b \neq 0$ and $d \neq 0$, we have $bd \neq 0$, so these operations make sense. We still have to show that they are well-defined, that is, a different choice of representatives would give the same result. For addition, this means that, if $(a, b) \sim (a', b')$ and $(c, d) \sim (c', d')$, then $(ad + bc, bd) \sim (a'd' + b'c', b'd')$. Translating, we have to show that

if $ab' = ba'$ and $cd' = dc'$, then $(ad + bc)b'd' = bd(a'd' + b'c')$,

a simple exercise. The proof for multiplication is similar.

Now we have some further work to do. We have to show that

- F, with addition and multiplication defined as above, is a field;
- The map θ defined by $a\theta = a/1$ is a homomorphism from R to F, with kernel $\{0\}$ (so that R is isomorphic to the subring $\{a/1 : a \in R\}$ of F).

These are fairly straightforward to prove, and their proof finishes the theorem.

EXERCISE 7.1

1. Defind a ring. Show that the set of all intgers is a ring with addition and multiplication.
2. Show that the set of even integers with zero is a ring with respect to ordinary addition and multiplication.
3. Show that set of real (complex) numbers is a ring with ordinary addition and multiplication.
4. Prove that the singleton $\{0\}$ is a ring with regard to addition and multiplication.
5. Prove that the set of the residue classes (mod 7) is a ring with regard to addition and multiplication (mod 7).
6. Prove that the set $I(\sqrt{-1})$ of complex numbers of the form $m + n\sqrt{-1}$, where m, n are integers is a ring with respect to addition and multiplication.
7. Prove that the set $R = \{ma : a \in I, m \text{ is a given integers}\}$ is a ring with respect to addition and multiplication.

8. If $a, b, \in$ ring $(R, t, .)$, then prove that
    (i) $(a + b)^2 = a^2 + a.b + b.a + b^2$ where $a^2 = a.a$,
    (ii) $a + a = 0, \forall\ a \in R$ if $a^2 = a, \forall\ a \in R$
    (iii) $a + b = 0 \Rightarrow a = b$, when $a.b \in (R, +, .)$
    (iv) $(R, +, .)$ is commutative ring if $a^2 = a, \forall\ a \in (R, +, .)$
    (v) Prove that a ring is commutative if and only if $(a + b)^2 = a^2 + 2ab + b^2\ \forall\ a, b \in R$.
9. Prove that the set of natural numbers is not a ring with addition and multiplication.
10. Prove that in ring R of integers mod 6 under addition and multiplication mod 6, $a.b = 0$ when $a \neq 0, b = 0, b \in R$.

EXERCISE 7.2

1. Prove that for $J = \{7x : x \in I\}$, $(J, +, .)$ is a prime ideal of ring of integers.
2. Prove that set I of all integers is a subring of the ring of numbers $a + ib$, where $a, b \in I$.
3. Prove that ideal $S = \{3x : x \in I\}$ is a maximal ideal of I.
4. Prove that ideal $S = \{7x : x \in I\}$ is a maximal ideal but ideal $J = \{14\ x : x \in I\}$ is not.
5. If $I \in S$ and S is an ideal of R, Prove that $S = R$.
6. If R is a commutative ring and $a \in R$. Prove that $aR = \{ar : r \in R\}$ is an indeal but not necessary a Prime ideal.
7. Prove that a necessary and sufficient condition that the element x is Euclidean ring is a unit that
$$f(x) = f(1).$$
8. Show that in a ring with unity a proper left zero division cannot have a left inverse.
9. If ϕ is an isomorphism of a ring R onto R', then prove that
    (i) $\phi(0) = 0'$, where $0 \in R, 0' \in R'$
    (ii) $\phi(-a) = -\phi(a), \forall\ a \in R$
    (iii) Isomorphic image of commutative ring with unity is a commutative ring with unity.
10. Show that the set of 2×2 matrics of the form.
$$\begin{bmatrix} a & 0 \\ b & 0 \end{bmatrix}$$
where a, b are integers is a left ideal but not a ring ideal in the ring of all 2×2 matrices with elements as integers.
11. Show that S is an ideal of $S + T$ where S is an ideal of the ring R and T any subring of R.
12. Find the characterisitics of ring of integers modulo a prime m. What will be the characteristic if m is a composite integer.

EXERCISE 7.3

1. Define a field and give examples of field.
2. Prime that the set {0, 1, 2} (mod 3) is a field with respect to '$+$', '$\cdot$' (mod 3).
3. Prove that set of residue classes (mod 7) is a field wide respect to '$+$' and '$\cdot$' (mod 7).
4. Show that the set of rational (real, complex) numbers is a field with respect to addition and multiplication.
5. Give an example of division ring which is not a field.
6. Prove that a finite commutative ring without zero divisions is a field.
7. Prove that the set $I/(n)$ with respect to addition and multiplication of residue classes (mode n), if n is not a prime, is not a field.
8. Show that the set $R \times R$ is a field for compositions defined by —
$$(a, b) + (c, d) = (a + c, b + d)$$
$$\text{and } (a, b) \cdot (c, d) = (ac + bd, ad + bc)$$
Prove that this is isomorphic to the field to complex numbers.
9. In a field F where $a, b, c, d \in F$ and are non zero, prove that —
   (i) $(a/b)^{-1} = b/a$;
   (ii) $a/b + c/b = \dfrac{a+c}{b}$
   (iii) $\dfrac{a}{b} - \dfrac{c}{b} = \dfrac{a-c}{b}$
   (iv) $(a^{-1})^{-1} = a$
   (v) $(-a)^{-1} = -(a^{-1})$.
10. If R is commutative ring with two or more elements and if $a, b \in R$, $a \neq 0$, there exist $x \in R$ such that $ax = b$, then prove that R is a field.
11. Prove that the quotient field of a finite integral domain coincides with itself.
12. The set of elements of the form $a + bc$ where c is an imaginary cube root of unity and a, b are rational is a subfield of the field of complex numbers.

❏❏

CHAPTER 8

POSETS, HASSE DIAGRAM AND LATTICES

8.1 INTRODUCTION

In this chapter, we shall discuss the ordered set, partially ordered set and lattices, which is a special kind of an ordered set.

Partially ordered set (on Poset)

A relation R on a set S is called a partial order on partial ordering, if it is

(i) *Reflexive* — For any $x \in S$, xRx or $(x, x) \in R$

(ii) *Anti symmetric* — If aRb and bRa then $a = b$

(iii) *Transitive* — If aRb and aRc then aRc.

A set S together with a partial orderning R is called a partially ordered set (or poset) and it is denoted by (S, R). For example, the relation $\leq$ (less than or equal to) on the set N of positive integer. The partial order relation is usually denoted by $\leq$ and $a \leq b$ is real as a precedes b.

Example 1: Show that the relation $\geq$ (greater than or equal to) is a partial orderning on the set of integers.

Solution : We have $a \geq a$ for every integer a. Therefore, the relation $\geq$ is reflexive. Also $a \geq b$ and $b \geq a$ implies $a = b$. Therefore, $\geq$ is antisymmetric. Finally $a \geq b$ and $b \geq c$ implies $a \geq c$. Therefore, is $\geq$ transitive. Hence $\geq$ is a partial ordering on the set of integers.

Example 2: Consider P(S) as the power set, *i.e.*, the set of all subsets of a given set S. Show that the inclusion relation $\subseteq$ is a partial ordering on the power set P(S).

Solution : Since

1. $A \subseteq A$ for all $A \subseteq S$, $\subseteq$ is reflexive.
2. $A \subseteq B$ and $B \subseteq A \Rightarrow A = B$, $\subseteq$ is antisymmetric.
3. $A \subseteq B$ and $B \subseteq A \Rightarrow A \subseteq C$, is $\subseteq$ transitive.

It follow that $\subseteq$ is a partial ordering on P(S) and (P(S), $\subseteq$) is a poset

Comparability : The elements a and b in a poset (S, $\leq$) are called comparable if either $a \leq b$ or $b \leq a$, then a and b are called incomparable.

Example : In the poset $(Z^+, 1)$ the integers 2 and 6 are comparable while 2 and 7 are incomparable.

8.2 TOTALLY ORDERED (OR LINEARLY ORDERED) SET

If every two elements of a poset $(S, \geq)$ are comparable then S is called a totally (or linearly) ordered set and the relation $\leq$ is called a total order or partial order. A totally ordered set is also called a chain.

Example: (a) Consider the set N of positive integers ordered by divisibility. Then 21 and 7 are comparable since 7/21 on the other hand, 3 and 5 are non comparable since neither 3/5 non 5/3. Thus N is hot line only ordered by divisibility. Observe that A = {2, 6, 12, 36} is a linearly ordered subset of N, since 2/6, 6/12 and 12/36.

(b) The set N of positive integers with the usual order $\leq$ is linearly ordered and hence every ordered subset of N is also line only ordered.

(c) The power set P(A) of a set A with two or bone elemens is not linearly ordered by set inclusion. For instance suppose a and b belong to A. Then {a} and {b} are non comparable. Observe that the empty set ϕ, {a} and A do form a linearly ordered subset of P(A), since $\phi \subseteq \{A\} \subseteq A$. Similarly, ϕ, {b} and A form a linearly ordered subset of P(A).

Well ordered set: A poset $(S, \leq)$ is called well ordered set if on is a total ordering and every non-empty subset of S has a least element.

Power sets and order: There are a number of ways to define an order relation on the cartesian product of given ordered sets. Two of these ways follow :

(a) *Product order* : Suppose S and T are ordered sets. Then the following, is an order relation on the product set $S \times T$ called the product order :

$$(a, b) \leq (a', b') \quad \text{if} \quad a \subseteq a' \text{ and } b \leq b'$$

(b) *Lexicographical order*: Suppose S and T are line only ordered sets. Then the following is an order relation on the product set $S \times T$, called the Lexicographical or dictionary order :

$$(a, b) < (a', b') \quad \text{if} \quad a < b \text{ on if } a = a' \text{ and } b < b'$$

This order can be extended to $S_1 \times S_2 \times ... \times S_n$ as follows :

$(a_1, a_2, ..., a_n) < (a'_1, a'_2, ..., a'_n)$ if $a_i = a'_1$ for $i = 1, 2, ..., K-1$ and $a_K < a'_K$.

Example: $(3, 5) < (4, 8)$ is lexicographic ordering constructed from the usual relation $\leq$ on Z.

Kleene closure and order: Let A be a (non empty) linearly ordered alphabet. Recall that A*, called the Kleene closure of A, cosists of all words W on A, and $|W|$ denotes the length W. Then the following are two order relations on A*.

(a) *Alphabetical (Lexicographical order)* : The order is no dobut familar with the usual alphabetical ordering of A*, i.e.,

(i) $\lambda < W$, where λ is the empty word and W is any non empty word.

(ii) Suppose $u = av'$ and $V = bv'$ are distinct non empty words where $a, b \in A$ and $u', v' \in A^*$, then :

$$u < v \quad \text{if} \quad |u| < |v| \text{ or if } |u| = |v| \text{ but } u \text{ precedes } v \text{ alphabetically.}$$

For example, "to" precedes "and" since $|to| = 2$ but $|and| = 3$. However, "an" precedes "to" since they have the same length; but "an" preceds "to" alphabetically. This order is also called the full semigroup order.

Minimal and Maximal elements: Let $(S, \leq)$ be a poset. An element a in S is called a minimal element of S if there is no element b in such that $a < b$. Similarly, an element a in S is called maximal element in S if there is no element b is S such that $b < a$. Minimal and maximal elements can be spotted easily by a Hasse diagram.

Supremum and infimum: Let A be a subset of a partially ordered set S. An element M in S is called an upper bound of A in M succeeds every element of A, *i.e.*, if for every x in A, we have $x \leq M$.

If an upper bound of A preceds every upper bound of A then it is called the supremum of A and it is denoted by sup (A).

Similarly an element M in a poset S is called a lower bound of a subset A of S if m preceds every element of A, *i.e.*, for every x in A we have :
$$M \leq x$$
If a lower bound of A succeeds every other lower bound of A then it is called the infimum of A and it is denoted by inf (A).

Example 3: Find the supremum and infimum of $\{1, 2, 4, 5, 10\}$ if thing exist in the poset $(Z^+, 1)$.

Solution 1: is the only lower bound for the given set. Therefore 1 is the infimum of the given set. 20 is the least upper bound of the given set. Therefore 20 is the supremum of the given set.

8.3 PROPERTIES OF POSETS

An element x of a poset (X, R) is called *maximal* if there is no element $y \in X$ satisfying $x < R\ y$. Dually, x is *minimal* if no element satisfies $y < R\ x$.

In a general poset there may be no maximal element, or there may be more than one. But in a finite poset there is always at least one maximal element, which can be found as follows: choose any element x; if it is not maximal, replace it by an element y satisfying $x < R\ y$; repeat until a maximal element is found. The process must terminate, since by the irreflexive and transitive laws the chain can never revisit any element. Dually, a finite poset must contain minimal elements.

An element x is an *upper bound* for a subset Y of X if $y \in R\ x$ for all $y \leq Y$. *Lower bounds* are defined similarly. We say that x is a *least upper bound* or *l.u.b.* of Y if it is an upper bound and satisfies $x < R\ x'$ for any upper bound x'. The concept of *a greatest lower bound* or *g.l.b.* is defined similarly.

A *chain* in a poset (X, R) is a subset C of X which is totally ordered by the restriction of R (that is, a totally ordered subset of X). An *antichain* is a set A of pairwise incomparable elements.

Infinite posets (such as $\mathbb{Z}$), as we remarked, need not contain maximal elements. *Zorn's Lemma* gives a sufficient condition for maximal elements to exist:

Let (X, R) be a poset in which every chain has an upper bound. Then X contains a maximal element.

As well known, there is no "proof of Zorn's Lemma, since it is equivalent to the Axiom of Choice (and so there are models of set theory in which it is true, and models in which it is false). Our proof of the existence of maximal elements in finite posets indicates why this should be so: the construction requires (in general infinitely many) choices of upper bounds for the elements previously chosen (which is form a chain by construction).

The *height* of a poset is the largest cardinality of a chain, and its *width* is the largest cardinality of an antichain. We denote the height and width of (X, R) by $h(X)$ and $w(X)$ respectively (suppressing as usual the relation R in the notation).

In a finite poset (X, R), a chain C and an antichain A have at most one element in common. Hence the least number of antichains whose union is X is not less than the size $h(X)$ of the largest chain in X. In fact there is a partition of X into $h(X)$ antichains. To see this, let A_1 be the set of maximal elements; by definition this is an antichain, and it meets every maximal chain. Then, let A_2 be the set of maximal elements in $X \setminus A_1$, and iterate this procedure to find the other antichains.

There is a kind of dual statement, harder to prove, known as *Dilworth's Theorem*:

Theorem 8.1: Let (X, R) be a finite poset. Then there is a partition of X into $w(X)$ chains.

An *up-set* in a poset (X, R) is a subset Y of X such that, if $y \in Y$ and $y \le R\, z$, then $z \in Y$. The set of minimal elements in an up-set is an antichain. Conversely, if A is an antichain, then

$$\uparrow (A) = \{x \in X : a < R\, x \text{ for some } a \in A\}$$

is an up-set. These two correspondences between up-sets and antichains are mutually inverse; so the numbers of up-sets and antichains in a poset are equal.

Down-sets are, of course, defined dually. The complement of an up-set is a down-set; so there are equally many up-sets and down-sets.

8.4 HASSE DIAGRAMS

Let x and y be distinct elements of a poset (X, R), We say that y covers x. if $[x, y]_R = \{x,y\}$, that is, $x < R\, y$ but no element z satisfies $x < R\, z < R\, y$. In general, there may be no pairs x and y such that y covers x (this is the case in the rational numbers, for example). However, locally finite posets are determined by their covering pairs:

Proposition 8.1: Let (X, R) be a locally finite poset, and $x, y \in X$. Then $x < R\, y$ if and only if there exist elements $z_0, ..., z_n$ (for some non-negative integer n) such that $z_0 = x$, $z_n = y$, and z_{i+1} covers z_{i+1} for $i = 0, ..., n-1$.

The *Hasse diagram* of a poset (X, R) is the directed graph whose vertex set is X and whose arcs are the covering pairs (x, y) in the poset. We usually draw the Hasse diagram of a finite poset in the plane in such a way that, if y covers x, then the point representing y is higher than the point representing x. Then no arrows are required in the drawing, since the directions of the arrows are implicit.

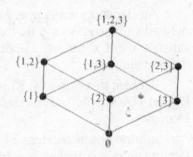

Figure 8.1: A Hasse diagram

For example, the Hasse diagram of the poset of subsets of {1, 2, 3} is shown in Figure 8.1.

Linear Extensions and Dimension

One view of a partial order is that it contains partial information about a total order on the underlying set. This view is borne out by the following theorem. We say that one relation *extends* another if the second relation (as a set of ordered pairs) is a subset of the first.

Theorem 8.2: Any partial order on a finite set X can be extended to a total order on X.

This theorem follows by a finite number of applications of the next result.

Proposition 8.2: Let R be a partial order on a set X, and let a, b be incomparable elements of X. Then there is a partial order R' extending R such that $(a, b) \in R'$ (that is, $a < b$ in the order R').

A total order extending R in this sense is referred to as a *linear extension* of R. (The term "linear order" is an alternative for "total order".)

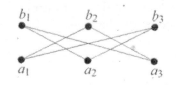

Figure 8.2. A crown

This proof does not immediately shows that every infinite partial order can be extended to a total order. If we assume Zorn's Lemma, the conclusion follows. It cannot be proved from the Zermelo-Fraenkel axioms alone (assuming their consistency), but it is strictly weaker than the Axiom of Choice, that is, the Axiom of Choice (or Zorn's Lemma) cannot be proved from the Zermelo-Fraenkel axioms and this assumption. In other words, assuming the axioms consistent, there is a model in which Theorem 8.2 is false for some infinite poset, and another model in which Theorem 8.2 is true for all posets but Zorn's Lemma is false.

The theorem gives us another measure of the size of a partially ordered set. To motivate this, we use another model of a partial order. Suppose that a number of products are being compared using several different attributes. We regard object a as below object b if b beats a on every attribute. If each beats the other on some attributes, we regard the objects as being incomparable. This defines a partial order (assuming that each attribute gives a total order). More precisely, given a set S of total orders on X, we define a partial order R on X by $x < Ry$ if and only if $x < Sy$ for every $s \in S$. In other words, R is the intersection of the total orders in S.

Theorem 8.3: Every partial order on a finite set X is the intersection of some set of total orders on X.

Now we define the *dimension* of a partial order R to be the smallest number of total orders whose intersection is R. In our motivating example, it is the smallest number of attributes which could give rise to the observed total order R.

The *crown* on $2n$ elements $a_1, ..., a_n, b_1, ..., b_n$ is the partial order defined as follows: for all indices $i \neq j$, the elements a_i and a_j are incomparable, the elements b_i and b_j are incomparable, but $a_i < b_j$, and for each i, the elements a_i and b_i are incomparable. Figure 8.2 shows the Hasse diagram of the 6-element crown.

Now we have the following result:

Proposition 8.3: The crown on $2n$ elements has dimension n.

8.6 THE MÖBIUS FUNCTION

Lef R be a partial order on the finite set X. We take any linear order extending R, and write $X = \{x_1, \ldots, x_n\}$, where $x_1 < \ldots < x_n$ (in the linear order S): this is not essential but is convenient later.

The *incidence algebra* $\mathcal{A}(R)$ of R is the set of all functions $f: X \times X \to \mathbb{R}$ which satisfy $f(x, y) = 0$ unless $x <$ R y holds. We could regard it as a function on R, regarded as a set of ordered pairs. Addition and scalar multiplication are defined pointwise; multiplication is given by the rule

$$(fg)(x, y) = \sum_z f(x, z) g(z, y)$$

If we represent f by the $n \times n$ matrix A_f with (i, j) entry $f(x_i, x_j)$, then this is precisely the rule for matrix multiplication. Also, if $x \not\leq$ Ry, then there is no point z such that $x \leq$ R z and $z \leq$ R y, and so $(fg)(x, y) = 0$. Thus, $\mathcal{A}(R)$ is closed under multiplication and does indeed form an algebra, a subset of the matrix algebra $M_n(\mathbb{R})$. Also, since f and g vanish on pairs not in R, the sum can be restricted to the interval $[x, y]_R = \{z : x <$ R $z <$ R $y\}$:

Incidentally, we see that the (i, j) entry of A_f is zero if $i > j$, and so consists of upper triangular matrices. Thus, an element $f \in \mathcal{A}(R)$ is invertible if and only if $f(x,x) \neq 0$ for all $x \in X$.

The *zeta-function* ζ_R is the matrix representing the relation R as defined earlier; that is, the element of $\mathcal{A}(R)$ defined by

$$\zeta_R(x, y) = \begin{cases} 1 & \text{if } x \leq \text{R} y, \\ 0 & \text{otherwise} \end{cases}$$

Its inverse (which also lies in $\mathcal{A}(R)$) is the *Möbius function* μ_R of R. Thus, we have, for all $(x,y) \in R$,

$$\sum_{x \in [x, y]_R} \mu(x, z) \begin{cases} 1 & \text{if } x = y, \\ 0 & \text{otherwise} \end{cases}$$

This relation allows the Möbius function of a poset to be calculated recursively. We begin with $\mu_R(x, x) = 1$ for all $x \in X$. Now, if $x <$ R y and we know the values of $\mu(x, z)$ for all $z \in [x, y]_R \setminus \{y\}$, then we have

$$\mu_R(x, y) = \sum_{z \in [x, y]_R \setminus \{y\}} \mu(x, z)$$

In particular, $\mu R(x, y) = -1$ if y covers x.

The definition of the incidence algebra and the Möbius function extend immediately to locally finite posets, since the sums involved are over intervals $[x, y]_R$.

The following are examples of Möbius functions.
- The subsets of a set:
$$\mu(A, B) = (-1)^{|B/A|} \text{ for } A \subseteq B;$$
- The subspaces of a vector space $V \subseteq GF(q)^n$:
$$\mu(U, W) = (-1)^k q^{\binom{k}{2}} \quad \text{for } U \subseteq W, \text{ where } k = \dim U - \dim W.$$
- The (positive) divisors of an integer n:
$$\mu(a, b) = \begin{cases} (-1)^r & \text{if } \dfrac{b}{a} \text{ is the product of } r \text{ distinct primes;} \\ 0 & \text{otherwise} \end{cases}$$

In number theory, the classical Möbius function is the function of one variable given by $\mu(n) = \mu(1, n)$ (in the notation of the third example above).

The following result is the *Möbius inversion* for locally finite posets. From the present point of view, it is obvious.

Theorem 8.4 $f = g\zeta \Leftrightarrow g = f\mu$. Similarly, $f = \zeta g \Leftrightarrow g = \mu f$.

Example 1. Suppose that f and g are functions on the natural numbers which are related by the identity $f(n) = \Sigma_{d|n} g(d)$. We may express this identity as $f = g\zeta$, where we consider f and g as vectors and where ζ is the zeta function for the lattice of positive integer divisors of n. Theorem 7 implies that $g = f\mu$, or

$$g(n) = \sum_{d|n} \mu(d, n) f(d) = \sum_{d|n} \mu\left(\frac{d}{n}\right) f(d)$$

which is precisely the classical Möbius inversion.

Example 2. Suppose that f and g are functions on the subsets of some fixed (countable) set X which are related by the identity $f(A) = \Sigma_{B \subseteq A} g(B)$. We may express this identity as $f = \zeta g$ where ζ, is the zeta function for the lattice of subsets of X. Theorem 7 implies that $g = \mu f$, or

$$g(A) = \sum_{B \supseteq A} \mu(A, B) f(B) = \sum_{B \supseteq A} (-1)^{|B \setminus A|} f(B)$$

which is a rather general form of the inclusion/exclusion principle.

8.7 LATTICES

A *lattice* is a poset (X, R) with the properties
- X has an upper bound 1 and a lower bound 0;
- for any two elements $x, y \in X$, there is a least upper bound and a greatest lower bound of the set $\{x, y\}$.

A simple example of a poset which is not a lattice is the poset

In a lattice, we denote the l.u.b. of $\{x, y\}$ by $x \vee y$, and the g.l.b. by $x \wedge y$. We commonly regard a lattice as being a set with two distinguished elements and two binary operations, instead of as a special kind of poset.

Lattices can be axiomatised in terms of the two constants 0 and 1 and the two operations $\vee$ and $\wedge$. The result is as follows, though the details are not so important for us. The axioms given below are not all independent. In particular, for finite lattices we do not need to specify 0 and 1 separately, since 0 is just the meet of all elements in the lattice and 1 is their join.

Properties of Lattices Let X be a set, $\wedge$ and $\vee$ two binary operations defined on X, and 0 and 1 two elements of X. Then $(X, \vee, \wedge, 0, 1)$ is a lattice if and only if the following axioms are satisfied:

- *Associative laws*: $x \wedge (y \wedge z) = (x \wedge y) \wedge z$ and $x \vee (y \vee z) = (x \vee y) \vee z$;
- *Commutative laws*: $x \wedge y = y \wedge x$ and $x \vee y = y \vee x$;
- *Idempotent laws*: $x \wedge x = x \vee x = x$;
- $x \wedge (x \wedge y) = x = x \vee (x \wedge y)$;
- $x \wedge 0 = 0$, $x \vee 1$.

A *sublattice* of a lattice is a subset of the elements containing 0 and 1 and closed under the operations $\vee$ and $\wedge$. It is a lattice in its own right.

The following are a few examples of lattices.

- The subsets of a (fixed) set:

$$A \wedge B = A \cap B$$
$$A \vee B = A \cup B$$

The subspaces of a vector space:

$$U \wedge V = U \cap V$$
$$U \vee V = \mathrm{span}\,(U \cup V)$$

The partitions of a set:

$$R \wedge T = R \cap T$$
$$R \vee T = \overline{R \cup T}$$

Here $\overline{R \cup T}$ is the partition whose classes are the connected components of the graph in which two points are adjacent if they lie in the same class of either R or T.

Distributive and Modular Lattices

A lattice is *distributive* if it satisfies the *distributive laws*

(D) $x \wedge (y \vee z) = (x \wedge y) \vee (x \wedge z)$ and $x \vee (y \wedge z) = (x \vee y) \wedge (x \vee z)$ for x, y, z.

A lattice is *modular* if it satisfies the *modular law*

(M) $x \vee (y \wedge z) = (x \vee y) \wedge z$ for all x, y, z such that $x \ne z$.

Figure 8.3 represents a lattice, N_5, which is not modular, as well as a modular lattice, M_3, which is not distributive.

Figure 8.3. Two lattices

Not only are N_5 and M_3 the smallest lattices with these properties, they are, in a certain sense, the only lattices with these properties. The following theorem states this more precisely.

Theorem 8.5: A lattice is modular if and only if it does not contain the lattice N_5 as a sublattice. A lattice is distributive if and only if it contains neither the lattice N_5 nor the lattice M_j as a sublattice.

The poset of all subsets of a set S (ordered by inclusion) is a distributive lattice: we have $0 = \theta$, $1 = S$, and l.u.b. and g.l.b. are union and intersection respectively. Hence every sublattice of this lattice is a distributive lattice.

Conversely, every finite distributive lattice is a sublattice of the lattice of sub-sets of a set. We describe how this representation works. This is important in that it gives us another way to look at posets.

Let (X, R) be a poset. Recall that an down-set in X is a subset Y with the property that, if $y \in Y$ and $z \leq R\ y$, then $z \in Y$.

Let L be a lattice. A non-zero element $x \in L$ is called *join-irreducible* if, whenever $x = y \vee z$, we have $x = y$ or $x = z$.

Theorem 8.6: (a) Let (X, R) be a finite poset. Then the set of down-sets. in X, with the operations of union and intersection and the distinguished elements $0 = \theta$ and $1 = X$, is a distributive lattice.

(b) Let L be a finite distributive lattice. Then the set X of non-zero join-irreducible elements of L is a sub-poset of L.

(c) These two operations are mutually inverse.

Meet-irreducible elements are defined dually, and there is of course a dual form of Theorem 10.

Example 4: Show that the inclusion $\subseteq$ relation is a partial ordering on the power set of a set S.

Solution : For each subset A of S we have $A \subseteq A$. Therefore $\subseteq$ is reflexive. Also $A \subseteq B$ and $B \subseteq A \Rightarrow A = B$.

Therefore $\subseteq$ is antisymmetric. Finally $A \subseteq B$ and $B \subseteq C$ implies $A \subseteq C$. Therefore $\subseteq$ is transitive. Hence $\subseteq$ is a partial ordering and so $(P(S), \subseteq)$ is a poset.

Example 5: Draw the Hasse diagram for the partial ordering $\{[A, B], A \subseteq B\}$ on the power set P(S) for S = {1, 2, 3}.

Solution: The required Hasse diagram can be obtained from the digraph of the given poset by deleting all the loops and all the edges that occur from transitive property, *i.e.*, (φ, {1, 23}), (φ, {1, 33}), (φ, (2, 33)), (φ {1, 2, 33}), ({1}{1, 2, 33}), ({2}, {1, 2, 33}) and deleting arrows which will be as given below:

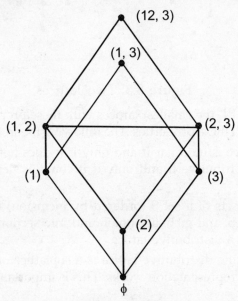

Figure 8.4

Example 6: Prerequisites in a college is a partial ordering of available classes. We say A < B if course A is prerequisite for course B. We consider the maths courses and third prerequisites as given below :

Course	Prerequisite
Math I	None
Math II	Math I
Math III	Math I
Math IV	Math III
Math V	Math II
Math VI	Math V
Math VII	Math II, Math III
Math VIII	Math VI, Math IV

Draw the Hasse diagram for the partial ordering of these courses.

Solution : We put Math I in the bottom of the diagram as it is the only course with no prerequisite. Math II and Math III only require Math I, so we have Math I << Math II and Math I << Math III.

Therefore, we draw lines starting upward from Math I to Math II and from Math I to Math III. Continuing this process we draw the complete Hasse diagram of the given partial ordering as shown below:

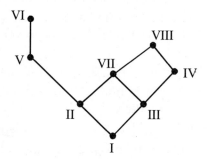

Example 7: Find the lower and upper bound of the subset {1, 2, 3} and {1, 3, 4, 6} in the poset with the Hasse diagram given below:

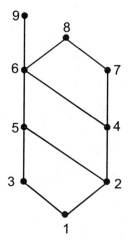

Solution: The upper bounds of such set {1, 2, 3} are 5, 6, 7, 8, 9 and its lower bound is only 1. The upper bound of {1, 3, 4, 6} are 6, 8 and 9 and its lower bound is only 1.

Example 8: Let N = {1, 2, 3, ...} be ordered by divisibility. State whether the following subsets are totally ordered—

(i) {2, 5, 24}
(ii) {3, 5, 15}
(iii) {5, 15, 30}
(iv) {2, 4, 8, 32}

Solution : (i) Here 2 divides 6 and 6 divides 24, therefore, the given set is totally ordered.

(ii) Here 3 and 5 are not comparable. Therefore, the given set is not totally ordered.

(iii) Here 5 divides 15 which divides 30. Therefore, the given set is totally ordered.

(iv) Here 2 divides 4, 4 divides 8 and 8 divides 32. Therefore, the given set is totally ordered.

Example 9: Show that every finite lattice L is bounded.

Solution: Let L = $\{a_1, a_2, ..., a_n\}$ is a finite lattice then $a_1 \wedge a_2 ... \wedge a_n$ and $a_1 \vee a_2 ... \vee a_n$ are lower and upper bounds of L respectively. Thus the lower and upper bounds of L exists. Hence the finite lattice is bounded.

Example 10: Show that the poset ({1, 2, 3, 4, 5} 1) is not a lattice.

Solution: 2 and 3 have no upper bound consequently they do not have a least upper bound. Hence the given set is not a lattice.

Example 11: Show that the poset ({1, 3, 6, 12, 24} 1) is a lattice.

Solution: Every two elements of the given poset have a least upper bound as well as a greatest lower bound which are longer and smaller elements respectively. Hence the given poset is a lattice.

Example 12: Determine whether the posets represented by the following Hasse diagrams are lattices :

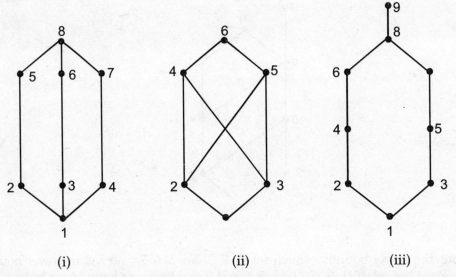

(i)   (ii)   (iii)

Solution: (*i*) The poset represented by the given Hasse diagram is a lattice as every pair of elements of it has a least upper bound as well as greatest lower bound.

(*ii*) The poset represented by the given Hasse diagram is not a lattice because the elements 2 and 3 have no least upper bound. The elements 4, 5 and 6 are the upper bounds but none of the sets three elements preceeds other two with respect to the ordering of the given poset.

(*iii*) The poset represented by the given Hasse diagram is a lattice as every pair of elements of it has a least upper bound as well as a greatest lower bound.

Example 13: Write the dual of the following:

(*i*) $(a \wedge b) \vee a = a \wedge (b \vee a)$

(*ii*) $(a \wedge b) \vee c = (a \vee c) \wedge (b \vee c)$

Solution: The dual will be obtained by replacing $\wedge$ by $\vee$ and $\vee$ by $\wedge$. Thus dual statements will be

(i) $(a \vee b) \wedge a = a \vee (b \wedge a)$

(ii) $(a \vee b) \wedge c = (a \wedge c) \vee (b \wedge c)$

Example 14: Give an example of an infinite lattice L with finite length.

Solution: Let $L = \{0, 1, a_1, a_2, ..., a_n, ...\}$ be ordered

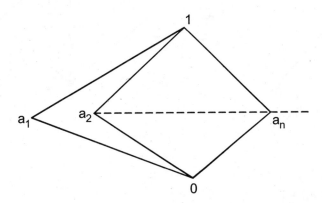

i.e., $0 < a_n < 1 ... n \in N$

Then N is an infinite lattice having finite length.

Example 15: Show that the posets given below are lattices obtain the Hasse diagrams.

(i) $(S_6, |)$ (ii) $(S_8, |)$ (iii) $(S_{24}, |)$ (iv) $(S_{30}, |)$

where S_n is the set of all divisions of n and $|$ denotes division.

Solution: We have $S_6 = \{1, 2, 3, 6\}$

If we take any two elements of S_6 then their lower bound and upper bound will also be in S_6. Therefore, $(S_6, 1)$ will be a lattice.

Similarly

$S_8 = \{1, 2, 4, 8\}$

$S_{24} = \{1, 2, 3, 4, 6, 8, 12, 24\}$

$S_{30} = \{1, 2, 3, 5, 6, 10, 15, 30\}$

We can show easily that there are lattices. The Hasse diagram of these lattices are given below :

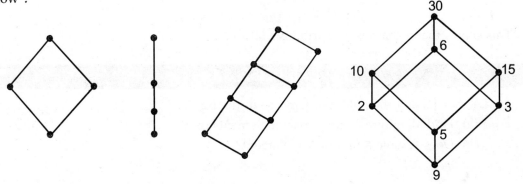

Example 16: Show that the lattices given by the following diagrams are not distributive.

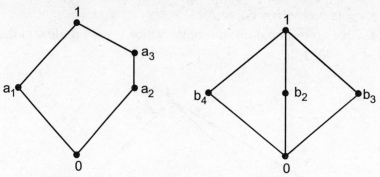

Solution : (i) We have
$$a_1 \wedge (a_2 \vee a_3) = (a_1 \wedge a_2) \wedge (a_1 \wedge a_3)$$
But
$$a_2 (a_1 \wedge a_3) = a_2 \wedge 1 = a_2$$
$$(a_2 \wedge a_1) \vee (a_2 \wedge a_3) = 0 \vee a_3 = a_3$$
Thus, $a_2 \wedge (a_1 \vee a_3) \neq (a_2 \wedge a_1) \vee (a_2 \wedge a_3)$
Hence the given lattices is not distributive.
(ii) We have
$$b_1 \wedge (b_2 \vee b_3) = b_1 \wedge 1 = b_1$$
$$(b_1 \wedge b_2) \vee (b_1 \wedge b_3) = 0 \vee 1 = 0$$
Thus, $b_1 \wedge (b_2 \vee b_3) \neq (b_1 \wedge b_2) \vee (b_1 \wedge b_3)$
Hence the given lattice is not distributive.

Example 17: Show that every chain and $a, b, c \in L$, we consider the cases :
(i) $a \leq b$ or $a \leq c$ and (ii) $a \geq b$ or $a \geq c$
Now we shall show that distributive law is satisfied by a, b, c :
For case (i) we have
$$a \wedge (b \vee c) = a \quad \text{and} \quad (a \wedge b) \vee (a \wedge c) = a$$
For case (ii), we have
$$a \wedge (b \vee c) = b \vee c \quad \text{and} \quad (a \wedge b) \vee (a \wedge c) = b \wedge c$$
Thus, we have
$$a \wedge (b \vee c) = (a \wedge b) \vee (a \wedge c)$$
This shows that a chain is a distributive lattice.

EXERCISE 8.1

1. Let $A = \{1, 2, 3, 4, 5, 6\}$ be ordered as in the figure.
   (a) Find all minimal and maximal elements of A.
   (b) Does A have a first or last element?

(c) Find all linearly ordered subsets of A, each of which contains at least three elements.

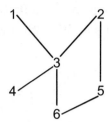

2. Let B = {a, b, c, d, e, f} be ordered as in the figure.
   (a) Find all minimal and maximal elements of B.
   (b) Does B have a first on last element?
   (c) List two and find the number of consistent environmentations of B into the set {1, 2, 3, 4, 5, 6}.

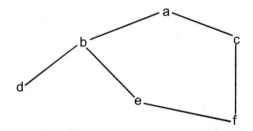

3. Let C = {1, 2, 3, 4} be ordered as shown in the figure. Let L(C) denote the collection of all non empty linearity ordered. Subsets of C ordered by set inclusion. Draw a diagram L(C)

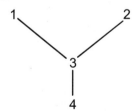

4. State whether each of the following is true or false and, if it is a false, give a counter example.
   (a) If a poset S has only one maximal element a then *a* is a last element.
   (b) If a finite poset S has only one maximal element *a*, then *a* is a last element.
   (c) If a linearly ordered set S has only one maximal element *a* then *a* is a last element.

5. Let S = {a, b, c, d, e} be ordered as in the figure.
   (a) Find all minimal and maximal elements of S.

(b) Does S have any first on last element?
(c) Find all subsets S in which C is a minimal element.
(d) Find all subsets of S in which C is a first element.
(e) List all linearly ordered subsets with three on more elements.

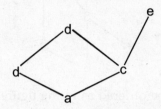

6. Let S = {a, b, c, d, e, f} be ordered as shown in the figure
   (a) Find all minimal and maximal elements of S.
   (b) Does S have any first on last element.
   (c) List all linearity ordered subsets with three on more elements

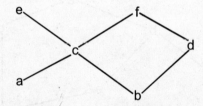

7. Let M = {2, 3, 4, ...} and let $M^2 = M \times M$ be ordered as follows :
   $(a, b) \leq (c, d)$ If a/c and $b \leq d$
   Find all minimal and maximal elements of M × M.

8. Consider the set R of real numbers with the usual order ≤. Let A = {x : x ∈ Q and 5 < x^3 < 27}.
   (a) Is A bounded above an below ?
   (b) Do Sup (A) and inf (A) exist ?

EXERCISE 8.2

1. Suppose the union S of the sets $A = \{a_1, a_2, a_3, ...\}$, $B = \{b_1, b_2, b_3, ...\}$ $C = \{c_1, c_2, c_3, ...,\}$ is ordered as follows :
   $$S = \{A; B; C\} = \{a_1, a_2, ..., b_1, b_2, ..., c_1, c_2, ...\}$$
   (a) Find all limit elements of S.
   (b) Show that S is not isomorphic to $N = \{1, 2, ...\}$ with the usual order $\leq$.

2. Let $A = \{a, b, c\}$ be linearly ordered by $a < b < c$ and let $N = \{1, 2, ...\}$ be given the usual order $\leq$.
   (a) Show that $S = \{A; N\}$ is isomorphic to N.
   (b) Show that $S' = \{N; A\}$ is not isomorphic to N.

3. Consider the lattice L in the figure.
   (a) Find all sublattices with five elements.
   (b) Find all join-irreducible elements and atoms.
   (c) Find complements of a and b, if they exist
   (d) Is L distributive? Complemented?

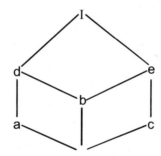

4. Consider the lattice M in the figure below :
   (a) Find join-irreducible elements
   (b) Find the atoms.
   (c) Find complements of a and b if exist
   (d) Is M distributive? Complemented?

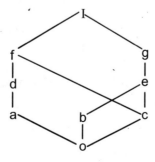

5. Show that the divisibility relation | is a partial ordering on the set of positive integers.
6. Find the maximal and minimal elements of the poset {2, 4, 5, 10, 12, 20, 25}
7. Find the greatest lower bound and least upper bound of the set {1, 2, 4, 5, 10} in the poset $(Z^+, 1)$.
8. Show that the poset {1, 2, 4, 8, 16} is a lattice.
9. Draw the Hasse diagram for divisibility on {1, 2, 3, 5, 7, 11, 13}.
10. Show that the poset given below are lattices.
    (i) $(S_9, 1)$, (ii) $(S_{12}, 1)$, (iii) $(S_{15}, 1)$
    Also obtain their Hasse diagram.

ANSWERS (EXERCISE 8.1)

1. (a) Minimal 4 and 6 maximal 1 and 2
   (b) First, none; last, none
   (c) {1, 3, 4}, {1, 3, 6} {2, 3, 4}, {2, 3, 6}, {2, 5, 6}
2. (a) Minimal d and f; maximal a
   (b) First, none; last, a
   (c) There are eleven: *dfebca, dfecba, dfceba, fdebca, fdecda, fdceba, fedbca, fcdeba, fecdba, fcedba*

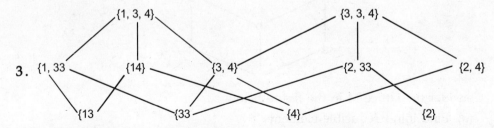

3.

4. (a) False, Example: N∪{a} where 1 << a, and N ordered by ≤.
   (b) True (c) True
5. (a) Minimal, a : maximal d and e
   (b) First a; last, none
   (c) Any subset which contains c and omits a; that is : c, cb, cd, ce, cbd, cbe, cde, cbde,
   (d) abd acd, ace
6. (a) Minimal a and b; maximal e and f. (b) First, none; last, none
   (c) ace, acf, bce, bcf, bdf
7. Minimal, (P, 2) where P is Prime, Maximal, none
8. (a) Both; (b) Sup (A) = 3; inf (A) = $\sqrt[3]{5}$

ANSWERS (EXERCISE 8.2)

1. (a) b_1, c (c) M has no limit points.
2. (a) Define $f : S \to N$ by $f(a) = 1, f(b) = 2, f(3) = 3, f(n) = n + 3$
   (b) The element a is a limit point of S', but M has no limit points.
3. (a) Six : $OabdI, OacdI, OadeI, ObceI, OaceI, OcdeI$
   (b) $a, b, e, o; b$ (ii) a, b, c
   (c) c and e are complements of a, b has no complement.
   (d) No, No
4. (a) a, b, c, g, o
   (b) a, b, c
   (c) $a; g; b;$ none
   (d) $I = avg, f = avb = avc, e = bvc, d = avc;$ none
   elements are join irreducible (e) No, No
6. Min. elements 2, 5 maximum elements 12, 20, 25, 2.
7. g. l. b = 1, l.v.b = 20

9.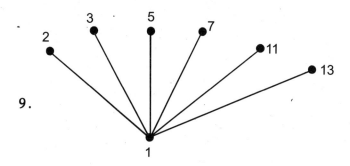

10.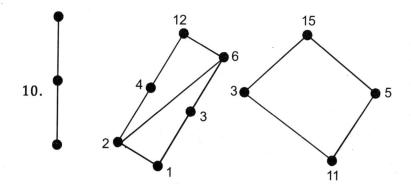

CHAPTER 9

BOOLEAN ALGEBRA

9.1 INTRODUCTION

We have already dealt extensively with the algebra of sets and mode mention of a related logical arithmetic. Now we will discuss Boolean algebra. Set algebra is a particular Boolean algebra and although different Boolean algebra are structurally very similar.

* Developed by famous mathematician **George Boole**.
* useful in analysis, design of electronic computers, dial telephone switching system and many kind of electronic control devices.

9.2 BOOLEAN ALGEBRA AND APPLICATIONS

A non empty set B together with two binary operations t, (known as addition and multiplication) and an separation t, i (complementation) on B is said to be Boolean algebra if it satisfied the following Axioms.

B_1 Commutativity : The operations are commutative i.e.,
$a + b = b + a$ and $a - b = b - a \; \forall \; ab \in B$

B_2 Distributivity : Each binary operations distributes over the other i.e.,
$a + (b.c) = (a + b).(a + c)$
$a.(b + c) = (a.b) + (a.c)$

B_3 Identity : B contains distinct identity elements 0 and 1 with respect to the operation + and. Respectively i.e.,
$a + 0 = a, a.1 = a \; \forall \; a \in B$.

B_4 Complementation : For every $a \in B$ there exist on element $a^1 \in B$ such that
$a + a^1 = 1$ and $aa^1 = 0$

* A boolean algebra is generally denoted by tuples
(B, +, 0, ', 0, 1) or B (B, +, 0, ') or by B.
* Instead of binary operations + and 0 we may use other symbols such as $\cup, \cap$ (Union Intersection) or $\wedge, \vee$ (join muss to denote these operations.)

Boolean Sub Algebra

A boolean sub algebra is a non empty subset S of a boolean algebra B such that $a, b \in S$ $\Rightarrow (a + b), a.b, a^1 \in S$.

Example 1. Let B • (0, a, b, 1) Define +, • and ' by the tables given below :

t	0	a	b	1
0	0	0	0	0
a	0	a	0	a
b	0	0	b	b
1	0	a	b	1

•	0	a	b	1
0	0	a	b	1
a	a	a	1	1
b	b	1	b	1
1	1	1	1	1

1	
0	1
a	b
b	a
1	0

Then B forms a boolean algebra under these operations

Example 2. Let B = [p, q, r, s] be a set addition and multiplication operations are defined on B as per table given below :

t	p	q	r	s
p	p	q	q	p
q	q	q	q	q
r	q	q	r	r
s	p	q	r	s

•	p	q	r	s
p	p	s	s	s
q	p	r	r	s
r	s	r	r	s
s	p	s	s	s

1. **+ and • are** binary operations since each element of the table is from the given set B.
2. **Commutativity :** From the table,
   (i) $p + q = q + p$
   (ii) $p.q = q.p \quad \forall \ pq \in B$
3. **Identity element :** From table it is clear that s is the additive identity
4. **Distributivity :**
   (i) $p.(q + r) = p.q = p$
   and $p.q + p.r = p + s = p$
   $\Rightarrow p(q + r) = p.q + p.r \ \forall \ p, q, r, s \in B$.
   (ii) We can also see that
   $p + qr = (p + q)(p + r) \ \forall \ p, q, r, \in B$
5. **Complement :** From table it can be verified that each element has get its complement such that $p + r = q$, $P.r = S \Rightarrow r$ is the complement of S. Hence B is a boolean algebra.

Example 3. Let B be a set of positive integers being divisions of 30 and operations ∨, ∧ on it are defined as

$a \vee b = c$ where c is the LCM of a, b.
$a \wedge b = d$ where d is the HCF of a, b. $\forall \ a, b, c, d \ \varepsilon \ B$

then show that B is a Boolean algebra.
B = (1, 2, 3, 5, 6, 10, 15, 30)

∨	1	2	3	5	6	10	15	30
1	1	2	3	5	6	10	15	30
2	2	2	6	10	6	10	30	30
3	3	6	3	15	6	30	15	30
5	5	10	15	5	30	10	15	30
6	6	6	6	30	6	30	30	30
10	1	2	3	10	30	10	30	30
15	15	30	15	15	30	30	15	30
30	30	30	30	30	30	30	30	30

∧	1	2	3	5	6	10	15	30
1	1	1	1	1	1	1	1	
2	1	2	1	1	2	2	1	2
3	1	1	3	1	3	1	3	3
5	1	1	1	5	1	5	5	5
6	1	2	3	1	6	2	3	6
10	1	2	1	3	2	10	5	10
15	1	1	3	5	3	5	13	13
30	2	2	3	5	6	10	15	30

9.3 THEOREMS OF BOOLEAN ALGEBRA

Theorem 9.1: In a boolean algebra

(I) Additive identity is unique.

(II) Multiplicative identity is unique.

(III) Complement of every element is unique.

Proof : Let (B, •+) is a boolean algebra. (I) Let if possible in B i_1 and i_2 be two additive identities then

When i_1 is additive identity $\Rightarrow a + i_1 = a$

and when i_2 is additive identity $\Rightarrow a + i_2 = a$

Hence, $\quad i_1 + i_2 = i_2$ $\hfill$ [Let i_1 is identity]

$\quad\quad\quad i_2 + i_1 = i_1$ $\hfill$ [Let i_2 is identity]

Since, $\quad i_1 + i_2 = i_2 + i$ $\hfill \Rightarrow i_i = i_2$

(II) Suppose it is possible e_1 and e_2 be two multiplicative identity in B.

$\quad\quad e_1 a = a$ and $e_2 a = a$

Also, $\quad e_1 e_2 = e_2$ $\hfill$ [When e_1 is identity]

$\quad\quad e_2 e_1 = e_1$ $\hfill$ [When e_2 is identity]

Since, $\quad e_1 e_2 = e_2 e_1$

$\Rightarrow \quad e_1 = e_2$

(III) Let if possible for $a \in B$. We have two different complement element b and c in b then

$\quad\quad a.b = 0$ and $a.c = 0$

$\quad\quad a + b = 1$ and $a + c = 1$

$\quad\quad b = b + 0$

$\quad\quad\quad = b + ac$

$\quad\quad\quad = (b + a)(b + c)$

$\quad\quad b = 1 . (b + c)$

$\quad\quad b = b + c$ $\hfill$ ...(1)

Similarly, $C = c + 0$
$= c + ab$
$= (c + a)(c + b)$
$= c + b$
$c = b + c$ ...(2)

(1) and (2) $\Rightarrow b = c$

Theorem 2 Idempotent law: If a be an element of a boolean Algebra then
(i) $a + a = a$ (ii) $a \cdot a = a$
$= [a + (b + c)]\, c + [a + (b + c)]\, c$
$= c\,[a + b(b + c)] + c_1\,[a + (b + c)]$
$= [ca + c(b + c)] + [c1\, a + c1\,(b + c)]$
$= (c a + c) +$

Theorem 9.2: Idempotent's law :

If a and b are arbitrary elements of a boolean algebra B then
(i) $(a + b)^1 = a^1 b^1$, (ii) $(ab)^1 = a^1 + b^1$

Proof : (i) Consider
$(a + b) + (a^1 b^1) = (aq + b + a^1)(a + b + b^1) = (a + a^1 + b)(a + b + b^1)$
$= (1 + b)(a + 1) = 1.1 = 1$ ...(1)
Again $(a + b) \cdot (a^1 b^1) = (a^1 \cdot b^1)(a + b) = (a^1 \cdot b^1)\, a + (a^1 \cdot b^1)\, b$
$= (b^1 a^1)\, a + (a^1 b^1)\, b$
$= b^1 (a^1 a) + a^1 (b^1 b)$
$= b^1.0 + a_1. 0$

1 and 2 $\Rightarrow$ Result $= 0 + 0 = 0$ ... (2)

(ii) $(a^1 + b^1) + (a \cdot b) = (a^1 + b^1 + a)(a^1 + b^1 + b) = (a^1 + a + b^1) \cdot (a^1 + b^1 + b)$
$= (1 + b^1)(a^1 + 1) = 1.1 = 1$ ...(1)
Also $ab\,(a^1 + b^1) = (ab)\, a^1 + (ab)\, b^1 = a^1 (ab) + (ab)d$
$= (a^1 a)\, b + a\,(bb) = 0.b + 9.0 = 0 + 0 = 0$ ...(2)

1 and 2 $\Rightarrow$ $(ab)^1 = (a^1 + b^1)$

Cancellation Law

Theorem 9.3: In a boolean algebra B of $b + a$ and $b + a^1 = c + a^1$ then $b = c$. Also if $ba = ca$ and $ba1 = ca^1$ then $b = c$

Proof: Assuming that $b + a = c + a$ and $b + a^1 = (+a)$
$b = b + 0 = b + aa_1$ Let, $ba = ca$ and $ba^1 = ca^1$
$= (b + a)(b + a^1)$ $b = b.1 = b\,(a + a^1) = ba + ba^1$
$= (c + a)(c + a^1)$ $= ca + ca^1 = c(a + a^1) = c.1$
$= c + aa^1 = c + 0$ $= c$.
$= c\; b = c$ $b = c$

9.4 LOGIC (DIGITAL) CIRCUITS AND BOOLEAN ALGEBRA

Some special type of net-works are used in digital computers for the processing of information in it. These net-works are represented by block diagrams. Logic circuits are structures which are built up from certain elementary circuits called logic gates.

Logic Gates

There are three basic logic gates.

(1) OR gate (OR)-Block used for OR gate is 'OR'. It converts two and more inputs into a single function given as follows:

Let x and y be two inputs, the output will be a function $(x \vee y)$ or $(x + y)$ as follows:

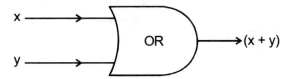

(2) AND gate (AG) block used for AG is 'AG'. If there be two or more inputs then the output will be a function of those input gives as below:

If x and y be two inputs then the output will be $(x \text{ and } y)$ i.e., $(x \wedge y)$ or $x \cdot y$ as follows:

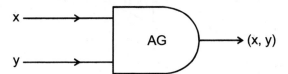

(3) NOT gate (or inverter). If the input be x then the output in converted into x' by an inverter.

Example : Write the Boolean function corresponding to the following network :

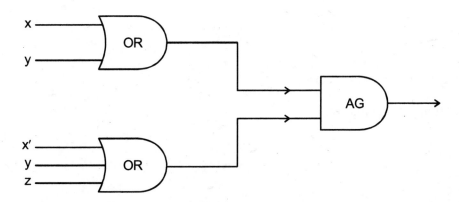

Required Boolean function is
$$(x + y)(x' + y + x')$$

Example : Design the gatting network for the function

(i) $x y + y z$ (ii) $(x + y)'$

Solution :

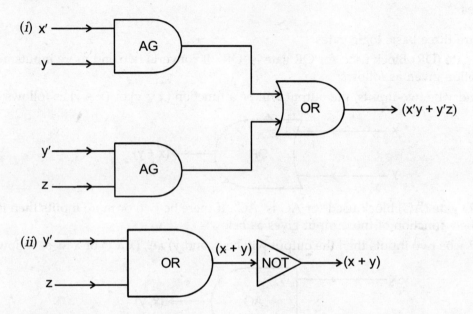

Many Terminal Networks

Any switching network having more than two terminate is called many terminal network.

If P, Q and R be the three terminals of any switching network then the transmittal of two terminals taken in pair will be denoted by T_{PQ}, T_{QR}, T_{RP}.

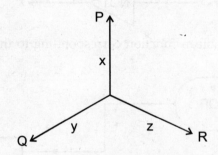

Four terminals network is

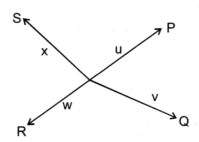

Symmetric Functions

Symmetric functions are those functions which remains unchanged by the interchange of any two variables

$$xy + x'y', x + y, x'y + xy' \quad \text{etc.}$$

OR Gate:

Input		Output
P	q	$r(P + q)$
1	1	1
1	0	1
0	1	1
0	0	0

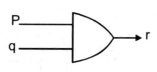

AND Gate:

Input		Output
P	q	P.q
1	1	1
1	0	0
0	1	0
0	0	0

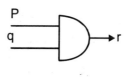

NOT Gate:

Input	Output
P	$r(\bar{P})$
1	0
0	1

Examples: 1. $(a + b) \cdot c$

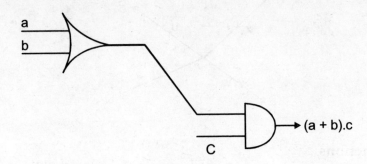

2. $(a\bar{b}) + (\bar{a}.b)$

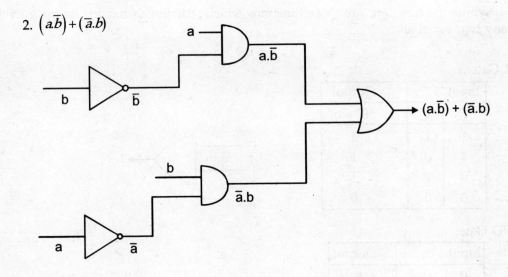

3. $(a+b).(\bar{a}+\bar{b})$

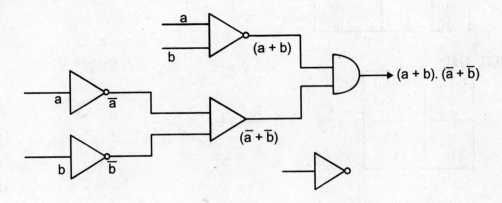

4. AND-to-OR logic network

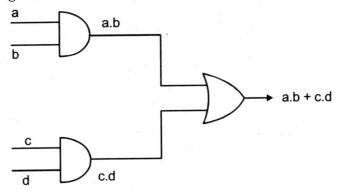

5. OR to AND logic network

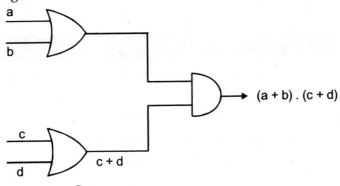

NAND Gate :

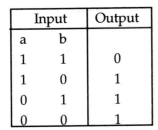

Input		Output
a	b	
1	1	0
1	0	1
0	1	1
0	0	1

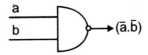

$\Leftrightarrow \overline{a}+\overline{b}$
$\Leftrightarrow a \uparrow b$

NOR Gate :

Input		Output
a	b	
1	1	0
1	0	0
0	1	0
0	0	1

$\rightarrow a \downarrow b$

$\Leftrightarrow \overline{a}+\overline{b}$

6. (a) $(P \wedge Q) \vee (\overline{7R} \wedge \overline{7P})$

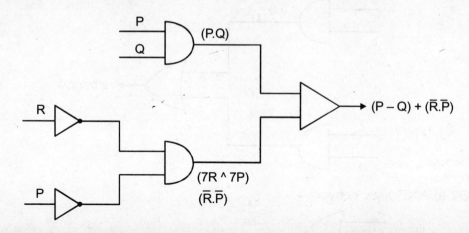

EXERCISE 9.1

1. If $[B + .]$ is a boolean algebra, prove that for all elements $a, b, c,$ of B
   (i) $ab + c(a^1 + b^1) = ab + c$
   (ii) $(a + b)(b^1 + c) + b(a^1 + c^1) = ab^1 + a + b$

True for Ordinary Algebra on Binary Algebra

2. Find whether the following statements are
   (i) $aa = a^2$ (ii) $aa = a$
   (iii) $0^1 = 1$ (iv) $1 + a = 1$
   (v) $a + a^1 = 1$ (vi) $0 + a = a$
   (vii) $aa^1 = 0$ (viii) $0 \, a = 0$
   (ix) $a(b + a) = ab + ac$ (x) $a + bc = (a + b)(a + c)$

3. Simplify the following products in boolean algebra :
   (i) $c(c + a)$ (ii) $S(S^1 + t^1)$
   (iii) $ab(a^1 + b^1)$ (iv) $ab(ab + a^1c)$
   (v) $(P + q^1)(P^1 + q)$ (vi) $a(a^1 + b + c)$
   (vii) $a^1(a + b^1 + c^1)$ (vii) $(a + b)(a^1 + b^1)$

4. Factories and simplify the following in boolean algebra.
   (i) $pq + pr$ (ii) $x^1 + x^1y$
   (iii) $cx^1e + cde$ (iv) $ab^1 + ab^1c^1$
   (v) $pq + qr + qr^1$ (vi) $ab(ab + bc)$
   (vii) $(cd^1 + d^1)(c + d)$ (viii) $(xy^1 + yz)(xz + yz^1)$
   (ix) $s^1(s + t)(s + t)$ (x) $(x + yz)(y^1 + x)(y^1 + z^1)$

5. By using the distractive law $a + bc = (a + b)(a + c)$ of boolean algebra factorise the following :
   (i) $x + yz$
   (ii) $ax + b$
   (iii) $t + Pqr$
   (iv) $ab^1 + a^1 b$
   (v) $x + x^1 y$
   (vi) $ab + ab^1$
   (vii) $a + bcd$
   (viii) $ab + cd$

6. Find the complement of the following
   (i) $p^1 + q$
   (ii) ab^1
   (iii) $xy^1 z$
   (iv) $c + d^1 + e$
   (v) $P + qx^1$
   (vi) $(P + q)(P + r^1)$

7. Find the complement of the following
   (i) $P^1 + q$
   (ii) ab^1
   (iii) $xy^1 z$
   (iv) $c + d^1 + e$
   (v) $P + qr^1$
   (vi) $(P + q)(P + r^1)$
   (vii) $ab^1 + ac + b^1 c$

8. Prove each of the following relation for elements $a, b, c,$ of a boolean algebra :
   (i) $c(a + b) + a^1 c + bc^1 = b + c$
   (ii) $a(a^1 + b) + (ab^1 + a) = a$
   (iii) $(a + b)^1 + (a + b^1)^1 = a^1$
   (iv) $[a^1 (a + b)]^1 + [b (b + a)]^1 + [b^1 (b^1 + a)] = 1$

Algebraic Manipulation of Boolean Expression

Example 1: If a, b, c are elements of any Boolean algebra then prove that
$$ba = ca \text{ and } ba' = ca' \Rightarrow b = c$$

Solution: Let $\quad ba = ca$ and $ba = ca$ ...(1)

Then, $\quad b = b.1 = b(a + a') = ba + ba$

$\quad\quad = ca + ca$ [from (1)]

$\quad\quad = c(a + a')$

$\quad\quad = c.1$

$\quad\quad = c$

$\Rightarrow \quad b = c$

Example 2: In any Boolean algebra B prove that
$$a + b = 1 \Leftrightarrow a + b = b \;\forall\; a, b, \in B$$

Solution: Let $\quad a + b = b$ ...(1)

Then, $\quad a' + (a + b) = (a' + a) + b$ [Associative law]

$\quad\quad = (a + a') + b$

$\quad\quad = 1 + b$

$\quad\quad = 1$

But, $\quad a' + (a + b) = a' + b$ [From (1)]

$\therefore \quad a' + b = 1$

Conversely: Let $\quad a' + b = 1$

Then, $\quad a + b = (a + b) \cdot 1$

$\qquad\qquad\quad = (a + b)(a' + b) \qquad\qquad [\because a' + b = 1]$

$\qquad\qquad\quad = (b + a)(b + a') \qquad\qquad$ [Commutative property]

$\qquad\qquad\quad = (b + aa')$

$\qquad\qquad\quad = b + a$

$\qquad\qquad\quad = b$

Hence, $a' + b = 1 \Leftrightarrow a + b = b$

Example 3: For any Boolean algebra B prove that
$$a + b = a + c \text{ and } ab = ac \Rightarrow b = c, \forall\ a, b, c, \in B.$$

Solution: $b = b.1 = b(1 + a) = b.1 + b.a \qquad\qquad [\because 1 + a = 1]$

$\qquad\qquad = b + ba$

$\qquad\qquad = bb + ba \qquad\qquad\qquad\qquad\qquad [\because bb = b]$

$\qquad\qquad = b(b + a)$

$\qquad\qquad = b(a + b)$

$\qquad\qquad = b(a + c) \qquad\qquad\qquad\qquad\qquad$ [Given]

$\qquad\qquad = b.a + b.c$

$\qquad\qquad = a.b + bc$

$\qquad\qquad = ac + bc \qquad\qquad\qquad\qquad\qquad\quad$ [Given]

$\qquad\qquad = (a + b).c$

$\qquad\qquad = c.(a + b)$

$\qquad\qquad = ca + c.c$

$\qquad\qquad = ca + c$

$\qquad\qquad = c(a + 1)$

$\qquad\qquad = c.1$

$\qquad\qquad = c$

Hence, $b = c$

Example 4: For any Boolean algebra B prove that
$(a + b) \cdot (b + c) \cdot (c + a) = a.b + b.c + c.a$

Solution: LHS $= (a + b) \cdot (b + c) \cdot (c + a)$

$\qquad\qquad = ((a + b)(a + c)(b + c)$

$\qquad\qquad = (a + b \cdot c)(b + a)$

$\qquad\qquad = (a + b \cdot c)b + (a + bc)c$

$\qquad\qquad = b(a + bc) + c(a + bc)$

$\qquad\qquad = b.a + bbc + ca + cbc$

$\qquad\qquad = a.b + (bb + bb')c + ca + ccb + cc'b$

$\qquad\qquad\qquad\qquad\qquad\qquad\qquad\qquad\qquad [bb' = 0;\ cc' = 0]$

$$= ab + b(b + b')c + ca + c(c + c')b$$
$$[b + b = 1 = c + c']$$
$$= a.b + bc + ca + cb$$
$$= ab + bc + ca = \text{RHS}$$

Example 5: If a, b, c are elements of any Boolean algebra then prove that
$$(a + b)(a' + c) = ac + a'b$$

Solution: LHS $= (a + b)(a' + c) = (a + b)a' + (a + b)c$
$$= a'(a + b) + c(a + b)$$
$$= aa' + a'b + c(a + a'b)$$
$$[\because a + a'b = (a + a')(a + b) = a + b]$$
$$= 0 + a'b + ca + ca'b$$
$$= a'b + ca'b + ca = ab + a'bc + ac$$
$$= a'b(a + c) + a.c = a'b.1 + a.c = a'b + ac$$
$$= ac + a'b = \text{RHS}$$

Example 6: If a, b, c are elements of a Boolean algebra then prove that
$$a + b.c = b(b + c)$$

Solution: $a + bc = (a + b)(a + c) = (b + a)(a + c) = (b + a).1.(a + c)$
$$= (b + a)(b + b')(a + c) = (b + ab')(a + c)$$
$$= (b + o)(a + c) \qquad [a \leq b \Rightarrow ab' = o]$$
$$= b(a + c) = \text{RHS}$$

Example 7: If P, q, r, a, b are elements of a Boolean algebra then prove that
(i) $Pqr + Pqr' + Pq'r + P'qr = Pq + qr + rP$
(ii) $ab + ab' + a'b + a'b' = 1$

Solution:
(i)
$$\text{LHS} = Pqr + Pqr' + Pq'r + P'qr$$
$$= Pq(r + r') + Pq'r + P'qr$$
$$= P.q.1 + Pq'r + P'qr = Pq + Pq'r + P'qr$$
$$= P(q + q'r) + P'qr = P(q + q')(q + r) + P'qr$$
$$= P.1(q + r) + P'qr = P(q + r) + P'qr = Pq + Pr + Pqr$$
$$= Pq(P + Pq)r = Pq + [(P + P').(P + q)]r$$
$$= Pq + [1.(P + q)]r = Pq + (P + q)r$$
$$= Pq + Pr + qr = Pq + qr + rP = \text{RHS}$$

(ii)
$$\text{LHS} = ab + a.b' + a'b + a'b' = a(b + b') + a'(b + b')$$
$$= a(1) + a'(1) = a + a' = 1 = \text{RHS}$$

Example 8: In a Boolean algebra $(B, +, \bullet, ')$, prove that
(i) $(a + b)' + (a + b')' = a'$
(ii) $ab + a'b' = (a' + b)(a + b')$

Solution : (*i*) We know by Demorgan's law that :
$$(ab)' = a' + b'$$
Hence, $(a + b)' + (a + b') = (a + b)(a + b')'$
$$= (a + bb') = (a + 0)'$$
$$= (a)' = a'$$
Hence, $(a + b)' + (a + b') = a'$

(*ii*) LHS $= ab + a'b' = (ab + a)(ab + b) = (a' + ab)(b' + ab)$
$$= (a' + a)(a' + b)(b' + a)(b' + b) = 1(a' + b)(a + b')$$
$$= (a' + b)(a + b') = \text{RHS}$$

Example 9: In a Boolean algebra B prove that
(i) $ab + [c(a' + b')] = ab + c$
(ii) $(a + a'b)(a' + ab) = b$

Solution : (*i*) LHS $= ab + [c(a' + b')] = ab + c(ab)' = (ab + c)(ab + (ab)')$
$$= (ab + c)1 = (ab + c) = \text{RHS}$$

(*ii*) LHS $= (a + a'b)(a' + ab) = (a + a')(a + b)(a' + a)(a' + b)$
$$= 1(a + b)1(a' + b) = (a + b)(a' + b) = (b + a)(b + a')$$
$$= b + aa'$$
$$= b + o = b = \text{RHS}$$

Example 10: In a Boolean algebra B prove that
(*i*) $[x'(x + y)]' + [y(y + x')]' + [y(y' + x)][y(y' + x)]' = 1$
(*ii*) $(x + yz)(y' + x)(y' + z') = x(y' + z')$

Solution : LHS $= [x'(x + y)]' + [y(y + x')] + [y(y' + x)]'$
$$= (x')' + (x + y) + (y)' + (y + x')' + y' + (y' + x')'$$
$$= x + x'y' + y' + y'x + y' + yx'$$
$$= x + x'y' + y' + y'x + yx'$$
$$= x + xy' + y' + y'x' + yx'$$
$$= x(1 + y') + y(1 + x') + yx'$$
$$= x + y' + yx' = x + x'y + y' = (x + x')(x + y) + y'$$
$$= x + y + y' = x + 1 = 1 = \text{RHS}$$

RHS $= (x + yz)(y' + x)(y' + z') = (x + y)(x + z)(x + y') + (y + z')$
$= (x + y)(x + y')(x + z)(y' + z') = (x + yy')(x + z)(y' + z')$
$= x(x + z)(y' + z') = (x + xz)(y' + z') = x(1 + z)(y' + z')$
$= x.1(y' + z') = x(y' + z') = \text{RHS}$

9.5 BOOLEAN FUNCTIONS, APPLICATIONS

Introduction. A *Boolean function* is a function from $\mathbb{Z}_n^2$ to $\mathbb{Z}_2$. For instance, consider the *exclusive-or* function, defined by the following table:

x_1	x_2	$x_1 \oplus x_2$
1	1	0
1	0	1
0	1	1
0	0	0

The exclusive-or function can interpreted as a function $\mathbb{Z}_2^2 \to \mathbb{Z}_2$ that assigns $(1, 1) \to 0$, $(1,0) \to 1$, $(0, 1) \to 1$, $(0, 0) \to 0$. It can also be written as a Boolean expression in the following way:

$$x_1 \oplus x_2 = (x_1 \cdot \overline{x_2}) + (\overline{x_2} \cdot x_2)$$

Every Boolean function can be written as a Boolean expression as we are going to see next.

Disjunctive Normal Form. We start with a definition. A *minterm* in the symbols $x_1, x_2, ..., x_n$ is a Boolean expression of the form $y_1 \cdot y_2 \cdots y_n$, where each y_i is either x_i or $\overline{x_i}$.

Given any Boolean function $f: \mathbb{Z}_2^n \to \mathbb{Z}_2$ that is not identically zero, it can be represented

$$f(x_1, ..., x_n) = m_1 + m_2 + ... + m_k,$$

where $m_1, m_2, ..., m_k$ are all the minterms $m_i = y_1 \cdot y_2 \cdots y_n$ such that $f(a_1, a_2, ..., a_n) = 1$, where $y_j = x_j$ if $a_j = 1$ and $y_j = \overline{x_j}$ if $a_j = 0$. That representation is called *disjunctive normal form* of the Boolean function f.

Example 9.1: We have seen that the exclusive-or can be represented $x_1 \oplus x_2 = (x_1 \cdot \overline{x_2})(\overline{x_1} \cdot x_2)$. This provides a way to implement the exclusive-or with a combinatorial circuit as shown in Figure below.

Conjunctive Normal Form. A *maxterm* in the symbols $x_1, x_2, ..., x_n$ is a Boolean expression of the form $y_1 + y_2 + ... + y_n$, where each y_i is either x_i or $\overline{x_i}$.

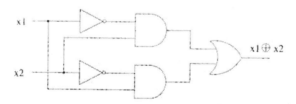

Figure 9.1. Exclusive-Or.

Given any Boolean function $f: \mathbb{Z}_2^n \to \mathbb{Z}_2$ that is not identically one, it can be represented

$$f(x_1, ..., x_n) = M_1 \cdot M_2 \cdots M_k.$$

where $M_1, M_2, ..., M_k$ are all the maxterms $M_i = y_1 + y_2 + ... + y_n$ such that $f(a_1, a_2, ..., a_n) = 0$, where $y_j = \overline{x_j}$ if $a_j = 0$ and $y_j = \overline{x_j}$ if $a_j = 1$. That representation is called *conjunctive normal form* of the Boolean function f.

Example: The conjunctive normal form of the exclusive-or is

$$x_1 \oplus x_2 = (x_1 + x_2) \bullet (\overline{x}_1 + \overline{x}_2).$$

Functionally Complete Sets of Gates. We have seen how to design combinatorial circuits using AND, OR and NOT gates. Here we will see how to do the same with other kinds of gates. In the following gates will be considered as functions from $\mathbb{Z}_2^n$ into $\mathbb{Z}_2$ intended to serve as building blocks of arbitrary boolean functions.

A set of gates $\{g_1, g_2, \ldots, g_k\}$ is said to be *functionally complete* if for any integer n and any function $f: \mathbb{Z}_2^n \to \mathbb{Z}_2$ it is possible to construct a combinatorial circuit that computes f using only the gates $g_1, g_2, \ldots, g_k$. *Example:* The result about the existence of a disjunctive normal form for any Boolean function proves that the set of gates {AND, OR, NOT} is functionally complete. Next we show other sets of gates that are also functionally complete.

1. The set of gates {AND, NOT} is functionally complete. Proof: Since we already know that {AND, OR, NOT} is functionally complete, all we need to do is to show that we can compute $x + y$ using only AND and NOT gates. In fact:

$$x + y = \overline{\overline{x} \cdot \overline{y}},$$

hence the combinatorial circuit of Figure 9.2 computes $x + y$.

Figure 9.2. OR with AND and NOT.

2. The set of gates {OR, NOT} is functionally complete. The proof is similar:

$$x \cdot y = \overline{\overline{x} + \overline{y}}$$

hence the combinatorial circuit of Figure 9.3 computes $x + y$.

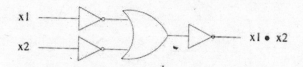

Figure 9.3. AND with OR and NOT.

3. The gate NAND, denoted $\uparrow$ and defined as

$$x_1 \uparrow x_2 = \begin{cases} 0 & \text{if } x_1 = 1 \text{ and } x_2 = 1 \\ 1 & \text{otherwise} \end{cases}$$

is functionally complete.

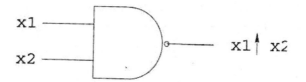

Figure 9.4. NAND gate.

Proof: Note that $x \uparrow y = \overline{x.y}$. Hence $x = \overline{x.y} = x_1 \uparrow x_2$, so the NOT gate can be implemented with a NAND gate. Also the OR gate can be implemented with NAND gates: $x + y = \overline{\overline{x}.\overline{y}} = (x_1 \uparrow x_2) \uparrow (y \uparrow y)$. Since the set {OR, NOT} is functionally complete and each of its element can be implemented with NAND gates, the NAND gate is functionally complete.

Minimization of Combinatorial Circuits. Here we address the problems of finding a combinatorial circuit that computes a given Boolean function with the minimum number of gates. The idea

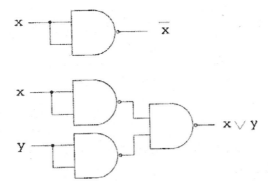

Figure 9.5. NOT and OR functions implemented with NAND gates.

is to simplify the corresponding Boolean expression by using algebraic properties such as (E . a) + (E . $\overline{a}$) = E and E + (E . a) = E, where E is any Boolean expression. For simplicity in the following we will represent a . b as a & b, so for instance the expressions above will look like this: Ea + E$\overline{a}$ = E and E + Ea = E.

Example: Let F(x, y, z) the Boolean function defined by the following table:

x	y	z	f(x, y, z)
1	1	1	1
1	1	0	1
1	0	1	0
1	0	0	1
0	1	1	0
0	1	0	0
0	0	1	0
0	0	0	0

Its disjunctive normal form is $f(x, y, z) = xyz + xy\overline{z} + x\overline{yz}$. This function can be implemented with the combinatorial circuit of Figure 9.6.

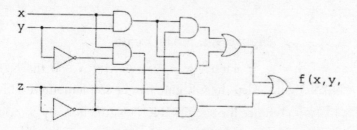

Figure 9.6. A circuit that computes $f(x, y, z) = xyz + xy\overline{z} + x\overline{yz}$.

But we can do better if we simplify the expression in the following way:

$$f(x, y, z) = \overbrace{xyz + xy\overline{z}}^{xy} + x\overline{yz}$$
$$= xy + x\overline{yz}$$
$$= x(y + \overline{yz})$$
$$= x(y + \overline{y})(y + \overline{z})$$
$$= x(y + \overline{z})$$

which corresponds to the circuit of Figure 9.7.

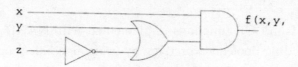

Figure 9.7. A simpler circuit that computes $f(x, y, z) = xyz + xy\overline{z} + x\overline{yz}$.

Multi-Output Combinatorial Circuits. *Example: Half-Adder.* A half-adder is a combinatorial circuit with two inputs x and y and two outputs s and c, where s represents the sum of x and y and c is the carry bit. Its table is as follows:

x	y	s	c
1	1	0	1
1	0	1	0
0	1	1	0
0	0	0	0

So the sum is $s = x \oplus y$ (exclusive-or) and the carry bit is $c = x \cdot y$. Figure 9.8 shows a half-adder circuit.

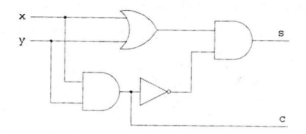

Figure 9.8. Half-adder circuit

EXERCISE 9.2

1. Reduce the following Boolean products to either 0 or a fundamental product:
   (a) $xyx'z$;
   (b) $xyzy$;
   (c) $xyz'yx$;
   (d) $xyz'yx'z'$.
2. Express each Boolean expression $E(x, y, z)$ as sum of products and then in its complete sum of products form:
   (a) $E = x(xy' + x'y + y'z)$;
   (b) $E = Z(x' + y) + y'$
3. Express $E(x, y, z) = (x' + y)' + x'y$ in its complete sum of products form.
4. Express each Boolean expression $E(x, y, z)$ as a sum of products form:
   (a) $E = y(x + yz)$;
   (b) $E = x(xy + y + x'y)$;
5. Express each set expression $E(A, B, C)$ involving sets A, B, C as a union of intersections.
   (a) $E = (A \cup B)^C \cap (C^C \cup B)$;
   (b) $E = (B \cap C)^C \cap (A^C \cap C)^C$
6. Let $E = xy' + xyz' + x'yz'$. Prove that:
   (a) $xz + E = E$;
   (b) $x + E \neq E$;
   (c) $Z' + E \neq E$
7. Let $E = xy' + xyz' + x'yz'$. Find
   (a) The prime implicants of E;
   (b) A minimum sum for E.
8. Let $E = xy + y't + x'yz' + xy'zt'$. Find
   (a) Prime implicants of E;
   (b) Minimal sum for E.
9. Express the output X as a Boolean expression in the inputs A, B, C for the logic circuits in fig. (a) and fig. (b)
   (a) The inputs to the first AND gate are A and B' and to the second AND gate are B and C. Thus $Y = AB' + B'C$.

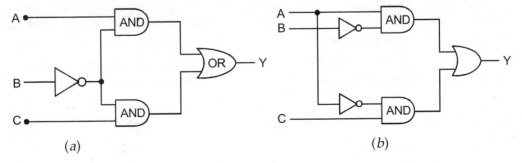

(a)   (b)

(b) The input to the first AND gate are A and B' to the second AND gate are A' and C. Thus X = AB' + A'C

10. Express the output Y as a Boolean expression in the inputs A, B, C for the logic circuit in the figure.

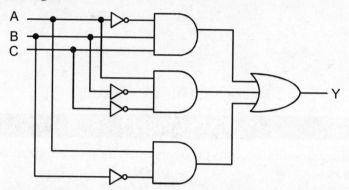

11. Express the output Y as a Boolean expression in the inputs A and B for the logic circuit in the figure.

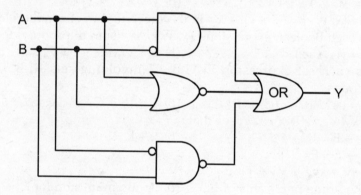

12. Find the output sequence Y for an AND gate with inputs A, B, C (or equivalently for Y = ABC) where :
    (a) A = 111001; B = 100101; C = 110011.
    (b) A = 11111100; B = 10101010; C = 00111100
    (c) A = 00111111; B = 11111100; C = 11000011

ANSWERS (EXERCISE 9.1)

2. (i) Only in ordinary algebra (ii) Only in Boolean algebra
   (iii) Only in Boolean algebra (iv) Only in Boolean algebra
   (vii) Only in Boolean algebra (viii) In the both algebras
   (ix) In the both algebras (x) In Boolean algebra
3. (i) c + cd; (ii) st'; (iii) O;

(iv) ab;   (v) $Pq + P'q'$;   (vi) $ab + ac$;
(vii) $a'b' + a'c'$;   (viii) $ab' + a'b$.

4. (i) $P(q + r)$;   (ii) x';   (iii) cl;
(iv) ab';   (v) q;   (vi) ab;
(vii) $cd + c'd'$;   (viii) xz;   (ix) o;
(x) $x(y' + z')$

5. (i) $(x + y)(x + z)$;   (ii) $(a + b)(x + b)$;   (iii) $(t + p)(t + q)(t + r)$;
(iv) $(a + b)(a' + b')$;   (v) $x + y$;   (vi) $(a + b')(b + a')$
(vii) $(a + b)(a + c)(a + d)$;   (vii) $(a + c)(b + c)(a + d)(b + d)$

6. (i) $a'b'c'$;   (ii) $a' + b' + c'$;   (iii) $P + q' + r$;
(iv) $(a' + b)c'$ (v) $(x' + y' + z)y + x'(y + z')$

7. (i) Pq';   (ii) $a' + b$;   (iii) $x' + y + z'$;
(iv) $c'de$; (v) $P'(q + r)$;   (vi) $P'q' + P'r = P'(q' + r)$; (vii) $(a' + b)(a' + c')(b + c')$

ANSWERS (EXERCISE 9.2)

1. (a) 0;   (b) xyz;   (c) xyz'
(d) 0;
2. (a) $xy'z + xy'z'$;   (b) $xyz + xy'z + x'yz + x'y'z + x'y'z'$
3. $xy'z + xy'z' + x'yz + x'yz'$
4. (a) $x'yz'$;   (b) $xyz + xyz' + xy'z + x'y'z'$
5. (a) $A^C \cap B^C \cap C^C$;   (b) $(A \cap B^C \cap C^C) \cup (A \cap C^C)$
7. (a) $xy' + xz' + yz'$;   (b) $xy' + yz'$
8. (a) $xy, y't, yz', xt, xz$ and $z't$ (b) $y't + xz + yz'$
9. (a) $Y = AB' + B'C$;   (b) $Y = AB' + A'C$;
10. $Y = A'BC + AB'C' + AB'$
11. (a) $Y = (A + BC)' + B$;   (b) $Y = (A'B)' + (A + C)'$
12. $Y = AB' + (A + B)' + (A'B)'$

CHAPTER 10

GRAPH THEORY

10.1 INTRODUCTION

The word *graph* refers to a specific mathematical structure usually represented as a diagram consisting of points joined by lines. In applications, the points may, for instance, correspond to chemical atoms, towns, electrical terminals or anything that can be connected in pairs. The lines may be chemical bonds, roads, wires or other connections. Applications of graph theory are found in communications, structures and mechanisms, electrical networks, transport systems, social networks and computer science.

10.2 GRAPHS—BASICS

A graph is a mathematical structure comprising a set of vertices, V, and a set of edges, E, which connect the vertices. It is usually represented as a diagram consisting of points, representing the **vertices** (or **nodes**), joined by lines, representing the **edges** (Figure 10.1). It is also formally denoted by G (V, E).

In a **labelled graph** the vertices have labels or names (Figure 10.2).

In a **weighted graph** each edge has a **weight** associated with it (Figure 10.3).

A **digraph (directed graph)** is a diagram consisting of points, called vertices, joined by directed lines, called **arcs** (Figure 10.4).

Figure 10.1

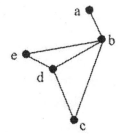
Figure 10.2

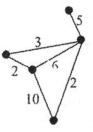
Figure 10.3

Figure 10.4

Two or more edges joined the same pair of vertices are called **multiple edges** (Figure 10.5). An edge joining a vertex to itself is a loop (Figure 10.5).

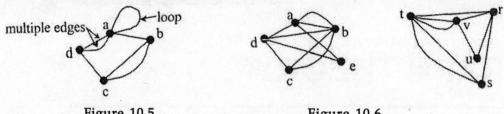

Figure 10.5   Figure 10.6

A graph with no multiple edges and no loops is a **simple graph** (e.g., Figure 10.1).

Two vertices joined by an edge are said to be **adjacent**.

Vertices are **incident** with the edges which joins them and an edge is **incident** with the vertices it joins.

Two graphs G and H are **isomorphic** if H can be obtained by relabelling the vertices of G, i.e. there is a one-one correspondence between the vertices and G and those of H such that the number of edges joining each pair of vertices in G is equal to the number of edges joining the corresponding pair of vertices in H (Figure 10.6 $a \leftrightarrow v, b \leftrightarrow t, c \leftrightarrow s, d \leftrightarrow r, e \leftrightarrow u$).

A **subgraph** of G is a graph all of whose vertices and edges are vertices and edges of G (Figure 10.7 shows a series of subgraph of G).

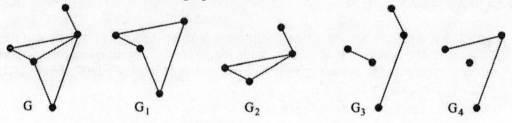

Figure 10.7

The **degree** of a vertex v is the number of edges incident with v, Loops count as 2.

The **degree sequence** of a graph G is the sequence obtained by listing, in ascending order with repeats, the degrees of the vertices of G (*e.g.* in Figure 10.7 the degree sequence of G is (1, 2, 2, 3, 4).)

The **Handshaking Lemma** states that the sum of the degrees of the vertices of a graph is equal to that twice the no. of edge this follow reality from the fact that each edge join two vertices necessarily distinct) and so contributes 1 to the degree of each of those vertices.

A walk of length k in a graph is a succession of k edges joining two vertices. NB Edges can occur more than once in a walk.

A trail is walk in which all the edges (but not necessarily all the vertices) are distinct.

A path is a walk in which all the edges and all the vertices are distinct.

So, in Figure 10.8, *abdcbde* is a walk of length 6 between *a* and *e*. It is not a trail (because edge *bd* is traversed twice). The walk *adcbde* is a trail length 5 between *a* and *e*. It is not a path (because vertex *d* is visited twice). The walk *abcde* is a path of length 4 between *a* and *e*.

A **connected graph** has a path between every pair of vertices. A **disconnected graph** is a graph which is not connected. e.g., Figure 10.7, G and the subgraphs G_1 and G_2 are connected whilst G_3 and G_4 are disconnected.

Figure 10.8

Every disconnected graph can be split into a number of connected subgraphs called its components. It may not be immediately obvious that a graph is disconnected. For instance Figure 10.9 shows 3 graphs, each disconnected and comprising 3 components.

  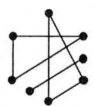

Figure 10.9

An edge in a connected graph is a **bridge** if its removal leaves a disconnected graph.

A **closed walk** or **closed trail** is a walk or trail starting and ending at the same vertex.

A **cycle** is a closed path, i.e. a path starting and ending at the same vertex.

Walks/trails/paths which are not closed are **open**.

In a **regular graph** all vertices have the same degree. If the degree is *r* the graph is *r*-**regular**.

If G is *r*-regular with *m* vertices it must have $1/2\ mr$ edges (from the Handshaking Theorem).

A **complete graph** is a graph in which every vertex is joined to every other vertex by exactly one edge. The complete graph with *m* vertices is denoted by K_m. K_m is $(m-1)$-regular and so has $1/2\ m(m-1)$ edges.

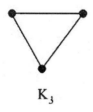 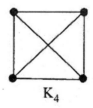 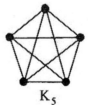

K_3 $\qquad\qquad K_4 \qquad\qquad K_5$

Figure 10.10

A **null graph** is a graph with no edges. The null graph with in vertices is denoted Nm is o-regular.

A **cycle graph** consines of a single cycle of vertices a edges. The cycle graph with m vertices is denoted cm

A **bipartite graph** is a graph whose vertices can be split into two subsets A and B in such a way that every edge of G joins a vertex in A with one in B. Figure 10.11 shows some bipartite graphs. Notice that the allocation of the nodes to the sets A and B can sometimes be done in several ways.

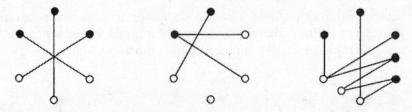

Figure 10.11

A **complete bipartite graph** is a bipartite graph in which every vertex in A is joined to every vertex in B by exactly one edge. The complete bipartite graph with r vertices in A and s vertices in B is denoted $K_{r,s}$. Figure 10.12 shows some complete bipartite graphs.

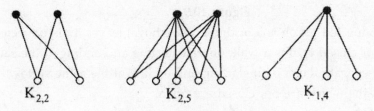

$K_{2,2}$ $K_{2,5}$ $K_{1,4}$

Figure 10.12

A **tree** is a connected graph with no cycles. In a tree there is just one path between each pair of vertices. Figure 10.13 shows some trees. Every tree is a bipartite graph. Start at any node, assign each node connected to that node to the other set, then repeat the process with those nodes!

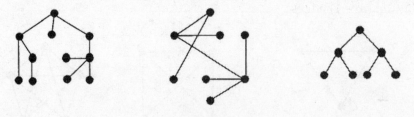

Figure 10.13

A **path graph** is a tree consisting of a single path through all its vertices. The path graph within vertices is denoted P_m. Figure 10.14 shows some path graphs.

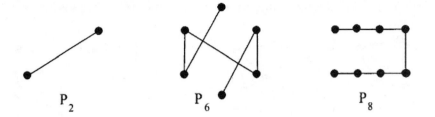

Figure 10.14

EXERCISE 10.1

1. Draw the graphs whose vertices and edges are as follows. In each case say if the graph is a simple graph.
   (a) V = {u, v, w, x}, E = {uv, vw, wx, vx}
   (b) V = {1, 2, 3 4, 5, 6, 7, 8}, E = {12, 22, 23, 34, 35, 67, 68, 78}
   (c) V = {n, p, q, r, s, t}, E = {np, nq, nt, rs, rt, st, pq}
2. Which of graph B, C and D are isomorphic to graph A? State the corresponding vertices in each isomorphic pair.

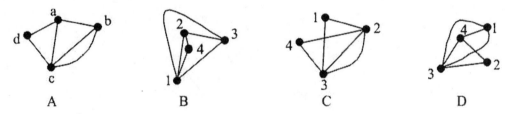

3. Which of the graphs P, Q, ... W are subgraphs of G?

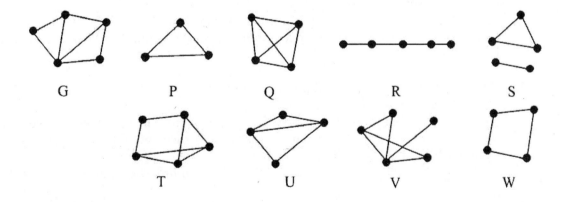

4. Write down the degree sequence of each of the following graphs:

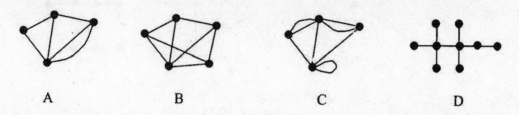

A   B   C   D

5. Graphs G_1 and G_2 have the same degree sequence—and they necessarily isomorphic? If your answer is no, give a counter example.
6. Graphs G_1 and G_2 are isomorphic. Do they necessarily have the same degree sequence? If your answer is no, give a counter example.
7. Draw simple connected graphs with the degree sequences:
   (a) (1, 1, 2, 3, 3, 4, 4, 6)
   (b) (3, 3, 3, 3, 3, 5, 5, 5)
   (c) (1, 2, 3, 3, 3, 4, 4)
8. Use the Handshaking Lemma to prove that the number of vertices of odd degree in any graph must be even.
9. Complete the following statements with walks/trail/path :

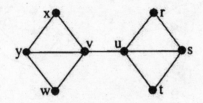

   (a) *wvyxvurs* is a .... of length .... between *w* and *s*
   (b) *vxvur* is a .... of length .... between *v* and *r*
   (c) *uvyxvw* is a .... of length .... between *u* and *w*
   (d) *ruvwy* is a .... of length .... between *r* and *y*
10. Write down all the paths between *x* and *t* in the graph in question 9. Which edge in this graph is a bridge?
11. Draw an *r*-regular graph with 6 vertices for $r = 3$ and $r = 4$.
12. Why are there no 3-regular graphs with 5 vertices ?
13. Draw the graphs K_5, N_5 and C_5.
14. Draw the complete bipartite graphs $K_{2,3}$, $K_{3,5}$, $K_{4,4}$. How many edges and vertices does each graph have? How many edges and vertices would you expect in the complete bipartite graphs $K_{r,s}$.
15. Under what conditions on *r* and *s* is the complete bipartite graph $K_{r,s}$ a regular graph ?
16. Show that, in a bipartite graph, every cycle has an even number of edges.
17. This is a more challenging question than the first sixteen.

The complement of a simple graph G is the graph obtained by taking the vertices of G (without the edges) and joining every pair of vertices which are not joined in G. For instance

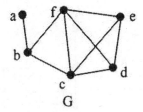

G

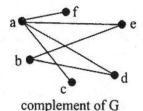

complement of G

(a) Verify that the complement of the path graph P_4 is P_4.
(b) What are the complement of K_4, $K_{3,3}$, C_5?
(c) What is the relationship between the degree sequence of a graph and that of its complement?
(d) Show that if a simple graph G is isomorphic to its complement then the number of vertices of G has the form $4k$ or $4k + 1$ for some integer k.
(e) Find all the simple graphs with 4 or 5 vertices which are isomorphic to their complements.
(f) Construct a graph with eight vertices which is isomorphic to its complement.

ANSWERS (EXERCISE 10.1)

1. 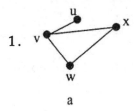 a   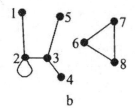 b   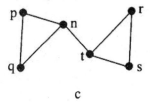 c

2. B is isomorphic to A with $1 \leftrightarrow c, 2 \leftrightarrow a, 3 \leftrightarrow b, 4 \leftrightarrow d$.
   C is not isomorphic to A.
   D is isomorphic to A with $1 \leftrightarrow b, 2 \leftrightarrow d, 3 \leftrightarrow c, 4 \leftrightarrow a$.

3. P, R, S, U, W are subgraphs of G.

4. Degree sequence of A is (2, 3, 3, 4).
   Degree sequence of B is (3, 3, 3, 3, 4).
   Degree sequence of C is (3, 3, 5, 5).
   Degree sequence of D is (1, 1, 1, 1, 1, 1, 2, 4, 4).

5. Both of the following graphs have degree sequence (3, 3) but they are not isomorphic.

6. If graphs G_1 and G_2 are isomorphic, then "there is a one-one correspondence between the vertices of G_1 and those of G_2 such that the number of edges joining each pair of

vertices in G_1 is equal to the number of edges joining the corresponding pair of vertices in G_2". Thus the degrees of the corresponding vertices in G_1 and G_2 are equal and so the degree sequences are the same.

7.

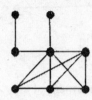

8. The Handshaking Lemma states that the sum of the degrees of all the vertices is twice the number of edges. Hence the degree sum is even. Since the sum of any number of even numbers is even and the sum of an even number of odd numbers is even, whilst the sum of an odd number of odd numbers is odd, the degree sum must be the sum of any number of even numbers and an even number of odd numbers. So the number of vertices of odd degree is even.

9. (a) *wvyxvurs* is a trail of length 7 between *w* and *s*.
    (b) *vxvur* is a walk of length 4 between *v* and *r*.
    (c) *uvyxvw* is a trail of length 5 between *u* and *w*.
    (d) *ruvwy* is a path of length 4 between *r* and *y*.

10. The possible paths are :

    | *xvut* | *xvust* | *xvurst* |
    | *xyvut* | *xyvust* | *xyvurst* |
    | *xywvut* | *xywvust* | *xywvurst* |

    The edge *vu* is a bridge since removing it produces a graph which comprises two disconnected sub graphs. There are no other bridges in the graph.

11.

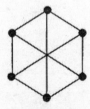 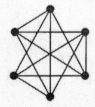

3-regular graph   4-regular graph

12. A 3-regular graph with 5 vertices would have a sum of vertex degree of 15, an odd number. But the sum of vertex degree is twice the number of edges and so is an even number. Hence there can be no 3-regular graph with 5 vertices.

13.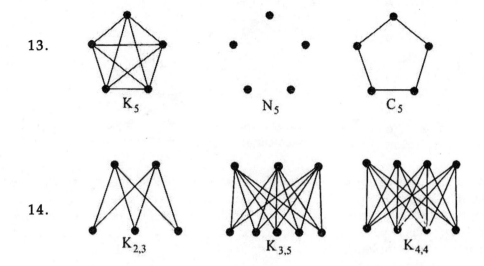

14.

The complete bipartite graphs $K_{2,3}$ has 5 vertices and 6 edges.
The complete bipartite graphs $K_{3,5}$ has 8 vertices and 15 edges.
The complete bipartite graphs $K_{4,4}$ has 8 vertices and 16 edges.
In general the complete bipartite graphs $K_{r,s}$ has $r + s$ vertices. Since each vertex in set A is connected to every vertex in set B we expect there to be rs edges in $K_{r,s}$.

15. In a complete bipartite graph every node is connected to every node in the other set so every node in set A has degree $|B|$ and every node in set B has degree $|A|$.

    This is a regular graph if $|A| = |B|$, i.e. $r = s$.

16. Taking any vertex in the cycle as a starting point. Suppose that this vertex is in set A, then the next vertex in the cycle is in set B and the following one in set A. In general every alternate vertex in the cycle is in set A. A cycle is a closed path so, eventually, the path leads back to the starting vertex. The vertex is in set A so the path length must be 2, 4, 6, 8, etc., i.e. the cycle has an even number of edges. If the starting vertex is in set B the argument is the same, substituting B for A.

17. (a) It can readily be seen that the complement of P_4 and P_4.

    P_4 complement of P_4

    (b) It can readily be seen that complement of K_4 is N_4, the complement of $K_{3,3}$ is a disjoint graph comprising a pair of C_3 s and the complement of C_5 is C_5.

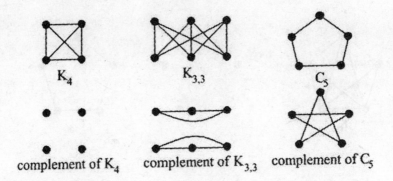

complement of K_4 complement of $K_{3,3}$ complement of C_5

(c) Each vertex in the complement of $G(V, E)$ is connected to every vertex in V to which it is not connected in G. Hence the sum of degree of a given vertex in G and its degree in the complement of G is one less than the number of vertices of G. So, if the graph has n vertices and degree sequence $(d_1, d_2, \dots d_n)$ then the degree sequence of the complement is $(n - 1 - d_n, n - 1 - d_{n-1}, \dots n - 1 - d_1)$.

(d) The edges of a graph G and of its complement make a complete graph so the sum of the vertex degrees of G and its complement is $n(n - 1)$. If a graph is isomorphic to its complement then the degree sequences of G and its complement are identical (see question 6) so the degree sums of G and its complement are equal. Hence the number of edges in G must be $n(n - 1)/4$. This must be an integer, k say, so we must have $n/4 = k$ or $(n - 1)/4 = k$, i.e. $n = 4k$ or $4k + 1$.

(e) For a graph with 4 vertices and a degree sequence (p, q, r, s) the degree sequences of the complement is $(3\text{-}s, 3\text{-}r, 3\text{-}q, 3\text{-}p)$. If the graph is isomorphic to its complement then the degree sequences are identical (see question 6) so we have $p = 3\text{-}s$ and $q = 3\text{-}r$. The degree sequences are in ascending order so the only possible sequences are $(0, 0, 3, 3)$, $(0, 1, 2, 3)$, $(1, 1, 2, 2)$. For $(0, 0, 3, 3)$ only 2 vertices have incident edges so there must be multiple edges or loops and the graph is not simple. For $(0, 1, 2, 3)$ only 3 vertices have incident edges so any vertex of degree greater than 2 is incident to multiple edges or a loop and so the graph is not simple. The only graph with degree sequence $(1, 1, 2, 2)$ is the path graph P_4 and we have already shown that P_4 is isomorphic to its complement (part (a)).

For a graph with 5 vertices and a degree sequence (p, q, r, s, t) the degree sequence of the complement is $(4\text{-}t, 4\text{-}s, 4\text{-}r, 4\text{-}q, 4\text{-}p)$. If the graph is isomorphic to its complement then the degree sequences are identical (see question 6) so we have $p = 4\text{-}t$, $q = 4\text{-}s$ and $r = 4\text{-}r$, that is $p + t = 4$, $q + s = 4$ and $r = 2$. The degree sequences are in ascending order so the only possible sequences are $(0, 0, 2, 4, 4)$, $(0, 1, 2, 3, 4)$, $(0, 2, 2, 2, 4)$, $(1, 1, 2, 3, 3)$, $(1, 2, 2, 2, 3)$, $(2, 2, 2, 2, 2)$. For $(0, 0, 2, 4, 4)$ only 3 vertices have incident edges so any vertex of degree greater than 2 incident to multiple edges or a loop and so the graph is not simple. For $(0, 1, 2, 3, 4)$ and $(0, 2, 2, 2, 4)$ only 4 vertices have incident edges so any vertex of degree greater than 3 is incident to multiple edges or a loop and so the graph is not simple. The only graph with degree sequence $(2, 2, 2, 2, 2)$ is the cycle graph C_5 and we have already shown that

C_5 is isomorphic to its complement (part (b)). The only simple graph with degree sequence (1, 1, 2, 3, 3) is A shown below and is isomorphic to its complement. The only simple graphs with degree sequence (1, 2, 2, 2, 3) are B and C shown below and the complement of B is C and, of course, vice versa.

So the only simple graph with 4 vertices which is isomorphic to its complement is P_4 and the only simple graphs with 5 vertices which are isomorphic to their complements are C_5 and the graph A below.

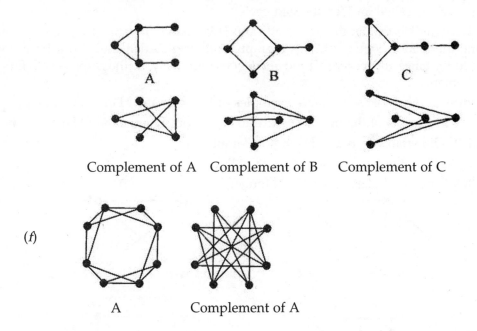

(f)

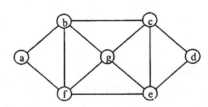

10.3 EULERIAN AND HAMILTONIAN GRAPHS

Consider the following map of 7 towns and the roads connecting them.

Figure 10.15

A highway engineer (E) wishes to inspect all the roads whilst an Egon Ronay inspector (H) wishes to dine in a restaurant in each town. Having studied Engineering Mathematics, each wishes to achieve their objective in as efficient a way as possible. So E states her aim as "I wish, if possible, to traverse every road once and only once and return to my starting point" whilst H says "I wish, if possible, to visit each town once and only once and return to my starting point".

A range of real problems give rise to versions of these two objectives, so graph theory has formalised them in the following way.

An **Eulerian graph** is a connected graph wishes contains a closed trail which includes every edge. The trail is called an **Eulerian trail**.

A **Hamiltonian graph** is a connected graph which contains a cycle which includes every vertex. The cycle is called an **Hamiltonian cycle**.

So E is saying "I want an Eulerian trail" and H is saying "I want a Hamiltonian cycle". Considering the map in Figure 10.15 as a graph, both an Eulerian trail and a Hamiltonian cycle exist, for instance *abcdecgefgbfa* and *abcdegfa* respectively. So the graph is both Eulerian and Hamiltonian.

In Figure 10.16 we see some more examples of Eulerian and Hamiltonian graphs.

Graph 1 (the graph of the map in Figure 10.16) is both Eulerian and Hamiltonian.

Graph 2 is Eulerian (*e.g. bcgefgb*) but not Hamiltonian.

Graph 3 is Hamiltonian (*e.g. bcgefb*) but not Eulerian.

Graph 4 is neither Eulerian nor Hamiltonian.

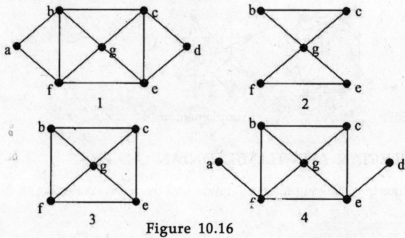

Figure 10.16

Eulerian Graphs

Theorem 10.1. A connected graph is Eulerian iff every vertex has even degree.

To prove this we first need a simpler theorem.

Theorem 10.2. If G is a graph all of whose vertices have even degree, then G can be split into cycles no two of which have an edge in common.

Proof. Let G be a graph all of whose vertices have even degree. Start at any vertex u and traverse edges in an arbitrary manner, without repeating any edge. Since every vertex has even degree it is always possible to find a different edge by which to leave a vertex. Since there is a finite number of vertices, eventually we must arrive at a vertex, v say, which we have already visited. The edges visited since the previous visit to v constitute a closed cycle, C_1 say. Remove all the edges in C_1 from the graph, leaving a subgraph G_1 say. Since we have removed a closed cycle of edges the vertices of G_1 will either have the same degree as the vertices of G or degrees 2 less than the equivalent vertices of G—either way G_1 is a graph all of whose vertices have been degree. We repeat the process with G_1, finding a cycle C_2, removing the edges in this cycle from G_1 and leaving G_2. Continue in this way until there are no edges left. Then we have a set of cycles, $C_1, C_2, C_3, \ldots$ which together include all edges of G and no two of which have an edge in common.

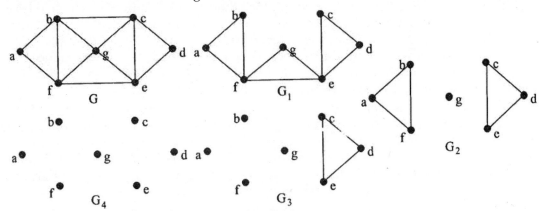

Figure 10.17

For instance, traverse *abcgb*—found cycle C_1 = *bcgb*—form G_1—traverse *degfe*—found cycle C_2 = *egfe*—form G_2—traverse *bafb*—found cycle C_3 = *bafb*—from G_3—traverse *edce*—found cycle C_4 = *edce*—from G_4—no edges left. The graph G can be split into cycles {*bcgb, egfe, bafb, edce*}. The decomposition is, of course, not unique. For instance we could equally decompose G into {*bcefb, gcdeg, abgfa*}.

Now we can prove **Theorem 2.1.** A connected graph is Eulerian iff every vertex has even degree.

Proof. First we prove "If a graph G is Eulerian then each vertex of G has even degree." Since G is Eulerian there exists an Eulerian trail. Every time the trail passes through a vertex it traverses two different edges incident with the vertex and so adds two to the degree of the vertex. Since the trail is Eulerian it traverses every edge, so the total degree of each vertex is 2 or 4 or 6 or ...., i.e. of the form $2k$, k = 1, 2, 3, ... Hence every has even degree.

Now we prove "If each vertex of a connected graph G has even degree then G is Eulerian." Since all the vertices of G have even degree, by theorem 10.2 G can be decomposed into a set of cycles no two of which have an edge in common. We will fit these cycles together to create an Eulerian trail. Start at any vertex of a cycle C_1. Travel round C_1 until we find a vertex which belongs also to another cycle, C_2 say. Travel round C_2 and then continue along C_1 until we reach the starting point. We have closed trail C_{12} which includes all the edges of C_1 and C_2. If this includes all the edges of G we have the required Eulerian trail, otherwise repeat the process starting at any point of C_{12} and travelling around it until we come to a vertex which is a member of another cycle, C_3 say. Travel round C_3 and then continue along C_{12} thus creating a closed trail C_{123}. Continue the process until we have a trail which includes all the edges of G and that will be an Eulerian trail in G.

We have proved both "If a graph G is Eulerian then each vertex of G has even degree" and "If each vertex of a connected graph G has even degree then G is Eulerian" and so we have "A connected graph is Eulerian iff every vertex has even degree".

A **semi-Eulerian graph** is a connected graph which contains an open trail which includes every edge. The trail is called a semi-Eulerian trail.

Theorem 10.3. A connected graph is semi-Eulerian iff exactly two vertices have odd degree.

Proof. (a) If G is a semi-Eulerian graph then there is an open trail which includes every edge. Let u and v be the vertices at the start and end of this trail. Add the edge uv to the graph. The graph is now Eulerian and so every vertex has even degree by theorem 10.1. If the added edge is now removed the degrees of the vertices u and v edge are reduced by one and so are odd, the degrees of all other vertices are unaltered and are even. So if G is semi-Eulerian it has exactly two vertices of odd degree.

(b) Suppose G is a connected graph with exactly two vertices of odd degree. Let those two vertices be u and v. Add an edge uv to the graph. Now every vertex of G has even degree and so G is Eulerian. The Eulerian trail in G includes every edge and so includes the edge uv. Now remove the edge uv, then there is a trail starting at vertex u and ending at vertex v (or vice versa) which includes every edge. Hence if G is a connected graph with exactly two vertices of odd degree then G is semi Eulerian.

Hence we see that a connected graph is semi-Eulerian iff exactly two vertices have odd degree.

Hamiltonian Graphs

No simple necessary and sufficient condition for a graph to be Hamiltonian is known—this is an open area of research in graph theory.

But we can identify some classes of graph which are Hamiltonian. Obviously the cycle graph C_n is Hamiltonian for all n. The complete graph K_n is Hamiltonian for all $n \geq 3$.—obvious because, if the vertices are denoted $\{v_1, v_2, ... v_n\}$ then the path $v_1 v_2 v_3 ... v_n v_1$ is a Hamiltonian path.

If we add an edge to a Hamiltonian graph then the resulting graph is Hamiltonian—the Hamiltonian cycle in the original graph is also a Hamiltonian cycle in the enhanced graph. Adding edges may make a non-Hamiltonian graph into a Hamiltonian graph but cannot convert a Hamiltonian graph into a non-Hamiltonian one so graphs with high vertex degrees are more likely to be Hamiltonian then graphs with small vertex degrees. Ore's theorem is one possible more precise statement relating Hamiltonian graphs and their vertex degrees.

Ore's Theorem (stated without proof) : If G is a simple connected graph with n vertices ($n \geq 3$) then G is Hamiltonian if deg (v) + deg $(w) \geq n$ for every non-adjacent pair of vertices v and w.

If G is a simple connected graph with n vertices ($n \geq 3$) then G is Hamiltonian if deg $(v) \geq n/2$ for every vertex v. This follows from Ore's theorem. From this we can determine that all the complete bipartite graphs $K_{p,p}$ are Hamiltonian (the degree of every vertex is p, the graph has $2p$ vertices, hence deg $(v) \geq n/2$ $(=p)$ for every vertex).

A **semi-Hamiltonian graph** is a connected graph which contains a path, but not a cycle, which includes every vertex. The path is called a **semi-Hamiltonian path**.

The Königsberg Bridges

Dr. Martin mentioned, in his introduction, the classic Königsberg Bridges problem. Königsberg lies at the confluence of two rivers. In the middle of the town lies an island, the Kneiphof. Spanning the rivers are seven bridges as shown in Figure 10.18.

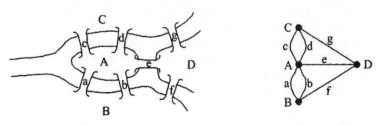

Figure 10.18   Figure 10.19

The citizens of Königsberg used to amuse themselves, during their Sunday afternoon strolls, trying to devise a route which crossed once only the returned to its starting point. Leonhard Euler finally proved (during the 1730s) that the task was impossible. We can see this readily using the result of theorem 10.1. Represent Königsberg and its bridges as the graph in Figure 10.19 where the vertices represent the different areas of land and the edges the bridges. What the citizens of Königsberg were seeking was an Eulerian trail through this graph. Theorem 2.1 tells us the graph is Eulerian iff every vertex has been degree. The degree sequence of the graph in Figure 10.19 is (3, 3, 3, 5) so it is not Eulerian and no Eulerian trail exists. If the citizens of Königsberg relaxed the requirement to finish up where they started (in other words to seek a semi-Eulerian trail) they would still fail since theorem 10.3 tells us that a graph a semi-Eulerian if every vertex bar two has even degree. Figure 10.19 fails this test also.

10.4 ISOMORPHISM

Two graphs G_1 and G_2 are said to be isomorphic to each other if there is a one to one correspondence between their vertices and between their edges so that the incidence relationship is maintained.

It means that if in graph G_1 an edge e_k is incident with vertices v_1 and v_j then in graph G_2 its corresponding edge e'_k must be incident with the vertices v'_i and v'_j that correspondent to the vertices v_i and v_j respectively.

The following two graphs G_1 and G_2 are isomorphic graphs.

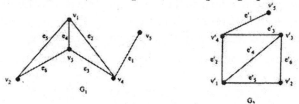

Figure 10.20.

Vertices v_1, v_2, v_3, v_4 and v_5 in G_1 corresponds to v'_1, v'_2, v'_3, v'_4 and v'_5 respectively in G_2. Edges e_1, e_2, e_3, e_4, e_5 and e_6 in G_1 corresponds to $e'_1, e'_2, e'_3, e'_4, e'_5$ and e'_6 respectively in G_2.

Here we can see that if any edge is incident with two vertices in G_1 then its corresponding edge shall be incident with the corresponding vertices in G_2 e.g. edges e_1, e_2 and e_3 are incident on vertex v_4, then the corresponding edges e'_1, e'_2 and e'_3 shall be incident on the corresponding vertex v'_4. In the way the incidence relationship shall be preserved.

In fact isomorphic graphs are the same graphs drawn differently. The difference is in the names or labels of their vertices and edges. The following two graphs are also isomorphic graphs in which vertices a, b, c, d, p, q, r, and s in G_1 corresponds to vertices $v_1, v_2, v_3, v_4, v_5, v_6, v_7$ and v_8 respectively in G_2 and edges $e_1, e_2, e_3, e_4, e_5, e_6, e_7, e_8, e_9, e_{10}, e_{11}$ and e_{12} in G_1 corresponds to edges $e'_1, e'_2, e'_3, e'_4, e'_5, e^x_6, e'_7, e'_8, e'_9, e'_{10}, e'_{11}$ and e'_{12} in G_2 to preserve the incidence relationship.

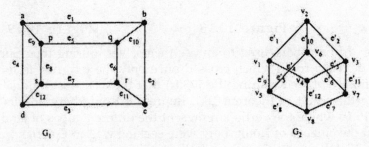

Figure 10.21

The incidence relationship between vertices and edges in between corresponding vertices and edges in G_2.

The following two graphs are not isomorphic

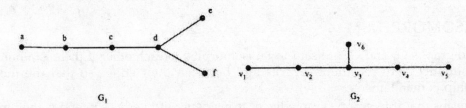

Figure 10.22

Vertex d in G_1 corresponds to vertex v_3 in G_2 as these are the only two vertices of degree 3.

In G_1, there are two pendant vertices adjacent to vertex d, while in G_2 there is only one pendant vertex adjacent to the corresponding vertex v_3. Thus the relationship of adjacency and incidence is not preserved and the two graphs are not isomorphic.

There is no simple and efficient criterion to identify isomorphic graphs.

10.5 ISOMORPHIC DIGRAPHS

Two digraphs are said to be isomorphic if,

(a) Their corresponding undirected graphs are isomorphic.

(b) Directions of the corresponding edges also agree.

The following two digraphs are not isomorphic because the directions of the two corresponding edges e_4 and e'_4 do not agree (although their corresponding undirected graphs are isomorphic).

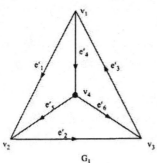

Figure 10.23

Isomorphism may also be defined as follows:

Isomorphism from a graph $G_1 = (V_1, E_1)$ to $G_2 = (V_2, E_2)$ is defined as mapping $f : V_1 \to V_2$ such that

(a) f is one–one and onto

(b) edge $v_i\, v_j \in E_1$ if and
only if $f(v_i) \cdot f(v_j) \in E_2$

where $f(v_1)$ and $f(v_j)$ are the images of v_i and v_j respectively in graph G_2

10.6 SOME PROPERTIES OF ISOMORPHIC GRAPHS

1. Number of vertices in isomorphic graphs is the same.

2. Number of edges in isomorphic graphs is also the same.

3. Each one of the isomorphic graphs has an equal number of vertices with a given degree.

This property is utilized in identifying two non-isomorphic graphs by writing down the degree sequence of their respective vertices.

Illustration: The degree sequence of graph G_1 is 4, 2, 2, 2, 2, 2, 1, 1 and that of G_2 is 3, 3, 2, 2, 2, 2, 1, 1 which are not the same. Therefore G_1 and G_2 are non-isomorphic

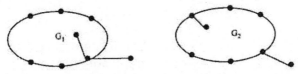

Figure 10.24

10.7 SOME OBSERVATION ON ISOMORPHISM

Let G_1 be a graph with vertices $v_0, v_1, v_2, \ldots v_n$ and f is an isomorphism from G_1 to G_2, then:

(i) G_2 will be connected only if G, is connected, because a path from v_i to v_j in G, induces a path from $f(v_i)$ to $f(v_j)$ in G_2.

(ii) If a Hamiltonian circuit exists in G_1, then a similar circuit shall exist in G_2, because if $v_0, v_1, v_2, \ldots, v_n, v_0$ is is a Hamiltonian circuit in G, then $f(v_0), f(v_1), f(v_2), \ldots, f(v_n), f(v_0)$ must be a Hamiltonian circuit in G_2 (as it is a circuit and $f(vi) \neq f(v_j$ for $0 <= i < j <= n)$.

(iii) Both G_1 and G_2 shall have the same number of isolated vertices, because if any vertex v_i is an isolated vertex in G_1 then its corresponding vertex $f(v_i)$ shall also be isolated in G_2.

(iv) If G_1 is bipartite, than G_2 shall also be bipartite.

(v) If G, has an Euler circuit, than G_2 shall also have an Euler circuit,

(vi) Both G_1 and G_2 shall have the same crossing number because they can be drawn the same way in the plane.

10.8 HOMEOMORPHISM GRAPHS

Let G_1 be a given graph. Another graph G_2 can be obtained from this graph by dividing an edge of G, with additional vertices.

Two graphs G_1 and G_2 are said to be Homeomorphic if these can be obtained from the same graph or isomorphic graphs by this method. Graphs G_1 and G_2, though not isomorphic, are Homeomorphic because each of them can be obtained from graph G by adding appropriate vertices.

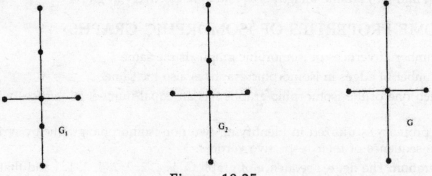

Figure 10.25

10.9 DIGRAPHS

Digraphs are similar to graphs except that the edges have a direction. To distinguish them from undirected graphs the edges of a digraph are called arcs. Much of what follows is exactly the same as for undirected graphs with 'arc' substituted for 'edge'. We will attempt to highlight the differences.

A **digraph** consists of a set of elements, V, called **vertices**, and a set of elements, A, called **arcs**. Each arc joins two vertices in a specified direction.

Two or more arcs joining the same pair of vertices in the same direction are called **multiple arcs** (see Figure 10.26). NB two arcs joining the same vertices in opposite directions are not multiple arcs.

An arc joining a vertex to itself is a loop (see Figure 10.26).

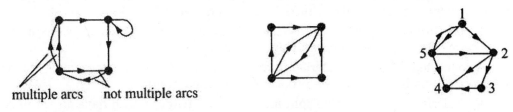

multiple arcs   not multiple arcs

Figure 10.26   Figure 10.27   Figure 10.28

A digraph with no multiple arcs and no loops is a **simple digraph** (e.g. Figure 10.27, Figure 10.28).

Two vertices joined by an arc in either direction are **adjacent**.

Vertices are **incident** to and form the arc which joins them and an arc is incident to and from the vertices its joins.

Two digraphs G and H are **isomorphic** if H can be obtained by relabelling the vertices of G, i.e. there is a one-one correspondence between the vertices of G and those of H such that the number and direction of the arcs joining each pair of vertices in G is equal to the number and direction of the arc joining the corresponding pair of vertices in H.

A **subdigraph** of G is a digraph all of whose vertices and arcs are vertices and arcs of G.

The underlying graph of a digraph is the graph obtained by replacing of all the arcs of the digraph by undirected edges.

The **out-degree** of a vertex v is the number of arcs incident from v and the **in-degree** of a vertex v is the number of arcs incident to v. Loops count as one of each.

The **out-degree sequence** and **in-degree sequence** of a digraph G are the sequences obtained by listing, in ascending order with repeats, the out-degrees and in-degrees of the of the vertices of G.

The **Handshaking Lemma** states that the sum of the out-degrees and of the in-degrees of the vertices of a graph are both equal to the number of arcs. This is pretty obvious since every arc contributes one to the out-degree of the vertex from which it is incident and one to the in-degree of the vertex to which it is incident.

A **walk of length k** in a digraph is a succession of k arcs joining two vertices.

A **trail** is a walk in which all the arcs (but not necessarily all the vertices) are distinct.

A **path** is a walk in which all the arcs and all the vertices are distinct.

A connected digraph is one whose underlying graph is a connected graph. A **disconnected digraph** is a digraph which is not connected. A digraph is **strongly connected** if there is a path between every pair of vertices. Notice here we have a difference between graphs and digraphs. The underlying graph can be connected (a path of edges exists between every pair of vertices) whilst the digraph is not because of the directions of the arcs (see Figure 10.28 for a graph which is connected but not strongly connected).

A **closed walk/trail** is a walk/tail starting and ending at the same vertex.

A **cycle** is a closed path, i.e. a path starting and ending at the some vertex.

An **Eulerian digraph** is a connected digraph which contains a closed trail which includes every arc. The trail is called an **Eulerian trail**.

A **Hamiltonian digraph** is a connected digraph which contains a cycle which includes every vertex. The cycle is called an **Hamiltonian cycle**.

A connected digraph is Eulerian iff for each vertex the out-degree equals the in-degree. The proof of this is similar to the proof for undirected graphs.

An Eulerian digraph can be split into cycles no two of which have an arc in common. The proof of this is similar to the proof for undirected graphs.

EXERCISE 10.2

1. Which of the following graphs are Eulerian and/or Hamiltonian. Give an Eulerian trail or a Hamiltonian cycle where possible.

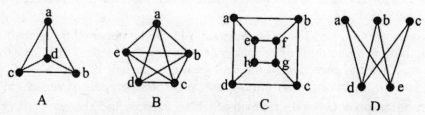

2. Which of the graphs K_8, $K_{4,4}$, C_6, $K_{2,5}$ are Eulerian graphs (use theorem 10.1 to decide). For those which are Eulerian, find an Eulerian trail.

3. Which of the following graphs are semi-Eulerian? Give a semi-Eulerian trail where possible. (Use theorem 10.3 to decide).

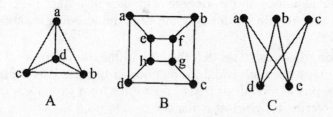

4. By finding a Hamiltonian cycle show that the complete bipartite graph $K_{3,3}$ is Hamiltonian. Show that the complete bipartite graph $K_{2,4}$ is not Hamiltonian. What condition on r and s is necessary for the complete bipartite graph $K_{r,s}$ to be Hamiltonian?

5. Which of the following graphs are semi-Hamiltonian? Give a semi-Hamiltonian path where possible.

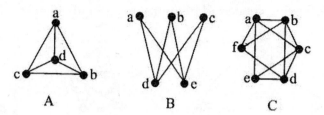

6. Check whether the conditions of Ore's theorem hold for these Hamiltonian graphs.

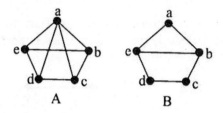

7. Draw the digraphs whose vertices and arcs are as follows. In each case say if the digraph is a simple digraph.
    (a) V = {u, v, w, x}, A = {vw, wu, wv, wx, xu}
    (b) V = {1, 2, 3, 4, 5, 6, 7, 8}, A = {12, 22, 23, 24, 34, 35, 67, 68, 78}
    (c) V = {n, p, q, r, s, t}, A = {np, nq, nt, rs, rt, st, pq, tp}

8. Which two of the following digraphs are identical, which one is isomorphic to the two identical ones and which is unrelated? Write down the in-degree sequence and the out-degree sequence for each digraph.

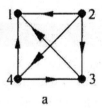 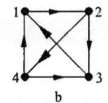 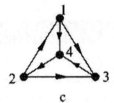 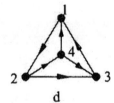

a b c d

9. Digraphs G_1 and G_2 have the same in-degree and out-degree sequence—are they necessarily isomorphic? If your answer is no, give a counter example.

10. Digraphs G_1 and G_2 are isomorphic. Do they necessarily have the same in-degree and out-degree sequences? If your answer is no, give a counter example.

11. In the digraph shown give (if possible)
    (a) a walk of length 7 from u to w,
    (b) cycles of length 1, 2, 3 and 4, and
    (c) a path of maximum length

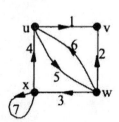

12. Which of the following connected digraphs are strongly connected?

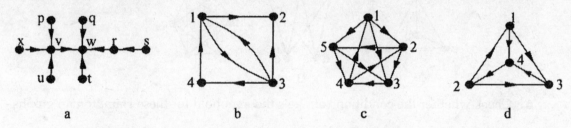

13. Are the following digraphs Eulerian/semi-Eulerian and/or Hamiltonian/semi-Hamiltonian? Give the Eulerian/semi-Eulerian trails and the Hamiltonian cycles/semi-Hamiltonian paths where they exist.

 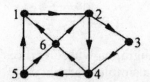

14. In the digraph shown, find
    (a) all the cycles of length 3, 4, and 5,
    (b) an Eulerian trail, and
    (b) a Hamiltonian cycle.

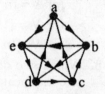

ANSWERS (EXERCISE 10.2)

1. (A) This graph is 3-regular with 4 vertices. It is not Eulerian because its vertices all have odd degree. It has a Hamiltonian cycle *abcda* and so is Hamiltonian.

   (B) This graph is 4-regular. All its vertices are of even degree so it is Eulerian. An Eulerian trail is *abecadbcdea*. It has a Hamiltonian cycle *abcdea* and so is Hamiltonian.

   (C) This graph is 3-regular. All its vertices are of odd degree so it is not Eulerian. It has a Hamiltonian cycle *abcdhgfea* and so is Hamiltonian.

   (D) This graph has 2 vertices of odd degree so it is not Eulerian. It is the complete bipartite graph $K_{2,3}$ so to construct a Hamiltonian cycle we necessarily have to visit vertices from the set {*a, b, c*} and the set {*d, e*} alternately. Start at any vertex in {*a, b, c*}, go to a vertex in {*d, e*} then to a different vertex in {*a, b, c*} then to the other vertex in {*d, e*} then to the only unvisited vertex in {*a, b, c*}. Now in order to get back to the starting vertex we must visit another vertex in {*d, e*}. But we have visited both of those already so we cannot return to the start without revisiting a vertex which is already in the walk. Thus no Hamiltonian cycle exists and the graph is not Hamiltonian.

2. K_8 is 7-regular, so all its vertices are of odd degree and it is not Eulerian.

$K_{4,4}$ is 4-regular, so all its vertices are of even degree and it is Eulerian. An Eulerian trail, referred to the diagram below, is *aebfcgdhcedfagbha*.

C_6 is 2-regular, so all its vertices are of even degree and it is Eulerian. Since it's a cycle graph the whole graph constitutes an Eulerian cycle.

$K_{2,5}$ has 5 vertices of degree 2 and 2 vertices of degree 5. Not all its vertices are of even degree so it is not Eulerian.

3. (A) This graph is 3-regular with 4 vertices. It is not semi-Eulerian because it has more than two vertices of odd degree.

   (B) This graph is 3-regular with 8 vertices. It is not semi-Eulerian because it has more than two vertices of odd degree.

   (C) This graph has 3 vertices of degree 2 and 2 vertices of degree 3 so it is semi-Eulerian (because exactly two vertices have odd degree). A semi-Eulerian trail is *daecdbe*.

4. A Hamiltonian cycle in $K_{3,3}$ is *adbecfa* (see figure below) so $K_{3,3}$ is Hamiltonian.

   To construct a Hamiltonian cycle is $K_{2,4}$ we need to visit vertices from the set B = {a, b, c, d} and the set A = {e, f} alternately. Start at any vertex in B, go to a vertex in A then to a different vertex in B then to the other vertex in A then to another vertex in B. Now in order to visit the remaining vertices in B and to get back to the starting vertex we must visit another vertex in A. But we have visited both of those already

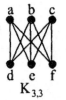
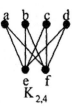

   so we cannot return to the start without re-visit a vertex which is already in the walk. Thus no Hamiltonian cycle exists and the graph is not Hamiltonian.

   A generalisation of this argument demonstrates that the only complete bipartite graphs, $K_{r,s}$, which are Hamiltonian are those with $r = s$.

5. (A) This graph is, in fact, Hamiltonian so it is not semi-Hamiltonian.

   (B) This graph is not Hamiltonian but *adbec* is a semi-Hamiltonian path so it is semi-Hamiltonian.

   (C) This graph is, in fact, Hamiltonian (*fabcedf* is a Hamiltonian cycle) so it is not semi-Hamiltonian.

6. (A) The non-adjacent vertex pairs are (c, e) and (b, d). We have n = 5, deg (c) + deg (e) = 6 and deg (b) + deg (d) = 6, so the conditions of Ore's theorem hold.

   (B) The non-adjacent vertex pairs are (a, d), (a, c), (c, e) and (b, d). We have n = 5, deg (a) + deg (d) = 4 = deg (a) + deg (c) and deg (c) + deg (e) = 5 = deg (b) + deg (d), so the conditions of Ore's theorem do not hold.

7.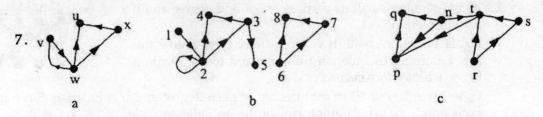

    a   b   c

8. Digraphs (b) and (d) are identical. Digraph (c) is isomorphic with the assignment $c(1) \to b(4)$, $c(2) \to b(2)$, $c(3) \to b(3)$, $c(4) \to b(1)$.
   (a) In-degree sequence is (0, 1, 2, 3) and out-degree sequence is (0, 1, 2, 3).
   (b) In-degree sequence is (1, 1, 2, 2) and out-degree sequence is (1, 1, 2, 2).
   (c) In-degree sequence is (1, 1, 2, 2) and out-degree sequence is (1, 1, 2, 2).
   (d) In-degree sequence is (1, 1, 2, 2) and out-degree sequence is (1, 1, 2, 2).

9. Both of the following digraphs have in-degree sequence (1, 2) and out-degree sequence (1, 2) but they are not isomorphic.

10. If digraphs G_1 and G_2 are isomorphic, then "there is a one-one correspondence between the vertices of G_1 and those of G_2 such that the arcs joining each pair of vertices in G_1 are identical in number and direction to the arcs joining the corresponding pair of vertices in G_2". Thus the in-degrees and out-degrees of the corresponding vertices in G_1 and G_2 are equal and so the in-degree and out-degree sequences of the digraphs are the same.

11. (a) 5374565, 5377745 are walks of length 7 from u to w,
    (b) 7 is a cycle of length 1, 56 is a cycle of length 2, 453 is a cycle of length 3 and there are no cycles of length 4.
    (c) 453, 452, 341, are all paths of length 3 which is the maximum length in this digraph.

12. (a) Not strongly connected. There is no path from, e.g. v to r.
    (b) Not strongly connected. There is no path from 2 to any other vertex.
    (c) Is strongly connected. Paths are (12, 13, 154, 15), (2341, 23, 234, 25), (341, 3412, 34, 3415), (41, 42, 423, 415), (541, 5412, 53, 54).
    (d) Is strongly connected. Paths are (142, 13, 14), (21, 23, 214), (3421, 342, 34), (412, 42, 423).

13. (a) In-degree is equal to out-degree for each vertex so the digraph is Eulerian. An Eulerian trail is 1532541. The digraph is not Hamiltonian but it is semi-Hamiltonian, a semi-Hamiltonian path is 41532.
    (b) In-degree is not equal to out-degree for each vertex so the digraph is not Eulerian. The digraph is Hamiltonian. A Hamiltonian cycle is 1234561.

14. (a) Each vertex has equal in-degree and out-degree so this digraph is Eulerian.
The cycles of length 3 are *aeda, aeca, abda, becb, bdcb*.
The cycles of length 4 are *aedca, abeda, abeca, abdca, bedcb*.
The cycles of length 5 are *abedca*.
(b) *abedcaecbda* is an Eulerian trail
(c) *abedca* is a Hamiltonian cycle.

10.10 MATRIX REPRESENTATIONS

The adjacency matrix A(G) of a graph G with n vertices is an $n \times n$ matrix with a_{ij} being the number of edges joining of edges joining the vertices i and j.

$$\begin{bmatrix} 0 & 1 & 0 & 1 \\ 1 & 1 & 1 & 2 \\ 0 & 1 & 0 & 1 \\ 1 & 2 & 1 & 0 \end{bmatrix}$$

The adjacency matrix A(D) of a digraph D with n vertices is an $n \times n$ matrix with a_{ij} being the number of arcs from vertex i to vertex j.

$$\begin{bmatrix} 0 & 1 & 0 & 1 \\ 0 & 1 & 0 & 2 \\ 0 & 1 & 0 & 0 \\ 0 & 0 & 1 & 0 \end{bmatrix}$$

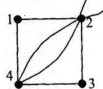

What is the number of walks of length 2 from vertex i to vertex j? There are, for instance, two walks of length 2 from 1 to 4. And so on. The matrix of these is

$$\begin{bmatrix} 0 & 1 & 1 & 2 \\ 0 & 1 & 2 & 2 \\ 0 & 1 & 0 & 2 \\ 0 & 1 & 0 & 0 \end{bmatrix}$$

and we see that this is just $A(D)^2$. In fact this generalises.

Theorem 10.4. The number of walks of length k from vertex i to vertex j in a digraph D with n vertices is given by the ijth element of the matrix A^k where A is the adjacency matrix of the digraph.

Proof. We prove this result by mathematical induction.

Assume that the result is true for $k \leq K - 1$. We will show that it is then also true for K. Consider any walk from vertex i to vertex j of length K. Such a walk consists of a walk of length $K - 1$ from vertex i to a vertex p which i adjacent to vertex j followed by a walk of length 1 from vertex p to vertex j. The number of such walks is $[A^{K-1}]_{ip} \times A_{pj}$. The total number of walks of length k from vertex i to vertex j will then be the sum of the walks through any p, i.e. $\sum_{p=1}^{n}\left[A^{K-1}\right]_{ip} A_{pj}$ but this is just the expression for the ijth element of the matrix $A^{K-1}A = A^K$ so the result is true for $k = K$. But the result is certainly true for walks of length 1, i.e. $k = 1$, because that is the definition of the adjacency matrix A. Hence the theorem is true for all k.

Now we can create a method of automating the determination of whether a given digraph is strongly connected or not. For the graph to be strongly connected there must be paths (of any length) from every vertex to every other vertex. The length of these paths cannot exceed $n - 1$ where n is the number of vertices in the graph (otherwise a path would be visiting at least one vertex at least twice). So the number of paths from a vertex i to a vertex j of any length from 1 to $n - 1$ is the sum of the ijth elements of the matrices A, A^2, A^3, ... A^{n-1}. So we introduce the matrix $B = A + A^2 + A^3 + \ldots A^{n-1}$ whose element B_{ij} represent the number of paths between all the vertices. If any off-diagonal element of B is zero then there are no paths from some vertex i to some vertex j. The digraph is strongly connected provided all the off-diagonal elements are non-zero!

Theorem 10.5. If A is the adjacency matrix of a digraph D with n vertices and B is the matrix $B = A + A^2 + A^3 + \ldots A^{n-1}$ then D is strongly connected iff each nondiagonal element of B is greater then 0.

Proof. To prove this theorem we must show both "if each non diagonal element of B is greater then 0 then D is strongly connected" and "if D is strongly connected then each non diagonal element of B is greater then 0".

Firstly, let D be a digraph and suppose that each non-diagonal element of the matrix B > 0, i.e. $B_{ij} > 0$ for all $i \neq j$. Then $[A^k]_{ij} > 0$ for some $k \in [1, n-1]$, i.e. there is a walk of some length k between 1 and $n - 1$ from every vertex i to every vertex j. So the digraph is strongly connected.

Secondly, suppose the digraph is strongly connected. Then, by definition, there is a path from every vertex i to every j. Since the digraph has n vertices the path is of length no more than $n - 1$.

Hence, for all $i \neq j$, $[A^k]_{ij} > 0$ for some $k \leq n - 1$. Hence, for all $i \neq j$, $B_{ij} > 0$.

Returning to the example of the digraph D in figure 4.2 above we have

$$A = \begin{bmatrix} 0 & 1 & 0 & 1 \\ 0 & 1 & 0 & 2 \\ 0 & 1 & 0 & 0 \\ 0 & 0 & 1 & 0 \end{bmatrix}, A^2 = \begin{bmatrix} 0 & 1 & 1 & 2 \\ 0 & 1 & 2 & 2 \\ 0 & 1 & 0 & 2 \\ 0 & 1 & 0 & 0 \end{bmatrix}, A^3 = \begin{bmatrix} 0 & 2 & 2 & 2 \\ 0 & 3 & 2 & 2 \\ 0 & 1 & 2 & 2 \\ 0 & 1 & 0 & 2 \end{bmatrix},$$

$$B = A + A^2 + A^3 = \begin{bmatrix} 0 & 4 & 3 & 5 \\ 0 & 5 & 4 & 6 \\ 0 & 3 & 2 & 4 \\ 0 & 2 & 1 & 2 \end{bmatrix}$$

so the graph is not strongly connected because we cannot get from vertex 2, vertex 3 or vertex 4 to vertex 1. Inspecting the digraph that is intuitively obviously! But, of course, this method is valid for large and complex digraphs which are less amenable to ad hoc analysis.

If we are only interested in whether there is at least one path from vertex i to vertex j (rather than wanting to know how many paths), then all of this can also be done using Boolean matrices. In this case the Boolean matrix (in which $a_{ij} = 1$ if there is at least one arc from vertex i to vertex j and 0 otherwise) of the graph is

$$\begin{bmatrix} 0 & 1 & 0 & 1 \\ 0 & 1 & 0 & 1 \\ 0 & 1 & 0 & 0 \\ 0 & 0 & 1 & 0 \end{bmatrix}$$

and the calculation of A^2, A^3 etc. is done using Boolean arithmetic (so × is replaced by $\wedge$ and + by $\vee$) so

$$A = \begin{bmatrix} 0 & 1 & 0 & 1 \\ 0 & 1 & 0 & 1 \\ 0 & 1 & 0 & 1 \\ 0 & 0 & 1 & 0 \end{bmatrix}, A^2 = \begin{bmatrix} 0 & 1 & 1 & 1 \\ 0 & 1 & 1 & 1 \\ 0 & 1 & 0 & 1 \\ 0 & 1 & 0 & 0 \end{bmatrix}, A^3 = \begin{bmatrix} 0 & 1 & 1 & 1 \\ 0 & 1 & 1 & 1 \\ 0 & 1 & 1 & 1 \\ 0 & 1 & 0 & 1 \end{bmatrix},$$

$$R = A \cup A^2 \cup A^3 = \begin{bmatrix} 0 & 1 & 1 & 1 \\ 0 & 1 & 1 & 1 \\ 0 & 1 & 1 & 1 \\ 0 & 1 & 1 & 1 \end{bmatrix}$$

A^2 is a matrix in which $(A^2)_{ij} = 1$ if there is at least one walk of length 2 from vertex i to vertex j and 0 otherwise and R is a matrix in which $R_{ij} = 1$ if there is at least one walk of length less than n from vertex i to vertex j and 0 otherwise. R is called the **reachability matrix**. In general R is easier and quicker to compute than B because it uses Boolean arithmetic. But, better, there is an even faster method called Warhsall's algorithm.

Let $D = (V, A)$ where $V = \{v_1, v_2, \ldots v_n\}$ and there are no multiple arcs. Warshall's algorithm computes a sequence of $n + 1$ matrices, $M_0, M_1, \ldots M_n$. For each $k \in [0 \ldots n]$, $[M_k]_{ij} = 1$ iff there is a path in G from v_i to v_j whose interior nodes come only from $\{v_1, v_2, \ldots v_k\}$. Warshall's algorithm is

    procedure Warshall (var M:n x n matrix);
    {initially M = A, the adjacency matrix of G}
    begin
        for $k := 1$ **to** n **do**
            for $i := 1$ **to** n **do**
                for $j := 1$ **to** n **do**
                      $M[i, j] := M[i, j] \vee (M[i, k] \wedge M[k, j])$;
    end;

Example. Find the reachability matrix of the digraph G using Warhall's algorithm.

Lets look at this in detail. The following table shows the derivation of the elements of M_1 from those of M_0. Notice that we use the updated elements of M as soon as they are available!

k	i	j		
1	1	1	$M_1[1,1] := M_0[1,1] \vee (M_0[1,1] \wedge M_0[1,1])$	$M_1[1,1] := 0 \vee (0 \wedge 0) = 0$
1	1	2	$M_1[1,2] := M_0[1,2] \vee (M_1[1,1] \wedge M_0[1,2])$	$M_1[1,2] := 1 \vee (0 \wedge 1) = 1$
1	1	3	$M_1[1,3] := M_0[1,3] \vee (M_1[1,1] \wedge M_0[1,3])$	$M_1[1,3] := 0 \vee (0 \wedge 0) = 0$
1	1	4	$M_1[1,4] := M_0[1,4] \vee (M_1[1,1] \wedge M_0[1,4])$	$M_1[1,4] := 1 \vee (0 \wedge 1) = 1$
1	2	1	$M_1[2,1] := M_0[2,1] \vee (M_0[2,1] \wedge M_1[1,1])$	$M_1[2,1] := 0 \vee (0 \wedge 0) = 0$
1	2	2	$M_1[2,2] := M_0[2,2] \vee (M_1[2,1] \wedge M_1[1,2])$	$M_1[2,2] := 1 \vee (0 \wedge 1) = 1$
1	2	3	$M_1[2,3] := M_0[2,3] \vee (M_1[2,1] \wedge M_1[1,3])$	$M_1[2,3] := 0 \vee (0 \wedge 0) = 0$
1	2	4	$M_1[2,4] := M_0[2,4] \vee (M_1[2,1] \wedge M_1[1,4])$	$M_1[2,4] := 1 \vee (0 \wedge 1) = 1$
1	3	1	$M_1[3,1] := M_0[3,1] \vee (M_0[3,1] \wedge M_1[1,1])$	$M_1[3,1] := 0 \vee (0 \wedge 0) = 0$
1	3	2	$M_1[3,2] := M_0[3,2] \vee (M_1[3,1] \wedge M_1[1,2])$	$M_1[3,2] := 1 \vee (0 \wedge 1) = 1$
1	3	3	$M_1[3,3] := M_0[3,3] \vee (M_1[3,1] \wedge M_1[1,3])$	$M_1[3,3] := 0 \vee (0 \wedge 0) = 0$
1	3	4	$M_1[3,4] := M_0[3,4] \vee (M_1[3,1] \wedge M_1[1,4])$	$M_1[3,4] := 0 \vee (0 \wedge 1) = 0$
1	4	1	$M_1[4,1] := M_0[4,1] \vee (M_0[4,1] \wedge M_1[1,1])$	$M_1[4,1] := 0 \vee (0 \wedge 0) = 0$
1	4	2	$M_1[4,2] := M_0[4,2] \vee (M_1[4,1] \wedge M_1[1,2])$	$M_1[4,2] := 0 \vee (0 \wedge 1) = 0$
1	4	3	$M_1[4,3] := M_0[4,3] \vee (M_1[4,1] \wedge M_1[1,3])$	$M_1[4,3] := 1 \vee (0 \wedge 0) = 1$
1	4	4	$M_1[4,4] := M_0[4,4] \vee (M_1[4,1] \wedge M_1[1,4])$	$M_1[4,4] := 0 \vee (0 \wedge 1) = 0$

$$M_0 = \begin{bmatrix} 0 & 1 & 0 & 1 \\ 0 & 1 & 0 & 1 \\ 0 & 1 & 0 & 0 \\ 0 & 0 & 1 & 0 \end{bmatrix}, M_1 = \begin{bmatrix} 0 & 1 & 0 & 1 \\ 0 & 1 & 0 & 1 \\ 0 & 1 & 0 & 0 \\ 0 & 0 & 1 & 0 \end{bmatrix}, M_2 = \begin{bmatrix} 0 & 1 & 0 & 1 \\ 0 & 1 & 0 & 1 \\ 0 & 1 & 0 & 1 \\ 0 & 0 & 1 & 1 \end{bmatrix},$$

$$M_3 = \begin{bmatrix} 0 & 1 & 0 & 1 \\ 0 & 1 & 0 & 1 \\ 0 & 1 & 0 & 1 \\ 0 & 1 & 1 & 1 \end{bmatrix}, M_4 = \begin{bmatrix} 0 & 1 & 1 & 1 \\ 0 & 1 & 1 & 1 \\ 0 & 1 & 1 & 1 \\ 0 & 1 & 1 & 1 \end{bmatrix}$$

The major advantage of Warshall's algorithm is that it is computationally more efficient. The number of operations for a digraph with n vertices is $O(n^3)$ whilst the number of operations required to compute R from $R = A \cup A^2 \cup A^3 \cup ... \cup A^{n-1}$ is $O(n^4)$.

To see this proceed thus :

Let an *and* take m and an *or* s msec. For the power method, to compute each element of A^2 takes $n*m + (n-1)*s$. A has n^2 elements to compute the whole of A^2 takes $n^2*(n*m + (n-1)*s)$. To compute A^3 we can multiply A times A^2 and this just takes the same time as computing A^2. For a graph with n vertices there are $n-2$ matrix multiplications and then an *or* of $n-1$ matrices, so the total time is $(n-1)*(n^2*(n*m+(n-1)*s))+n^2*(n-2)*s = (n^4-2*n^3)*(m+s) = O(n^4)$.

For Warshall's algorithm, each basic operation takes one *and* and one *or*. The triple loop means there are n^3 basic operations so the total time taken is $n^3*(m+s) = O(n^3)$.

Overall therefore, for a small graph we can compute the reachability matrix by hand using the power method relatively quickly. But if we are looking at a larger graph (think of 50 vertices and then consider the situation for 500 or 5000 vertices), we need computational help and Warshall's algorithm will take $1/n$ of the compute time taken by the power method.

EXERCISE 10.3

1. Write down the adjacency matrices of the following graphs and digraphs.

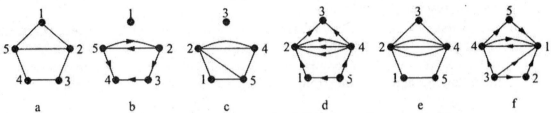

    a   b   c   d   e   f

2. Draw the graphs corresponding to the following adjacency matrices.

$$\begin{bmatrix} 0 & 1 & 1 & 1 & 0 \\ 1 & 0 & 0 & 0 & 1 \\ 1 & 0 & 0 & 0 & 1 \\ 1 & 0 & 0 & 0 & 1 \\ 0 & 1 & 1 & 1 & 0 \end{bmatrix} \quad \begin{bmatrix} 0 & 2 & 0 & 1 & 1 \\ 2 & 0 & 0 & 1 & 1 \\ 0 & 0 & 0 & 0 & 0 \\ 1 & 1 & 0 & 0 & 2 \\ 1 & 1 & 0 & 2 & 0 \end{bmatrix} \quad \begin{bmatrix} 0 & 1 & 1 & 1 & 0 & 0 \\ 1 & 0 & 0 & 1 & 0 & 0 \\ 1 & 0 & 0 & 1 & 0 & 0 \\ 1 & 1 & 1 & 0 & 0 & 0 \\ 0 & 0 & 0 & 0 & 0 & 1 \\ 0 & 0 & 0 & 0 & 1 & 0 \end{bmatrix}$$

3. Draw the digraphs corresponding to the adjacency matrices.

$$\begin{bmatrix} 0 & 1 & 0 & 0 & 1 \\ 1 & 0 & 0 & 1 & 0 \\ 0 & 0 & 0 & 0 & 0 \\ 1 & 0 & 1 & 0 & 0 \\ 0 & 1 & 0 & 2 & 0 \end{bmatrix} \quad \begin{bmatrix} 0 & 0 & 0 & 0 & 1 \\ 0 & 0 & 0 & 0 & 1 \\ 1 & 0 & 0 & 0 & 1 \\ 1 & 1 & 0 & 0 & 0 \\ 1 & 0 & 0 & 1 & 0 \end{bmatrix} \quad \begin{bmatrix} 0 & 0 & 1 & 0 & 1 \\ 1 & 0 & 0 & 1 & 0 \\ 0 & 0 & 1 & 0 & 0 \\ 1 & 0 & 0 & 0 & 0 \\ 0 & 1 & 0 & 1 & 0 \end{bmatrix}$$

4. Determine the in-degree and the out-degree of each vertex in the following digraph and hence determine if it is Eulerian. Draw the digraph and determine an Eulerian trail.

$$\begin{bmatrix} 0 & 1 & 0 & 0 & 0 \\ 0 & 0 & 1 & 0 & 0 \\ 1 & 0 & 0 & 0 & 1 \\ 0 & 0 & 1 & 0 & 0 \\ 0 & 0 & 0 & 1 & 0 \end{bmatrix}$$

5. Write down the adjacency matrices of the digraph shown, calculate the matrices A^2, A^3 and A^4, and hence find the number of walks of length 1, 2, 3 and 4 from w to u. Is there a walk of length 1, 2, 3 or 4 from u to w? Find the matrix B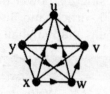
 (= $A + A^2 + A^3 + A^4$) for the digraph and hence say whether it is strongly connected. Write down the Boolean matrix for the digraph and use Warshall's algorithm to find the reachability matrix.

6. Use Warshall's algorithm to determine if the digraph defined by the adjacency matrix in question 4 is strongly connected.

7. Write a Maple procedure to implement Warshall's algorithm. Use it to determine if the following digraphs are strongly connected.

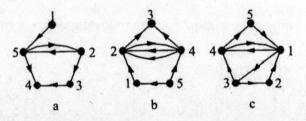

a b c

ANSWERS (EXERCISE 10.3)

1. (a) $A = \begin{bmatrix} 0 & 1 & 0 & 0 & 1 \\ 1 & 0 & 1 & 0 & 1 \\ 0 & 1 & 0 & 1 & 0 \\ 0 & 0 & 1 & 0 & 1 \\ 1 & 1 & 0 & 1 & 0 \end{bmatrix}$ (b) $A = \begin{bmatrix} 0 & 0 & 0 & 0 & 0 \\ 0 & 0 & 1 & 0 & 1 \\ 0 & 0 & 0 & 1 & 0 \\ 0 & 0 & 0 & 0 & 0 \\ 0 & 1 & 0 & 1 & 0 \end{bmatrix}$ (c) $A = \begin{bmatrix} 0 & 1 & 0 & 0 & 1 \\ 1 & 0 & 0 & 2 & 1 \\ 0 & 0 & 0 & 0 & 0 \\ 0 & 2 & 0 & 0 & 1 \\ 1 & 1 & 0 & 1 & 0 \end{bmatrix}$

(d) $A = \begin{bmatrix} 0 & 1 & 0 & 0 & 0 \\ 0 & 0 & 1 & 1 & 0 \\ 0 & 0 & 0 & 0 & 0 \\ 0 & 2 & 1 & 0 & 0 \\ 1 & 0 & 0 & 1 & 0 \end{bmatrix}$ (c) $A = \begin{bmatrix} 0 & 1 & 0 & 0 & 1 \\ 1 & 0 & 1 & 3 & 0 \\ 0 & 1 & 0 & 1 & 0 \\ 0 & 3 & 1 & 0 & 1 \\ 1 & 0 & 0 & 1 & 0 \end{bmatrix}$ (e) $A = \begin{bmatrix} 0 & 0 & 0 & 1 & 0 \\ 1 & 0 & 0 & 0 & 0 \\ 1 & 1 & 0 & 1 & 0 \\ 1 & 0 & 0 & 0 & 1 \\ 1 & 0 & 0 & 0 & 0 \end{bmatrix}$

2.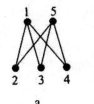

3.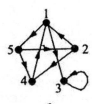

4. The out-degree of each vertex is given by adding the entries in the corresponding row and the in-degree by adding the entries in the corresponding column.

   Vertex 1 has out-degree 1 and in-degree 1
   Vertex 2 has out-degree 1 and in-degree 1
   Vertex 3 has out-degree 2 and in-degree 2
   Vertex 4 has out-degree 1 and in-degree 1
   Vertex 5 has out-degree 1 and in-degree 1

   Hence each vertex has out-degree equal to in-degree and the digraph is Eulerian. The digraph is as below and an Eulerian trail is, for instance, 1235431.

5. The adjacency matrix is

$$\begin{array}{c c c c c c} & u & v & w & x & y \end{array}$$
$$\begin{bmatrix} 0 & 1 & 0 & 0 & 1 \\ 0 & 0 & 0 & 1 & 1 \\ 1 & 1 & 0 & 0 & 0 \\ 1 & 0 & 1 & 0 & 0 \\ 0 & 0 & 1 & 1 & 0 \end{bmatrix}$$

so the powers of A are

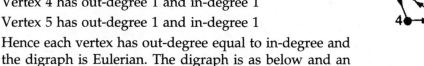

The number of walks from w to u is given by the "31" element of each matrix so there is 1 walk of length 1, 0 walks of length 2, 1 walk of length 3 and 6 walks of length 4 from w to u. Walks from u to w are given by the "13" element of each matrix so there are 0 walks of length 1, 1 walk of length 2, 3 walks of length 3 and 1 walk of length 4 from u to w. The matrix B for this digraph is

$$B = A + A^2 + A^3 + A^4 = \begin{bmatrix} 7 & 8 & 5 & 4 & 6 \\ 5 & 7 & 4 & 6 & 8 \\ 8 & 6 & 7 & 5 & 4 \\ 6 & 4 & 8 & 7 & 5 \\ 4 & 5 & 6 & 8 & 7 \end{bmatrix}$$

and the digraph is therefore strongly connected (all the off diagonal elements of B are non-zero). The Boolean matrix of the digraph is the same as the adjacency matrix A and the successive steps of Warshalls algorithm yield.

$$M_0 = \begin{bmatrix} 0 & 1 & 0 & 0 & 1 \\ 0 & 0 & 0 & 1 & 1 \\ 1 & 1 & 0 & 0 & 0 \\ 1 & 0 & 1 & 0 & 0 \\ 0 & 0 & 1 & 1 & 0 \end{bmatrix}, \quad M_1 = \begin{bmatrix} 0 & 1 & 0 & 0 & 1 \\ 0 & 0 & 0 & 1 & 1 \\ 1 & 1 & 0 & 0 & 1 \\ 1 & 1 & 1 & 0 & 1 \\ 0 & 0 & 1 & 1 & 0 \end{bmatrix}, \quad M_2 = \begin{bmatrix} 0 & 1 & 0 & 1 & 1 \\ 0 & 0 & 0 & 1 & 1 \\ 1 & 1 & 0 & 1 & 1 \\ 1 & 1 & 1 & 1 & 1 \\ 0 & 0 & 1 & 1 & 0 \end{bmatrix},$$

$$M_3 = \begin{bmatrix} 0 & 1 & 0 & 1 & 1 \\ 0 & 0 & 0 & 1 & 1 \\ 1 & 1 & 0 & 1 & 1 \\ 1 & 1 & 1 & 1 & 1 \\ 1 & 1 & 1 & 1 & 1 \end{bmatrix}, \quad M_4 = \begin{bmatrix} 1 & 1 & 1 & 1 & 1 \\ 1 & 1 & 1 & 1 & 1 \\ 1 & 1 & 1 & 1 & 1 \\ 1 & 1 & 1 & 1 & 1 \\ 1 & 1 & 1 & 1 & 1 \end{bmatrix} = M_5 = R$$

so the reachability matrix also (as we would hope) indicates that the graph is strongly connected.

6. The Boolean matrix of the graph is the same as the adjacency matrix A and the successive steps of Warshall's algorithm yield.

$$M_0 = \begin{bmatrix} 0 & 1 & 0 & 0 & 0 \\ 0 & 0 & 1 & 0 & 0 \\ 1 & 0 & 0 & 0 & 1 \\ 0 & 0 & 1 & 0 & 0 \\ 0 & 0 & 0 & 1 & 0 \end{bmatrix}, \quad M_1 = \begin{bmatrix} 0 & 1 & 0 & 0 & 0 \\ 0 & 0 & 1 & 0 & 0 \\ 1 & 1 & 0 & 0 & 1 \\ 0 & 0 & 1 & 0 & 0 \\ 0 & 0 & 0 & 1 & 0 \end{bmatrix}, \quad M_2 = \begin{bmatrix} 0 & 1 & 1 & 0 & 0 \\ 0 & 0 & 1 & 0 & 0 \\ 1 & 1 & 1 & 0 & 1 \\ 0 & 0 & 1 & 0 & 0 \\ 0 & 0 & 0 & 1 & 0 \end{bmatrix},$$

$$M_3 = \begin{bmatrix} 1 & 1 & 1 & 0 & 1 \\ 1 & 1 & 1 & 0 & 1 \\ 1 & 1 & 1 & 0 & 1 \\ 1 & 1 & 1 & 0 & 1 \\ 0 & 0 & 0 & 1 & 0 \end{bmatrix}, M_4 = \begin{bmatrix} 1 & 1 & 1 & 0 & 1 \\ 1 & 1 & 1 & 0 & 1 \\ 1 & 1 & 1 & 0 & 1 \\ 1 & 1 & 1 & 0 & 1 \\ 1 & 1 & 1 & 1 & 1 \end{bmatrix}, M_5 = \begin{bmatrix} 1 & 1 & 1 & 1 & 1 \\ 1 & 1 & 1 & 1 & 1 \\ 1 & 1 & 1 & 1 & 1 \\ 1 & 1 & 1 & 1 & 1 \\ 1 & 1 & 1 & 1 & 1 \end{bmatrix}$$

so the digraph is strongly connected.

7. A suitable Maple worksheet is

> *itob* : proc (*a*) ; if *a* = 1 then true else false and if and proc :

> *btoi* = proc (*a*) ; if *a* = true then 1 else 0 end if the proc :

> warshall : = proc (A, *n*) local *i, j, k;*

for *k* from 1 to *n* do

for *j* from 1 to *n* do

for *i* from 1 to *n* do

A[*i, j*] : = A[*i, j*] or (A[*i, k*] and A[*k, j*])

    end do;

    end do;

    M| | *k;* = map (btoi,A) ;

    end do

    print (seq (M| | *k, k*=1 . . 5))

    end proc;

> A : =matrix ([0, 1, 0, 0, 1], [0, 0, 0, 1, 1], [1, 1, 0, 0, 0], [1, 0, 1, 0, 0], [0, 0, 1, 1, 0]) ;

> A : =map (itob,A) :warshall (A, 5) ;

NB: I wrote the Maple code above in a hurry. It certainly works, and it is reasonably transparent (in that it reflects the logic of the explanation I give of Warshall's algorithm in the lectures). But it is not as elegant as I might wish, so if anyone (staff or student) thinks they have a better solution I would be pleased to know about it.

Using this procedure we find that digraph (a) is not strongly connected

$$M_0 = \begin{bmatrix} 0 & 0 & 0 & 0 & 1 \\ 0 & 0 & 1 & 0 & 1 \\ 0 & 0 & 0 & 1 & 0 \\ 0 & 0 & 0 & 0 & 0 \\ 0 & 1 & 0 & 1 & 0 \end{bmatrix}, M_1 = \begin{bmatrix} 0 & 0 & 0 & 0 & 1 \\ 0 & 0 & 1 & 0 & 1 \\ 0 & 0 & 0 & 1 & 0 \\ 0 & 0 & 0 & 0 & 0 \\ 0 & 1 & 0 & 1 & 0 \end{bmatrix}, M_2 = \begin{bmatrix} 0 & 0 & 0 & 0 & 1 \\ 0 & 0 & 1 & 0 & 1 \\ 0 & 0 & 0 & 1 & 0 \\ 0 & 0 & 0 & 0 & 0 \\ 0 & 1 & 1 & 1 & 1 \end{bmatrix},$$

$$M_3 = \begin{bmatrix} 0 & 0 & 0 & 0 & 1 \\ 0 & 0 & 1 & 1 & 1 \\ 0 & 0 & 0 & 1 & 0 \\ 0 & 0 & 0 & 0 & 0 \\ 0 & 1 & 1 & 1 & 1 \end{bmatrix}, M_4 = \begin{bmatrix} 0 & 0 & 0 & 0 & 1 \\ 0 & 0 & 1 & 1 & 1 \\ 0 & 0 & 0 & 1 & 0 \\ 0 & 0 & 0 & 0 & 0 \\ 0 & 1 & 1 & 1 & 1 \end{bmatrix}, M_5 = \begin{bmatrix} 0 & 1 & 1 & 1 & 1 \\ 0 & 1 & 1 & 1 & 1 \\ 0 & 0 & 0 & 1 & 0 \\ 0 & 0 & 0 & 0 & 0 \\ 0 & 1 & 1 & 1 & 1 \end{bmatrix}$$

and that digraph (b) is also not strongly connected

$$M_0 = \begin{bmatrix} 0 & 1 & 0 & 0 & 0 \\ 0 & 0 & 1 & 1 & 1 \\ 0 & 0 & 0 & 0 & 0 \\ 0 & 1 & 1 & 0 & 0 \\ 1 & 0 & 0 & 1 & 0 \end{bmatrix}, M_1 = \begin{bmatrix} 0 & 1 & 0 & 0 & 0 \\ 0 & 0 & 1 & 1 & 1 \\ 0 & 0 & 0 & 0 & 0 \\ 0 & 1 & 1 & 0 & 0 \\ 1 & 1 & 0 & 1 & 0 \end{bmatrix}, M_2 = \begin{bmatrix} 0 & 1 & 1 & 1 & 1 \\ 0 & 0 & 1 & 1 & 1 \\ 0 & 0 & 0 & 0 & 0 \\ 0 & 1 & 1 & 1 & 1 \\ 1 & 1 & 1 & 1 & 1 \end{bmatrix},$$

$$M_3 = \begin{bmatrix} 0 & 1 & 1 & 1 & 1 \\ 0 & 0 & 1 & 1 & 1 \\ 0 & 0 & 0 & 0 & 0 \\ 0 & 1 & 1 & 1 & 1 \\ 1 & 1 & 1 & 1 & 1 \end{bmatrix}, M_4 = \begin{bmatrix} 0 & 1 & 1 & 1 & 1 \\ 0 & 1 & 1 & 1 & 1 \\ 0 & 0 & 0 & 0 & 0 \\ 0 & 1 & 1 & 1 & 1 \\ 1 & 1 & 1 & 1 & 1 \end{bmatrix}, M_5 = \begin{bmatrix} 1 & 1 & 1 & 1 & 1 \\ 1 & 1 & 1 & 1 & 1 \\ 0 & 0 & 0 & 0 & 0 \\ 1 & 1 & 1 & 1 & 1 \\ 1 & 1 & 1 & 1 & 1 \end{bmatrix}$$

which digraph (c) is strongly connected

$$M_0 = \begin{bmatrix} 0 & 0 & 1 & 1 & 0 \\ 1 & 0 & 0 & 0 & 0 \\ 0 & 1 & 0 & 1 & 0 \\ 1 & 0 & 0 & 0 & 1 \\ 1 & 0 & 0 & 0 & 0 \end{bmatrix}, M_1 = \begin{bmatrix} 0 & 0 & 1 & 1 & 0 \\ 1 & 0 & 1 & 1 & 0 \\ 0 & 1 & 0 & 1 & 0 \\ 1 & 0 & 1 & 1 & 1 \\ 1 & 0 & 1 & 1 & 0 \end{bmatrix}, M_2 = \begin{bmatrix} 0 & 0 & 1 & 1 & 0 \\ 1 & 0 & 1 & 1 & 0 \\ 1 & 1 & 1 & 1 & 0 \\ 1 & 0 & 1 & 1 & 1 \\ 1 & 0 & 1 & 1 & 0 \end{bmatrix},$$

$$M_3 = \begin{bmatrix} 1 & 1 & 1 & 1 & 0 \\ 1 & 1 & 1 & 1 & 0 \\ 1 & 1 & 1 & 1 & 0 \\ 1 & 1 & 1 & 1 & 1 \\ 1 & 1 & 1 & 1 & 0 \end{bmatrix}, M_4 = \begin{bmatrix} 1 & 1 & 1 & 1 & 1 \\ 1 & 1 & 1 & 1 & 1 \\ 1 & 1 & 1 & 1 & 1 \\ 1 & 1 & 1 & 1 & 1 \\ 1 & 1 & 1 & 1 & 1 \end{bmatrix}, M_5 = \begin{bmatrix} 1 & 1 & 1 & 1 & 1 \\ 1 & 1 & 1 & 1 & 1 \\ 1 & 1 & 1 & 1 & 1 \\ 1 & 1 & 1 & 1 & 1 \\ 1 & 1 & 1 & 1 & 1 \end{bmatrix}$$

10.11 TREE STRUCTURES

A tree is a connected graph which has no cycler

Trees are a relatively simple type of graph but they are also very important. Many applications use trees as a mathematical representation, e.g. decision trees in OR, some utility networks, linguistic analysis, family trees, organisation trees.

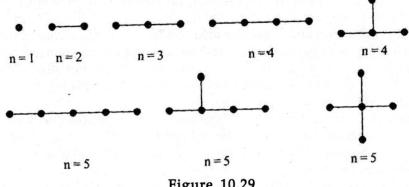

Figure 10.29

Figure 10.29 shows all the possible unlabelled trees with up to five vertices. Every tree with $n + 1$ vertices can be formed by adding a new vertex joined by a new edge to one of the vertices of one of the n vertex trees. For instance if we take the second 5-vertex tree we would obtain the trees shown in Figure 10.30.

Figure 10.30

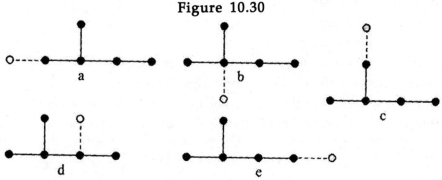

Of course (a) and (c) are isomorphic. If we complete this process with all three 5-vertex trees and eliminate the isomorphic duplicates we obtain the six trees with 6 vertices shown in Figure 10.31.

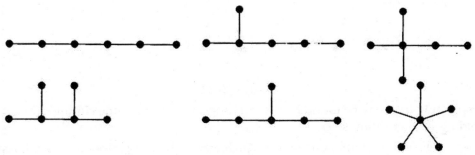

Figure 10.31

Trees have a number of special properties as follows :
(a) It is obvious, from the constructive process of building all possible trees step by step from the simplest tree (one vertex, no edges) that a tree with n vertices has exactly $n - 1$ edges.
(b) When a new vertex and edges is added to a tree, no cycle is created (since the new edge joins an existing vertex to a new vertex) and the tree remains connected.
(c) There is exactly one path from any vertex in a tree to any other vertex—if there were two or more paths between any two vertices then the two paths would form a cycle and the graph would not be a tree.
(d) Because there is exactly one path between any two vertices then there is one (and only one) edge joining any two adjacent vertices. If this edge is removed, the graph is no longer connected (and so is not a tree). So the removal of any edge from a tree disconnects the graph.
(e) Since there is a path between any two vertices, if an edge is added to the tree joining two existing vertices then a cycle is formed comprising the existing path between the two vertices together with the new edge.

All these properties can be used to define a tree. If T is a graph with n vertices then the following are all equivalent definitions of a tree :

T is connected and has no cycles.

T has $n - 1$ edges and has no cycles.

T is connected and has $n - 1$ edges.

Any two vertices of T are connected by exactly one path.

T is connected and the removal of any edge disconnects T.

T contains no cycles but the addition of any new edge creates a cycle.

A **spanning tree** in a connected graph G is a subgraph which includes every vertex and a tree. For instance Figure 10.32 below shows the complete graph K_5 and several possible spanning trees. Large and complex graphs may have very many spanning trees.

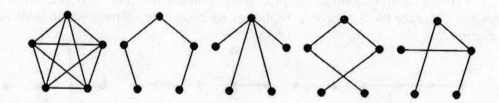

Figure 10.32

A spanning tree may be found by the building-up method or the cutting-down method. The building-up algorithm is

Select edges of the graph, one by one, such a way that no cycles are created; repeating until all vertices are included and the cutting-down method is

Choose any cycle in the graph and remove one of its edges; repeating until no cycles remain.

For instance, in Figure 10.33 below a spanning tree in the graph G is built up by selecting successively edges *ab* (1st diagram), then *ce* (2nd diagram), then *bg*, then *ge* (3rd diagram), then *gf* and finally *de* (final diagram).

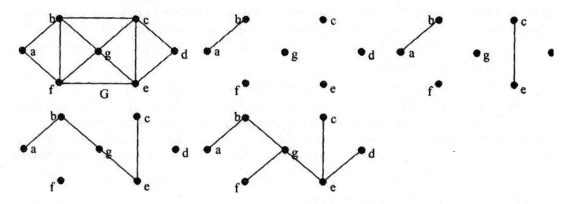

Figure 10.33

In Figure 10.34, a spanning tree in the graph G of Figure 10.33 is derived by the cutting down method, by successively finding cycles and removing edges. First cycle is *abcdefa*—remove *bc* (1st diagram).

Cycle *bgfb*—remove *fb* (2nd diagram). Cycle *gedcg*—remove *cd*. Cycle *cgec*—remove *gc*. Cycle *gfabg*—removed *ab*. Cycle *gfeg*—remove *fe*. No more cycles so this is a spanning tree.

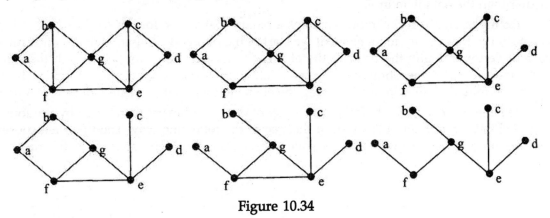

Figure 10.34

A **rooted tree** is a tree in which one vertex is selected as a root and all other edges branch out from the root vertex. Any given tree can be drawn as a rooted tree in a variety of ways depending on the choice of root vertex (see Figure 10.35).

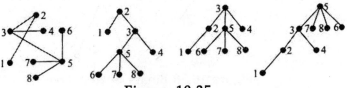

Figure 10.35

10.12 BINARY TREES

A *binary tree* is a rooted tree in which each vertex has at most two children, designated as *left child* and *right child*. If a vertex has one child, that child is designated as either a left child or a right child, but not both. A *full binary tree* is a binary tree in which each vertex has exactly two children or none. The following are a few results about binary trees:

1. If T is a full binary tree with i internal vertices, then T has $i+1$ terminal vertices and $2i+1$ total vertices.
2. If a binary tree of height h has t terminal vertices, then $t < 2^h$.

More generally we can define a *m-ary tree* as a rooted tree in which every internal vertex has no more than m children. The tree is called a *full m-ary tree* if every internal vertex has exactly m children. An *ordered rooted tree* is a rooted tree where the children of each internal vertex are ordered. A binary tree is just a particular case of *m-ary* ordered tree (with $m = 2$).

10.13 BINARY SEARCH TREES

Assume S is a set in which elements (which we will call "data") are ordered; e.g., the elements of S can be numbers in their natural order, or strings of alphabetic characters in lexicographic order. A *binary search tree* associated to S is a binary tree T in which data from S are associate with the vertices of T so that, for each vertex u in T, each data item in the left subtree of v is less than the data item in v, and each data item in the right subtree of v is greater than the data item in v.

Example: Figure 10.36 below, contains a binary search tree for the set S = {1, 2, 3, 4, 5, 6, 7, 8, 9, 10}. In order to find a element we start at the root and compare it to the data in the current vertex (initially the root). If the element is greater we continue through the right child, if it is smaller we continue through the left child, if it is equal we have found it. If we reach a terminal vertex without founding the element, then that element is not present in S.

Making a Binary Search Tree. We can store data in a binary search tree by randomly choosing data from S and placing it in the tree in the following way: The first data chosen will be the root of the tree. Then for each subsequent data item, starting at the root we

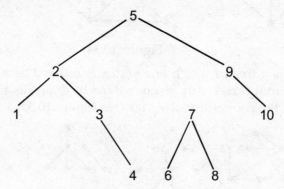

Figure 10.36 Binary Search Tree

compare it to the data in the current vertex v. If the new data item is greater than the data in the current vertex then we move to the right child, if it is less we move to the left child. If there is no such child then we create one and put the new data in it. For instance, the tree in Figure 10.37 below has been made from the following list of words choosing them in the order they occur: "IN A PLACE OF LA MANCHA WHOSE NAME I DO NOT WANT TO REMEMBER".

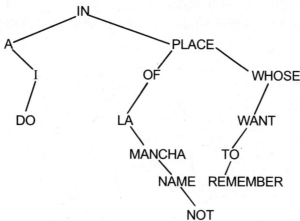

Figure 10.37 Another binary search tree.

10.14 TRAVERSAL OF A BINARY TREE

Tree traversal is one of the most common operations performed on tree data structures. It is a way in which each node in the tree is visited exactly once in a systematic manner. There are many applications that essentially require traversal of binary trees. For example, a binary tree could be used to represent an arithmetic expression as shown in Figure 10.38.

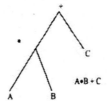

Figure 10.38

The full binary tree traversal would produce a linear order for the nodes in a binary tree. There are three ways of binary tree traversal,

(i) In-order traversal

(ii) Pre-order traversal

(iii) Post-order traversal

Inorder Traversal

The in-order traversal of a non-empty tree is defined as follows:

(i) Traverse the left subtree inorder (L).

(ii) Visit the root node (N)

(iii) Traverse the right subtree in order (R).

Illustration: In Fig. 10.39 inorder traversal of a binary tree is DBFEGAC.

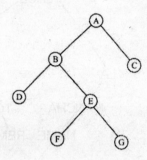

Figure 10.39

Pre-order Traversal

The pre-order traversal of a non-empty binary tree is defined as follows:

(i) Visit the root node (N).

(ii) Traverse the left subtree in pre-order (L)

(iii) Traverse the right subtree in pre-order (R)

Illustration; In Fig. 15 the preorder traversal of a binary tree is ABDEFGC.

Postorder Traversal

The postorder traversal of non-empty binary tree is defined as follows:

(i) Traverse the left subtree in postorder (L).

(ii) Traverse the right subtree in postorder (R).

(iii) Visit the root node (N).

Example: In Fig. 10.39 the postorder traversal of a binary tree is DFGEBC A.

Level of a vertex in a full binary tree

In a binary tree the distance of a vertex v_i from the root of the tree is called the level of v_i and is denoted by Ii. Thus level of the root is zero. Levels of the various vertices in the following tree have been denoted by numbers written adjacent to the respective vertices.

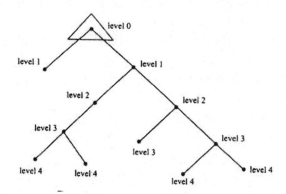

Figure 10.40. Twelve vertex binary tree of level 4

Number of vertices of different levels in a binary tree

In a binary tree there will be two edges adjacent to the root vertex v_0. Let these edges be v_0u_1 and v_0v_1. Levels of each of the vertices u_1 and v_1 is 1. So maximum number of vertices of level 0 is $1(=2^0)$ and maximum number of vertices with level 1 is $=2^1$.

Again there can be either 0 or 2 edges adjacent to each of the vertices u_1 and v_1. Let these edges be u_1u_2, u_1u_3, v_1v_2 and v_1v_3. Levels of each of the four vertices u_2, u_3, v_2, v_3 is 2. So maximum number of vertices of level 2 is $4(=2^2)$. In a similar way the levels of the 8 vertices that will be obtained by adding two edges to each of the four vertices u_2, u_3, v_2, v_3 shall be 3. So maximum number of vertices each of level 3 is $8(=2^3)$. Not more than two edges can be added to any of the vertices so obtained to keep the degree of that vertex as 3.

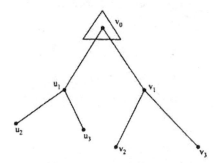

Figure 10.41

Proceeding in this way we see that the maximum number of vertices in a n level binary tree at levels 0, 1, 2, 3,... shall be 2^0, 2^1, 2^2, 2^3,... respectively.

whose sum $= 2^0 + 2^1 + 2^2 + 2^3 + ... + 2^n$.

The maximum level of any vertex in a binary tree (denoted by l_{max}) is called *height of the tree*. The minimum possible height of an n vertex binary tree is $[\log_2(n+1) - 1]$ which is equal *to* the smallest integer $\geq [\log_2(n+1) - 1]$ and Max. $l_{max} = \dfrac{n-1}{2}$

Example 3: If a tree T has n vertices of degree 1, 3 vertices of degree 2, 2 vertices of 3 and 2 vertices of degree 4 find the value of n.

Solution: Let $|E|$ denote the number of edges in the graph T and $|V|$ denote the number of vertices in the same graph T.

Sum of degrees of all vertices in T = $2|E|$

$$n.1 + 3.2 + 2.3 + 2.4 = 2|E|$$

or $\qquad n + 6 + 6 + 8 = 2(|V| - 1)$

or $\qquad n + 20 = 2[(n + 3 + 2 + 2) - 1]$

or $\qquad n + 20 = 2n + 12$

or $\qquad n = 8.$

Theorem 10.6: To prove that in every non-trivial tree there is at least one vertex of degree one.

Proof: Let us start at vertex v_1. If $d(v_1) = 1$, then the theorem is already proved. If $d(v_1) > 1$, then we can move to a vertex say v_2 that is adjacent to v_1. Now if $d(v_2) > 1$, again move to another vertex say v_3 that is adjacent to v_2. In this way we can continue to produce a path $v_1, v_2, v_3 \ldots$ (without repetition of any vertex, in order to avoid formation of circuit as the graph is a tree). As the graph is finite, this path must end at some vertex whose degree shall be one because we shall only enter this vertex and cannot exit from it.

10.15 PLANAR GRAPHS

A graph G is **planar** if it can be drawn on a plane in such a way that is no two edges meet except at a vertex with which they both incident. Any such drawing is a **plane drawing** of G. A graph is non-planar if no plane drawing of it exists. Figure 10.42 shows a common representation of the graph K_4 and Figure 10.43 shows three possible plane drawings of K_4.

Figure 10.42   Figure 10.43

The complete bipartite graph $K_{3,3}$ and the complete graph K_5 are important simple examples of non-planar graphs. (You will prove this in Graph Theory Exercises 4!).

Any plane drawing of a planar graph G divides the set of points of the plane not lying on G into regions called **faces**. The region outside the graph is of infinite extent and is called the **infinite face**.

The **degree** of a face f of a connected planar graph G, denoted deg (f), is the number of edges encountered in a walk around the boundary of f. If all faces have the same degree, g, then G is **face-regular of degree g**.

In Figure 10.43 each plane drawing of the graph K_4 has 4 faces (including the infinite face) each face being of degree 3 so K_4 is face-regular of degree 3.

In any plane drawing of a planar graph, the sum of all the face degrees is equal to twice the number of edges.

In Figure 10.43 the plane drawing of the graph K_4 has 4 face (including the infinite face) each face being of degree 3 so K_4 is face-regular of degree 3. The sum of the face degrees is therefore 12 whilst the number of edges is 6.

Euler's formula for planar graphs. If n, m and f denote respectively the number of vertices, edges and faces in a plane drawing of a connected plane graph G then $n - m + f = 2$.

In Figure 10.43 we have $n = 4$, $m = 6$ and $f = 4$ satisfying Euler's formula.

Proof of Euler's formula : A plane drawing of any connected planar graph G can be constructed by taking a spanning tree of G and adding edges to it, one at a time, until a plane drawing of G is obtained.

We prove Euler's formula by showing that

(a) for any spanning tree G, $n - m + f = 2$ and

(b) adding an edge to the spanning tree does not change the value of $n - m + f$.

Let T be any spanning tree of G. We may draw T in a plane without crossings. T has n vertices $n - 1$ edges and 1 face (the infinite face). Thus $n - m + f = n - (n - 1) + 1 = 2$ so we have shown (a).

Now if we add an edge to T either it joins two different vertices, or it joins a vertex to itself (it is a loop). In either case it divides some face into two faces, so adding one face. Hence we have increased m, the number of edges, and f, the number of faces, by one each. The value of the expression $n - m + f$ is unchanged. We add more edges, one at a time, and at each addition the value of $n - m + f$ is unchanged. Hence we have shown (b) and so proved Euler's theorem.

It is useful to be able to test a graph for planarity. There are a variety of algorithms for determining planarity, mostly quite complex. Here we will describe a simple test, the **cycle method**, which determines the planarity of a Hamiltonian graph.

First we identify a Hamiltonian cycle, C, in the graph. Now list the remaining edges of the graph, and then try to divide those edges into two disjoint sets A and B such that

A is a set of edges which can be drawn inside C without crossings and

B is a set of edges which can be drawn outside C without crossings.

Example. Determine whether the graph in Figure 10.44 is planar.

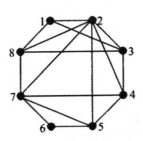
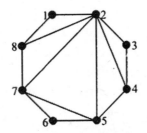
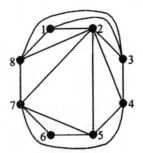

Figure 10.44   Figure 10.45   Figure 10.46

First find a Hamiltonian cycle. 123456781 will do. Draw a plane drawing of the cycle. The remaining edges are {24, 25, 27, 28, 31, 38, 47, 57}. Take the first edge, 24, and put it in set A. Take the second edge, 25, and put it in set A if compatible—it is so put it in set A. Consider the next edge, 27—it is comparable with set A so add it to set A. At'this point we have A = {24, 25, 27}, B = { }. The next edge is 28—it is compatible with set A so add it to set A (A = {24, 25, 27, 28}). The next edge is 31 which is not compatible with set A so put it in set B(B = {31}). The next edge is 38 which is not compatible with set A so put it in set B(B = {31, 38}). The next edge is 47 which is not compatible with set A so put it in set B (B = {31, 38, 47}). The next edge 57 which is compatible with set A so put it in set A(A = 24, 25, 27, 28, 57}). Figure 10.45 shows the Hamiltonian cycle 123456781 and the edges in set A drawn inside the cycle. Now if we can add the edges from set B, all outside the cycle, without crossings then we have a plane drawing of the graph and it will be planar. Figure 10.46 shows that the edges in set B can be drawn in that way so the graph is planar and Figure 10.46 is a plane drawing.

EXERCISE 10.4

1. By adding a new edge in every possible way to each unlabelled tree with 6 vertices, draw the 11 unlabelled trees with 7 vertices.

2. Give an example of a tree with eight vertices and
   (a) exactly 2 vertices of degree 1,
   (b) exactly 4 vertices of degree 1, and
   (c) exactly 7 vertices of degree 1.

3. Use the Handshaking Lemma to show that every tree with n vertices, where $n \geq 2$, has at least 2 vertices of degree 1.

4. (a) Find 3 different spanning trees in the graph shown using the building up method.
   (b) Find 3 more spanning trees (different from those found in part (a) using the cutting down method.

5. Find all the spanning trees in each of the two graphs shown.

   Hint : How many edges does a tree with 5 vertices have?

6. (a) How many spanning trees has $K_{2,3}$?
   (b) How many spanning trees has $K_{2,100}$?
   (c) How many spanning trees has $K_{2,n}$?

7. Find a planar drawing of each of the following graphs.

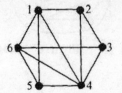 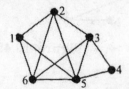 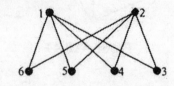

8. If G is a simple, connected, planar graph with $n(\geq 3)$ vertices and m edges, and if g is the length of the shortest cycle in G, show that

   $m \leq g(n-2)/(g-2)$

   Hint: The edges around a face in a plane drawing of a planar graph are a cycle. Find a lower bound on the face degree sum of G then use the Handshaking Lemma and Euler's formula.

9. What is the smallest cycle length in the complete bipartite graph $K_{3,3}$? Hence use the result from question 8 to show that the $K_{3,3}$ is not planar.

10. What is the smallest cycles in the complete graph K_5? Hence use the result from question 8 to show that the K_5 is not planar.

11. Use the cycle method to determine whether each of the following graphs is planar. If it is give a plane drawing.

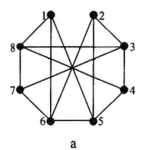
a

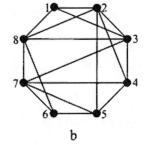
b

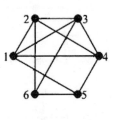
c

ANSWERS (EXERCISE 10.4)

1.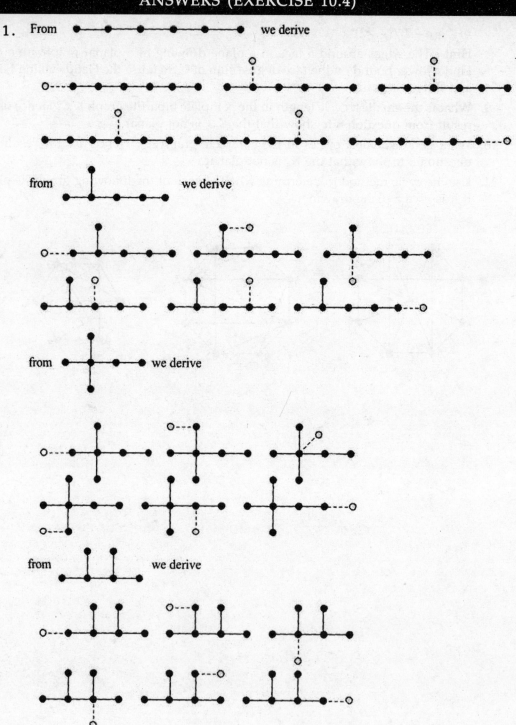

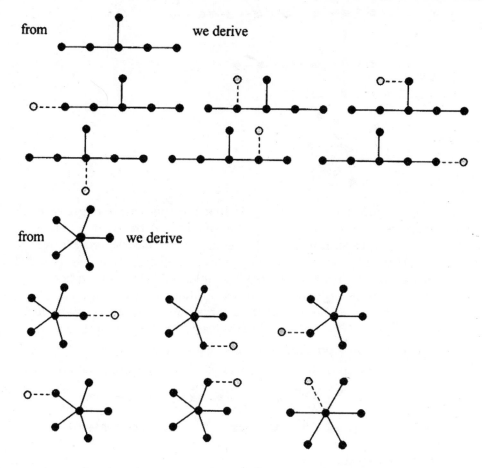

There are many isomorphic trees in there collections—the eleven genuinely distinct ones are shown below

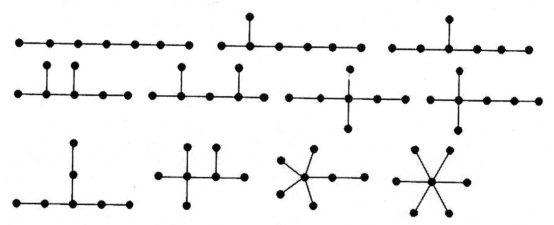

2. (a)

(b)

(c)

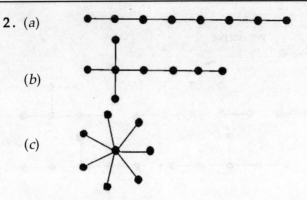

3. The Handshaking Lemma states that the sum of the degrees of the vertices of a graph is equal to twice the number of edges. A tree with n vertices has $n-1$ edges so the degree sum is $2(n-1)$. If every vertex has degree ≥ 2 then the degree sum would be $\geq 2n$. So there must be at least one vertex with degree < 2. There could be at least one vertex with degree 0 or at least 2 vertices with degree 1. A vertex with degree 0 is an unconnected vertex. In that case the graph would be unconnected. But a tree is a connected graph so there cannot be vertices with degree 0. Hence the only possibility is that there are at least 2 vertices with degree 1.

4. There are many possible solutions to this question—here are a few of them
   (a) Build up (i) *xu, uy, yw, wz*; (ii) *xu, uw, wx, uy*; (iii) *xu, uw, uy, yz*

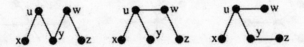

   (b) Cut down
       (i) Remove *xy* from *xuyx*, remove *uy* from *uywu*, remove *wy* from *wyzw*,
       (ii) Remove *xy* from *xuyx*, remove *uw* from *uywu*, remove *wz* from *wzyw*,
       (iii) Remove *uy* from *xuyx*, remove *uw* from *uwyxu*, remove *wy* from *wyzw*.

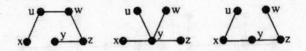

5. In a tree with n vertices there are $n-1$ edges. The first graph has 6 edges and 5 vertices so every spanning tree is derived from the graph by removing 2 edges. There are $^6C_2 = 15$ ways of removing two edges. Of course some of these result in a disconnected graph. The 15 possibilities are:

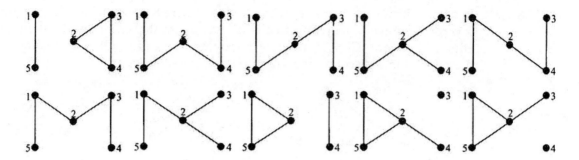

Six of these are disconnected graphs so there a 9 possible spanning trees:

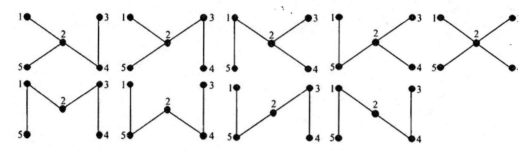

The second graph has 5 edges and 4 vertices so again every spanning tree is derived from the graph by removing 2 edges. There are $^5C_2 = 10$ ways of removing two edges.

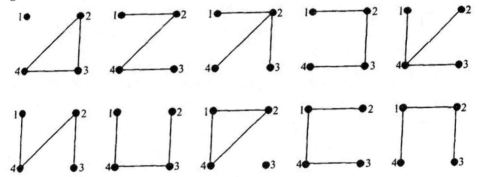

Two of these are disconnected graphs so there a 8 possible spanning trees:

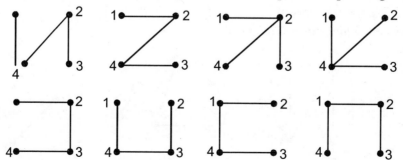

6. (a) In a tree with n vertices there are $n-1$ edges. There $K_{2,3}$ graph has 6 edges and 5 vertices so every spanning tree is derived from the graph by removing 2 edges. There are ${}^6C_2 = 15$ ways of removing two edges. Of course some of these result in a disconnected graph. The 15 possibilities are:

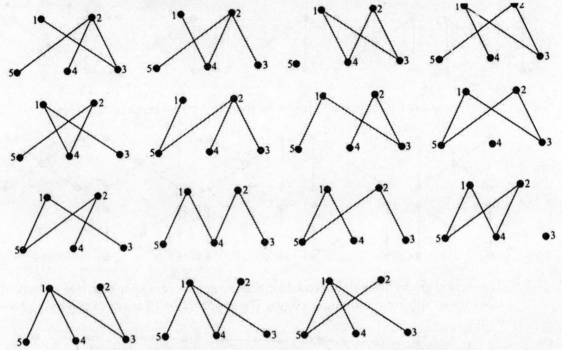

Three of these (3, 8 and 12) are disconnected graphs so there are 12 possible spanning trees of $K_{2,3}$.

(b) The $K_{2,100}$ graph has 200 edges and 102 vertices so a spanning tree has 101 edges. Each of the 100 vertices in set B must be connected to one or other of the two vertices in set A. The 101st edge then connects one vertex in set B to the vertex in set A to which it is not yet connected. There are 100 ways of choosing the vertex in set B which is connected to both vertices of set A. The other 99 vertices in set B must be connected to one or other vertex in set A. There are 2^{99} ways of doing that. So there are 100×2^{99} possible spanning trees in $K_{2,100}$.

(c) In general, by the same argument as part (b), there are $n \times 2^{n-1}$ possible spanning trees in $K_{2,n}$.

7. Possible planar drawing are:

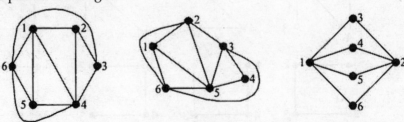

8. In a plane drawing of a planar graph the edges surrounding a face are a cycle. Thus, if g is the length of the shortest cycle in a planar graph, the degree of each face in the plane drawing is $\geq g$. Therefore the sum of the face degrees is $\geq gf$. The Handshaking Lemma tells us that the sum of the face degrees is twice the number of edges $= 2m$, so $2m \geq gf$. Now Euler's formula tells us that $n - m + f = 2$ so $f = m + 2 - n$. Hence $2m \geq gf = g(m + 2 - n)$. Hence, we have $2m \geq g(m + 2 - n)$. Hence $g(n - 2) \geq (g - 2)m$, i.e. $g(n - 2)/(g - 2) > m$.

9. In $K_{3,3}$ the shortest cycle is length 4 and the number of vertices is 6. Hence, if $K_{3,3}$ is planar we must have $m \leq g(n - 2)/(g - 2) = 8$. But, in $K_{3,3}$, the number of edges, is 9. Hence $K_{3,3}$ cannot be planar.

10. In K_5 the shortest is length 3 and the number of vertices is 5. Hence, if K_5 is planar we must have $m \leq g(n - 2)/(g - 2) = 9$. But, in K_n, m, the number of edges, is $1/2 n(n - 1)$ so K_5 has 10 edges. Hence K_5 cannot be planar.

11. (a) 154326781 is a Hamiltonian cycle. The remaining edges are {56, 38, 37, 48, 16, 25}. Edges {38, 37, 48} can be added inside the cycle and {56 16, 25} outside thus. The graph is planar.

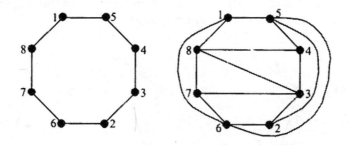

(b) 123456781 is a Hamiltonian cycle. The remaining edges are {13, 24, 25, 28, 38, 37, 47, 57, 68}. Edges {13, 38, 37, 47, 57} can be added inside the cycle and {24, 25, 28, 68} outside thus. The graph is planar.

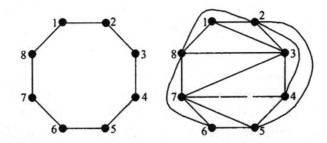

(c) 1236541 is a Hamiltonian cycle. The remaining edges are {13, 15, 24, 26, 34}. Edges {13, 15} can be added inside the cycle leaving B = {24, 26, 34} to go outside. But the set of edge in B are incompatible for a plane drawing (we can add 24 and 26 but 34 cannot be added without crossing 26) so the graph is not planar.

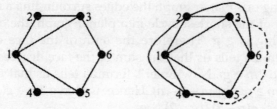

10.16 THE TRAVELLING SALESMAN PROBLEM

The **travelling salesmen problem** is the problem of finding the minimum weight Hamiltonian cycle in a weighted complete graph. It is called the travelling salesman problem because it is essentially the problem of working out a route between a given set of locations (vertices) travelling along roads (edges) of known lengths (weights) in such a way as to minimise the total distance travelled. We tackle the problem in stages. First of all we need the concept of a minimum spanning tree.

A **minimum spanning tree** or **minimum connector** is a spanning tree in a connected weighted graph which has minimum total weight.

Figure 10.47 shows a number of possible spanning tree for the graph shown. How do we know if any of them are minimum spanning trees?

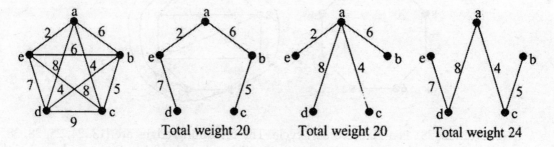

Figure 10.47

There are two algorithms for constructing minimum spanning trees.

10.17 KRUSKAL'S ALGORITHM

Start with a finite set of vertices, each pair joined by a weighted edge.
1. List all the weights in ascending order.
2. Draw the vertices and weighted edge corresponding to the first weight in the list provided that, in so doing, no cycle is formed. Delete the weight from the list.
3. Repeat step 2 until all vertices are connected, then stop. The weighted graph obtained is a minimum connector, and the sum of weights on its edges is the total weight of the minimum connector.

To construct a minimum spanning tree for the graph of Figure 10.47 we proceed thus. The weights are

    2 (ae) 4 (ac) 4 (ce) 5 (bc) 6 (ab) 6 (be) 7 (de) 8 (ad) 8 (bd) 9 (cd)

so the steps of Kruskal's algorithm will give the following : choose *ae*, choose *ac*, reject *ce* (would from cycle area), choose *bc*, reject *ab* (would from cycle *acba*), reject *be* (would from cycle *acbea*), choose *de*. All vertices now connected. The total weight is 18.

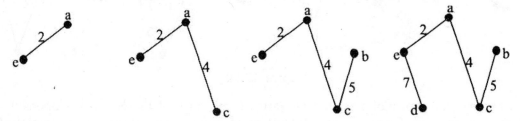

Figure 10.48

Although, in the above example, the subgraph was a constructed subgraph at each stage this is not necessarily the case. For instance suppose edge *bc* had weigh 3 instead of 5. Now the weights are

2 (*ae*)   3 (*bc*)   4 (*ac*)   4 (*ce*)   6 (*ab*)   6 (*be*)   7 (*de*)   8 (*ad*)   8 (*bd*)   9 (*cd*)

so the steps of Kruskal's algorithm will give the following : choose *ae*, choose *bc*, choose *ac*, reject *ce* (would from cycle), reject *ab* (would from cycle), reject *be* (would from cycle), choose *de*. All vertices now connected. The total weight is 16.

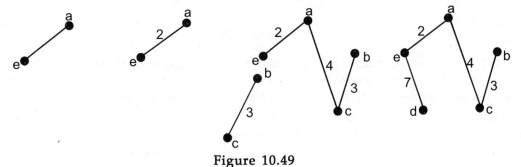

Figure 10.49

10.18 PRIM'S ALGORITHM

Start with a finite set of vertices, each pair joined by a weighted edge.
1. Chose and draw any vertex
2. Find the edge of least weight joining a drawn vertex to a vertex not currently drawn. Draw this weighted edge and the corresponding new vertex.

Repeat step 2 until all vertices are connected, then stop.

To construct a minimum spanning tree for the graph of Figure10.42 using Prim's algorithm we proceed thus. Choose vertex *a* to start, minimum weight edge is *ae* (weight 2), minimum weight edge from {*a,e*} to new vertex is *ac* (weight 4), minimum weight edge from {*a, c, e*} to new vertex is *cb* (weight 5), minimum weight edge from {*a, b, c, e*} to new vertex is *ed* (weight 7), all vertices now connected. Total weight is 18.

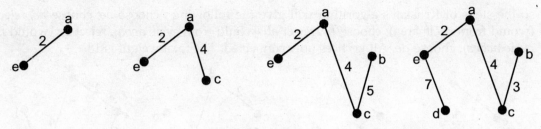

Figure 10.50

Of course the choice of a as starting vertex is arbitrary—let's see what happens if we start somewhere else. Choose vertex b to start, minimum weight edge is bc (weight 5), minimum weight edge from {b, c} to new vertex is ca or ce (both weight 4), choose ce, minimum weight edge from {b, c, e} to new vertex is ea (weight 2), minimum weight edge from {a, b, c, e} to new vertex is ed (weight 7), all vertices now connected. Total weight is 18.

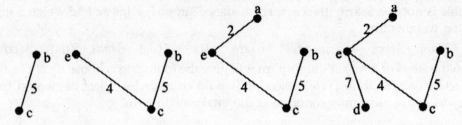

Figure 10.51

In this case we see that we get a different minimum spanning tree (if we had chosen ca instead of ce at stage 2 we would have got the same minimum spanning tree as previously). This is because, in this graph, there is more than one minimum spanning tree.

Now, the **travelling salesmen problem** is the problem of finding the minimum weight Hamiltonian cycles in a weighted complete graph.

No simple algorithm to solve the travelling salesmen problem is known. We can seek approximate solutions which provide upper and lower bounds for the solution of the problem. Obviously if we can find upper and lower bounds which are close enough together then we have a solution which may be good enough for practical purposes and the closer they are together, the better. So we will try to find the smallest upper bound possible and the largest lower bound.

Heuristic algorithm for an upper bound on the travelling salesmen problem

1. Start with a finite set of 3 or more vertices, each pair joined by a weight edge.
2. Choose any vertex and find a vertex joined to it by an edge of least weight.
3. Find the vertex which is joined to either of the two vertices identified in step (2) by an edge of least weight. Draw these three vertices and the three edges joining them to form a cycle.
4. Find a vertex, not currently drawn, joined by an edge of least weight to any of the vertices already drawn. Label the new vertex v_n and the existing vertex to which it is joined by the edge of least weight as v_k. Label the two vertices which are joined

to v_k in the existing cycle as v_{k-1} and v_{k+1}. Denote the weight of an edge $v_p v_q$ by $w(v_p v_q)$. If $w(v_n v_{k-1}) - w(v_k v_{k-1}) < w(v_n v_{k+1}) - w(v_k v_{k+1})$ then replace the edge $v_k v_{k-1}$ in the existing cycle with the edge $v_n v_{k-1}$ otherwise replace the edge $v_k v_{k+1}$ in the existing cycle with the edge $v_n v_{k+1}$. See Figure 10.52

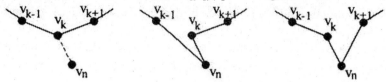

Figure 10.52

5. Repeat step 4 until all vertices are joined by a cycle, then stop. The cycle of obtained is an upper bound for the solution to the travelling salesmen problem.

Applying this to our previous problem we proceed thus. Start by choosing vertex a, then the edge of least weight incident to a is ae and the vertex adjacent to $\{a, e\}$ joined by an edge of least weight is c. From the cycle $aeca$. Now the vertex adjacent to $\{a, c, e\}$ joined by an edge of least weight is b (edge bc) and the vertices adjacent to c in the existing cycle are a and e. In this case $w(be) - w(ce) = w(ba) - w(ca)$ so we can create the new cycle by replacing either of edges ca or ce. We will choose ca. Now the vertex adjacent to $\{a, b, c, e\}$ joined by an edge of least weight is d (edge ed) and the vertices adjacent to e in the existing cycle are a and c. Now $w(da) - w(ea) > w(dc) - w(ec)$ so we create the new cycle by replacing edge ec with edge dc. The cycle is $abcdea$ and the weight is 29.

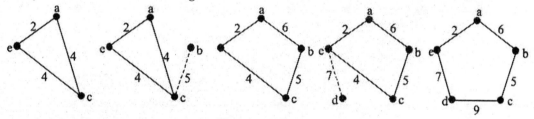

Figure 10.53

What would have happened if we had chosen a different starting vertex? Let us try it. Start by choosing vertex b, then the edge of least weight incident to b is bc and the vertex adjacent to $\{b, c\}$ joined by an edge of least weight is a or e. Choose a and from the cycle $abca$. Now the vertex adjacent to $\{a, b, c\}$ joined by an edge of least weight is e (edge ae) and the vertices adjacent to a in the existing cycle are b and c. In this case $w(eb) - w(ab) < w(ec)$ so we can create the new cycle by replacing edge ab with eb, creating the cycle $aebca$. Now the vertex adjacent to $\{a, b, c, e\}$ joined by an edge of least weight is d (edge de) and the vertices adjacent to e in the existing cycle are a and b. Now $w(da) - w(ea) > w(db) - w(eb)$ so we create the new cycle by replacing edge eb with edge db. The cycle is $acbdea$ and the weight is 26.

Figure 10.54

So different starting points and different choices will give different upper bounds! It is a heuristic algorithm and there is no theory to tell us how to make choices which result in a least upper bound, trial and error is the only way forward!

Algorithm for a lower bound on the travelling salesmen problem

An algorithm which produces a lower bound for the travelling salesman problem is as follows.

1. Start with a finite set of vertices, each pair joined by a weighted edge.
2. Choose a vertex v_s and remove it from the graph.
3. Find a minimum spanning tree connecting the remaining vertices and calculate its total weight w.
4. Find the two smallest weights, w_1 and w_2, of the edges incident with v_s.

Then $w + w_1 + w_2$ is a lower bound for the solution of the travelling salesmen problem.

For instance, remove vertex a from the graph G. The resulting graph G_1 has a minimum spanning tree G_2 so the lower bound graph is G_3 with total weight 22.

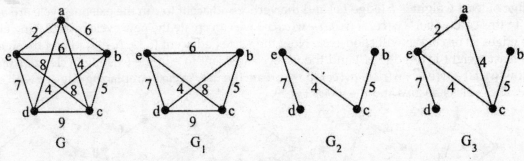

Figure 10.55

Of course we will get a different lower bound for each different choice of starting vertex. If we start by removing vertex d then we get the following sequence (although there are two points at which we could make alternative, equal weight choices) and the lower bound is 26. Since we have already found an upper bound of 26 we know this is, in fact, the solution to this travelling salesman problem!

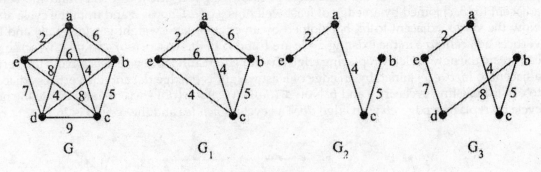

Figure 10.56

We can also point out that this lower bound algorithm will produce an actual solution (rather than a lower bound on a solution) to the travelling salesman problem whenever the following is true :
(a) The minimum spanning tree, T_G, produced at step 2 is a path graph P_{n-2} and
(b) The two edges incident with v_s which have smallest weight connect v_s to v_a to v_b, the vertices at either end of the minimum spanning tree T_G (which is a path graph).

This is evidently the case in the last example so the graph which provides the lower bound is actually a Hamiltonian cycle and so is a solution of the travelling salesman problem (since it is a lower bound on the solution and is a solution, then the minimum Hamiltonian cycle is this cycle).

EXERCISE 10.5

1. Find a minimum weight spanning tree in the following graphs using Kruskal's algorithm.

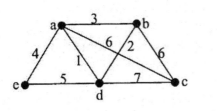
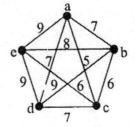

G   H

2. Find a minimum weight spanning tree in the graphs from question 1 using Prim's algorithm.

3. How do you think you might modify Kruskal's algorithm to find a maximum weight spanning tree. Find maximum weight spanning trees for the two graphs in question 1.

4. The following tables gives the distances between 6 European cities (in hundreds of km). Find an upper bound on the solution of the travelling salesman problem for these cities starting at (a) Moscow and (b) Seville. Which is the better bound? Find a lower bound to the solution by removing (a) Moscow and (b) Seville. Which is the better lower bound?

	Berlin	London	Moscow	Paris	Rome	Seville
Berlin		8	11	7	10	15
London	8	—	18	3	12	11
Moscow	11	18	—	19	20	27
Paris	7	3	19	—	9	8
Rome	10	12	20	9	—	13
Seville	15	11	27	8	13	—

5. The following table gives the distances between English cities (in miles). Find an upper bound on the solution of the travelling salesman problem for these cities

starting at (*a*) Bristol and (*b*) Leeds. Which is the better bound? Find a lower bound to the solution by removing (*a*) Exeter and (*b*) York. Which is the better lower bound?

	Bristol	Exeter	Hull	Leeds	Oxford	York
Bristol	—	84	231	220	74	225
Exeter	84	—	305	271	154	280
Hull	231	305	—	61	189	37
Leeds	220	271	61	—	169	24
Oxford	74	154	189	169	—	183
York	225	280	37	24	183	—

ANSWERS (EXERCISE 10.5)

1. To construct a minimum spanning tree for the graph G we proceed thus. The weights are

   1 (*ad*)   2 (*db*)   3 (*ab*)   4 (*ae*)   5 (*ed*)   6 (*ac*)   6 (*bc*)   7 (*cd*)

   So the steps of Kruskal's algorithm will give the following : choose *ad*, choose *db*, reject *ab* (would form cycle *adba*), choose *ae*, reject *ed* (would form cycle *aeda*), choose *ac*. All vertices now connected. The total weight is 13.

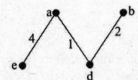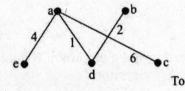

Construct a minimum spanning tree for the graph H we proceed thus. The weights are :

   5 (*ac*)   6 (*db*)   6 (*bc*)   7 (*ab*)   7 (*ad*)   7 (*cd*)   8 (*be*)   9 (*ae*)   9 (*de*)   9 (*ce*)

   so the steps of Kruskal's algorithm will give the following : choose *ac*, choose *db*, choose *bc*, reject *ab* (would form cycle *abca*), reject *ad* (would form cycle *acbda*), reject *cd* (would form cycle *cbdc*), choose *be*. All vertices now connected. The total weight is 25.

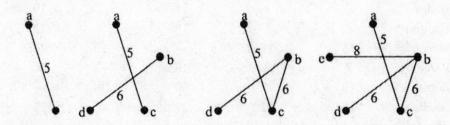

2. To construct a minimum spanning tree for the graph G using Prim's algorithm we

proceed thus. Choose vertex a to start, minimum weight edge is *ad* (weight 1), minimum weight edge from {*a, d*} to new vertex is *db* (weight 2), minimum weight edge from {*a, b, d*} to new vertex is *ae* (weight 4), minimum weight edge from {*a, b, d, e*} to new vertex is *bc* or *ac* (both weight 6), choose *bc*, all vertices now connected. Total weight is 13.

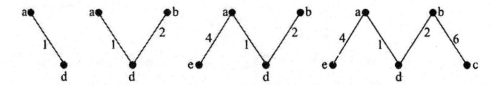

To construct a minimum spanning tree for the graph H using Prim' algorithm we proceed thus. Choose vertex e to start, minimum weight edge is *eb* (weight 8), minimum weight edge from {*b, e*} to new vertex is *bd* or *bc* (both weight 6), choose *bd*, minimum weight edge from {*b, d, e*} to new vertex is *bc* (weight 6), minimum weight edge from {*b, c, d, e*} to new vertex is *ca* (weight 5), all vertices now connected. Total weight is 25.

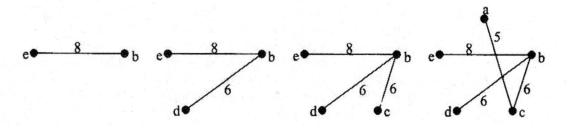

3. To construct a maximum spanning tree we could list the edges in descending order (or, equivalently, process them from the other end of the list) instead of listing the edges in ascending order. We would then use the same algorithm!

   To construct a maximum spanning tree for the graph G we proceed thus. The weights are :

   1 (*ad*)   2 (*db*)   3 (*ab*)   4 (*ae*)   5 (*ed*)   6 (*ac*)   6 (*bc*)   7 (*cd*)

   so the steps of Kruskal's algorithm will give the following : choose *cd*, choose *bc*, choose *ac*, choose *ed*. All vertices now connected. The total weight is 24.

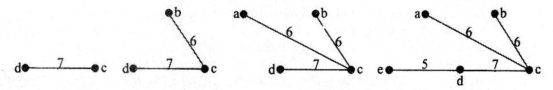

   To construct a maximum spanning tree for the graph H we proceed thus. The weights are :

5 (ac)   6 (db)   6 (dc)   7 (ab)   7 (ad)   7 (cd)   8 (be)   9 (ae)   9 (de)   9 (ce)

So the steps of kruskal's algorithm will give the following : choose, ce, choose de, choose ae, choose be. All vertices now conncerid. The total weight is 35.

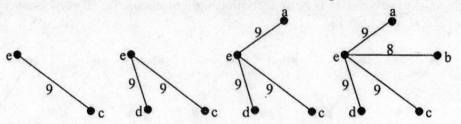

4. The graph of the connections and distances is as shown here.

To find an upper bound we use the Heuristic algorithm. Starting at M we find the edge of least weight is MB (weight 11) and the least weight edge joined to {B, M} is BP (weight 7) so we establish the cycle MBPM. The least weight edge joined to {B, M, P} is LP (weight 3). We have $w(LB) - w(PB) = 1$ and $w(LM) - w(PM) = -1$ so we insert L in PM and establish cycle MBPLM. The least weight edge joined to {B, L, M, P} is PS (weight 8). We have $w(SB) - w(PB) = 8$ and $w(SL) - w(PL) = 8$ so we can insert P either in PB or PL. Choose PL and establish cycle MBPSLM. Finally the least weight edge joined to {B, L, M, P, S} is RP (weight 9). We have $w(RB) - w(PB) = 3$ and $w(RS) - w(PS) = 5$ so we insert R in PB and establish cycle MBRPSLM with total weight $11 + 10 + 9 + 8 + 11 + 18 = 67$.

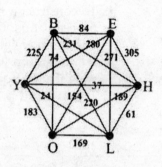

Alternatively starting at S we find the edge of least weight is SP (weight 8) and the least weight edge joined to {P, S} is LP (weight 3) so we establish the cycle SPLS. The least weight edge joined to {L, P, S} is BP (weight 7). We have $w(BS) - w(PS) = 7$ and $w(BL) - w(PL) = 5$ so we insert B in PL and establish cycle SPBLS. The least weight edge joined to {B, L, P, S} is RP (weight 9). We have $w(RS) - w(PS) = 5$ and $w(RB) - w(PB) = 3$ so we insert R in PB and establish cycle SPRBLS. Finally the least weight edge joined to {B, L, P, R, S} is MB (weight 11). We have $w(MR) - w(BR) = 10$ and $w(ML) - w(BL) = 10$ so we can insert M in BR or BL. Choose BR and establish cycle SPRMBLS with total weight $8 + 9 + 20 + 11 + 8 + 11 = 67$. So the upper bounds found starting at Moscow and at Seville a

Removing Moscow from the graph we have the graph shown below. A minimum spanning tree is also shown with weight 27. The two least weight edges from M are MB and ML. So $w(T_{min}) + w(MB) + w(ML) = 27 + 11 + 18 = 56$ is a lower bound on the solution of the travelling salesman problem.

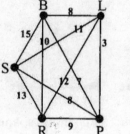 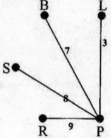

Removing Seville from the graph we have the graph shown below. A minimum spanning tree is also shown with weight 30. The two least weight edges from S are SP and SL. So $w(T_{min}) + w(SP) + w(SL) = 30 + 8 + 11 = 49$ is a lower bound on the solution of the travelling salesman problem. So the better (greater) lower bound is found by removing Moscow from the graph.

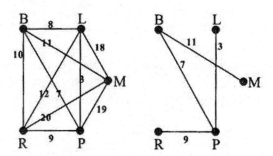

The solution of this travelling salesman problem has a total distance between 56 and 67. In fact I think the solution is probably BMLPSRB for a total weight of 63. So the heuristic algorithm gets within 7% of the optimum!

5. The graph of the connections and distances is as shown here.

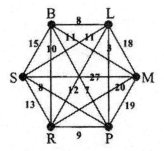

To find an upper bound we use the heuristic algorithm. Starting at B we find the edge of least weight is BO (weight 74) and the least weight edge joined at {B, O} is BE (weight 84) so we establish the cycle EBOE. The least weight edge joined to {B, E, O} is LO (weight 169). We have $w(LB) - w(OB) = 146$ and $w(LE) - w(OE) = 116$ so we insert L in OE and establish cycle EBOLE. The least weight edge joined to {B, E, L, O} is YL (weight 24). We have $w(YE) - w(LE) = 9$ and $w(YO) - w(LO) = 14$ so we insert Y in LE and establish cycle EBOLYE. Finally the least weight edge joined to {B, E, L, O, Y} is HY (weight 37). We have $w(HL) - w(YL) = 37$ and $w(HE) - (YE) = 25$ so we insert H in YE and establish cycle EBOLYHE with total weight $84 + 74 + 169 + 24 + 37 + 305 = 693$.

Starting at L we find the edge of least weight is YL (weight 24) and the least weight edge joined to {L, Y} is YH (weight 37) so we establish the cycle YLHY. The least weight edge joined to {H, L, Y} is OL (weight 169). We have $w(OH) - w(LH) = 128$ and $w(OY) - w(LY) = 159$ so we insert O in LH and establish cycle YLOHY. The least weight edge joined to {H, L, O, Y} is BO (weight 74). We have $w(BL) - w(OL) = 51$ and $w(BH) - w(OH) = 42$ so we insert B in OH and establish cycle YLOBHY. Finally the least weight edge joined to {B, H, L, O, Y} is BE (weight 84). We have $w(EO) - w(BO) = 80$ and $w(EH) - w(BH) = 74$ so we inset E in BH and establish cycle YLOBEHY with total weight $24 + 169 + 74 + 84 + 305 + 37 = 693$.

Removing Exeter from the graph we have the graph shown below. A minimum spanning tree is also shown with weight 304. The two least weight edges from E are EB and EO. So $w(T_{min}) + w(EB) + w(EO) = 304 + 84 + 154 = 542$ is a lower bound on the solution of the travelling salesman problem.

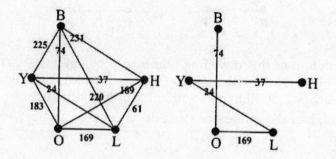

Removing York from the graph we have the graph shown below. A minimum spanning tree is also shown with weight 378. The two least weight edges from E are YL and YH. So $w(T_{min}) + w(YL) + w(YH) = 378 + 24 + 37 = 439$ is a lower bound on the solution of the travelling salesman problem.

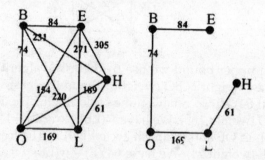

The solution of this travelling salesman problem has a total distance between 542 and 693.

CHAPTER 11

MATHEMATICAL LOGIC

11.1 PROPOSITION LOGIC

Well-defined formula of Proposition:
"A proposition is a statement that is either true or false but not both".
Example 1: "Washington D.C is the capital of the United States of America"
2. Toronto is the capital of Canada
3. $1 + 1 = 2$
4. $2 + 2 = 3$
5. What time is it?
6. Real this carefully
7. $x + 1 = 2$
8. $x + 7 = z$

(a) Delhi is the capital of India
(b) Haryana is a country
(c) 7 is a prime number.
(d) $2 + 3 = 5$
(e) Wish you a happy new year.
(f) How are you?
(g) Please wait.
(h) Take one disprin.

S → Statements 1 and 3 are true, whereas 2 and 4 are false, so they are propositions. Sentences 1 and 2 are not propositions because they are not statements. Sentences 3 and 4 are not propositions because they are neither true nor false.

11.2 OPERATORS OF PROPOSITION

1. Negation operator: Let P be a proposition. The statement
"It is not the case that P" is another proposition, called the negation of P. The negation of P is denoted by ΓP.

Example: Find the negation of the proposition
"Today is Friday"
and express this in simple English.

Solution : The negation is
" It is not the case that today is Friday"

This negation can be more simply expressed by "Today is not Friday"

11.3 TRUTH TABLE

"A truth table displays the relationship between the truth values of propositions".

The truth table for the negation of a proportion :

P	⌐P
T	F
F	T

2. Conjunction operation: Let P and q be two propositions. The proposition "P and 'V'" denoted by P ∧ q, is the proposition that is true when both P and q are true and is false otherwise and denoted by "∧".

Example 1: Find the conjunction of the propositions P and q where P is the proposition "Today is Friday" and q is the proposition " It is raining today".

Solution: The conjunction of these Propositions P and q where P ∧ q is the proposition "Today is Friday and it is raining today". This is true on rainy Fridays and is false on any day that is not a Friday and on Fridays when it does not rain.

3. Disjunction operation: Let P and q be propositions. The propositions "P or q" denoted by P ∨ q, is the proposition that is false when P and q are both false and true otherwise the proposition P ∨ q is called the disjunction of P and q.

The truth table for the conjunction of two propositions

P	q	P ∧ q
T	T	T
T	F	F
F	T	F
F	F	F

The truth table for the disjunction of two propositions

P	q	P ∨ q
T	T	T
T	F	T
F	T	T
F	F	F

4. Exclusive OR operation: Let P and q be propositions. The exclusive or of P and q, denoted by P ⊕ q, is the proposition that is true when exactly one of P and q is true and is false otherwise.

The truth table for the exclusive or of two propositions

P	q	P ⊕ q
T	T	F
T	F	T
F	T	T
F	F	F

5. Implication operation: Let P and q be propositions. The implication $P \to q$ is the proposition that is false, when P is true and q is false and true otherwise. In this implication P is called the hypothesis (or antecedent or premise) and q is called conclusion (or consequence).

The truth table for the Implication $P \to q$

P	q	$P \to q$
T	T	T
T	F	F
F	T	T
F	F	T

Some of the common ways of expressing this implication are

* "If P the q",   * "P is sufficient for q"
* "P implies q",   * "q if P"
* " If P, q"   * "q whenever P"
* "P only if"   * "q is necessary for P"

6. Biconditional operator: Let P and q be propositions the biconditional $P \to q$ is the proposition that is true when P and q have the same truth values and is false otherwise.

"P if and only if q"

The truth table for the
Biconditional $P \leftrightarrow q$

P	q	$P \leftrightarrow q$
T	T	T
T	F	F
F	T	F
F	F	T

11.4 LAWS

1. Law of Identity: Under this law the symbol used will carry the same sense through out the specified problems.

2. Law of the Excluded middle: Which express that a statement is either true or false.

3. Law of Non-contradiction: It expresses that no statement is true and false simultaneously.

11.5 TYPES OF STATEMENTS

1. Atomic statement (simple statements): A statement which has no logical connective in it is called an atomic sentences for example:

(a) This is my body (true)

(b) Delhi is the capital of U.P (False).

2. Molecular or compound statement: The sentence formed by two simple sentences using a logical connective is called a compound sentences. For example

(a) U.P. is in India and Lucknow is the capital of U.P.

(b) Hari will study he will play.

Example: Construct the truth table for
$$\sim (P \wedge q) \Leftrightarrow (\sim P \vee \sim q)$$

P	q	P∧q	~(P∧q)	~q	~q	~P∨~q	~(P∧q) ⇔ (~P∨~q)
T	T	T	F	F	F	F	F
T	F	F	T	F	T	T	T
F	T	F	T	T	F	T	T
F	F	F	T	T	T	T	T

3. Contingency and Satisfiable

A statement formula (expression involving prepositional variable) that is neither a tautology nor a contradiction is called a contingency

If the resulting truth value of a statement formula A $(p_1, p_2, ... p_n)$ is true for at least one combination of truth values of the variables $p_1, p_2,..., p_n$ then A is said to be satisfiable

Example 2: Prove that the sentence 'It is wet or it is not wet' is a tautology.

Solution: Sentence can be symbolized by $p \vee \sim p$ where p symbolizes the sentence 'It is wet'. Its Truth table is given below:

p	~p	p ∨ ~p
T	F	T
F	T	T

As column 3 contains T everywhere, therefore the given proposition is tautology. This tautology is called the **Law of the Excluded Middle.**

Example 3: Prove that the proposition $\sim [p \wedge (\sim p)]$ is a tautology.

Solution: The truth table is given below:

P	~q	P ∧ (~p)	~[p ∧ (~p)]
T	F	F	T
F	T	F	T

Since last column contains 'T' everywhere therefore the given proposition is a tautology.

Example 4: Show that $((p \vee \sim q) \wedge (\sim p \vee \sim q)) \vee q$ is a tautology be definition

Solution: $((p \vee \sim q) \wedge (\sim p \vee \sim q)) \vee q$

$\Rightarrow \quad ((p \wedge \sim p) \vee \sim q) \vee q$ (distributive law)

$\Rightarrow \quad (F \vee \sim q) \vee q$

$\Rightarrow \quad (\sim q) \vee q \quad$ (as $p \wedge \sim p = F$)

$\Rightarrow \quad T$

Example 5: Prove that $(\sim p \vee q) \wedge (p \wedge \sim q)$ is a contradiction.

Solution: Truth table for the given proposition

p	q	$\sim p$	$\sim q$	$\sim p \vee q$	$p \wedge \sim q$	$(\sim p \vee q) \wedge (p \wedge \sim q)$
T	T	F	F	T	F	F
T	F	F	T	F	T	F
F	T	T	F	T	F	F
F	F	T	T	T	F	F

Since F appears in every row of the last column, therefore the given proposition is contradiction.

4. Equivalence of formulae: Two statements A and B in variable $P_1 \ldots P_n$ ($n \geq U$) are said to be equivalent if they acquire the same truth values for all interpretation, i.e., they have identical truth values

Tautology: A statement that is true for all possible values of its propositional variable is called a tautology. A statement that is always false is called a contradiction and a statement that can be either true or false depending on the truth values of its propositional variables is called a contingency.

Example:

P	$\sim P$	$P \vee (\sim P)$	$P \wedge (\sim P)$
T	F	T	F
F	T	T	F

Hence $P \vee (\sim P)$ is a tautology and $P \wedge (\sim P)$ is a contradiction

11.6 LOGICAL EQUIVALENCE

Two propositions $P(p, q, \ldots)$ and $Q(p, q, \ldots)$ are said to be **logically equivalent** or simply equivalent or **equal** if they have identical truth tables and is written as

$$P(p, q, \ldots) \equiv Q(p, q, \ldots)$$

As shown below truth tables of $\neg (p \wedge q)$ and $\neg p \vee q$ are identical. Hence $\neg (p \wedge q) = \neg p \vee \neg q$

P	q	$p \wedge q$	$\sim (p \wedge q)$	P	q	$\neg p$	$\neg q$	$\neg p \vee \neg q$
T	T	T	F	T	T	F	F	F
T	F	F	T	T	F	F	T	T
F	T	F	T	F	T	T	F	T
F	F	F	T	F	F	T	T	T

$\sim (p \wedge q)$ $\qquad\qquad\qquad\qquad$ $\neg p \vee \neg q$

Also when a conditional $p \Rightarrow q$ & its converse $q \Rightarrow p$ are both ture, the statement p & q are said to be logically equivalent. Consider the following table

P	F	p ∧ F	p ∨ F	T	p ∧ T	p ∨ T
T	F	F	T	T	T	T
T	F	F	T	T	T	T
F	F	F	F	T	F	T
F	F	F	F	T	F	T

$$p \wedge F \equiv F \quad p \vee F \equiv p \quad \quad p \wedge T \equiv p \quad p \vee T \equiv T$$

Hence T is the **identity** element for the operation $\wedge$ (**conjunction**) and F is the **identity** element for the operation $\vee$ (**disjunction**).

Duality : Any two formulas A and A* are said to be duals of each other if one can be obtained from the other by replaying $\wedge$ by $\vee$ and $\vee$ by $\wedge$.

Example : (a) The dual of $(P \wedge q) \vee r$ is $(P \wedge q) \wedge r$.

(b) The dual of $(P \vee q) \wedge r$ is $(P \wedge q) \vee r$.

Functionally complete Set : "A set of connectives is called functionally complete if every formula can be expressed in terms of an equivalent formula containing the connectives from this set".

$$\wedge, \vee, \sim, \Rightarrow, \Leftrightarrow$$

Example: Consider $P \Leftrightarrow q$

It is equivalent to $(P \Rightarrow q) \wedge (q \Rightarrow P)$

11.7 PREDICATED LOGIC

Well formed formula of Predicate:

The statement "x is greater than >" has two parts the first part the variable x is the subject of the statement the second part the predicate "is greater then >" refers to a property that the subject of the statement can have.

$P(x) \rightarrow$ Proposition function P at q.

Quantifiers: When all the variables in a propositional function are assigned values, the resulting statement has a truth value. However, there is another important way called quantification to create a proposition from a propositional function. There are two types of quantification namely **universal quantification and existential quantification.**

Many mathematically statement assert that a property is true for all values of a variable in a particular domain, called the universe of discourse. Such a statement is expressed using a universal quantification. The universal quantification of a propositional function is the proposition that asserts that $P(x)$ is true for all values of x in the universe of discourse. The universe of discourse specifies the possible values of the variable x.

Arguments: An argument (denoted by the symbol $\vdash$ which is called trunstile) is a sequence of propositions that purport to imply another propositions.

The sequence of propositions serving as evidence will be called, the **premises,** and the proposition inferred will be called the **conclusion.**

An arguments is valid if and only if whenever the conjunction of the premises is true, the conclusion is also true. If we let p_1, p_2, p_3 be the premises and p_4 the conclusion then argument $p_1, p_2, p_3, \dashv p_4$ will be valid if and only if whenever $p_1 \wedge p_2 \wedge p_3$ is true, p_4 is also. We can reduce this to the conditional $\Rightarrow$ as follows:

Def^n : If $p_1, p_2, ... , p_n$ are premises and p is a conclusion then the argument $p_1, p_2, ..., p_n \vdash p$ is valid if and only if $p_1 \wedge p_2 \wedge ... p_n \Rightarrow p$ is true for all combinations of truth values of $p_1 p_2 ... p_n$ and p.

In other word in order to decide whether an argument is valid, use the conjunction of evidences as the antecedent of conditional of which the conclusion of the argument is the consequent and see whether or not a tautology results.

If $p_1 \wedge p_2 \wedge ... \wedge p_n \Rightarrow p$ is not a tautology then the argument $p_1 ... p_n \vdash$ is invalid.

Example 6: "If the labour market is perfect then the wages all persons in a particular employment will be equal. But it is always the case that wages for such persons are not equal therefore the labour market is not perfect". Test the validity of this argument.

Solution: In the given case let

p_1 : "The labour market is perfect"

p_2 : "Wages of all persons in a particular employment will be equal"

$\sim p_2$: Wages for such persons are not equal.

$\sim p_1$: The labour market is not perfect.

The premise are $p_1 \Rightarrow p_2, \sim p_2$ and the conclusion is $\sim p_1$

The argument $p_1 \Rightarrow p_2 \sim p_2 \vdash \sim p_1$

valid if and only if $(p_1 \Rightarrow p_2) \Rightarrow \sim p_2 \Rightarrow \sim p_1$ is a tautology.

We construct the truth tables as below :

$p_1\ p_2$	$\sim p_1$	$\sim p_2$	$p_1 \Rightarrow p_2$	$p_1 \Rightarrow p_2 \wedge \sim p_2$	$p_1 \Rightarrow p_2 \wedge \sim p_2 \Rightarrow \sim p_1$
T T	F	F	T	F	T
T F	F	T	F	F	T
F T	T	F	T	F	T
F F	T	T	T	T	T

is follows that $p_1 \Rightarrow p_2 \wedge \sim p_2 \Rightarrow \sim p_1$ is a tautology

Hence the argument is valid.

Example 7: Test the validity of the following argument. "If Ashok wins then Ram will be happy. If Kamal wins Raju will be happy. Either Ashok will win or Kamal will win. However if Ashok wins, Raju will not be happy and if Kamal wins Ram will not be happy. So Ram will be happy if and only if Raju is not happy."

Solution : Here let

p_1 : Ashok wins

p_2 : Ram is happy

p_3 : Kamal wins

p_4 : Raju is happy

The premises are $p_5 : P_1 \Rightarrow P_2$, $p_6 : P_3 \rightarrow P_4$, $p_7 : P_1 \vee P_2$
$p_8 : P_1 \Rightarrow \sim P_4$
$p_9 : P_3 \Rightarrow \sim P_2$
The conclusion is $p_1 : P_2 \Leftrightarrow \sim P_4$
The above argument is valid if $p_1 : P_2 \Leftrightarrow \sim P_4$ is a tautology so we construct the truth table.

1	2	3	4	5	6	7	8	9	10	11	12	13
P_1	P_2	P_3	P_4	$P_1 \Rightarrow P_2$	$P_3 \Rightarrow P_4$	$P_1 \vee P_3$ or	$\sim P_4$	$P_1 \Rightarrow \sim P_4$	$P_3 \Rightarrow \sim P_2$	$P_2 \Rightarrow \sim P_4$	$P_1 \Rightarrow P_2 \wedge P_3 \Rightarrow P_4 \wedge (P_1 \wedge P_3) \Rightarrow P_1 \Rightarrow \sim P_4 \wedge P_3 \Rightarrow P_2$	$(12) \Rightarrow \Rightarrow (13)$
T	T	T	T	T	T	F	F	F	F	F	F	T
T	T	T	T	T	T	F	F	T	F	F	F	T
T	F	T	F	F	F	T	T	T	T	F	F	T
T	F	F	F	F	T	T	T	T	T	F	F	T
F	T	T	T	T	T	T	F	T	F	F	F	T
F	T	F	T	T	T	F	F	T	T	F	F	T
F	F	T	F	T	F	T	F	T	T	F	F	T
F	F	F	F	T	F	F	T	T	T	F	F	T

Since the given statement is a tautology. Hence the argument is valid.

11.8 INFERENCE THEORY

In the following, we list some of the rules in logic that are used to infer results from the given arguments.

- Rule P : A premise may be introduced at any point in the derivative.
- Rule T : A formula S may be introduced in the derivation if S is tautologically implied by any one or more of the preceding formulas.
- Modus' ponens (method of affirming)

$$p, q \rightarrow q \Rightarrow q.$$

- Modus tolens (method of denying)

$$\neg q, p \rightarrow q \Rightarrow \neg p.$$

Example 8: Show that r is a valid inference from the premises $p \to q$, $p \to r$ and p.
Solution:

$p \to q$ Rule P
p Rule P
q Rule T, modus ponens
$q \to r$ Rule P
r Rule T, modus ponens.

Example 9: Show that $r \vee s$ follows logically from the premises $c \vee d$, $(c \vee d) \to \neg h$, $\neg h \to (a \wedge \neg b)$ and $(a \wedge \neg b) \to (r \vee s)$.
Solution:

$(c \vee d) \to \neg h$ Rule P
$\neg h \to (a \wedge \neg b)$ Rule P
$(c \vee d) \to (a \wedge \neg b)$ Rule T
$(a \wedge \neg b) \to (r \vee s)$ Rule P
$(c \vee d) \to (r \vee s)$ Rule T
$(c \vee d)$ Rule P
$(r \vee s)$ Rule T, modus ponens

Example 10: Show that $s \vee r$ is tautologically valid by $(p \vee q) \wedge (p \to r) \wedge (q \to s)$.
Solution:

$p \vee q$ Rule P
$\neg p \to q$ Rule T
$q \to s$ Rule P
$\neg p \to s$ Rule T
$\neg s \to p$ Rule T
$p \to r$ Rule P
$\neg s \to r$ Rule T
$s \vee r$ Rule T.

Inference Theory in Predicate Calculus

The following are some of the rules that are further needed in the inference theory of predicate calculus.

De-Morgans Laws
1. $\neg (\forall x) A(x) \Leftrightarrow (\exists x)(\neg A(x))$
1. $\neg (\exists x) A(x) \Leftrightarrow (\forall x)(\neg A(x))$

Additional rules in Predicate Calculus
1   **Rule US (Universal Specification):** $(\forall x) A(x) \Rightarrow A(y)$

If a property is true for all members of the universe of discourse, then it is true for a particular member of the universe of discourse.

2. **Rule ES (Existential Specification):** $(\exists x)\ A(x) \to A(y)$
   If a property is true for a variable x, then it can be substituted with y.
3. **Rule EG (Existential Generalisation):** $A(x) \Leftrightarrow (\exists y)\ A(y)$
   If a property is true for a variable x, then there exists y with the same property.
4. **Rule UG (Universal Generalisation):** $A \Leftrightarrow (\forall y)\ A(y)$
   If a property is true for an arbitrary variable x, then it is true for all y.

Example 11: Show the validity of the Socrates' argument 'All men are mortal. Socrates is a man. Therefore Socrates is mortal'.

Solution: Let us assume: $H(x) : x$ is a man, $M(x) : x$ is mortal and s : Socrates. Hence in these notations, the given statements takes the form

$$(\forall x)\ (H(x) \to M(x)) \wedge H(s) \Rightarrow (\neg A(x))$$

To prove this,

$(\forall x)\ (H(x) \to M(x))$ Rule P
$H(s) \to M(s))$ US
$H(s)$ Rule P
$M(s)$ Rule T, modus ponens

Example 12: Show that $(\forall x)(P(x) \to Q(x)) \wedge (\forall x)(P(x) \to Q(x)) \Rightarrow (\forall x)(P(x) \to R(x))$.

Solution:

$(\forall x)(P(x) \to Q(x))$ Rule P
$P(y) \to Q(y)$ Rule US
$(\forall x)(Q(x) \to R(x))$ Rule P
$Q(y) \to R(y)$ Rule US
$P(y) \to R(y)$ Rule T
$(\forall x)(P(x) \to R(x))$ Rule UG.

Example 13: Show that $(\exists x)M(x)$ follows logically from $(\forall x)(H(x) \to M(x))$ and $(\exists x) H(x)$

Solution:

$(\exists x)H(x)$ Rule P
$H(y))$ Rule ES
$(\forall x)(H(x) \to M(x))$ Rule P
$H(y) \to M(y))$ Rule ES
$M(y) \to M(y))$ Rule T, modus poneus
$(\exists x)M(x)$ Rule EG.

Example 14: Show that $(\forall x)(P(x) \wedge Q(x)) \Rightarrow (\forall x)P(x) \vee (\exists x)Q(x)$.

Solution: Here we apply the indirect method of proof which means that inorder to prove the result $p \to g$, it is sufficient to prove $\neg q \to \neg p$.

$\neg((\forall x)P(x) \vee (\exists x)Q(x))$	Rule P
$\neg((\forall x)P(x) \wedge \neg((\exists x)Q(x))$	De Morgans' Law P
$\neg((\forall x)P(x))$	Rule T
$(\exists x)\neg P(x)$	De Morgans' Law
$\neg P(y)$	Rule ES
$\neg((\exists x)Q(x))$	RuleT
$(\forall x)\neg Q(x)$	De Morgans' Law
$\neg Q(y)$	Rule US
$\neg P(y) \wedge \neg Q(y)$	Rule T
$(\exists x)\neg P(x) \wedge \neg Q(y)$	RuleES
$\neg(\forall x)(P(x) \vee Q(x))$	De Morgans' Law

11.9 ALGEBRA OF PROPOSITIONS

We can easily verify by truth tables that every proposition satisfy the following laws. If P and Q are propositions and F are tautology and contradiction respectively.

1. **Associative Laws**
   (a) $(P \vee Q) \vee R = P \vee (Q \vee R)$
   (b) $(P \wedge Q) \wedge R = P \wedge (Q \vee R)$

2. **Commutative Laws**
   (a) $P \vee Q = Q \wedge P$
   (b) $P \wedge Q = Q \wedge P$

3. **Distributive Laws**
   (a) $P \vee (Q \wedge R) = (P \vee Q) \wedge (P \vee R)$
   (b) $P \wedge (Q \vee R) = (P \wedge Q) \vee (P \wedge R)$

4. **DeMorgan's Laws**
   (a) $\sim (P \vee Q) \equiv \sim P \wedge \sim Q$
   (b) $\sim (P \wedge Q) \equiv \sim P \vee \sim Q$
   (c) $\sim (P \Rightarrow Q) \equiv \sim P \wedge \sim Q$
   (d) $\sim (P \Leftrightarrow Q) \equiv P \Leftrightarrow \sim Q \sim P \Leftrightarrow Q$

5. **Identity Laws**
   (a) $P \vee F = P$
   (b) $P \vee \sim P \equiv F$
   (c) $\sim \sim P \equiv P$
   (d) $\sim T \equiv F$
   (e) $\sim F \equiv T$

6. Idempotent Laws
(a) $P \vee P \equiv P$
(b) $P \wedge P \equiv P$

7. Compliment Laws
(a) $P \vee \sim P = T$
(b) $P \wedge \sim P \equiv F$
(c) $\sim \sim P \equiv P$
(d) $\sim T \equiv F$
(e) $\sim F \equiv T$

Examples on truth tables

Example 15: Prove that $p \vee (q \wedge r)$ and $(p \vee q) \wedge (p \vee r)$ have the same truth table i.e. they are equivalent.

Solution:

1	2	3	4	5	6	7	8
p	q	r	$q \wedge r$	$p \vee q$	$p \vee r$	$p \vee (q \wedge r)$	$(p \vee q) \wedge (q \vee r)$
T	T	T	T	T	T	T	T
T	T	F	F	T	T	T	T
T	F	T	F	T	T	T	T
F	T	T	T	T	T	T	T
F	F	T	T	F	T	T	F
F	T	F	F	T	F	F	F
T	F	F	F	T	T	T	T
F	F	F	F	F	F	F	F

Since the truth values of the columns (7) and (8) are the same for all values of the component of sentences. Hence the given functions are equivalent.

Example 16: Using, truth table prove the DeMorgan's laws
(i) $\sim (p \wedge q) \equiv \sim p \vee \sim q$
(ii) $\sim (p \vee q) \equiv \sim p \wedge \sim q$

Solution (1) Consider the truth table

1	2	3	4	5	6	7
p	q	$p \wedge q$	$\sim p$	$\sim q$	$\sim (p \wedge q)$	$(\sim p \vee \sim q)$
T	T	T	F	F	F	F
T	F	F	F	T	T	T
F	T	F	T	F	T	T
F	F	F	T	T	T	F

Since the truth values of the columns (6) and (7) are same law both functions are equivalent.

Similarly we can prove (U) result.

Example 17: Prove that $(p \Rightarrow q) \vee r \equiv (p \vee r) \Rightarrow (q \vee r)$

Solution. Consider the truth table

1	2	3	4	5	6	7	8
p	q	r	$p \Rightarrow q$	$(p \Rightarrow q) \vee r$	$p \vee r$	$q \vee r$	$(p \vee r) \Rightarrow (q \vee r)$
T	T	T	T	T	T	T	T
T	T	F	T	T	T	T	T
T	F	T	F	T	T	T	T
F	T	T	T	T	T	T	T
T	F	F	F	F	T	F	F
F	F	T	T	T	T	F	T
F	T	F	T	T	F	T	T
F	F	F	T	T	F	F	T

Since the truth values of the columns (5) and (8) are same hence the given sentences are logically equivalent.

Example 18: Write the Converse, Inverse and Contrapositive of the following simplications :

(a) If PQRS is a rectangle then PQRS is a square.

(b) If $3x - 2 = 10$ then $x = 4$.

Solution. (a) Let p = PQRS is a rectangle

q = PQRS is a square.

then $p \Rightarrow q$ is given.

Its converse will be $q \Rightarrow p$ i.e. if PQRS is a square then PQRS is a rectangle.

Its inverse will be $\sim p \Rightarrow \sim q$ if PQRS is not rectangle then PQRS is not a square.

Its contrapositive will be $\sim q \Rightarrow \sim p$ i.e. if PQRS is not a square then PQRS is not a rectangle.

(b) **Consider** $p = 3x - 2 = 10$

$q = x = 4$

then given that $p \Rightarrow q$

Its converse will be $q \Rightarrow p$

i.e., If $x = 4$ then $3x - 2 \neq 10$

Its inverse will be

$\sim p \Rightarrow \sim q$

i.e., If $3x - 2 \neq 10$ then $x \neq 4$

Its contrapositive will be $\sim q \Rightarrow \sim p$

i.e., If $x \neq 4$ then $3x - 2 \neq 10$

Example 19: Simplify the following propositions:

(a) $P \vee (P \wedge Q)$

(b) $(P \vee Q) \wedge \sim P$

(c) $\sim (P \vee Q) \vee (\sim P \wedge Q)$

Solution.

(a) $P \vee (P \wedge Q)$

$$\equiv (P \wedge T) \vee (P \wedge Q) \quad \text{(since } P \wedge T = P\text{)}$$
$$\equiv P \wedge (T \vee Q)$$
$$\equiv P \quad \text{(since } T \vee Q = T\text{)}$$

(b) $(P \vee Q) \wedge \sim P$

$$\equiv \sim P \wedge (P \vee Q) \quad \text{(Commutative Law)}$$
$$\equiv (\sim P \wedge P) \vee (\sim P \vee Q) \quad \text{(Distributive Law)}$$
$$\equiv F \vee (\sim P \vee Q) \quad \text{(Complement Law)}$$
$$\equiv (\sim P \vee Q) \quad \text{(Identity Law)}$$
$$= \sim P \vee Q$$

(c) $\sim(P \vee Q) \vee (\sim P \wedge Q)$

$$\Rightarrow (\sim P \wedge \sim Q) \vee (\sim P \wedge Q)$$
$$\equiv \sim P(\sim Q \vee Q) \quad \text{(By distributive law)}$$
$$\equiv \sim P \wedge T \quad (\because \sim Q \vee Q = T)$$
$$\equiv \sim P \quad (\because P \wedge T = P)$$

Example 20: Construct truth table for the following functions and check whether it is a tautology or contradiction.

(a) $[(p \wedge q) \vee (q \wedge r) \vee (r \wedge p)] \Leftrightarrow [(p \vee q) \wedge (q \vee r) \wedge (r \vee p)]$

Solution : Truth table corresponding to the given function is given below :

p	q	r	$p \wedge q$	$q \wedge r$	$r \wedge p$	$p \vee q$	$q \vee r$	$r \vee p$	$(p \wedge q)$ $\vee (q \wedge r)$ $\vee (r \wedge p)$	$(p \vee q)$ $\wedge (q \vee r)$ $\wedge (r \vee p)$
T	T	T	T	T	T	T	T	T	T	T
T	T	F	T	F	F	T	T	T	T	T
T	F	T	F	F	T	T	T	T	T	T
F	T	T	F	T	F	T	T	T	T	T
T	F	F	F	F	F	T	F	T	F	F
F	T	F	F	F	F	T	T	F	F	F
F	F	T	F	F	F	F	T	T	F	F
F	F	F	F	F	F	F	F	F	F	F

Since the last two columns of the above table are the same. Hence given function represents a tautology,

(b) $(p \Rightarrow q) \Leftrightarrow (\sim p \vee q)$

p	q	~p	p⇒q	~p∨q	p⇒q⇔(~p∨q)
T	T	F	T	T	T
T	T	F	F	F	T
F	F	T	T	T	T
F	F	F	T	T	T

Since the last column contains all truth values T. Hence given function represents a tautology.

(c) $(p \wedge q) \Rightarrow p] \Rightarrow [q \wedge \sim q]$

p	q	~q	p∧q	(p∧q)⇒p	q∧~q	(p∧q)⇒p⇒[q∧~q]
T	T	F	T	T	F	F
T	F	T	F	T	F	F
F	T	F	F	T	F	F
F	F	T	F	T	F	F

Since the last column contains all truth values F. Hence the given function represents a contradiction.

(d) $[(p \wedge q) \Rightarrow r] \Leftrightarrow [(p \Rightarrow r) \vee (q \Rightarrow r)]$

Solution:

p	q	r	p∧q	(p⇒q)	q⇒r	(p∧q)⇒r	(p⇒q)∨(q⇒r)	(d)
T	T	T	T	T	T	T	T	T
T	T	F	T	T	F	F	T	F
T	F	T	F	F	T	T	T	T
F	T	T	F	T	T	T	T	T
T	F	F	F	F	T	T	T	T
F	T	F	F	T	F	T	T	T
F	F	T	F	T	T	T	T	T
F	F	F	F	T	T	T	T	T

Since the last column of above table does not contain all truth values T, so, it is not a tautology.

Example 21: Establish the equivalence using truth tables.

(a) $(p \Rightarrow q) \vee (p \Rightarrow r) \equiv p \Rightarrow (q \vee r)$

Solution : Truth table for the given statement is

p	q	r	$q \vee r$	$p \Rightarrow q$	$p \Rightarrow r$	$(p \Rightarrow q) \vee (p \Rightarrow r)$	$p \Rightarrow (q \vee r)$
T	T	T	T	T	T	T	T
T	T	F	T	T	F	T	T
T	F	T	T	F	T	T	T
F	T	T	T	T	T	T	T
T	F	F	F	F	F	F	F
F	T	F	T	T	T	T	T
F	F	T	T	T	T	T	T
F	F	F	F	T	T	T	T

Since the last two columns of the above table have the same truth values. Hence the given statements are logically equivalent.

(b) $(p \Rightarrow q) \Rightarrow (p \wedge q) = (\sim p \Rightarrow q) \wedge (q \Rightarrow p)$

Solution : We construct the truth table

p	q	$\sim p$	$p \wedge q$	$p \wedge q$	$\sim p \Rightarrow q$	$q \Rightarrow p$	$(p \wedge q) \Rightarrow (p \wedge q)$	$(\sim p \Rightarrow q) \wedge (q \Rightarrow p)$
T	T	F	T	T	T	T	T	T
T	F	F	F	F	T	T	T	T
F	T	T	T	F	T	F	F	F
F	F	T	T	F	F	T	F	F

Since the last two columns have the same truth values. Hence given statements are logically equivalent.

(c) $(p \vee q) (p \wedge q) \equiv p$

Solution : Truth table is

p	q	$p \vee q$	$p \wedge q$	$(p \vee q) \vee (p \wedge q)$
T	T	T	T	T
T	F	T	F	T
F	T	T	F	T
F	F	F	F	F

Since the first and last columns of the given statement do not contain same truth values. Hence given statements are not logically equivalent.

(d) $p \Rightarrow (q \wedge r) \equiv (p \Rightarrow q) \wedge (p \Rightarrow r)$

Solution : Truth Tables given by

p	q	r	q∧r	p⇒q	p⇒r	p⇒(q∧r)	(p⇒q)∧(p⇒r)
T	T	T	T	T	T	T	T
T	T	F	F	T	F	F	F
T	F	T	F	F	T	F	F
F	T	T	T	T	T	T	T
T	F	F	F	F	F	F	F
F	T	F	F	T	T	T	T
F	F	T	F	T	T	T	T
F	F	F	F	T	T	T	T

Since the last two columns of the above table have the same truth values hence the given statements are logically equivalent.

Example 22: Establish the equivalence analytically and write dual also of the given statement

$$\sim p \wedge (\sim q \wedge r) \vee (q \wedge r) \vee (p \wedge r) \equiv r.$$

Solution: Consider L.H.S.

$\sim p \wedge (\sim q \wedge r) \vee (q \wedge r) \vee (p \wedge r)$
$\equiv [(\sim p \wedge \sim q) \wedge r] \vee [[(q \wedge r) \vee p] \vee [(q \wedge r) \vee r]$
$\equiv [(\sim p \wedge \sim q) \wedge r] \vee [\{(q \wedge r) \vee p\} \vee r]$
$\equiv [(\sim p \wedge \sim q) \wedge r] \vee [\{(q \wedge r) \wedge r\} \vee (p \wedge r)]$
$\equiv [(\sim p \wedge \sim q) \wedge r] \vee [\{(q \wedge r) \vee (p \wedge r)\}]$
$\equiv [(\sim p \wedge \sim q) \wedge r] \vee [(q \wedge p) \vee r]$
$\equiv [(\sim p \wedge \sim q)] \vee [(q \vee p)] \vee r$
$\equiv [\sim(p \vee q) \vee (p \vee q)] \wedge r$
$\equiv r$ R.H.S.

Dual of the given statement can be obtained by interchanging $\wedge$ and $\vee$ and is $\sim p \vee (\sim q \vee r) \wedge (q \vee r) \wedge (p \vee r) \equiv r.$

EXERCISE 11.1

1. Let P be "It is cold" and let q be "It is training". Give a simple verbal sentence which describes each of the following statements.
   (a) $\neg P$
   (b) $s(s^1 + 1)$
   (c) $P \vee q$
   (d) $q \vee \neg P$

2. Let P be "Ram reads Hindustan Times", let q be "Ram reads Times of India," and let r be "Ram reads NBT! write each of the following in symbolic form :
   (a) Ram reads Hindustan Times on Times of India not NBT.
   (b) Ram reads Hindustan Times and Times of India, on he does not read Hindustan Times and NBT.
   (c) It is not true that Ram reads Hindustan Times but not NBT.

(d) It is not true that Ram reads NBT on Times of India but not Hindustan Times.
3. Determine the truth value of each of the following statements :
   (a) 4 + 2 = 5 and 6 + 3 = 9
   (b) 3 + 2 = 5 and 6 + 1 = 7
   (c) 4 + 5 = 9 and 1 + 2 = 4
   (d) 3 + 2 = 5 and 4 + 7 = 11
4. Find the truth table of $\neg P \wedge q$.
5. Verify that the propostion $P \vee \neg (P \vee q)$ is tautalogy.
6. Show that the propostions $\neg (P \wedge q)$ and $\neg P \vee \neg q$ are logically equivalent.
7. Rewrite the following statements without using the conditonal :
   (a) If it is cold, he means a hat.
   (b) If productivity increases, then wages rise.
8. Determine the contraposition of each statement :
   (a) If John is a poet, then he is poor;
   (b) Only if he studies more then he will pass the test.
9. Consider the conditional proposition $P \downarrow q$. The simple propositions $q \rightarrow P$, $\neg P \rightarrow \neg q$ and $\neg q \rightarrow \neg P$ are called respectively the converse inverse and contrapositive of the conditional $P \rightarrow q$. Which if any of these proposition are logically equivalent to $P \rightarrow q$?
10. Write the negation of each statement as simple as possible.
    (a) If she works, she will down money.
    (b) He swims if and only if the water is warm.
    (c) If it shows, then they do not drive the can.
11. Show that the following arguments is a fallacy :
    $P \rightarrow q, \neg P + \neg q$.
12. Write the negation of each statement as simply as possible.
    $P \rightarrow q, \neg q + \neg P$.

EXERCISE 11.2

1. Let A = {1, 2, 3, 4, 5}. Determine the truth table value of each of the following statements :
   (a) $(\exists x \in A)(x + 3 = 10)$
   (b) $(\forall x \in A)(x + 3 > 10)$
   (c) $(\exists x \in A)(x + 3 < 5)$
   (d) $(\forall x \in A)(x + 3 \leq 7)$
2. Determine the truth table value of each of the following statements where U = {1, 2, 3} is the universal set :
   (a) $\exists x \forall y, x^2 < y + 1$;
   (b) $\forall x \exists y, x^2 + y^2 < 12$
   (c) $\forall x \forall y, x^2 + y^2 < 12$
3. Negate each of the following statements :
   (a) $\exists x \forall y, P(x, y)$
   (b) $\exists x \forall y, P(x, y)$
   (c) $\exists x \exists y \exists z, P(x, y, z)$
4. Let P(x) denote the sentence "x + 2 > 5". State whether on not P(x) is a propositional function on each of the following sets :

(a) N, the set of positive integers;
(b) M = (− 1, − 2, − 3, ...)
(c) C, the set of complex numbers.

5. Negative each of the following statements :
   (a) All students live in the domintonics
   (b) All Mathematics majors are values.
   (c) Some students are 25 (years) on older.

6. Let P denote the is rich and let q denote "He is happy". Write each statement in symbolic form using P and q. Note that "He is poor" and he is unhappy" are equivalent to. $\neg$P and $\neg q$ respectively.
   (a) If he is rich, then he is unhappy
   (b) He is neither rich nor happy
   (c) It is necessary to be poor in order to be happy.
   (d) To be poor is to be unhappy.

7. Find the truths tables for :
   (a) $P \vee \neg q$ (b) $P \wedge \neg q$

8. Verify that the propostion $(P \wedge q) \wedge \neg (P \vee q)$ is a contradiction.

9. Show that :
   (a) $P \wedge q$ logically implies $P \leftrightarrow q$
   (b) $P \rightarrow \neg q$ does not logically imply $P \leftrightarrow q$.

10. Let A = {1, 2, ..., 9, 10}. Consider each of the following sentences. If it is a statement, then determine its truth value. If it is a propositional function, determine its truth set.
    (a) $(\forall x \in A)(\forall y \in A)(x + y < 14)$ (b) $(\forall y \in A)(x + y < 14)$
    (c) $(\forall x \in A)(\forall y \in A)(x + y < 14)$ (d) $(\exists y \in A)(x + y < 14)$

11. Negative each of the following statement :
    (a) If the teacher is absent, then some students do not complete their homework.
    (b) All the students completed their homework and the teacher is present.
    (c) Some of the students did not compute their homework on teacher is absent.

ANSWERS (EXERCISE 11.1)

1. In each case, translate $\wedge$, $\vee$ and ~ to read "and", "or", and "It is false that" on "hat", respectively and then simplify the English sentence.
   (a) It is not cold (b) It is cold and training
   (c) It is cold on it is training (d) It is training or it is cold.

2. Vise $\vee$ for "an", $\vee$ for "and", (or, its logical equivalent, "but"), and $\neg$ for "not" (negation).

(a) $(P \vee q) \wedge \neg r$ (b) $(P \wedge q) \vee \neg (P \wedge r)$

(c) $\neg (P \wedge \neg r)$ (d) $\neg [(r \vee q) \wedge \neg P]$

3. The statement "P and q" is true only when both substatements are true. Thus
   (a) False   (b) True
   (c) False   (d) True

4.

P	q	$\neg P$	$\neg P \wedge q$
T	T	F	F
T	F	F	F
F	T	T	T
F	F	T	F

5. Construct the truth table of $P \vee \neg (P \wedge q)$. Since the truth values of $P \vee \neg (P \wedge q)$ is T for all values of P and q, the proposition is a tautology.

P	q	$P \vee q$	$\neg (P \wedge q)$	$P \vee \neg (P \vee q)$
T	T	T	F	T
T	F	F	T	T
F	T	F	T	T
F	F	F	T	T

6. Construct the truth tables for $\neg (P \wedge q)$ and $\neg P \vee \neg q$. Since the truth tables are the same, the propostions $\neg (P \wedge q)$ and $\neg P \wedge q$ are logically equivalent and we can write

$$\neg (P \wedge q) \equiv P \vee \neg q$$

P	q	$P \wedge q$	$\neg (P \wedge q)$	P	q	$\neg P$	q	$\neg P \wedge \neg q$
T	T	T	F	T	T	F	F	F
T	F	F	T	T	F	F	T	T
F	T	F	T	F	T	T	F	T
F	F	F	T	F	F	T	T	T

7. Recall that "If P then q" is equivalent to "Not P or q"; that is, $P \to q \equiv \neg P \vee q$. Hence,
   (a) It is not cold on he means a hot.
   (b) Productivity does not increase on wages rise.

8. (a) The contrapositive of $P \to q$ is $\neg q \to \neg P$. Hence the contrapositive of the given statement is

   If John is not poor, then he is not a poet.

   (b) The given statement is equivalent to "If Marc Passes the test, then he studied." Hence its contrapostive is

   If Marc does not study, then he will not pass the test.

9. Construct their truth tables as given below : Only the contrapositive $\neg q \to \neg P$ is logically equivalent to the original conditional proposition $P \to q$.

P	q	¬P	¬q	Conditional $P \to q$	Converse $q \to P$	Inverse $\neg P \to \neg q$	Contrapositive
T	T	F	F	T	T	T	T
T	F	F	T	F	T	T	F
F	T	T	F	T	F	F	T
F	F	T	T	T	T	T	T

10. (a) Note that $\neg(P \to q) \equiv P \wedge \neg q$; hence the negative of the statement follows :

   She works and she will not earn money.

   (b) Note that $\neg (P \to q) \equiv P \leftrightarrow \neg q \equiv \neg P \leftrightarrow q$. Hence the negation of the statement is either of the following :

   He swims if and only if the water is not warm.

   He does not swim if and only if the water is warm.

   (c) Note that $\neg (P \to \neg q) \equiv P \wedge \neg \neg q \equiv P \wedge q$. Hence the negation of the statement follows :

   It shows and they drive the can.

11. Construct the truth table for $[(P \to q) \wedge \neg P] \to \neg q$.

   Since the proposition $[(P \to q) \wedge \neg P] \to \neg q$ is not a tautology, the argument is a fallacy. Equivalently, the argument is a fallacy since in third line of the truth table $P \to q$ and $\neg P$ are true but $\neg q$ is false.

P	q	$P \to q$	¬P	$(P \to q) \wedge \neg P$	¬q	$[(P \to q) \wedge \neg P] \to \neg q$
T	T	T	F	F	F	T
T	F	F	F	F	T	T
F	T	T	T	T	F	F
F	F	T	T	T	T	T

11. Construct the truth table for $[(P \to q) \wedge \neg q] \to P$. Since the proposition $[(P \to q) \wedge \neg q] \to \neg P$ is a tautalogy, the argument is valid.

P	q	$(P \to q)$	$\wedge$	$\neg q]$	$\to$			¬P		
T	T	T	F	F	T	T	F	T		
T	F	F	F	T	T	F	T	F	T	
F	T	T	F	F	T	T	T	T	F	
F	F	T	T	T	F	T	F	T	T	F
Step		1	2	1	3	2	1	4	2	

12. Construct the truth tables of the premises and conclusion as shown below. Now, $P \to \neg q$, $r \to q$, and r and true simultaneously only in the fifth now of the table, where P is also true. Hence, the argument is valid.

	P	q	r	P → q	r → q	q
1	T	T	T	F	T	F
2	T	T	F	F	T	F
3	T	F	T	T	F	F
4	T	F	F	T	T	F
5	F	T	T	T	T	T
6	F	T	F	T	T	T
7	F	F	T	T	T	T
8	F	F	F	T	T	T

ANSWERS (EXERCISE 11.2)

1. (a) False. For no number in A is a solution to $x + 3 = 10$.
   (b) True. For every number in A satisfies $x + 3 < 10$.
   (c) True. For if $x_0 = 1$, then $x_0 + 3 <$, i.e., 1 is a solution.
   (d) False. For if $x_0 = 5$, then $x_0 + 3$ is not less than on equal.
   (e) In other words, 5 is not a solution to the given condition.

2. (a) True. For if $x = 1$, then 1, 2 and 3 are all solutions to $1 < y + 1$.
   (b) True. For each x_0, let $y = 1$; then $x_0^2 + 12$ is a true statement.
   (c) False. For if $x_0 = 2$ and $y_0 = 3$, then $x_0^2 + y_0^2 < 12$ is not a true statement.

3. Use $\neg \forall x\, P(x) \equiv \exists x\, \neg P(x)$ and $\neg \exists x\, P(x) \equiv \forall x\, \neg P(x)$;
   (a) $\neg (\exists x, \forall y, P(x, y)) \equiv \forall x, \exists y, \neg P(x, y)$
   (b) $\neg (\forall x, \forall y, P(x, y)) \equiv \exists x, \exists y, \neg P(x, y)$
   (c) $\neg (\exists y \exists x \forall z, P(x, y, z)) \equiv \forall y, \forall x \exists z, \neg P(x, y, z)$

4. (a) Yes.
   (b) Although $P(x)$ is false for every element in M, $P(x)$ is. still al propostional function on M.
   (c) No. Note that $2i + 2 > 5$ does not have meaning in other words in equalities are not defined for complex number.

5. (a) At least one student does not live in the donmitonic.
   (b) At least one Mathematics major is female.
   (c) None of the students is 25 on older.

6. (a) P

7. (a)

P	q	~q	P ∨ ~q
T	T	F	T
T	F	T	T
F	T	F	F
F	F	T	T

(b)

P	q	¬P	¬q	¬P ∨ ¬q
T	T	F	F	F
T	F	F	T	F
F	T	T	F	F
F	F	T	T	T

8. It is contradiction since its truth table is false fortable values of p and q.

P	q	P∧q	P∨q	⌐(P∨q)	(P∨q)∧⌐(P∨q)
T	T	T	T	F	F
T	F	F	T	F	F
F	T	F	T	F	F
F	F	F	F	T	F

9. (a) The open sentence in two variables is preceded by two quantifiers; hence it is a statement. Moreover, the statement is true.

(b) The open sentence is preceded by one quantifier; hence it is a propositional function of the other variable. Note that for every $y \in A$, $x_0 + y < 14$ if and only if $x_0 = 1, 2$, or 3. Hence the truth set is $\{1, 2, 3\}$.

(c) It is a statement and it is false : if $x_0 = 8$ and $y_0 = 9$, then $x_0 + y_0 < 14$ is not true.

(d) It is an open sentence in x. The truth set is A itself.

10. (a) The teacher is absent and all the students completed their homework.

(b) Some of the students did not complete their homework and the teacher is absent.

(c) All the students completed their homework and the teacher is present.

CHAPTER 12

COMBINATORICS

12.1 BASIC COUNTING TECHNIQUES

The Rule of Sum: If a task can be performed in m ways, while another task can be performed in n ways, and the two tasks cannot be performed simultaneously, then performing either task can be accomplished in $m + n$ ways.

Set theoretical version of the rule of sum: If A and B are disjoint sets ($A \cap B = \emptyset$) then
$$|A \cup B| = |A| + |B|$$
More generally, if the sets $A_1, A_2, \ldots, A_n$ are pairwise disjoint, then:
$$|A_1 \cup A_2 \cup \ldots \cup A_n| = |A_1| + |A_2| + \ldots + |A_n|$$
For instance, if a class has 30 male students and 25 female students, then the class has $30 + 25 = 45$ students.

The Rule of Product: If a task can be performed in m ways and another independent task can be performed in n ways, then the combination of both tasks can be performed in mn ways.

Set theoretical version of the rule of product: Let $A \times B$ be the Cartesian product of sets A and B. Then:
$$|A \times B| = |A| \cdot |B|$$
More generally:
$$|A_1 \times A_2 \times \ldots \times A_n| = |A_1| \cdot |A_2| \ldots |A_n|$$
For instance, assume that a license plate contains two letters followed by three digits. How many different license plates can be printed?

Answer: each letter can be printed in 26 ways, and each digit can be printed in 10 ways, so $26 \cdot 26 \cdot 10 \cdot 10 \cdot 10 = 676000$ different plates can be printed.

Exercise: Given a set A with m elements and a set B with n elements, find the number of functions from A to B.

The Inclusion-Exclusion Principle. The inclusion-exclusion principle generalizes the rule of sum to non-disjoint sets.

In general, for arbitrary (but finite) sets A, B:

$$|A \cup B| = |A| + |B| - |A \cup B|.$$

Example 1: Assume that in a university with 1000 students, 200 students are taking a course in mathematics, 300 are taking a course in physics, and 50 students are taking both. How many students are taking at least one of those courses?

Solution: If U = total set of students in the university, M = set of students taking Mathematics, P = set of students taking Physics, then:

$$|M \cup P| = |M| + |P| - |M \cup P| = 300 + 200 - 50 = 450$$

students are taking Mathematics or Physics.

For three sets the following formula applies:

$$|A \cup B \cup C| = |A| + |B| + |C| - |A \cap B| - |A \cap C| - |B \cap C| + |A \cap B \cap C|, \text{and}$$

for an arbitrary union of sets:

$$|A_1 \cup A_2 \cup \ldots \cup A_n| = s_1 - s_2 + s_3 - s_4 + \ldots \pm s_n,$$

where s_k = sum of the cardinalities of all possible k-fold intersections of the given sets.

12.2 COMBINATORICS

Permutations. Assume that we have n objects. Any arrangement of any k of these objects in a given order is called a *permutation* of size k. If $k = n$ then we call it just a *permutation* of the n objects. For instance, the permutations of the letters a, b, c are the following: $abc, acb, bac, bca, cab, cba$. The permutations of size 2 of the letters a, b, c, d are: $ab, ac, ad, ba, be, bd, ca, cb, cd, da, db, dc$.

Note that the order is important. Given two permutations, they are considered equal if they have the same elements arranged in the same order.

We find the number $P(n, k)$ of permutations of size k of n given objects in the following way: The first object in an arrangement can be chosen in n ways, the second one in $n - 1$ ways, the third one in $n - 2$ ways, and so on, hence :

$$P(n, k) = n \times (n-1) \times \overbrace{\cdots \times (n-k+1)}^{(k \text{ factors})} =$$

$$\underbrace{(k \text{ factors})}$$

where $n! = 1 \times 2 \times 3 \times \cdots \times n$ is called "n factorial".

The number $P(n, k)$ of permutations of n objects is

$$P(n, n) = n!.$$

By convention $0! = 1$.

For instance, there are $3! = 6$ permutations of the 3 letters a, b, c. The number of permutations of size 2 of the 4 letters a, b, c, d is $P(4, 2) = 4 \times 3 = 12$.

Exercise: Given a set A with m, elements and a set B with n elements, find the number of one-to-one functions from A to B.

Combinations. Assume that we have a set A with n objects. Any subset of A of size r is called a *combination of n elements taken r at a time*. For instance, the combinations of the letters a, b, c, d, e taken 3 at a time are: $abc, abd, abe, acd, ace, ade, bed, bee, bde, cde$, where two combinations are considered identical if they have the same elements regardless of their order.

The number of subsets of size r in a set A with n elements is:

$$C(n, r) = \frac{n!}{r!(n-r)!}$$

The symbol $\binom{n}{r}$ (read "n choose r") is often used instead of $C(n, r)$.

One way to derive the formula for $C(n, r)$ is the following. Let A be a set with n objects. In order to generate all possible permutations of size r of the elements of A we 1) take all possible subsets of size r in the set A, and 2) permute the k elements in each subset in all possible ways. Task 1) can be performed in $C(n, r)$ ways, and task 2) can be performed in $P(r, r)$ ways. By the product rule we have $P(n, r) = C(n, r) \times P(r, r)$, hence

$$C(n-r)! = \frac{P(n,r)}{P(r,r)} = \frac{n!}{r!(n-r)!}$$

Generalized Permutations and Combinations

Permutations with Repeated Elements. Assume that we nave an alphabet with k letters and we want to write all possible words containing n_1 times the first letter of the alphabet, n_2 times the second letter, ..., n_k times the kth letter. How many words can we write? We call this number $P(n; n_1, n_2, ..., n_k)$, where $n = n_1 + n_2 + ... + n_k$.

Example: With 3 a's and 2 b's we can write the following 5-letter words:

aaabb, aabab, abaab, baaab, aabba, ababa, baaba, abbaa, babaa, bbaaa.

We may solve this problem in the following way, as illustrated with the example above. Let us distinguish the different copies of a letter with subscripts: $a_1 a_2 a_3 b_1 b_2$. Next, generate each permutation of this five elements by choosing 1) the position of each kind of letter, then 2) the subscripts to place on the 3 a's, then 3) these subscripts to place on the 2 b's. Task 1) can be performed in $P(5; 3, 2)$ ways, task 2) can be performed in $3!$ ways, task 3) can be performed in $2!$. By the product rule we have $5! = P(5; 3, 2) \times 3! \times 2!$, hence $P(5; 3, 2) = 5!/3!\, 2!$.

In general the formula is:

$$P(n; n_1, n_2, ..., n_k) = \frac{n!}{n_1!\, n_2!...n_k!}$$

Combinations with Repetition. Assume that we have a set A with n elements. Any selection of r objects from A, where each object can be selected more than once, is called a *combination of n objects taken r at a time with repetition*. For instance, the combinations of the letters a, b, c, d taken 3 at a time with repetition are: *aaa, aab, aac, aad, abb, abc, abd, ace, acd, add, bbb, bbc, bbd, -bcc, bcd, bdd, ccc, ccd, cdd, ddd*. Two combinations with repetition are considered identical if they have the same elements repeated the same number of times, regardless of their order.

Note that the following are equivalent:

1. The number of combinations of n objects taken r at a time with repetition.

2. The number of ways r identical objects can be distributed among n distinct containers.

3. The number of nonnegative integer solutions of the equation:
$$x_1 + x_2 + \ldots + x_n = r$$

Example: Assume that we have 3 different (empty) milk containers and 7 quarts of milk that we can measure with a one quart measuring cup. In how many ways can we distribute the milk among the three containers? We solve the problem in the following way. Let x_1, x_2, x_3 be the quarts of milk to put in containers number 1, 2 and 3 respectively. The number of possible distributions of milk equals the number of non negative integer solutions for the equation $x_1 + x_2 + x_3 = 7$. Instead of using numbers for writing the solutions, we will use strokes, so for instance we represent the solution $x_1 = 2, x_2 = 1, x_3 = 4$, or $2 + 1 + 4$, like this: $||+|+||\,|$. Now, each possible solution is an arrangement of 7 strokes and 2 plus signs, so the number of arrangements is $P(9; 7, 2) = 9!/7!\,2! = \binom{9}{7}$.

The general solution is:

$$P(n+r-1; r, n-1) = \frac{(n+r-1)!}{r!(n-1)!} = \binom{n+r-1}{r}$$

Binomial Coefficients

Binomial Theorem. The following identities can be easily checked:

$$(x+y)^0 = 1$$
$$(x+y)^1 = x+y$$
$$(x+y)^2 = x^2 + 2xy + y^2$$
$$(x+y)^3 = x^3 + 3x^2y + 3xy^2 + y^3$$

They can be generalized by the following formula, called the *Binomial Theorem*:

$$(x+y)^n = \sum_{k=0}^{n} \binom{n}{k} x^{n-k} y^k .$$

$$= \binom{n}{0} x^n + \binom{n}{1} x^{n-1} y + \binom{n}{2} x^{n-2} y^2 + \ldots + \binom{n}{n-1} x y^{n-1} + \binom{n}{n} y^n$$

We can find this formula by writing

$$(x+y)^n = \overset{(k\ \text{factors})}{(x+y) \times (x+y) \times \cdots \times (x+y)},$$

expanding, and grouping terms of the form $x^a y^b$. Since there are n factors of the form $(x+y)$, we have $a + b = n$, hence the terms must be of the form $x^{n-k} y^k$. The coefficient of $x^{n-k} y^k$ will be equal to the number of ways in which we can select the y from any k of the factors (and the x from the remaining $n-k$ factors), which is $C(n, k) = \binom{n}{k}$. The expression $\binom{n}{k}$ is often called binomial *coefficient*.

Exercise: Prove

$$\sum_{k=0}^{n}\binom{n}{k}=2^n \quad \text{and} \quad \sum_{k=0}^{n}(-1)^k\binom{n}{k}=0.$$

Hint: Apply the binomial theorem to $(1 + I)^2$ and $(1 - I)^2$.

Properties of Binomial Coefficients. The binomial co-efficients have the following properties :

1. $\binom{n}{k}=\binom{n}{n-k}$

2. $\binom{n+1}{k+1}=\binom{n}{k}+\binom{n}{k+1}$

The first property follows easily from $\binom{n}{k}=\dfrac{n!}{k!(n-k)!}$

The second property can be proved by choosing a distinguished element a in a set A of $n + 1$ elements. The set A has $\binom{n+1}{k+1}$ subsets of size $k + 1$. Those subsets can be partitioned into two classes: that of the subsets containing a, and that of the subsets not containing a. The number of subsets containing a equals the number of subsets of A – {a} of size k, i.e., $\binom{n}{k}$. The number of subsets not containing a is the number of subsets of A – {a} of size $k + 1$, i.e., $\binom{n}{k+1}$. Using the sum principle we find that in fact $\binom{n+1}{k+1}=\binom{n}{k}+\binom{n}{k+1}$.

Pascal's Triangle. The properties shown in the previous section allow us to compute binomial coefficients in a simple way. Look at the following triangular arrangement of binomial coefficients:

$$\binom{0}{0}$$

$$\binom{1}{0} \quad \binom{1}{1}$$

$$\binom{2}{0} \quad \binom{2}{1} \quad \binom{2}{2}$$

$$\binom{3}{0} \quad \binom{3}{1} \quad \binom{3}{2} \quad \binom{3}{3}$$

$$\binom{4}{0} \quad \binom{4}{1} \quad \binom{4}{2} \quad \binom{4}{3} \quad \binom{4}{4}$$

We notice that each binomial coefficient on this arrangement must be the sum of the two closest binomial coefficients on the line above it. This together with $\binom{n}{0} = \binom{n}{n} = 1$, allows us to compute very quickly the values of the binomial coefficients on the arrangement:

```
              1
           1     1
         1    2    1
      1    3     3    1
   1    4     6     4    1
```

This arrangement of binomial coefficients is called *Pascal's Triangle*.

12.3 THE PIGEONHOLE PRINCIPLE

The Pigeonhole Principle: The *pigeonhole principle* is used for proving that a certain situation must actually occur. It says the following : *If n pigeonholes are occupied by m piegeons and m > n, then at least one pigeonhole is occupied by more than one pigeon.*

Example: In any given set of 13 people at least two of them have their birthday during the same month.

Example 2: Let S be a set of eleven 2-digit numbers. Prove that S must have two elements whose digits have the same difference (for instance in S = {10, 14, 19, 22, 26, 28, 49, 53, 70, 90, 93}, the digits of the numbers 28 and 93 have the same difference : 8 – 2 = +, 9 – 3 = 6.)

Solution : The digits of a two-digit number can have 10 possible differences (from 0 to 9). So, in a list of 11 numbers there must be two with the same difference.

Example 3: Assume that we choose three different digits from 1 to 9 and write all permutations of those digits. Prove that among the 3-digit numbers written that way there are two whose difference is a multiple of 500.

Solution: There are 9 . 8 . 7 = 504 permutations of three digits. On the other hand if we divide the 504 numbers by 500 we can get only 500 possible remainders, so at least two numbers give the same remainder, and their difference must be a multiple of 500.

Proposition PHP1. (The Pigeonhole Principle, simple version.) If $k + 1$ or more pigeons are distributed among k pigeonholes, then at least one pigeonhole contains, two or more pigeons.

Proof

The contrapositive of the statement is: If each pigeonhole contains at most one pigeon, then there are at most k pigeons. This is easily seen to be true.

The same argument can be used to prove a variety of different statements. We prove the general version of the Pigeonhole Principle and leave the others as exercises.

Proposition PHP2. (The Pigeonhole Principle.) If n or more pigeons are distributed among $k > 0$ pigeonholes, then at least one pigeonhole contains at least $\left\lceil \dfrac{n}{k} \right\rceil$ pigeons.

Proof

Suppose each pigeonhole contains at most $\left\lceil \frac{n}{k} \right\rceil - 1$ pigeons. Then, the total number of pigeons is at most $k\left(\left\lceil \frac{n}{k} \right\rceil - 1\right) < k\left(\frac{n}{k}\right) = n$ pigeons (because $\left\lceil \frac{n}{k} \right\rceil - 1 < \frac{n}{k} \leq \left\lceil \frac{n}{k} \right\rceil$).

Exercises. Prove:

(a) If n objects are distributed among $k > 0$ boxes, then at least one box contains at most $\left\lceil \frac{n}{k} \right\rceil$ objects.

(b) Given $t > 0$ pigeonholes $h_1, h_2, ..., h_t$ and t integers $n_1, n_2, ..., n_t$, if $\left[\sum_{i=1}^{t}(n_i - 1)\right] + 1$ pigeons are distributed among these t pigeonholes, then there exists at least one i such that pigeonhole h_i contains at least n_i pigeons.

(c) Given a collection of n numbers, at least one of the numbers is at least as large as the average.

(d) Given a collection of n numbers, at least one of the numbers is no larger than the average.

Example PHP1. Prove that if seven distinct numbers are selected from $\{1, 2, ..., 11\}$, then some two of these numbers sum to 12.

Solution: Let the pigeons be the numbers selected. Define six pigeonholes corresponding to the six sets: $\{1, 11\}, \{2, 10\}, \{3, 9\}, \{4, 8\}, \{5, 7\}, \{6\}$. (Notice that the numbers in each of the first five sets sum to 12. and that there is no pair of distinct numbers containing 6 that sum to 12.) When a number is selected, it gets placed into the pigeonhole corresponding to the set that contains it. (This means at most one number can go into the pigeonhole corresponding to $\{6\}$. This does not cause trouble.) Since seven numbers are selected and placed in six pigeonholes, some pigeonhole contains two numbers. By the way the pigeonholes were defined, these two numbers sum to 12.

Another way to write up the above proof is: *Since seven numbers are selected, the Pigeonhole Principle guarantees that two of them, are selected from, one of the six sets* (1, 11), (2, 10), (3, 9), (4, 8), (5, 7), (6). *These two numbers sum to* 12.

In Example PHP1, the quantity seven is the best possible in the sense that it is possible to select six numbers from $\{1, 2, ..., 11\}$ so that no two of the numbers selected sum to 12. One example of six such numbers is 1, 2, 3, 4, 5, 6.

Proving things with the Pigeonhole Principle. There are four steps involved. First, decide what the pigeons are. They will be the things that you'd like several of to have some special property. Second, set-up the pigeonholes. You want to do this so that when you get two pigeons in the same pigeonhole, they have the property you want. To use the Pigeonhole Principle, it is necessary to set things up so that there are fewer pigeonholes than pigeons. Sometimes the way to do this relies on some astute observation. Third, give a rule for assigning the pigeons to the pigeonholes. It is important to note that the conclusion of the Pigeonhole Principle holds for any assignment of pigeons to pigeonholes, so it holds for any assignment you describe. Pick the rule so that when "enough" pigeons occupy the same pigeonhole, that collection has the property you want. Fourth, apply the Pigeonhole Principle to your set-up and get the desired conclusion.

Exercise. Prove the general version of Example PHP1: If $n + 1$ numbers are selected from $\{1, 2, ..., 2n - 1\}$, then some two of these numbers sum to $2n$. Show that it is possible to select n numbers so that no two of them sum to $2n$. Formulate and prove similar statements for collections of numbers selected from $\{1, 2, ..., 2n\}$.

Example PHP2. Prove that if five points are selected from the interior of a 1×1 square, then there are two points whose distance is less than $\sqrt{2}/2$.

Let the pigeons be the five points selected. Define the four pigeonholes corresponding to the four $1/2 \times 1/2$ subsquares obtained by joining the midpoints of opposite sides of the square. When a point is selected is is placed into a pigeonhole according to the subsquare that contains it, and points on the boundary of these subsquares (and interior to the whole square) can be assigned arbitrarily. Since five points are selected and placed in four pjgeonholes, some pigeonhole contains two points. Since these points are on the interior of the square, the distance between them is less than the length of a diagonal of a subsquare, which is $\sqrt{2}/2$.

Exercises. Prove that if four points are selected from the interior of a unit circle, then there are two points whose distance apart is less than $\sqrt{2}$. How many points must be selected from the interior of an equilateral triangle of side two in order to guarantee that there are two points whose distance apart is less than one?

Example PHP3. For a subset $X \subseteq \{1, 2, ..., 9\}$, define $\sigma(X) = \sum_{x \in X} x$ (For example $\sigma(\{1, 6, 8\}) = 1 + 6 + 8 = 15$.) Prove that among any 26 subsets of $\{1, 2, ..., 9\}$, each having size at most three, there are subsets A and B such that $\sigma(A) = \sigma(B)$.

For $X \subseteq \{1, 2, ..., 9\}$ with $|X| \leq 3$, the possible values of $\sigma(X)$ lie between 0 (corresponding to $X = \emptyset$) and 24 (corresponding to $X = \{7, 8, 9\}$). Since there are 25 possible values for $\sigma(X)$ and 26 subsets are selected, we have by the Pigeonhole Principle that the selection contains subsets A and B such that $\sigma(A) = \sigma(B)$.

Exercise. Write up Example PHP3 to make the pigeons and pigeonholes explicit.

Exercise. How many subsets of $\{1, 2, ..., 10\}$, each containing at least three and at most five elements, must be selected in order to guarantee that the selection contains subsets A, B and C such that $\sigma(A) = \sigma(B) = \sigma(C)$?

Example PHP4. Prove that if 10 integers are selected from $\{1, 2, ..., 18\}$. the selection includes integers a and b such that $a \,|\, b$ (that is, a divides b - there exists an integer k such that $ak = b$).

Let the pigeons be the 10 integers selected. Define nine pigeonholes corresponding to the odd integers 1, 3, 5, 7, 9, 11, 13, 15, and 17. Place each integer selected into the pigeonhole coresponding to its largest odd divisor (which must be one of 1, 3, 5, ..., 17). Notice that if x gets placed in the pigeonhole corresponding to the odd integer m, then $x = 2^k m$ for some integer $k \geq 0$. Since 10 integers are selected and placed in nine pigeonholes, some pigeonhole contains two integers a and b, where $a < b$. Suppose this pigeonhole corresponds to the odd integert. Then, $a = 2^r t$ and $b = 2^s t$, where $r < s$, so that $a 2^{s-r} = b$. Since $s - r$ is a positive integer, it follows that $a \mid b$.

Here is an alternative write up. The largest odd divisor of an integer between 1 and 18 is one of the nine numbers 1. 3, 5, 7, 9, 11, 13, 15, 17. Since 10 integers are selected, the Pigeonhole Principle guarantees that some two of them have the same largest odd divisor, t. Let these two numbers be a and b, where $a < b$. Then, $a = 2^r t$ and $b = 2^s t$, where $r < s$, so that $a 2^{s-r} = b$. Since $s - r$ is a positive integer, it follows that $a \mid b$.

Exercise. Prove that if $n + 1$ integers arc selected from $\{1, 2, ..., 2n\}$, then the selection includes integers a and b such that $a \mid b$.

Example PHP5. Prove that if 11 integers are selected from among $\{1, 2, ..., 20\}$, then the selection includes integer a and b such that $a - b = 2$.

Let the pigeons be the 11 integers selected. Define 10 pigeonholes corresponding to the sets $\{3, 1\}, \{4, 2\}, \{7, 5\}, \{8, 6\}, \{11, 9\}, \{12, 10\}, \{15, 13\}, \{16, 14\}, \{19, 17\}, \{20, 18\}$. Place each integer selected into the pigeonhole corresponding to the set that contains it. Since 11 integers are selected and placed into 10 pigeonholes, some pigeonhole contains two pigeons. By the way the pigeonholes were defined, these two integers differ by two.

Exercise. Prove an alternative write up (as above) of Example PHP5.

Exercise. Prove that if 11 integers are selected from among $\{1, 2, ..., 20\}$, then the selection includes integers a and b such that $b = a + 1$.

Exercise. Prove that if $n + 1$ integers are selected from among $\{1, 2, ..., 2n\}$, then the selection includes integers a and b such that $b = a + 1$. This implies that if $n + 1$ integers are selected from among $\{1, 2, ... ,2n\}$, then the selection includes integer a and b such that $\gcd(a, b) = 1$. Why is that?

Example PHP6. Over a 44 day period, Gary will train for triathlons at least once per day, and a total of 70 times in all. Show that there is a period of consecutive days during which he trains exactly 17 times.

For $i = 1, 2, ... , 44$, let x_i be the number of times Gary trains up to the end of day i. Then $1 < x_1 < x_2 < x_3 < ... < x_{44} = 70$. We need to find subscripts i and j such that $x_i + 17 = x_j$. This implies that Gary trains exactly 17 times in the period of days $i + 1, i + 2, ... ,j$. Therefore, we want one of $x_1, x_2, ..., x_{44}$ to be equal to one of $x_1 + 17, x_2 + 17, ..., x_{44} + 17$. Using the inequality for the x_i s it follows that $18 < x_1 + 17 < x_2 + 17 < ... < x_{44} + 17 = 87$. Thus, the 88 numbers $x_1, x_2, ..., x_{44}, x_1 + 17, x_2 + 17, ... , x_{44} + 17$ can take on at most 87 different values. Hence, by the Pigeonhole Principle, some two of them must be equal. The inequalities imply that one of $x_1, x_2, ..., x_{44}$ must equal one of $x_1 + 17, x_2 + 17, ..., x_{44} + 17$, which is what we wanted.

Exercise. Over a 30 day period, Rick will walk the dog at least once per day, and a total of 45 times in all. Prove that there is a period of consecutive days in which he walks the dog exactly 14 times.

Example PHP7. A party is attended by $n \geq 2$ people. Prove that there will always be two people in attendance who have the same number of friends at the party. (Assume that the relation "is a friend of" is symmetric, that is, if x is a friend of y then y. is a friend of x.)

Each person either is, or is not, a friend of each of the the other $n-1$ people in attendance. Thus, the possible values for the number of friends a person can have in attendance at the party are $0, 1, \ldots, n-1$. However, it cannot be the case that there is someone at the party with 0 friends and someone else with $n-1$ friends: if a person is friends with everyone then (since "is a friend of" is symmetric) everyone at the party has at least one friend there. Thus, the possible values for the number of friends a person can have in attendance at the party are $0, 1, \ldots, n-2$ or $1, 2, \ldots, n-1$. In either case, there are n numbers (of friends among the people in attendance) that can take on at most $n-1$ different values. By the Pigeonhole Principle, two of the numbers are equal. Thus, some two people in attendance who have the same number of friends at the party.

Exercise. Ten baseball teams are entered in a round-robin tournament (meaning that every team plays every other team exactly once) in which ties are not allowed. Prove that if no team loses all of its games, then some two teams finish the tournament with the same number of wins.

Example PHP8. We prove that any collection of eight distinct integers contains distinct integers x and y such that $x - y$ is a multiple of 7,

By the Division Algorithm, every integer n can be written as $n = 7q + r$, where $0 \le r \le 6$. Since there are eight integers in the collection but only seven possible values for the remainder r on division by 7, the Pigeonhole Principle asserts that the collection contains integers x and y that leave the same remainder on division by 7, that is, there exists s with $0 \le s \le 6$ such that $x = 7q_1 + s$ and $y = 7q_2 + s$. For this x and y we have $x - y = 7q_1 + s - (7q_2 + s) = 7(q_1 - q_2)$. Since $(q_1 - q_2)$ is an integer, x is a multiple of 7.

Exercise. Prove that in and collection of $n + 1$ distinct integers, there are distinct integers x arid y such that $x - y$ is a multiple of n.

12.4 RECURRENCE RELATIONS

Here we look at recursive definitions under a different point of view. Rather than definitions they will be considered as equations that we must solve. The point is that a recursive definition is actually a definition when there is one and only one object satisfying it, *i.e.*, when the equations involved in that definition have a unique solution. Also, the solution to those equations may provide a *closed-form* (explicit) formula for the object defined.

The recursive step in a recursive definition is also called a *recurrence, relation*. We will focus on *kth-order linear recurrence relations,* which are of the form

$$C_0 x_n + C_1 x_{n-1} + C_2 x_{n-2} + \ldots + C_k x_{n-k} = b_n.$$

where $C_0 \ne 0$. If $b_n = 0$ the recurrence relation is called *homogeneous*. Otherwise it is called *non-homogeneous*.

The basis of the recursive definition is also called *initial conditions* of the recurrence. So, for instance, in the recursive definition of the Fibonacci sequence, the recurrence is

$$F_n = F_{n-1} + F_{n-2}$$

or

$$F_n - F_{n-1} - F_{n-2} = 0,$$

and the initial conditions are
$$F_0 = 0, F_1 = 1.$$

One way to solve some recurrence relations is by *iteration*, i.e., by using the recurrence repeatedly until obtaining a explicit close-form formula. For instance consider the following recurrence relation :
$$x_n = r x_{n-1} \ (n > 0); \qquad x_0 = A$$
By using the recurrence repeatedly we get:
$$x_n = r x_{n-1} = r^2 x_{n-2} = r^3 x_{n-3} = \ldots = r^n x_0 = A r^n,$$ hence the solution is $x_n = A r^n$.

In the following we assume that the coefficients $C_0, C_1, \ldots, C_k$ are constant.

First Order Recurrence Relations. The homogeneous case can be written in the following way:
$$x_n = r x_{n-1} \ (n > 0); \qquad x_0 = A$$
Its general solution is
$$x_n = A r^n,$$
which is a *geometric sequence* with *ratio* r.

The non-homogeneous case can be written in the following way:
$$x_n = r x_{n-1} + c_n \ (n > 0); \qquad x_0 = A$$
Using the summation notation, its solution can be expressed like this:
$$x_n = A r^n + \sum_{k=1}^{n} c_k r^{n-k}$$

We examine two particular cases. The first one is
$$x_n = r x_{n-1} + c \ (n > 0); \qquad x_0 = A$$
where c is a constant. The solution is
$$x_n = A r^n + c \sum_{k=1}^{n} r^{n-k} = A r^n + c \frac{r^n - 1}{r - 1} \qquad \text{if } r \neq 1,$$
and
$$x_n = A + c n \qquad \text{if } r = 1.$$

Example 4: Assume that a country with currently 100 million people has a population growth rate (birth rate minus death rate) of 1% per year, and it also receives 100 thousand immigrants per year (which are quickly assimilated and reproduce at the same rate as the native population). Find its population in 10 years from now. (Assume that all the immigrants arrive in a single batch at the end of the year.)

Solution: If we call x_n = population in year n from now, we have:
$$x_n = 1.01 \, x_{n-1} + 100{,}000 \qquad (n > 0); \qquad x_0 = 100{,}000{,}000.$$
This is the equation above with $r = 1.01$, $c = 100{,}000$ and $A = 100{,}000{,}00$, hence:

$$x_n = 100{,}000{,}000 \cdot 101^n + 100{,}000 \, \frac{1.01^n - 1}{1.01 - 1}$$

$$x_n = 100{,}000{,}000 \cdot 1.01^n + 1000 \, (1.01^n - 1)$$

So:

$$x_{10} = 110{,}462{,}317.$$

The second particular case is for $r = 1$ and $c_n = c + dn$, where c and d are constant (so c_n is an arithmetic sequence):

$$x_n = x_{n-1} + c + dn \quad (n > 0); \quad x_0 = A.$$

The solution is now

$$x_n = A + \sum_{k=1}^{n}(c + dk) = A + cn + \frac{dn(n+1)}{2}$$

Second Order Recurrence Relations. Now we look at the recurrence relation

$$C_0 x_n + C_1 x_{n-1} + C_2 x_{n-2} = 0$$

First we will look for solutions of the form $x_n = cr^n$. By plugging in the equation we get:

$$C_0 \, cr^n + C_1 \, cr^{n-1} + C_2 \, cr^{n-2} = 0$$

hence r must be a solution of the following equation, called the *characteristic equation* of the recurrence:

Let r_1, r_2 be the two (in general complex) roots of the above equation. They are called *characteristic roots*. We distinguish three cases:

1. *Distinct Real Roots.* In this case the general solution of the recurrence relation is

$$x_n = c_1 r_1^n + c_2 r_2^n$$

where c_1, c_2 are arbitrary constants.

2. *Double Real Root.* If $r_1 = r_2 = r$, the general solution of the recurrence relation is

$$x_n = c_1 r^n + c_2 \, n \, r^n,$$

where c_1, c_2 are arbitrary constants.

3. *Complex Roots.* In this case the solution could be expressed in the same way as in the case of distinct real roots, but in order to avoid the use of complex numbers we write $r_1 = r \, e^{-ai}$, $r_2 = r e^{-ai}$, $k_1 = c_1 + c_2$, $k_2 = (c_1 - c_2) \, i$, which yields:[1]

$$x_n = k_1 \, r^n \cos n\alpha + k_2 \, r^n \sin n\alpha$$

Example 5: Find a closed-form formula for the Fibonacci sequence defined by:

$$F_{n+1} = F_n + F_{n-1} \quad (n > 0); \quad F_0 = 0, F_1 = 1.$$

Solution: The recurrence relation can be written

$$F_n - F_{n-1} - F_{n-2} = 0.$$

The characteristic equation is

$$r^2 - r - 1 = 0.$$

Its roots are:

$$r_1 = \phi = \frac{1+\sqrt{5}}{2}; \quad r_2 = -\phi^{-1} = \frac{1+\sqrt{5}}{2}$$

They are distinct real roots, so the general solution for the recurrence is:

$$F_n = c_1 \phi^n + c_2 (-\phi^{-1})^n$$

Using the initial conditions we get the value of the constants:

$$\begin{cases}(n=0) & c_1 + c_2 = 0 \\ (n=1) & c_1\phi + c_2(-\phi^{-1}) = 1\end{cases} \Rightarrow \begin{cases}c_1 = 1/\sqrt{5} \\ c_2 = -1/\sqrt{5}\end{cases}$$

Hence:

$$F_n = \frac{1}{\sqrt{5}} \{\phi^n - (-\phi)^{-n}\}$$

Suppose $a_0, a_1, a_2, \ldots$ is a sequence. A *recurrence relation* for a_n is a formula (*i.e.*, function) giving a_n in terms of some or all previous terms (*i.e.*, $a_0, a_1, \ldots, a_{n-1}$). To completely describe the sequence, the first few values are needed, where "few" depends on the recurrence. These are called the *initial conditions*.

If you are given a recurrence relation and initial conditions, then you can write down as many terms of the sequence as you please: just keep applying the recurrence. For example, $f_0 = f_1 = 1, f_n = f_{n-1} + f_{n-2}, n \geq 2$, defines the *Fibonacci Sequence* 1, 1, 2, 3, 5, 8, 13, ... where each subsequent term is the sum of the preceding two terms. On the other hand, if you are given a sequence, you may or may not be able to determine a recurrence relation with inital conditions which describes it. For example, 0, 1, 2, 3, 0, 2, 4, 6, 0, 4, 8, 12, ... satisfies $a_n = 2a_{n-4}, a_0 = 0. a_1 = 1, a_2 = 2, a_3 = 3$, but I can't think of a recurrence relation and initial conditions that describes the sequence $\{p_n\}$ of prime numbers.

It is typical to want to derive a recurrence relation with initial conditions (abbreviated to RR with 1C from now on) for the number of objects satisfying certain conditions. The basic idea is to give a counting argument for the number of objects of "size" n in terms of the number of objects of smaller size. This may involve an analysis consisting of several cases.

Fibonacci numbers. *Assume you start with one pair of newborn rabbits (one of each gender), and in each subsequent month each pair of rabbits which are 'more than 1 -month old gives birth to a new pair of rabbits, one of each gender. Determine a RR with 1C for f_n, the number of pairs of rabbits present at the end of n months.* The statement tells us that $f_0 = 1$. Also, $f_1 = 1$ because the original pair of rabbits is not yet old enough to breed. At the end of two months, we have our pair from before, plus one new pair. At the end of 3 months, we have the f_2 pairs from before, and f_1 of them are old enough to breed, so we have $f_3 = f_2 + f_1 = 3$ pairs. Consider what happens at the end of n months. We still have the f_{n-1} pairs from the month before. The number of pair's old enough to breed is the number alive two months ago, or f_{n-2} so we get f_{n-2} new pairs. Thus, $f_n = f_{n-1} + f_{n-2}, n \geq 2$. and $f_0 = f_1 = 1$. Using the RR and 1C yields the sequence 1, 1, 2, 3, 5, 8. ... which agrees with our initial counting.

Try to derive a RR with 1C as above if each pair of rabbits give birth to two pairs each month (once they are old enough) and die after living for 4 months. The answer is $f_0 = f_1 = 1$, $f_2 = 3$, $f_3 = 5$, and $f_n = f_{n-1} + 2f_{n-2} - f_{n-4}$, $n \geq 4$.

Counting sequences. Suppose you want to derive a recurrence for a_n, the number of sequences with a certain property. Breaking into cases based on the first (or last) entry is often successful.

Example 6: Derive a RR with 1C for u_n, the number of sequences (strings) of upper case letters that do not contain ZZ.

Solution: By counting, $u_0 = 1$, $u_2 = 26$, $u_2 = 26^2 - 1 = 675$, $u_3 = 25(26^2 - 1) + 1 \cdot 25 \cdot 26 = 17525$. The method used to derive u_3 suggests breaking the counting into cases depending on the first letter in the sequence. Let's call a string *valid* if it does not contain ZZ. Consider a valid string of length n. Suppose the first letter is not Z (25 choices). Then, the remaining $n - 1$ letters can be any valid string of length $n - 1$. Since there u_{n-1} of these, by the Rule of Product there are $25u_{n-1}$ valid string in which the first letter is not Z. Now suppose the first letter is Z (1 choice). Since the string is valid, the second letter is not Z (25 choices), and then the remaining $n - 2$ letters can be any valid string of length $n - 2$. Since there u_{n-2} of these, by the Rule of Product there are $25u_{n-2}$ valid string in which the first letter is Z. Therefore, by the Rule of Sum, $u_n = 25u_{n-1} + 25u_{n-2}$, $n \geq 2$. As a check on the work, computing with the recurrence, and $U_0 = 1$ and $u_1 = 25$ gives $u_2 = 25 \cdot 26 + 25 \cdot 1 = 675$ and $u_3 = 25 \cdot 675 + 25 \cdot 26 = 17525$. Since these agree with what was obtained in step 1, there is some evidence that the RR with IC is correct.

Similar methods often work when you are considering the number of ways to accomplish a sequence of steps, where at each step one of a few things happens. For example, a RR with IC for the number of ways to climb n stairs where on each stride you climb 1, 2, or 3 stairs is $a_0 = a_1 = 1$, $a_2 = 2$ and for $n \geq 3$, $a_n = a_{n-1} + a_{n-2} + a_{n-3}$.

Sometimes one or more of the cases do not involve the number of objects of a smaller size. For example, let's derive a recurrence for b_n, the number, of bit strings of length n that contain 00. By counting, $b_0 = b_1 = 0$, $b_2 = 1$, and $b_3 = 3$. Call a bit string *good* if it contains 00. Consider a good bit string of length n. If it starts with a 1 the remaining $n - 1$ bits can be any good bit string of length $n - 1$, so there are b_{n-1} of these. Suppose it starts with a 0. If the next bit is 1, then the remaining $n - 2$ bits can be any good bit string of length $n - 2$, so there are b_{n-2} good bit strings that begin 01. On the other hand, if the second bit is 0, then we have 00 already, so the remaining $n - 2$ bits can be *any* bit string of length $n - 2$, and there are 2^{n-2} of these. Thus, by the Rule of Sum, $b_n = b_{n-1} + b_{n-2} + 2^{n-2}$, $n \geq 2$. Plugging in the initial conditions and computing yields $b_2 = 1$ and $b_3 = 3$, which agrees with what was obtained directly.

Another thing to remember is that if a_n counts the number of objects you want, then the number that you don't want equals the total number of object minus a_n. For example, let's derive a RR with IC for z_n, the number of sequences of A's, B's and C's that contain an odd number of C's. By counting, $z_0 = 0$, $z_1 = 1$, $z_2 = 4$ and $z_3 = 13$. Let's call a sequence *odd if* it contains an odd number of C's, arid *even* otherwise. Consider an odd sequence of length n. If the first character is A or B (2 choices), then the remaining $n - 1$ characters form an odd sequence of length $n - 1$, so there are $2z_{n-1}$ of these. If the first character is (7, then the remaining $n - 1$ characters form an even sequence of length $n - 1$, and the number of these is $3^{n-1} - z_{n-1}$. Thus, by the Rule of Sum, $z_n = 2z_{n-1} + 3^{n-1} - z_{n-1} = z_{n-1} + 3^{n-1}$ $n \geq 1$. Plugging in the initial conditions and computing yields $z_1 = 1$, $z_2 = 4$. and $z_3 = 13$, which agrees with our initial counting.

Catalan Numbers. Let c_n be the number of well formed strings of n left brackets and n right brackets. Since the empty sequence is well formed, $c_0 = 1$. By counting, $c_1 = 1$, $c_2 = 2$, and $c_3 = 3$. Consider a well formed sequence of n pairs of brackets. The first element in the sequence is a left bracket. Consider the corresponding right bracket. This pair encloses some well formed string of k brackets, for some k between 0 and $n-1$, and is followed by a well formed string of $n-1-k$ brackets. Hence, the number of such well formed sequences is $c_k c_{n-1-k}$. Thus, by the Rule of Sum $c_n = c_0 c_{n-1} + c_1 c_{n-2} + \ldots + c_{n-1} c_0 = \sum_{k=0}^{n-1} c_k c_{n-1}$, $n \geq 1$. Computing using this recurrence and $c_0 = 1$ yields $c_1 = 1$, $c_2 = 2$. and $c_3 = 3$, as above. It can be shown that the function $\binom{2n-1}{n} - \binom{2n-1}{n-1} = \frac{1}{n+1}\binom{2n}{n}$ satisfies this recurrence and agrees with the initial condition, so it equals c_n. These numbers also count many other things.

Counting subsets. If you are trying to derive a RR with IC that involves subsets of $\{x_1, x_2, \ldots, x_n\}$ with certain properties, try breaking into cases depending on what happens to x_1.

Example: Derive a RR with IC for p_n, the number of ways to partition an n-set into subsets. (Remember that the subsets in a partition are non-empty and pairwise disjoint). By counting, $p_1 = 1$. $p_2 = 2$, $p_3 = 7$. Let's agree that $p_0 = 1$; we'll see why this is helpful in a moment. Consider a partition of the n-set $\{x_1, x_2, \ldots, x_n\}$. Then x_1 belongs to one of the subsets in the partition, call it S and let $k = |S|$. Then $1 \leq k \leq n$. Deleting S leaves a partition of the $(n-k)$-set formed by the remaining elements. Since there are $\binom{n-1}{k-1}$ choices for a k-subset containing x_1, and for each of these there are p_{n-k} partitions of the remaining elements, by the Rule of Product the number of partitions of $\{x_1, x_2, \ldots, x_n\}$ in which x_1 belongs to a subset of size k is $\binom{n-1}{k-1} p_{n-k}$. (Note: it is this line that would lead to wanting to define $p_n = p_{n-n} = 1$.) Thus, by the Rule of Sum, $p_n = \sum_{k=1}^{n} \binom{n-1}{k-1} p_{n-k}$ $n \geq 1$.

Computing using this recurrence leads to $p_2 = \binom{1}{0} p_1 + \binom{1}{1} p_0 = 2$, and $p_3 = \binom{2}{0} p_2 + \binom{2}{1} p_1 + \binom{2}{2} p_0 = 7$, which agrees with what we obtained initially.

Summary. A good way to start is to work out the first few cases directly. For the general case, try to organise the counting into cases depening on what happens on the first move, or first step, or to the first object. Or, replace first by n-th.

12.5 GENERATING FUNCTIONS

If you take $f(x) = 1/(1 - x - x^2)$ and expand it as a power series by long division, say, then you get $f(x) = 1/(1 - x - x^2) = 1 + x + 2x^2 + 3x^3 + 5x^4 + 8x^6 + 13x^6 + \ldots$ It certainly seems as though the coefficient of x^n in the power series is the n-th Fibonacci number. This is not difficult to prove directly, or using techniques discussed below. You can think of the process of long division as *generating* the coefficients, so you might want to call $1/(1 - x - x^2)$ the generating function for the Fibonacci numbers. The actual definition of generating function is a bit more general. Since the closed form and the power series represents the same function (within the circle of convergence), we will regard either one as being *the* generating function.

Definition of generating function. The *generating function* for the sequence $a_0, a_1,$ ... is defined to be the function $f(x) = \sum_{n=0}^{\infty} a_n x^n$.

That is, the generating function for the sequence $a_0, a_1, ...$ is the function whose power series representation has a_n as the coefficient of x^n. We'll call $a_0, a_1, ...$ the *sequence generated* by $f(x)$. We will not be concerned with matters of convergence, and instead treat these as *formal* power series. Perhaps "symbolic" would be a better word than formal.

When determining the sequence generated by a generating function, you will want to get a formula for the n-the term (that is, for the coefficient of x^n), rather than just computing numerical values for the first few coefficients.

Useful Facts

- If $x \neq 1$ then $1 + x + x^2 + ... + x^r = \dfrac{x^{r+1} - 1}{x - 1}$ (although it is often best not to do anything with sums of only a few terms).

- $\dfrac{1}{1 - x} = 1 + x + x^2 + ...$

Both of these facts can be proved by letting S be the sum, calculating $xS - S$, and doing a little bit of algebra.

The second fact above says that $\dfrac{1}{1-x}$ is the generating function for the sequence 1, 1, 1, 1, ... It also lets you determine the sequence generated by many other functions. For example:

- $\dfrac{1}{1 - ax} = 1 + a + a^2 x^2 + a^3 x^3 + ... = \sum_{n=0}^{\infty} a^n x^n$. To see this, substitute $y = ax$ into the series expansion of $\dfrac{1}{1-y}$. Thus $\dfrac{1}{1 - ax}$ is the generating function for the sequence $a^0, a^1, a^2, ...$ The coefficient of x^n is a^n.

- $\dfrac{1}{1 - ax} = 1 - a + a^2 x^2 - a^3 x^3 + ... = \sum_{n=0}^{\infty} (-1)^n a^n x^n$. To see this substitute $y = (-a)x$ into the series expansion of $\dfrac{1}{1-y}$. Thus $\dfrac{1}{1 - ax}$ is the generating function for the sequence $a^0, -a^1, a^2, ...$ The coefficient of x^n is $(-a)^n = (-1)^n x^n$.

More Useful Facts

- $\left(\dfrac{1}{1+x}\right)^k = \sum_{n=0}^{\infty} \binom{n+k+1}{n} x^n$.

- $\left(\dfrac{1}{1-ax}\right)^k = \sum_{n=0}^{\infty} \binom{n+k+1}{n}^n a^n x^n$.

- $\left(\dfrac{1}{1+ax}\right)^k = \sum_{n=0}^{\infty} (-1)^n \binom{n+k+1}{n}^n a^n x^n$

To see the first of these, consider the LHS, $(1 + x + x^2 + \ldots)^k$, multiplied out, but not simplified. Each term is a product of k (possibly different) powers of x (remembering that $x^0 = 1$). By the rules of exponents, $x^{n_1} x^{n_2} \ldots x^{n_k} = x^{n_1 + n_2 + \ldots + n_k}$. Thus, after simplifying, there is a term x^n for every way of expressing n as a sum of k numbers each of which is one of 0, 1, 2, …. That is, the number of terms x^n equals the number of integer solutions to $n_1 + n_2 + \ldots + n_k = n$, $0 \leq n_i$ $\forall i$. We know that this number is $\binom{n+k-1}{k-1}$ (count using "bars and stars").

To see the second fact, let $y = ax$ in $\left(\dfrac{1}{1-y}\right)^k$. (This is the same idea as before.) To see the third fact, let $y = -ax$ in $\left(\dfrac{1}{1-y}\right)^k$.

Multiplying a generating function by a constant, multiplies every coefficient by that constant. For example, $\dfrac{3}{1+5x} = 3 \dfrac{1}{1+5x} = 3(1 - 5x + 5^2 x^2 - 5^3 x^3 + \ldots) = 3 - 3 \cdot 5x + 3 \cdot 5^2 x^2 - 3 \cdot 5^3 x^3 + \ldots = \sum_{n=0}^{\infty} 3 \cdot (-1)^n 5^n x^n$. The coefficient of x^n is $3 \cdot (-1)^n 5^n$.

Multiplying a generating function by x^k "shifts" the coefficients by k. This has the effect of introducing k zeros at the start of the sequence generated. For example,

$$\left(\dfrac{x}{1+7x}\right)^4 = x^4 \left(\dfrac{1}{1+7x}\right)^4$$

$$= x^4 \sum_{n=0}^{\infty} (-1)^n \binom{n+4-1}{n} 7^n x^n$$

$$= \sum_{n=0}^{\infty} (-1)^n \binom{n+4-1}{n} 7^n x^{n+4}$$

$$= \sum_{n=0}^{\infty} (-1)^{t-4} \binom{t-1}{3} 7^{t-4} x^t$$

(Let $t = n + 4$ in the second last sum. If $n = 0$ then $t = 4$, so that gives the lower limit for the sum. If $n \to \infty$, so does t, which gives the upper limit. And $n = t - 4$.) Since $(-1)^{t-4} = (-1)^t$, the last sum equals: $\sum_{t=4}^{\infty} (-1)^t \binom{t-1}{3} 7^{t-4} x^t$. The coefficient of x^t is zero if $t < 4$, and $(-1)^t \binom{t-1}{3} 7^{t-4}$ if $t \geq 4$.

If you add two generating functions together, the coefficient of x^n in the sum is what you would expect, the sum of the coefficients of x^n in the summands. For example, adding the generating functions from the above two paragraphs gives:

$$\frac{3}{1+5x} + \left(\frac{x}{1+7x}\right)^4$$

$$= \sum_{n=0}^{\infty} 3(-1)^n 5^n x^n + \sum_{n=4}^{\infty} (-1)^n \binom{n-1}{3} 7^{n-4} x^n$$

$$= 3 - 3(5^1)x + 3(5^2)x^2 - 3(5^3)x^3 + \sum_{n=4}^{\infty} (3(-1)^n 5^n + (-1)^n \binom{n-1}{3} 7^{n-4}) x^n$$

$$= 3 - 3(5^1)x + 3(5^2)x^2 - 3(5^3)x^3 + \sum_{n=4}^{\infty} (-1)^n (3 \cdot 5^n + \binom{n-1}{3} 7^{n-4}) x^n$$

Thus, if $n \leq 3$, the coefficient of x^n is $3(-1)^n 5^n$, and if $n \geq 4$ it is $(-1)^n \left(3 \cdot 5^n + \binom{n-1}{3} 7^{t-4} \right)$.

The above describes most of the basic tools you need. *When trying to determine what sequence is generated by some generating function, your goal will be to write it as a sum of known generating functions, some of which may be multiplied by constants, or constants times some power of x.* Once you have done this, you can use the techniques above to determine the sequence. Most of the time the known generating functions are among those described above. Occasionally you may need to directly compute the product of two generating functions:

Cauchy Product

$$\left(\sum_{n=0}^{\infty} a_n x^n \right) \left(\sum_{n=0}^{\infty} b_n x^n \right)$$

$$= a_0 b_0 + (a_0 b_1 + a_1 b_0)x + (a_0 b_2 + a_1 b_1 + a_2 b_0)x^2 + \ldots$$

$$= \sum_{n=0}^{\infty} \left(\sum_{n=0}^{n} a_k b_{n-k} \right) x^n$$

To see this, carry out the multiplication and simplify.

The last tool you will need allows you to write a, product of (known) generating functions a» a sum of known generating functions, some of which may be multiplied by constants, or constants times some power of x. It is the method of **partial fractions,** which you should have learned while studying Calculus. We will do a couple of examples in the course of this section, but won't give a comprehensive treatment of the method. For more info, look in (almost) any Calculus textbook.

Example 7: What sequence is generated by $\dfrac{2+x}{1-x-8x^2+12x^3}$?

Solution: The first task is to factor the denominator. We want to write it as a product of terms of the form $(1 - ax)$. Since the constant term is 1, if the a's. are all integers then they are among the divisors of the coefficient of the highest; power of x. and their negatives. (This is similar to the hints for finding integer roots of polynomials discussed in the section on solving recurrences. To see it, imagine such a factorization multiplied out and notice how the coefficient of the highest power of x arises.) Here, we find that $1 - x - 8x^2 + 12x^3 = (1 + 3x)(1 - 2x)^2$. Thus, we want to use partial fractions to write $\dfrac{2+x}{(1+3x)(1-2x)^2}$ as a sum of "known" generating functions. These arise from the factors in the denominator. Using partial fractions

$$\frac{2+x}{(1+3x)(1-2x)^2} = \frac{A}{1+3x} + \frac{B}{1-2x} + \frac{C}{(1+2x)^2}.$$

Thus, $2 + x = A(1 - 2x)^2 + B(1 + 3x)(1 - 2x) + C(1 + 3x)$. Expanding the RHS and equating like powers of x on the LHS and RHS gives

$$A + B + C = 2, \quad -4A + B + 3C = 1, \quad 4A - 6B = 0$$

The solution is $A = 3/5$, $B = 2/5$, $C = 1$. Therefore,

$$\frac{2+x}{(1+3x)(1-2x)^2} = (3/5)\frac{1}{1+3x} + (2/5)\frac{B}{1-2x} + \left(\frac{1}{1+2x}\right)^2.$$

The coefficient of x^n on the RHS is $(3/5)(-1)^n 3^n + (2/5)2^n + \binom{n+1}{n} 2^n = (3/5)(-1)^n 3^n + (2/5)2^n + (n+1)2^n$. Thus the sequence generated is $(3/5)(-1)^n 3^n + (2/5)2^n + (n+1)2^n$.

12.6 DERIVING GENERATING FUNCTIONS FROM RECURRENCES

If you are given a sequence defined by a recurrence relation and initial conditions, you can use these to get a generating function for the sequence. Having done that, you can then apply the facts and methods above to get a formula for the n-th term of the sequence. This is **another method of solving recurrences.** We'll illustrate the method first, and then try to give a description of it below. It would be wise to work through the example twice: once before reading the description of the method, and once after. That way you should be able to recognize (and understand) the major steps.

Example 8. Find the generating function for the sequence a_n defined by $a_n = a_{n-1} + 8a_{n-2} - 12a_{n-3}$, $n > 3$, with initial conditions $a_0 = 2$, $a_1 = 3$, and $a_3 = 19$.

Solution: Let $g(x) = \sum_{n=0}^{\infty} a_n x^n$

$= 2 + 3x + 19x^2 + \sum_{n=3}^{\infty} a_n x^n$

$= 2 + 3x + 19x^2 + \sum_{n=3}^{\infty} (a_{n-1} + 8a_{n-2} - 12a_{n-3}) x^n$

$= 2 + 3x + 19x^2 + x \sum_{n=3}^{\infty} a_{n-1} x^n + \sum_{n=3}^{\infty} 8 a_{n-2} x^n - \sum_{n=3}^{\infty} 12 a_{n-3} x^n$

$= 2 + 3x + 19x^2 + x \sum_{n=3}^{\infty} a_{n-1} x^{n-1} + 8x^2 \sum_{n=3}^{\infty} a_{n-2} x^{n-2} - 12x^3 \sum_{n=3}^{\infty} a_{n-3} x^{n-3}$

$= 2 + 3x + 19x^2 + \sum_{k=2}^{\infty} a_k x^k + 8x^2 + \sum_{k=1}^{\infty} a_k x^k - 12x^3 \sum_{k=0}^{\infty} a_k x^k$

$= 2 + x + xg(x) + \left(x \sum_{k=0}^{\infty} a_k x^k \right) - x(2 + 3x) + 8x^2 \left(\sum_{k=0}^{\infty} a_k x^k \right) - 8x^2(2) - 12x^3 \sum_{k=0}^{\infty} a_k x^k$

$= 2 + x + xg(x) + 8x^2 g(x) - 12x^3 g(x)$.

Therefore $g(x) - xg(x) - 8x^2 g(x) - 12x^2 g(x) = 2 + x$. That is, $g(x)(1 - x - 8x^2 - 12x^3) = 2 + x$,

so $g(x) = \dfrac{2+x}{1-x-8x^2-12x^3}$ is the generating function. You can now apply the method in

Example 9: (since this *is* the generating function from Example 1) to find that $a_n = (3/5)(-1)^n 3^n + (2/5)2^n + (n+1)2^n$.

Solution: The main idea is to let $g(x)$ be the generating function, say $g(x) = \sum_{n=0}^{\infty} a_n x^n$: and then use the recurrence, initial conditions, manipulation of the sum(s), and algebra to convert the RHS into an expression which is a sum of terms some of which involve $g(x)$ itself. The following steps are usually involved (and in this order):

(1) Split the terms corresponding to the initial values from the sum on the RHS. The sum should then run from the first integer for which the recursive definition takes effect.

(2) Substitute the recursive definition of a_n into the summation.

(3) Split the sum into several sums, and in each one factor out the constant times a high enough power of x so that the subscript on the coefficient and the exponent of x are equal.

(4) Make a change of variable in each summation so each is the sum of $a_n x^n$ from some value to infinity. One of the sums should start at 0, another at one, and so on, until finally one starts at the subscript of the "last" initial value (assuming consecutive initial values are given). Also, the sum that starts at 0 should be multiplied by a constant (may be 1), the one that starts at 1 should be multiplied by a constant (may

be 1) times x, and in general each sum that starts at t should be multiplied by a constant (may be 1) times x^t.

(5) By adding the "missing" (first few) terms, and subtracting them off outside the sum (don't forget to multiply by whatever is in front of the summation!), convert each sum so that it goes from 0 to infinity. That is, do algebra so that each sum involving a term of the recurrence equals $g(x)$. If the recurrence is non-homogeneous, you may need to do the same sort of thing to convert the sum(s) arising from the non-homogeneous term(s) into known sums.

(6) Collect all terms involving $g(x)$ on the LHS, factor $g(x)$ out (if possible), then divide (or do whatever algebra is needed), to obtain a closed form for the generating function.

You might have noticed that the characteristic equation for the recurrence in Example 2 is $x^3 - x^2 - 8x + 12$, while the denominator for the generating function is $1 - x - 8x^2 - 12x^3$. It is always true that *if the characteristic equation for a LHRRWCC is $a_0 x^k + a_1 x^{k-1} + \ldots + a_{k-1}$, then the denominator for (some form of) the generating function is $a_0 + a_1 x + a_2 x^2 + \ldots + a_{k-1} x^k$*. This follows from the general method described above.

Once you have the generating function, you can factor the denominator and (hopefully) use partial fractions to write it as a sum of multiples of known generating functions. By finding the coefficient of x^n in each sum and, you get a formula for a_n.

Example 10. Solve $a_n = -3a_{n-1} + 10a_{n-2} + 3 \cdot 2^n$, $n \geq 2$ with initial conditions $a_0 = 0$ and $a_1 = 6$.

Solution: Let $g(x) = \sum_{n=0}^{\infty} a_n x^n$

$$= a_0 + a_1 x + \sum_{n=0}^{\infty} a_n x^n$$

$$= 0 + 6x + \sum_{n=2}^{\infty} (-3a_{n-1} + 10a_{n-2} + 3 \cdot 2^n) x^n$$

$$= 6x - 3x \sum_{n=2}^{\infty} a_{n-1} x^{n-1} + 10x^2 \sum_{n=2}^{\infty} a_{n-2} x^{n-2} + 3 \sum_{n=2}^{\infty} 2^n x^n$$

$$= 6x - 3x \sum_{k=1}^{\infty} a_k x^k + 10x^2 \sum_{k=0}^{\infty} a_k x^k + 3 \sum_{n=2}^{\infty} 2^n x^n$$

$$= 6x - 3x \left(\sum_{k=1}^{\infty} a_k x^k \right) - (-3x)(0) + 10x^2 \sum_{k=0}^{\infty} a_k x^k + 3 \sum_{n=2}^{\infty} 2^n x^n$$

$$= 6x - 3xg(x) + 10x^2 g(x) + 3 \left(\sum_{n=2}^{\infty} 2^n x^n \right) - 3(1 + 2x)$$

Therefore,

$$g(x)(1 + 3x - 10x^2) = -3 + \frac{3}{1-2x}, \text{ so } g(x) = \frac{-3}{1+3x-10x^2} + \frac{3}{(1-2x)(1+3x-10x^2)}$$

We need to expand the RHS using partial fractions.

$$\frac{6x}{(1-2x)^2(1+5x)} = \frac{A}{1-2x} + \frac{B}{(1-2x)^2} + \frac{C}{1+5x} = \frac{A(1-2x)(1+5x) + B(1+5x) + C(1-2x)^2}{(1-2x)^2(1+5x)}$$

Thus $6x = A(1-2x)(1+5x) + 5(1+5x) + C(1-2x)^2$. Expanding the RHS and then equating coefficients of like powers of x yields:

$$A + B + C = 0, \quad 3A + 5B - 4C = 6, \quad -10A + 4C = 0$$

The solution is $A = -12/49$, $B = 6/7$, $C = -30/49$. Thus, $g(x) = (-12/49)\frac{1}{1-2x} + (6/7)$

$\frac{1}{(1-2x)^2} + (-30/49)\frac{1}{1+5x}$, so $a_n = (-12/49)2^n + (6/7)\binom{n+1}{1}2^n - (30/49)(-1)^n 5^n = (-12/49)2^n + (6/7)(n+1)2^n + (30/49)(-1)^{n+1}5^n$.

Using generating functions to solve counting problems. In showing that $(1 + x + x^2 + \ldots)^k = \sum_{n=0}^{\infty} \binom{n+k-1}{k-1} x^n$ we argued that the coefficient of x^n equals the number of ways of writing n as a sum of k integers, each of which occurs as an exponent in one of the factors on the LHS. Stated slightly differently, $\left(\frac{1}{1-x}\right)^k$ is the generating function for the number of ways to write n as a sum of k non-negative integers.

Continuing this line of reasoning, the generating function for the number of ways to write n as a sum of five odd integers would be $(x + x^3 + x^5 + \ldots)^5 = \{x(1 + x^2 + x^4 + \ldots)\}^5 = \left(\frac{x}{1-x^2}\right)^5$. Think about what happens with the exponents when the LHS is multiplied out.)

Similarly, the generating function for the number of ways to write n as a sum of two odd integers and an even integer would be $(x + x^3 + x^5 + \ldots)^2(1 + x^2 + x^4 + \ldots) = \frac{x^2}{(1-x)^3}$.

In general, suppose you want the generating function for the number of ways to write n as an (ordered) sum of k terms, some of which may be restricted (as in being odd, for example). Then the generating function will be a product of k factors each of which is a sum of powers of x. The exponents of x in the first factor will be the possibilities for the first term in sum (that adds to n), the exponents of x in the second factor will be the possibilities for the second term in the sum, and so on until, finally, the exponents in the n-th term are the possibilities for the A;-th term in the sum.

Remember that $x^0 = 1$, so a factor has a 1 if and only if zero is one of the possibilities for the corresponding term in the sum. Similarly for $x^1 = x$, and so on.

Example 11: Determine the generating function for the number of ways to distribute a large number of identical candies to four children so that the first two children receive an odd number of candies, the third child receives at least three candies, and the fourth child receives at most two candies.

Solution: Suppose the number of candies distributed is n. Then we are looking for the generating function for the number of ways to write n as a sum of four integers, the first two of which are odd, the third of which is at least 3. and the fourth of which is at most 2. Following the reasoning from above, the generating function is a product of four factors, and the exponents of x in each factor give the possible values for the corresponding term in the sum. Thus, the generating function is $(x + x^3 + x^5 + \ldots)^2(x^3 + x^4 + \ldots)(1 + x + x^2) =$

$$\left(\frac{x^2}{1-x^2}\right)^2 \frac{x^3}{1-x}(1 + x + x^2)$$

Once you have the generating function for the number of ways to do something, you can apply the methods you know (i.e. write the generating function as a sum of known generating functions) to determine the coefficient of x^n, and hence the number of ways.

(This gets a touch ugly at the end, but it is a good illustration of the methods, so hang with it.) Use the generating function to determine the number of ways.

The generating function is

$$g(x) = \left(\frac{x^2}{1-x^2}\right)^2 \frac{x^3}{1-x}(1 + x + x^2) = (x^2 + x^3 + x^4)\left(\frac{1}{(1-x^2)^2(1-x)}\right)$$

$$= (x^2 + x^3 + x^4)\left(\frac{1}{(1-x)^2(1-x)^3}\right)$$

An exercise in partial fractions shows that :

$$\frac{1}{(1-x)^2(1-x)^3} = (1/8)\left(\frac{1}{1-x}\right)^2 + (3/16)\left(\frac{1}{1-x}\right) + (1/4)\left(\frac{1}{1+x}\right)^3 + (1/4)\left(\frac{1}{1+x}\right)^2$$

$$+ (3/16)\left(\frac{1}{1+x}\right)$$

Since these are all known generating functions, the coefficient of x^n in *this* expression (which is not $g(x)$ is $(1/8) + (3/16)(n + 1) + (1/4)(-1)^n \binom{n+3-1}{3-1} + (1/4)(-1)^n(n + 1) + (3/16)$

$\left(\frac{1}{1+x}\right)$.

We need the coefficient of x^n in $g(x)$, which equals $\dfrac{1}{(1-x)^2(1-x)^3}$ multiplied by $(x^2 + x^3 + x^4)$. Thus, the coefficient of x^n in $g(x)$ is the sum of the coefficient of x^{n-2} in $\dfrac{1}{(1-x)^2(1-x)^3}$

the coefficient of x^{n-3} in $\dfrac{1}{(1-x)^2(1-x)^3}$ and the coefficient of x^{n-4} in $\dfrac{1}{(1-x)^2(1-x)^3}$. This equals $a + b + c$, where

$a = (1/8) + (3/16)(n-1) + (1/4)(-1)^{n-2}\dbinom{n}{2} + (1/4)(-1)^{n-2}(n-1) + (3/16)(-1)^{n-2}$,

$b = (1/8) + (3/16)(n-2) + (1/4)(-1)^{n-2}\dbinom{n-1}{2} + (1/4)(-1)^{n-3}(n-2) + (3/16)(-1)^{n-3}$,

$c = (1/8) + (3/16)(n-3) + (1/4)(-1)^{n-4}\dbinom{n-2}{2} + (1/4)(-1)^{n-4}(n-3) + (3/16)(-1)^{n-4}$.

No doubt this expression can be simplified.

Example 12: Use generating functions to find b_n, the number of ways that $n > 0$ identical candies can be distributed among 4 children and 1 adult so that each child receives an odd number of candies, and the adult receives 1 or 2 candies.

Solution: Following the method outlined above, the generating function is

$$g(x) = (x + x^3 + x^5 + \ldots)^4(x + x^2) = \left(\dfrac{x}{1-x^2}\right)^4 (x + x^2)$$

$$= (x^5 + x^6) \sum_{k=0}^{\infty} \binom{k+4-1}{3} x^{2k}$$

$$= \sum_{k=0}^{\infty} \binom{k+3}{3} x^{2k+5} + \sum_{k=0}^{\infty} \binom{k+3}{3} x^{2k+6}$$

The first sum contains all terms with *odd* exponents, and the second sum contains all terms with even exponents. Thus, if $n \leq 4$, $b_n = 0$. If $n \geq 5$ and odd, $b_n = \dbinom{n+1/2}{3}$ (let $n = 2k + 5$ in the first sum). If $n \geq 6$ and even, $b_n = \dbinom{n/2}{3}$ (let $n = 2k + 6$ in the second sum). These cases can all be described by the single expression $b_n = \dbinom{[n/2]}{3}$.

Example 13: In a certain game it is possible to score 1, 2, or 4 points on each turn. Find the generating functions for the number of ways to score n points in a game in which there are at least two turns where 4 points are scored.

Solution: Here we are looking for the number of ways to write n as an ordered sum of three terms. The first term represents the number of points obtained from turns where 1 point was scored. The second term represents the number of points obtained from turns where 2 points were scored (and thus is a multiple of 2). The third term represents the number of points obtained from turns where 4 points were scored (and thus is a multiple of 4, and at least 8). Thus, the generating function will be a product of three factors, where the exponents in each factor correspond to the possibilities just discussed. Therefore,

$$g(x) = (1 + x + x^2 + \ldots)(1 + x^2 + x^4 + \ldots)(x^8 + x^{12} + x^{16} + \ldots) = \frac{1}{1-x}\frac{1}{1-x^2}\frac{x^8}{1-x^4}.$$

If you want an exercise in partial fractions, continue with Example 6 and determine the number of ways by writing $g(x)$ as a sum of known generating functions and finding the coefficient of x^n. This involves a 6×6 linear system (unless you makle an astute observation that is eluding me right now).

12.7 RECURRENCE RELATIONS

(a) Recursive Definitions

A definition such that the object defined occurs in the definition is called a *recursive definition*. For*instance the *Fibonacci sequence*

$$0, 1, 1, 2, 3, 5, 8, 13, \ldots$$

is defined as a sequence whose two first terms are $F_0 = 0$, $F_1 = I$ and each subsequent term is the sum of the two previous ones: $F_n = F_{n-1} + F_{n-2}$ (for $n \geq 2$).

Other examples:
- Recursive definition of *factorial*:
  (1) $0! = 1$
  (2) $n! = n \cdot (n-1)!$ $\quad (n \geq 1)$
- Recursive definition of *power*:
  (1) $a^0 = 1$
  (2) $a^n = a^{n-1} a$ $\quad (n \geq 1)$

In all these examples we have:
(1) A *Basis*, where the function is explicitly evaluated for one or more values of its argument.
(2) A *Recursive Step*, stating how to compute the function from its previous values.

(b) Recurrence Relations

When we considerer a recursive definition as an equation to be solved we call it *recurrence relation*. Here we will focus on *kth- order linear recurrence relations*, which are of the form

$$C_0 x_n + C_1 x_{n-1} + C_2 x_{n-2} + \ldots + C_k x_{n-k} = b_n,$$

where $C_0 \neq 0$. If $b_n = 0$ the recurrence relation is called *homogeneous*. Otherwise it is called *non-homogeneous*. The coefficients C_i may depend on n, but here we will assume that they are constant unless stated otherwise.

The basis of the recursive definition is also called *initial conditions* of the recurrence. So, for instance, in the recursive definition of the Fibonacci sequence, the recurrence is

$$F_n = F_{n-1} + F_{n-2}$$

or
$$F_n - F_{n-1} - F_{n-2} = 0,$$
and the initial conditions are
$$F_0 = 0, F_1 = 1.$$

1. Solving Recurrence Relations. A solution of a recurrence relation is a sequence x_n that verifies the recurrence.

An important property of *homogeneous* linear recurrences ($b_n = 0$) is that given two solutions x_n and y_n of the recurrence, any linear combination of them $z_n = rx_n + sy_n$, where r, s are constant, is also a solution of the same recurrence, because

$$\sum_{i=0}^{k} C_i (rx_{n-i} + sy_{n-i}) = r \sum_{i=0}^{k} C_i x_{n-i} + s \sum_{i=0}^{k} C_i y_{n-i} = r \cdot + s \cdot 0 = 0$$

For instance, the Fibonacci sequence $F_n = 0, 1, 1, 2, 3, 5, 8, 13, \ldots$ and the Lucas sequence $L_n = 2, 1, 3, 4, 7, 11, \ldots$ verify the same recurrence $x_n = x_{n-1} + x_{n-2}$, so any linear combination of them $aF_n + bL_n$, for instance their sum $F_n + L_n = 2, 2, 4, 6, 10, 16, \ldots$, is also a solution of the same recurrence.

If the recurrence is non-homogeneous then we have that the difference of any two solutions is a solution of the homogeneous version of the recurrence, i.e., if $\sum_{i=0}^{k} C_i x_{n-i} = b_n$ and $\sum_{i=0}^{k} C_i y_{n-i} = b_n$ then obviously $z_n = x_n - y_n$ verifies:

Some recurrence relations can be solved by *iteration*, i.e., by using the recurrence repeatedly until obtaining a explicit close-form formula. For instance consider the following recurrence relation:

$$x_n = r x_{n-1} \quad (n > 0); \qquad x_0 = A.$$

By using the recurrence repeatedly we get:

$$x_n = r x_{n-1} = r^2 x_{n-2} = r^3 x_{n-3} = \ldots = r^n x_0 = A r^n,$$

hence the solution is $x_n = A r^n$.

Next we look at two particular cases of recurrence relations, namely first and second order recurrence relations, and their solutions.

2. First Order Recurrence Relations. The homogeneous case can be written in the following way:

$$x_n = r x_{n-1} \quad (n > 0); \qquad x_0 = A.$$

Its general solution is

$$x_n = A r^n$$

which is a *geometric sequence* with *ratio* r.

The non-homogeneous case can be written in the following way:

$$x_n = r x_{n-1} + c_n \quad (n > 0); \qquad x_0 = A.$$

Using the summation notation, its solution can be expressed like this:

$$x_n = Ar^n + \sum_{k=1}^{n} c_k r^{n-k}$$

We examine two particular cases. The first one is
$$x_n = rx_{n-1} + c \quad (n > 0); \qquad x_0 = A.$$
where c is a constant. The solution is
$$x_n = Ar^n + c\sum_{k=1}^{n} r^{n-k} = Ar^n + c\frac{r^n - 1}{r-1} \quad \text{if } r \neq 1,$$
and
$$x_n = A + cn \qquad \text{if } r = 1.$$

The second particular case is for $r = 1$ and $c_n = c + dn$, where c and d are constant (so c_n is an arithmetic sequence):
$$x_n = x_{n-1} + c + dn \quad (n > 0); \qquad x_0 = A.$$
The solution is now
$$x_n = A + \sum_{k=1}^{n}(c + dk) = A + cn\frac{dn(n+1)}{2}$$

3. Second Order Recurrence Relations. Now we look at the recurrence relation
$$C_0 x_n + C_1 x_{n-1} + C_2 x_{n-2} = 0$$
First we will look for solutions of the form $x_n = cr^n$. By plugging in the equation we get:
$$C_0 cr^n + C_1 cr^{n-1} + C_2 cr^{n-2} = 0,$$
hence r must be a solution of the following equation, called the *characteristic equation* of the recurrence:
$$C_0 r^2 + C_1 r + C_2 = 0$$
Let r_1, r_2 be the two (in general complex) roots of the above equation. They are called *characteristic roots*. We distinguish three cases:

(1) **Distinct Real Roots.** In this case the general solution of the recurrence relation is
$$x_n = c_1 r_1^n + c_2 r_2^n$$
where c_1, c_2 are arbitrary constants.

(2) **Double Real Root.** If $r_1 = r_2 = r$, the general solution of the recurrence relation is
$$x_n = c_1 r^n + c_2 n r^n$$
where c_1, c_2 are arbitrary constants.

(3) **Complex Roots.** In this case the solution could be expressed in the same way as in the case of distinct real roots, but in order to avoid the use of complex numbers we write $r_1 = re^{\alpha i}, r_2 = re^{-\alpha i}$ $k_1 = c_1 + c_2, k_2 = (c_1 - c_2) i$, which yields:

$$x_n = k_1 r^n \cos n\alpha + k_2 r^n \sin n\alpha.$$

Example 14: Find a closed-form formula for the Fibonacci sequence defined by:
$$F_{n+1} = F_n + F_{n-1} \ (n > 0); \quad F_0 = 0, F_1 = 1$$

Solution: The recurrence relation can be written
$$F_n - F_{n-1} - F_{n-2} = 0.$$

The characteristic equation is
$$r^2 - r - 1 = 0.$$

Its roots are:
$$r_1 = \phi = \frac{1+\sqrt{5}}{2}; \quad r_2 = -\phi^{-1} = \frac{1-\sqrt{5}}{2}$$

They are distinct real roots, so the general solution for the recurrence is:
$$F_n = c_1 \phi^n + c_2 (-\phi^{-1})^n.$$

Using the initial conditions we get the value of the constants:
$$\begin{cases} (n=0) & c_1 + c_2 = 0 \\ (n=1) & c_1 \phi + c_2(-\phi^{-1}) = 1 \end{cases} \Rightarrow \begin{cases} c_1 = 1/\sqrt{5} \\ c_1 = -1/\sqrt{5} \end{cases}$$

Hence:
$$F_n = \frac{1}{\sqrt{5}} \{\phi^n - (-\phi)^{-n}\}$$

(c) Solving Recurrence Relations

We want to solve the recurrence relation
$$a_n = A a_{n-1} + B a_{n-2}$$

where A and B are real numbers. The solutions depend on the nature of the roots of the characteristic equation
$$s^2 - As - B = 0 \qquad \ldots(1)$$

We consider three cases for the roots of (1).

1. If we have two distinct real roots s_1 and s_2, then
$$a_n = \alpha s_1^n + \beta s_2^n$$

2. If we have exactly one real root s, then
$$a_n = \alpha s^n + \beta n s^n.$$

3. If we have two complex conjugate roots in polar form $s_1 = r \angle \theta$ and $s_2 = r \angle (-\theta)$, then
$$a_n = r^n (\alpha \cos(n\theta) + \beta \sin(n\theta)).$$

In all cases, the numbers α and β can be determined if we are given the values of a_0 and a_1.

Example : Consider the recurrence relation
$$a_n = 5 a_{n-1} - 6 a_{n-2} \qquad \ldots(2)$$

with initial conditions $a_0 = 1$ and $a_1 = 4$. The characteristic equation is
$$s^2 - 5s + 6 = (s-2)(s-3) = 0.$$
Since the roots are $s = 2$ and $s = 3$, any solution of (2) has the form $a_n = \alpha 3^n + \beta 2^n$. Therefore,
$$a_0 = \alpha + \beta = 1$$
$$a_1 = 3\alpha + 2\beta = 4.$$
Solving this linear system, we get $\alpha = 2$ and $\beta = -1$. The solution of (2) with the given initial conditions is then
$$a_n = 2 \cdot 3^n - 2^n$$

Example : Consider the recurrence relation
$$a_n = 6a_{n-1} - 9a_{n-2} \qquad \ldots (3)$$
with initial conditions $a_0 = 4$ and $a_1 = 6$. The characteristic equation is
$$s^2 - 6s + 9 = (s-3)^2 = 0.$$
Since $s = 3$ is the only root, any solution of (3) has the form $a_n = \alpha 3^n + \beta n 3^n$. Therefore,
$$a_0 = \alpha = 4$$
$$a_1 = 3\alpha + 3\beta = 6.$$
Solving this system, we get $\alpha = 4$ and $\beta = -2$. The solution of (3) with its initial conditions is then
$$a_n = 4 \cdot 3^n - 2n3^n$$

Example : Consider the recurrence relation
$$a_n = 2a_{n-1} - 2a_{n-2} \qquad \ldots (4)$$
with initial conditions $a_0 = 1$ and $a_1 = 3$. The characteristic equation is
$$s^2 - 2s + 2 = (s-1)^2 + 1 = 0.$$
We have two complex conjugate roots $s_1 = 1 + i$ and $s_1 = 1 - i$. In polar form $s_1 = r\angle\theta$ with $r = \angle 2$ and $\theta = \frac{\pi}{4}$. Any solution of (4) has the form $a_n = (\sqrt{2})^n \left(\alpha \cos\left(n\frac{\pi}{4}\right) + \beta \sin\left(n\frac{\pi}{4}\right) \right)$. Therefore,
$$a_0 = \alpha = 1$$
$$a_1 = \sqrt{2}\left(\alpha \frac{1}{\sqrt{2}} + \beta \frac{1}{\sqrt{2}} \right) = 3$$
Solving this system, we get $\alpha = 1$ and $\beta = 2$. The solution of (4) with its initial conditions is then
$$a_n = (\sqrt{2})^n \left(\cos\left(n\frac{\pi}{4}\right) + 2\sin\left(n\frac{\pi}{4}\right) \right)$$

Example : Consider the sequence of Fibonacci numbers that satisfy the recurrence relation

$$f_n = f_{n-1} + f_{n-2}$$

with initial conditions $f_0 = 0$ and $f_1 = 1$. The characteristic equation is
$$s^2 - s - 1 = 0.$$

The roots are $s = (1+\sqrt{5})/2$ and $s = (1-\sqrt{5})/2$. Then, any solution of (5) has the form Therefore,

$$f_0 = \alpha + \beta = 0$$

$$f_1 = \alpha\left(\frac{1+\sqrt{5}}{2}\right) + \beta\left(\frac{1-\sqrt{5}}{2}\right) = 1$$

Solving this linear system, we get $\alpha = 1/\sqrt{5}$ and $\beta = -1/\sqrt{5}$. Therefore, the Fibonacci numbers are obtained by the following formula, commonly known as Binet's formula.

$$f_n = \frac{1}{\sqrt{5}}\left(\frac{1+\sqrt{5}}{2}\right)^n - \frac{1}{\sqrt{5}}\left(\frac{1-\sqrt{5}}{2}\right)^n$$

$$a_n = Aa_{n-1} + Ba_{n-2} + F(n)$$

for $F(n)$ not identically zero is said to be **nonhomogeneous.** Its associated **homogeneous** recurrence relation is

$$a_n = Aa_{n-1} + Ba_{n-2}.$$

Theorem: If $a_n^{(p)}$ is a particular solution of the nonhomogeneous recurrence relation

$$a_n = Aa_{n-1} + Ba_{n-2} + F(n), \qquad \ldots(6)$$

then every solution of (6) is of the form

$$a_n = a_n^{(h)} + a_n^{(p)}$$

where $a_n^{(h)}$ is a solution of the associated homogeneous recurrence relation.

To apply this theorem, we need to find a particular solution of (6). This is a difficult problem in general but a standard technique exists for simple types of $F(n)$ such as

- Polynomial,   e.g. $F(n) = 5n^2 - 2n + 1$.
- Exponential,   e.g. $F(n) = 3^n$.
- Exponential × Polynomial,   e.g. $F(n) = 2^n(5n^2 + 3n + 1)$.

Theorem: *Consider the nonhomogeneous recurrence relation*

$$a_n = Aa_{n-1} + Ba_{n-2} + F(n),$$

where $F(n) = t^n$ (Polynomial of degree N). If t is not a root of $s^2 - As - B = 0$, then there is a particular solution of the form

$$a_n^{(p)} = t^n(p_0 + p_1 n + p_2 n^2 + \ldots + p_N n^N).$$

If t is a root of $s^2 - As - B = 0$ of multiplicity m, then there is a particular solution of the form

$$a_n^{(p)} = t^n n^m (p_0 + p_1 n + p_2 n^2 + \ldots + p_N n^N).$$

Note that if $F(n)$ is simply a polynomial like $F(n) = 5n^2 - 2n + 1$, then $t = 1$ in the above theorem.

Example Consider the recurrence relation

$$a_n = 5a_{n-1} - 6a_{n-2} + F(n).$$

The characteristic equation of its associated homogeneous equation is

$$s^2 - 5s + 6 = (s-2)(s-3) = 0.$$

1. If $F(n) = 2n^2$, then a particular solution has the form $a_n^{(p)} = An^2 + Bn + C$.

2. If $F(n) = 5^n(3n^2 + 2n + 1)$, then $a_n^{(p)} = 5^n(An^2 + Bn + C)$.

3. If $F(n) = 5^n$, then $a_n^{(p)} = 5^n A$.

4. If $F(n) = 3^n$, then $a_n^{(p)} = 3^n An$.

5. If $F(n) = 2^n(3n + 1)$, then $a_n^{(p)} = 2^n n(An + B)$.

The values of the constants A, B, and C can be found by substituting $a_n^{(p)}$ in the recurrence relation.

Example. For the recurrence relation

$$a_n = 6a_{n-1} - 9a_{n-2} + F(n),$$

the characteristic equation of its associated homogeneous equation is

$$s^2 - 6s + 9 = (s-3)^2 = 0.$$

1. If $F(n) = 3^n$, then $a_n^{(p)} = 3^n An^2$.

2. If $F(n) = 3^n(5n + 1)$, then $a_n^{(p)} = 3^n n^2(An + B)$.

3. If $F(n) = 2^n(5n + 1)$, then $a_n^{(p)} = 2^n(An + B)$.

Example. For the recurrence relation

$$a_n = 3a_{n-1} - 2a_{n-2} + F(n),$$

the characteristic equation of its associated homogeneous equation is

$$s^2 - 3s + 2 = (s-1)(s-2) = 0.$$

If $F(n) = 3n + 1$, then a particular solution has the form $a_n^{(p)} = n(An + B)$.

To see this, observe that $F(n) = 1^n(3n + 1)$ and $s = 1$ is a root of multiplicity one of the characteristic equation.

Finally, here are two basic recurrence relations.

Theorem (*Arithmetic sequence*) If $a_n = a_{n-1} + d$ and $a_0 = \alpha$, then $a_n = dn + \alpha$.

Theorem (*Geometric sequence*) If $a_n = ka_{n-1}$ and $a_0 = a$, then $a_n = \alpha k^n$.

EXERCISE 12.1

1. Simplify:
   (a) $\dfrac{(h+1)!}{h!}$;
   (b) $\dfrac{h!}{(h-2)!}$;
   (c) $\dfrac{(h-1)!}{(h+2)!}$;
   (d) $\dfrac{(h-r+1)!}{(h-r-1)!}$

2. Evaluate:
   (a) $\binom{5}{2}$;
   (b) $\binom{7}{3}$;
   (c) $\binom{14}{2}$;
   (d) $\binom{6}{4}$;
   (e) $\binom{20}{17}$;
   (f) $\binom{18}{15}$

3. How many automobile license plates can be made if each. Plate contains two different let ten followed by two different digits.

4. There are six roads between A and B and four roads between B and C. Find the number of ways that one can drive : (a) from A to C by way of B; (b) round-trip from A to C by way of B; (c) round-trip from A to C by way of B without using the same road more than once.

5. Find the number of ways in which six people can ride a toboggan if one of a subset of three must drive.

6. Find the number of ways in which five persons can site in a row.

7. Find the number of permutations that can be formed from the letters of the word ELEVEN.
   (a) How many of them begin and end with E?
   (b) How many of them have the three ES together?
   (c) How many begin with E and end with N?

8. A woman has 11 close friends.
   (a) In how many ways can she invite five of them to differ?
   (b) In how many ways if two of the friends are married and will not attend separately?
   (c) In how many ways if two of them are not on speaking terms and will not attend together?

9. A woman has 11 close friends of whm six are also woman.
   (a) In how many ways can she invite three on bone to a party?
   (b) In how many ways can she invite three on more of them if she wants the same number of men as women. (including herself)?
10. A student is to answer 10 out of 13 questions on an exam.
    (a) How many choices has he?
    (b) How many if he must answer the first two questions?
    (c) How many if he must answer the first or second question but not bath?
    (d) How many if he must answer exactly three out of the first five questions?
    (e) How many if he must answer at least three of the first five questions?

EXERCISE 12.2

1. In how many ways can 10 students be divided into three teams, one containing four students and the other three?
2. In how many ways can 14 people be partitioned into six committees when two of the committees contains three members and the other two?
3. Assuming that a cell can be empty, in how many ways can a set with three elements be partitioned into
   (i) three ordered cells
   (ii) three unordered cells
4. In how many ways can a set with found elements be partitioned into
   (i) three ordered cells
   (ii) three unordered cells.
5. Let a be a number function, such that—

$$an = \begin{cases} 2 & 0 \leq r \leq 3 \\ 2^{-r} + 5 & r \geq 4 \end{cases}$$

   (i) Determine $S^2 a$ and $S^{-2} a$.
   (ii) Determine Δa and ∇a.
6. Let $C = ab$, show that
   $$\Delta C r = a_{r+1} + \Delta br + br \Delta a_r$$
7. Let a and b be two numeric functions such that $a_r = r + 1$ and $br = a^r$ for all $r \geq 0$. Determine ab.
8. Determine the discrete numeric function of $G(x) = \dfrac{1}{1-x^2}$
9. Let a_r denote the number of ways a sum of r can be obtained when two indistinguishable dice are ralled. Determine $G(x)$.
10. In how many ways can $3r$ balls be selected from $2r$ red balls, $2r$ blue balls, and $2r$ white balls?
11. Let $G(x) = (1 + x)^{2n} + n(1 + x)^{2n-1} + \ldots + x^n (1 + x)^h$ find a_r.
12. Find first five terms of the sequence defined by
    $$a_r = a^2_{r-1}, a_1 = 2$$

13. For the discrete function $a_r = 2r + 3$. Find the recurrence relation.

14. Find the solution to the recurrence relation
$$a_r + 3a_{r-1} + 3a_{r-2} + a_{r-3} = 0$$

15. Find all solutions of the recurrence relation
$$a_r = 5a_{r-1} - 6a_{r-2} + 7r$$

ANSWERS (EXERCISE – 12.1)

1. (a) $h + 1$; (b) $h(n - 1) = n^2 - n$; (c) $1/[h(h + 1)(h + 2)]$; (d) $(h - r)(h - r + 1)$
2. (a) 10; (b) 35; (c) 91; (d) 15; (e) 1140; (f) 816;
3. $26.25.10.9.8 = 468000$
4. (a) 24; (b) 576; (c) 360;
5. 360;
6. 120;
7. (a) 120 (b) 24; (c) 24; (d) 12;
8. (a) 462; (b) 210; (c) 252;
9. $2^{11} - 1 - \binom{11}{2} - \binom{11}{2} = 1981$ or $\binom{11}{3} + \binom{11}{4} + \ldots + \binom{11}{11} = 1981$

   (b) $\binom{5}{5}\binom{6}{4} + \binom{5}{4}\binom{6}{3} + \binom{5}{3}\binom{6}{2} + \binom{5}{2}\binom{6}{1} = 325$

10. (a) 286; (b) 165; (c) 110; (d) 80; (e) 276

ANSWERS (EXERCISE – 12.2)

1. $\dfrac{10!}{4!3!3!} \cdot \dfrac{1}{2!} = 2100$ or $\binom{10}{4}\binom{5}{2} = 2100$

2. $\dfrac{14!}{3!3!2!2!2!2!} \cdot \dfrac{1}{2!4!} = 3153150$

3. (i) $3^3 = 27$; (ii) 5
4. $3^4 = 81$; (ii) 14
14. $a_r = (1 + 3r - 2r^2)(-1)^r$
15. $a_r = \alpha_1 3^r + \alpha_2 \cdot 2^r + \left(\dfrac{49}{20}\right)^r$

CHAPTER 13

KARNAUGH MAP

13.1 INTRODUCTION

Why learn about *Karnaugh* maps? The Karnaugh map, like Boolean algebra, is a simplification tool applicable to digital logic. See the "Toxic waste incinerator" in the Boolean algebra chapter for an example of Boolean simplification of digital logic. The Karnaugh Map will simplify logic faster and more easily in most cases.

Boolean simplification is actually faster than the Karnaugh map for a task involving two or fewer Boolean variables. It is still quite usable at three variables, but a bit slower. At four input variables, Boolean algebra becomes tedious. Karnaugh maps are both faster and easier. Karnaugh maps work well for up to six input variables, are usable for up to eight variables. For more than six to eight variables, simplification should be by *CAD* (computer automated design).

Recommended logic simplification vs number of inputs			
Variables	Boolean algebra	Karnaugh map	Computer automated
1-2	X		?
3	X		?
4	?	X	?
5-6		X	X
7-8		?	X
>8			X

In theory any of the three methods will work. However, as a practical matter, the above guidelines work well. We would not normally resort to computer automation to simplify a three input logic block. We could sooner solve the problem with pencil and paper. However, if we had seven of these problems to solve, say for a *BCD* (Binary Coded Decimal) to *seven segment decoder*, we might want to automate the process. A BCD to seven segment decoder generates the logic signals to drive a seven segment LED (light emitting diode) display.

Examples of computer automated design languages for simplification of logic are PALASM, ABEL, CUPL, Verilog, and VHDL. These programs accept a *hardware descriptor language* input file which is based on Boolean equations and produce an output file describing a *reduced* (or simplified) Boolean solution. We will not require such tools in this chapter. Let's move on to Venn diagrams as an introduction to Karnaugh maps.

13.2 VENN DIAGRAMS AND SETS

Mathematicians use *Venn diagrams* to show the logical relationships of *sets* (collections of objects) to one another. Perhaps you have already seen Venn diagrams in your algebra or other mathematics studies. If you have, you may remember overlapping circles and the *union* and *intersection* of sets. We will review the overlapping circles of the Venn diagram. We will adopt the terms OR and AND instead of union and intersection since that is the terminology used in digital electronics.

The Venn diagram bridges the Boolean algebra from a previous chapter to the Karnaugh Map. We will relate what you already know about Boolean algebra to Venn diagrams, then transition to Karnaugh maps.

A *set* is a collection of objects out of a universe as shown below. The *members* of the set are the objects contained within the set. The members of the set usually have something in common; though, this is not a requirement. Out of the universe of real numbers, for example, the set of all positive integers {1,2,3...} is a set. The set {3,4,5} is an example of a smaller set, or *subset* of the set of all positive integers. Another example is the set of all males out of the universe of college students. Can you think of some more examples of sets?

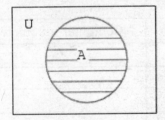 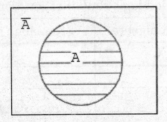 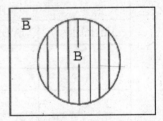

Above left, we have a Venn diagram showing the set A in the circle within the universe U, the rectangular area. If everything inside the circle is A, then anything outside of the circle is not A. Thus, above center, we label the rectangular area outside of the circle A as A-not instead of U. We show B and B-not in a similar manner.

What happens if both A and B are contained within the same universe? We show four possibilities.

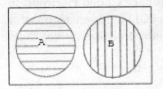 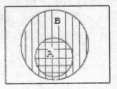 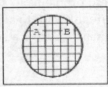 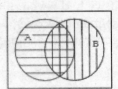

Let us take a closer look at each of the the four possibilities as shown above.

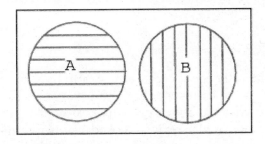

The first example shows that set A and set B have nothing in common according to the Venn diagram. There is no overlap between the A and B circular hatched regions. For example, suppose that sets A and B contain the following members:

set A = {1,2,3,4}
set B = {5,6,7,8}

None of the members of set A are contained within set B, nor are any of the members of B contained within A. Thus, there is no overlap of the circles.

In the second example in the above Venn diagram, Set A is totally contained within set B How can we explain this situation? Suppose that sets A and B contain the following members:

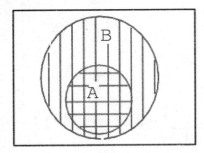

set A = {1,2}
set B = {1,2,3,4,5,6,7,8}

All members of set A are also members of set B. Therefore, set A is a subset of set B. Since all members of set A are members of set B, set A is drawn fully within the boundary of set B.

There is a fifth case, not shown, with the four examples. Hint: it is similar to the last (fourth) example. Draw a Venn diagram for this fifth case.

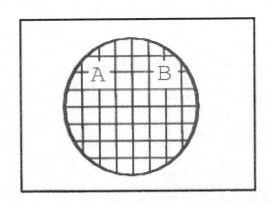

The third example above shows perfect overlap between set A and set B. It looks like both sets contain the same identical members. Suppose that sets A and B contain the following:

set A = {1,2,3,4} set B = {1,2,3,4}

Therefore,

Set A = Set B

Sets A and B are identically equal because they both have the same identical members. The A and B regions within the corresponding Venn diagram above overlap completely. If there is any doubt about what the above patterns represent, refer to any figure above or below to be sure of what the circular regions looked like before they were overlapped.

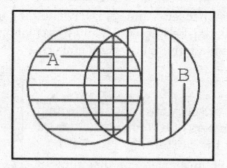

The fourth example above shows that there is something in common between set A and set B in the overlapping region. For example, we arbitrarily select the following sets to illustrate our point:

set A = {1,2,3,4}

set B = {3,4,5,6}

Set A and Set B both have the elements 3 and 4 in common. These elements are the reason for the overlap in the center common to A and B. We need to take a closer look at this situation.

13.3 BOOLEAN RELATIONSHIPS ON VENN DIAGRAMS

The fourth example has **A** partially overlapping **B**. Though, we will first look at the whole of all hatched area below, then later only the overlapping region. Let us assign some Boolean expressions to the regions above as shown below. Below left there is a red horizontal hatched area for **A**. There is a blue vertical hatched area for **B**.

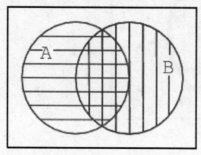 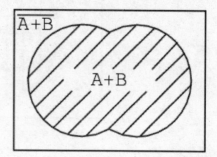

If we look at the whole area of both, regardless of the hatch style, the sum total of all hatched areas, we get the illustration above right which corresponds to the inclusive **OR** function of A, B. The Boolean expression is **A+B**. This is shown by the 45° hatched area. Anything outside of the hatched area corresponds to **(A+B)-not** as shown above. Let's move on to next part of the fourth example.

The other way of looking at a Venn diagram with overlapping circles is to look at just the part common to both **A** and **B**, the double hatched area below left. The Boolean expression for this common area corresponding to the **AND** function is **AB** as shown below right. Note that everything outside of double hatched **AB** is **AB-not**.

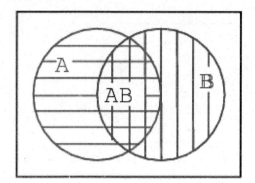
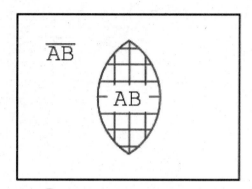

Note that some of the members of **A**, above, are members of **(AB)'**. Some of the members of **B** are members of **(AB)'**. But, none of the members of **(AB)'** are within the doubly hatched area **AB**.

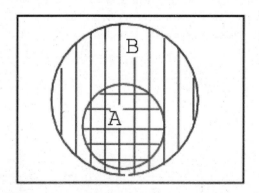
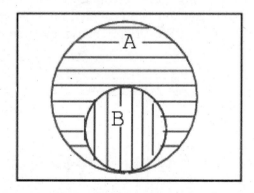

We have repeated the second example above left. Your fifth example, which you previously sketched, is provided above right for comparison. Later we will find the occasional element, or group of elements, totally contained within another group in a Karnaugh map.

Next, we show the development of a Boolean expression involving a complemented variable below.

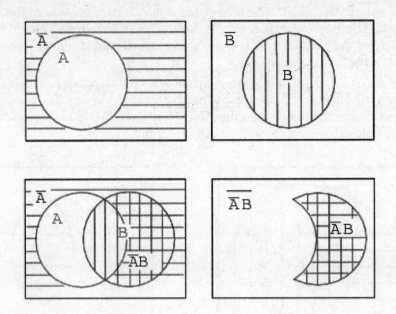

Example 1: (above)

Show a Venn diagram for **A'B** (A-not AND B).

Solution: Starting above top left we have red horizontal shaded **A'** (A-not), then, top right, **B**. Next, lower left, we form the AND function **A'B** by overlapping the two previous regions. Most people would use this as the answer to the example posed. However, only the double hatched **A'B** is shown far right for clarity. The expression **A'B** is the region where both **A'** and **B** overlap. The clear region outside of **A'B** is **(A'B)'**, which was not part of the posed example.

Let's try something similar with the Boolean **OR** function.

Example 2: Find **B'+A**

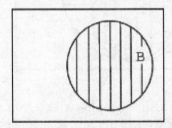 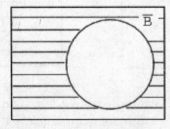 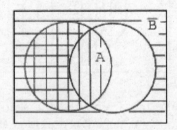

Solution: Above right we start out with **B** which is complemented to **B'**. Finally we overlay **A** on top of **B'**. Since we are interested in forming the **OR** function, we will be looking for all hatched area regardless of hatch style. Thus, **A+B'** is all hatched area above right. It is shown as a single hatch region below left for clarity.

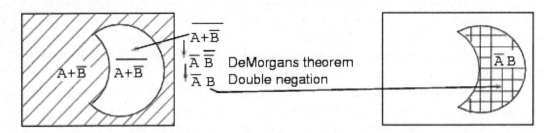

Example 3: Find (A+B')'

Solution: The green 45° **A+B'** hatched area was the result of the previous example. Moving on to a to,**(A+B')'** ,the present example, above left, let us find the complement of **A+B'**, which is the white clear area above left corresponding to **(A+B')'**. Note that we have repeated, at right, the **AB'** double hatched result from a previous example for comparison to our result. The regions corresponding to **(A+B')'** and **AB'** above left and right respectively are identical. This can be proven with DeMorgan's theorem and double negation.

This brings up a point. Venn diagrams don't actually prove anything. Boolean algebra is needed for formal proofs. However, Venn diagrams can be used for verification and visualization. We have verified and visualized DeMorgan's theorem with a Venn diagram.

Example 4: What does the Boolean expression **A'+B'** look like on a Venn Diagram?

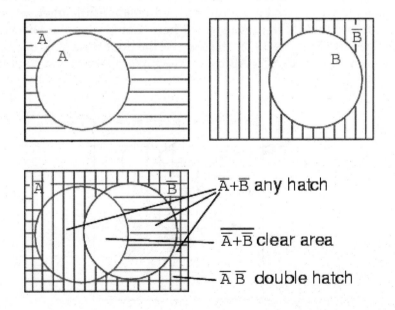

Solution: above figure

Start out with red horizontal hatched **A'** and blue vertical hatched **B'** above. Superimpose the diagrams as shown. We can still see the **A'** red horizontal hatch superimposed on the other hatch. It also fills in what used to be part of the **B** (B-true) circle, but only that part of the **B** open circle not common to the **A** open circle. If we only look at the **B'** blue vertical hatch, it fills that part of the open **A** circle not common to **B**. Any region with any hatch at all, regardless of type, corresponds to **A'+B'**. That is, everything but the open white space in the center.

Example 5: What does the Boolean expression **(A'+B')'** look like on a Venn Diagram?

Solution: Above figure, lower left Looking at the white open space in the center, it is everything **NOT** in the previous solution of **A'+B'**, which is **(A'+B')'**.

Example 6: Show that **(A'+B') = AB**

Solution: Below figure, lower left

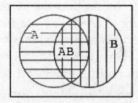 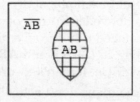 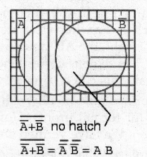

We previously showed on the above right diagram that the white open region is **(A'+B')'**. On an earlier example we showed a doubly hatched region at the intersection (overlay) of **AB**. This is the left and middle figures repeated here. Comparing the two Venn diagrams, we see that this open region , **(A'+B')'**, is the same as the doubly hatched region **AB** (A AND B). We can also prove that **(A'+B')'=AB** by DeMorgan's theorem and double negation as shown above.

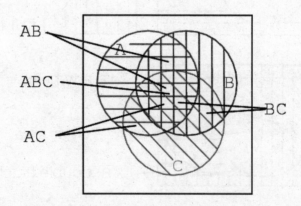

Three variable venn diagram

We show a three variable Venn diagram above with regions **A** (red horizontal), **B** (blue vertical), and, **C** (green 45°). In the very center note that all three regions overlap representing

Boolean expression **ABC**. There is also a larger petal shaped region where **A** and **B** overlap corresponding to Boolean expression **AB**. In a similar manner **A** and **C** overlap producing Boolean expression **AC**. And **B** and **C** overlap producing Boolean expression **BC**.

Looking at the size of regions described by AND expressions above, we see that region size varies with the number of variables in the associated AND expression.

- **A**, 1-variable is a large circular region.
- **AB**, 2-variable is a smaller petal shaped region.
- **ABC**, 3-variable is the smallest region.
- The more variables in the AND term, the smaller the region.

13.4 MAKING A VENN DIAGRAM LOOK LIKE A KARNAUGH MAP

Starting with circle **A** in a rectangular **A' universe** in figure (a) below, we morph a Venn diagram into almost a Karnaugh map.

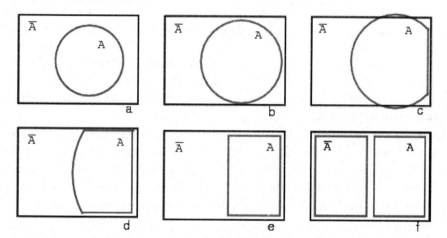

We expand circle **A** at (b) and (c), conform to the rectangular **A' universe** at (d), and change **A** to a rectangle at (e). Anything left outside of **A** is **A'**. We assign a rectangle to **A'** at (f). Also, we do not use shading in Karnaugh maps. What we have so far resembles a 1-variable Karnaugh map, but is of little utility. We need multiple variables.

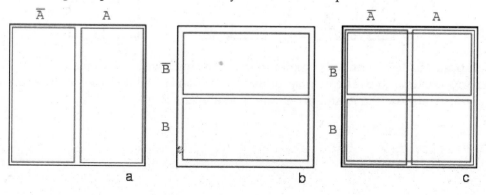

Figure (a) above is the same as the previous Venn diagram showing **A** and **A'** above except that the labels **A** and **A'** are above the diagram instead of inside the respective regions. Imagine that we have go through a process similar to figures (a-f) to get a "square Venn diagram" for **B** and **B'** as we show in middle figure (b). We will now superimpose the diagrams in Figures (a) and (b) to get the result at (c), just like we have been doing for Venn diagrams. The reason we do this is so that we may observe that which may be common to two overlapping regions— say where **A** overlaps **B**. The lower right cell in figure (c) corresponds to **AB** where **A** overlaps **B**.

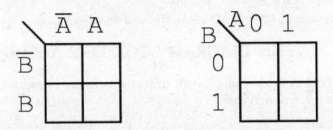

We don't waste time drawing a Karnaugh map like (c) above, sketching a simplified version as above left instead. The column of two cells under **A'** is understood to be associated with **A'**, and the heading **A** is associated with the column of cells under it. The row headed by **B'** is associated with the cells to the right of it. In a similar manner **B** is associated with the cells to the right of it. For the sake of simplicity, we do not delineate the various regions as clearly as with Venn diagrams.

The Karnaugh map above right is an alternate form used in most texts. The names of the variables are listed next to the diagonal line. The **A** above the diagonal indicates that the variable **A** (and **A'**) is assigned to the columns. The **0** is a substitute for **A'**, and the **1** substitutes for **A**. The **B** below the diagonal is associated with the rows: **0** for **B'**, and **1** for **B**

Example 7: Mark the cell corresponding to the Boolean expression **AB** in the Karnaugh map above with a **1**

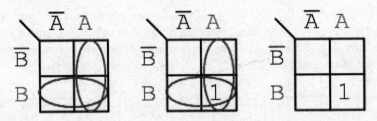

Solution: Shade or circle the region corresponding to **A**. Then, shade or enclose the region corresponding to **B**. The overlap of the two regions is **AB**. Place a **1** in this cell. We do not necessarily enclose the **A** and **B** regions as at above left.

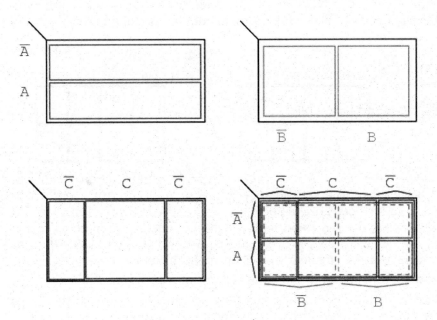

We develop a 3-variable Karnaugh map above, starting with Venn diagram like regions. The universe (inside the black rectangle) is split into two narrow narrow rectangular regions for **A'** and **A**. The variables **B'** and **B** divide the universe into two square regions. **C** occupies a square region in the middle of the rectangle, with **C'** split into two vertical rectangles on each side of the **C** square.

In the final figure, we superimpose all three variables, attempting to clearly label the various regions. The regions are less obvious without color printing, more obvious when compared to the other three figures. This 3-variable *K-Map* (Karnaugh map) has $2^3 = 8$ *cells*, the small squares within the map. Each individual cell is uniquely identified by the three Boolean Variables (**A, B, C**). For example, **ABC'** uniquely selects the lower right most cell(*), **A'B'C'** selects the upper left most cell (x).

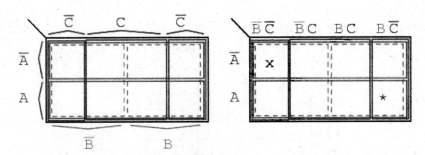

We don't normally label the Karnaugh map as shown above left. Though this figure clearly shows map coverage by single boolean variables of a 4-cell region. Karnaugh maps are labeled like the illustration at right. Each cell is still uniquely identified by a 3-variable *product term*, a Boolean **AND** expression. Take, for example, **ABC'** following the **A** row across to the right and the **BC'** column down, both intersecting at the lower right cell **ABC'**. See (*) above figure.

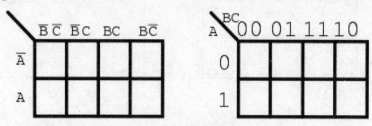

The above two different forms of a 3-variable Karnaugh map are equivalent, and is the final form that it takes. The version at right is a bit easier to use, since we do not have to write down so many boolean alphabetic headers and complement bars, just **1**s and **0**s Use the form of map on the right and look for the the one at left in some texts. The column headers on the left **B'C', B'C, BC, BC'** are equivalent to **00, 01, 11, 10** on the right. The row headers **A, A'** are equivalent to **0, 1** on the right map.

II.4 Karnaugh maps, truth tables, and Boolean expressions

Maurice Karnaugh, a telecommunications engineer, developed the Karnaugh map at Bell Labs in 1953 while designing digital logic based telephone switching circuits.

Now that we have developed the Karnaugh map with the aid of Venn diagrams, let's put it to use. Karnaugh maps *reduce* logic functions more quickly and easily compared to Boolean algebra. By reduce we mean simplify, reducing the number of gates and inputs. We like to simplify logic to a *lowest cost* form to save costs by elimination of components. We define lowest cost as being the lowest number of gates with the lowest number of inputs per gate.

Given a choice, most students do logic simplification with Karnaugh maps rather than Boolean algebra once they learn this tool.

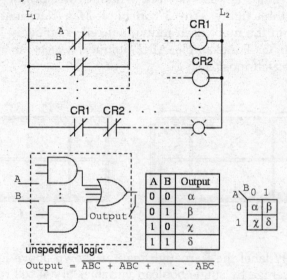

Karnaugh Map

We show five individual items above, which are just different ways of representing the same thing: an arbitrary 2-input digital logic function. First is relay ladder logic, then logic gates, a truth table, a Karnaugh map, and a Boolean equation. The point is that any of these are equivalent. Two inputs **A** and **B** can take on values of either **0** or **1**, high or low, open or closed, True or False, as the case may be. There are $2^2 = 4$ combinations of inputs producing an output. This is applicable to all five examples.

These four outputs may be observed on a lamp in the relay ladder logic, on a logic probe on the gate diagram. These outputs may be recorded in the truth table, or in the Karnaugh map. Look at the Karnaugh map as being a rearranged truth table. The Output of the Boolean equation may be computed by the laws of Boolean algebra and transferred to the truth table or Karnaugh map. Which of the five equivalent logic descriptions should we use? The one which is most useful for the task to be accomplished.

The outputs of a truth table correspond on a one-to-one basis to Karnaugh map entries. Starting at the top of the truth table, the A = 0, B = 0 inputs produce an output α. Note that this same output α is found in the Karnaugh map at the A = 0, B = 0 cell address, upper left corner of K-map where the A = 0 row and B = 0 column intersect. The other truth table outputs α, β, δ from inputs AB = 01, 10, 11 are found at corresponding K-map locations.

Below, we show the adjacent 2-cell regions in the 2-variable K-map with the aid of previous rectangular Venn diagram like Boolean regions.

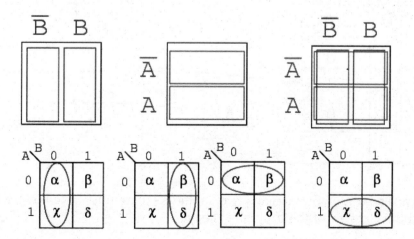

Cells α and χ are adjacent in the K-map as ellipses in the left most K-map below. Referring to the previous truth table, this is not the case. There is another truth table entry (β) between them. Which brings us to the whole point of the organizing the K-map into a square array, cells with any Boolean variables in common need to be close to one another so as to present a pattern that jumps out at us. For cells α and χ they have the Boolean variable **B'** in common. We know this because **B = 0** (same as **B'**) for the column above cells α and χ. Compare this to the square Venn diagram above the K-map.

A similar line of reasoning shows that β and δ have Boolean **B** (B = 1) in common. Then, α and β have Boolean **A'** (A = 0) in common. Finally, χ and δ have Boolean **A** (A = 1) in common. Compare the last two maps to the middle square Venn diagram.

To summarize, we are looking for commonality of Boolean variables among cells. The Karnaugh map is organized so that we may see that commonality. Let's try some examples.

A	B	Output
0	0	0
0	1	1
1	0	0
1	1	1

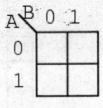

Example 8: Transfer the contents of the truth table to the Karnaugh map above.

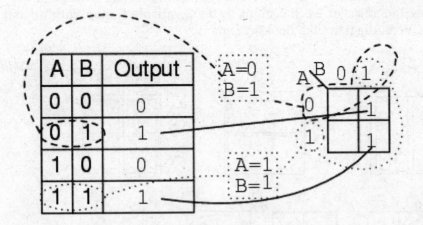

Solution: The truth table contains two **1**s. the K- map must have both of them. locate the first **1** in the 2nd row of the truth table above.

- note the truth table AB address
- locate the cell in the K-map having the same address
- place a **1** in that cell

Repeat the process for the **1** in the last line of the truth table.

Example 9: For the Karnaugh map in the above problem, write the Boolean expression. Solution is below.

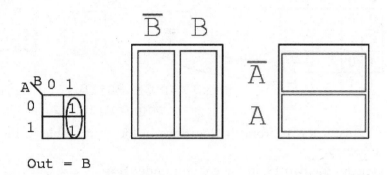

Out = B

Solution: Look for adjacent cells, that is, above or to the side of a cell. Diagonal cells are not adjacent. Adjacent cells will have one or more Boolean variables in common.

- Group (circle) the two **1**s in the column
- Find the variable(s) top and/or side which are the same for the group, write this as the Boolean result. It is **B** in our case.
- Ignore variable(s) which are not the same for a cell group. In our case A varies, is both 1 and 0, ignore Boolean A.
- Ignore any variable not associated with cells containing 1s. B' has no ones under it. Ignore B'
- Result **Out = B**

This might be easier to see by comparing to the Venn diagrams to the right, specifically the **B** column.

Example 10: Write the Boolean expression for the Karnaugh map below.

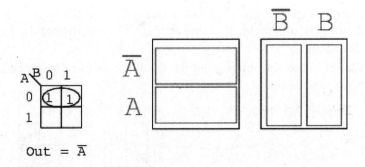

Out = $\overline{A}$

Solution: (above)
- Group (circle) the two **1's** in the row
- Find the variable(s) which are the same for the group, **Out = A'**

Example 11: For the Truth table below, transfer the outputs to the Karnaugh, then write the Boolean expression for the result.

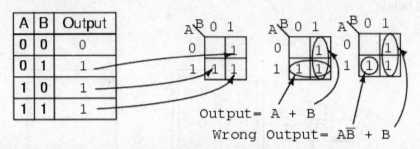

Solution: Transfer the 1s from the locations in the Truth table to the corresponding locations in the K-map.
- Group (circle) the two 1's in the column under **B=1**
- Group (circle) the two 1's in the row right of **A=1**
- Write product term for first group = **B**
- Write product term for second group = **A**
- Write Sum-Of-Products of above two terms **Output = A+B**

The solution of the K-map in the middle is the simplest or lowest cost solution. A less desirable solution is at far right. After grouping the two 1s, we make the mistake of forming a group of 1-cell. The reason that this is not desirable is that:
- The single cell has a product term of **AB'**
- The corresponding solution is **Output = AB' + B**
- This is not the simplest solution

The way to pick up this single **1** is to form a group of two with the **1** to the right of it as shown in the lower line of the middle K-map, even though this **1** has already been included in the column group (**B**). We are allowed to re-use cells in order to form larger groups. In fact, it is desirable because it leads to a simpler result.

We need to point out that either of the above solutions, Output or Wrong Output, are logically correct. Both circuits yield the same output. It is a matter of the former circuit being the lowest cost solution.

Example 12: Fill in the Karnaugh map for the Boolean expression below, then write the Boolean expression for the result.

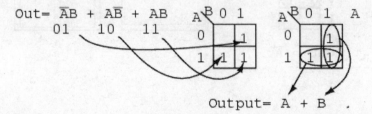

Solution: (above)

The Boolean expression has three product terms. There will be a **1** entered for each product term. Though, in general, the number of **1s** per product term varies with the number of variables in the product term compared to the size of the K-map. The product term is the address of the cell where the **1** is entered. The first product term, **A'B**, corresponds to the **01** cell in the map. A **1** is entered in this cell. The other two P-terms are entered for a total of three **1s**

Next, proceed with grouping and extracting the simplified result as in the previous truth table problem.

Example 13: Simplify the logic diagram below.

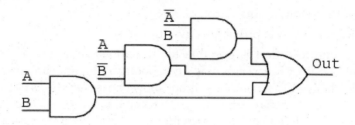

Solution: (Figure below)
- Write the Boolean expression for the original logic diagram as shown below
- Transfer the product terms to the Karnaugh map
- Form groups of cells as in previous examples
- Write Boolean expression for groups as in previous examples
- Draw simplified logic diagram

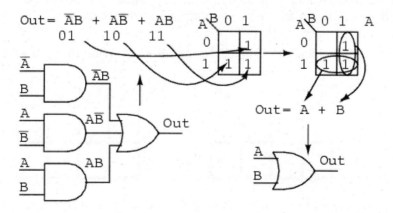

Example 14: Simplify the logic diagram below.

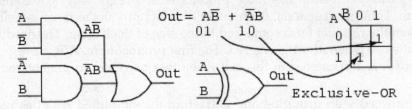

Solution: Write the Boolean expression for the original logic diagram shown above
- Transfer the product terms to the Karnaugh map.
- It is not possible to form groups.
- No simplification is possible; leave it as it is.

No logic simplification is possible for the above diagram. This sometimes happens. Neither the methods of Karnaugh maps nor Boolean algebra can simplify this logic further. We show an Exclusive-OR schematic symbol above; however, this is not a logical simplification. It just makes a schematic diagram look nicer. Since it is not possible to simplify the Exclusive-OR logic and it is widely used, it is provided by manufacturers as a basic integrated circuit (7486).

13.6 LOGIC SIMPLIFICATION WITH KARNAUGH MAPS

The logic simplification examples that we have done so could have been performed with Boolean algebra about as quickly. Real world logic simplification problems call for larger Karnaugh maps so that we may do serious work. We will work some contrived examples in this section, leaving most of the real world applications for the Combinatorial Logic chapter. By contrived, we mean examples which illustrate techniques. This approach will develop the tools we need to transition to the more complex applications in the Combinatorial Logic chapter.

We show our previously developed Karnaugh map. We will use the form on the right

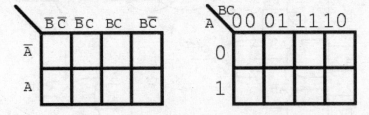

Note the sequence of numbers across the top of the map. It is not in binary sequence which would be **00, 01, 10, 11**. It is **00, 01, 11 10**, which is Gray code sequence. Gray code sequence only changes one binary bit as we go from one number to the next in the sequence, unlike binary. That means that adjacent cells will only vary by one bit, or Boolean variable. This is what we need to organize the outputs of a logic function so that we may view commonality. Moreover, the column and row headings must be in Gray code order, or the map will not work as a Karnaugh map. Cells sharing common Boolean variables would no longer be adjacent, nor show visual patterns. Adjacent cells vary by only one bit because a Gray code sequence varies by only one bit.

If we sketch our own Karnaugh maps, we need to generate Gray code for any size map that we may use. This is how we generate Gray code of any size.

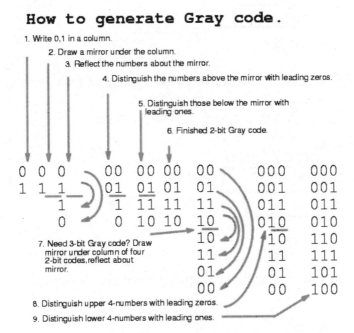

Note that the Gray code sequence, above right, only varies by one bit as we go down the list, or bottom to top up the list. This property of Gray code is often useful in digital electronics in general. In particular, it is applicable to Karnaugh maps.

Let us move on to some examples of simplification with 3-variable Karnaugh maps. We show how to map the product terms of the unsimplified logic to the K-map. We illustrate how to identify groups of adjacent cells which leads to a Sum-of-Products simplification of the digital logic.

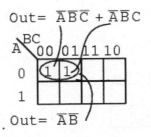

Above we, place the 1's in the K-map for each of the product terms, identify a group of two, then write a *p-term* (product term) for the sole group as our simplified result.

Out= $\overline{A}\overline{B}\overline{C} + \overline{A}\overline{B}C + \overline{A}BC + \overline{A}B\overline{C}$

```
  BC
A \ 00 01 11 10
0    1  1  1  1
1
```

Out= $\overline{A}$

Mapping the four product terms above yields a group of four covered by Boolean **A'**

Out= $\overline{A}\overline{B}C + \overline{A}BC + A\overline{B}C + ABC$

```
  BC
A \ 00 01 11 10
0       1  1
1       1  1
```

Out= C

Mapping the four p-terms yields a group of four, which is covered by one variable **C**.

Out= $\overline{A}\overline{B}\overline{C} + \overline{A}\overline{B}C + \overline{A}BC + \overline{A}B\overline{C} + ABC + AB\overline{C}$

```
  BC
A \ 00 01 11 10
0    1  1  1  1
1          1  1
```

Out= $\overline{A}$ + B

After mapping the six p-terms above, identify the upper group of four, pick up the lower two cells as a group of four by sharing the two with two more from the other group. Covering these two with a group of four gives a simpler result. Since there are two groups, there will be two p-terms in the Sum-of-Products result **A'+B**

Out= $\overline{A}BC + ABC$

```
  BC
A \ 00 01 11 10
0       1
1       1
```

Out= B C

The two product terms above form one group of two and simplifies to **BC**

Out = $\overline{A}BC + \overline{A}B\overline{C} + ABC + AB\overline{C}$

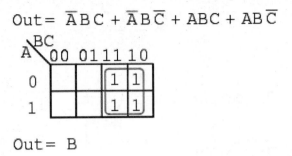

Out = B

Mapping the four p-terms yields a single group of four, which is **B**

Out = $\overline{A}\,\overline{B}\,\overline{C} + A\overline{B}\,\overline{C} + \overline{A}B\overline{C} + AB\overline{C}$

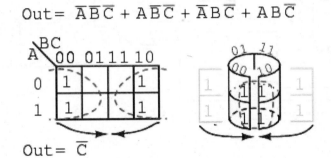

Out = $\overline{C}$

Mapping the four p-terms above yields a group of four. Visualize the group of four by rolling up the ends of the map to form a cylinder, then the cells are adjacent. We normally mark the group of four as above left. Out of the variables A, B, C, there is a common variable: C'. C' is a 0 over all four cells. Final result is **C'**.

Out = $\overline{A}\,\overline{B}\,\overline{C} + \overline{A}B\overline{C} + \overline{A}BC + \overline{A}\,B\,\overline{C} + A\overline{B}\,\overline{C} + AB\overline{C}$

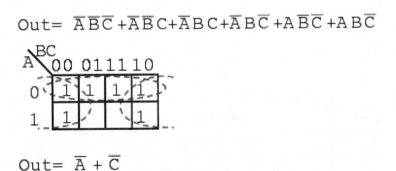

Out = $\overline{A} + \overline{C}$

The six cells above from the unsimplified equation can be organized into two groups of four. These two groups should give us two p-terms in our simplified result of **A' + C'**.

Below, we revisit the Toxic Waste Incinerator from the Boolean algebra chapter. See

Boolean algebra chapter for details on this example. We will simplify the logic using a Karnaugh map.

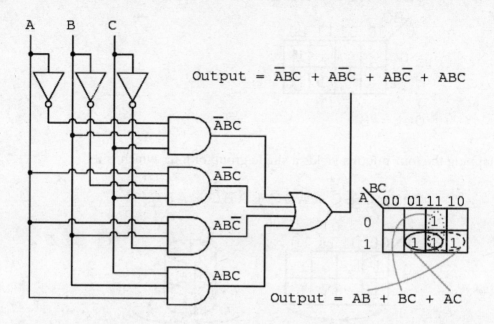

The Boolean equation for the output has four product terms. Map four 1's corresponding to the p-terms. Forming groups of cells, we have three groups of two. There will be three p-terms in the simplified result, one for each group. See "Toxic Waste Incinerator", Boolean algebra chapter for a gate diagram of the result, which is reproduced below.

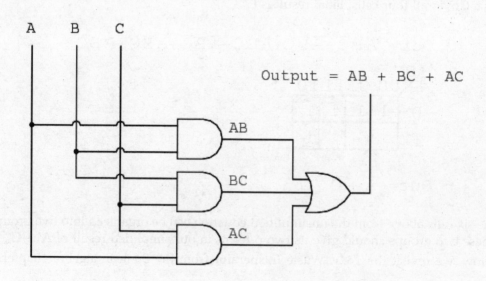

Below we repeat the Boolean algebra simplification of Toxic waste incinerator for comparison.

$$\overline{A}BC + A\overline{B}C + AB\overline{C} + ABC$$

↓ Factoring BC out of 1st and 4th terms

$$BC(\overline{A} + A) + A\overline{B}C + AB\overline{C}$$

↓ Applying identity $A + \overline{A} = 1$

$$BC(1) + A\overline{B}C + AB\overline{C}$$

↓ Applying identity $1A = A$

$$BC + A\overline{B}C + AB\overline{C}$$

↓ Factoring B out of 1st and 3rd terms

$$B(C + A\overline{C}) + A\overline{B}C$$

↓ Applying rule $A + \overline{A}B = A + B$ to the $C + A\overline{C}$ term

$$B(C + A) + A\overline{B}C$$

↓ Distributing terms

$$BC + AB + A\overline{B}C$$

↓ Factoring A out of 2nd and 3rd terms

$$BC + A(B + \overline{B}C)$$

↓ Applying rule $A + \overline{A}B = A + B$ to the $B + \overline{B}C$ term

$$BC + A(B + C)$$

↓ Distributing terms

$$BC + AB + AC$$

or   Simplified result

$$AB + BC + AC$$

Below we repeat the Toxic waste incinerator Karnaugh map solution for comparison to the above Boolean algebra simplification. This case illustrates why the Karnaugh map is widely used for logic simplification.

A\BC	00	01	11	10
0			1	
1		1	1	1

Output = AB + BC + AC

The Karnaugh map method looks easier than the previous page of Boolean algebra.

13.7 LARGER 4-VARIABLE KARNAUGH MAPS

Knowing how to generate Gray code should allow us to build larger maps. Actually, all we need to do is look at the left to right sequence across the top of the 3-variable map, and copy it down the left side of the 4-variable map. See below.

```
      CD
  AB\ 00 01 11 10
  00 |  |  |  |  |
  01 |  |  |  |  |
  11 |  |  |  |  |
  10 |  |  |  |  |
```

The following four variable Karnaugh maps illustrate reduction of Boolean expressions too tedious for Boolean algebra. Reductions could be done with Boolean algebra. However, the Karnaugh map is faster and easier, especially if there are many logic reductions to do.

Out = $\bar{A}\bar{B}CD + \bar{A}BCD + ABCD + A\bar{B}CD + AB\bar{C}\bar{D} + AB\bar{C}D + ABC\bar{D}$

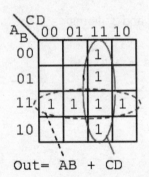

Out = AB + CD

The above Boolean expression has seven product terms. They are mapped top to bottom and left to right on the K-map above. For example, the first P-term **A'B'CD** is first row 3rd cell, corresponding to map location **A=0, B=0, C=1, D=1**. The other product terms are placed in a similar manner. Encircling the largest groups possible, two groups of four are shown above. The dashed horizontal group corresponds the the simplified product term **AB**. The vertical group corresponds to Boolean CD. Since there are two groups, there will be two product terms in the Sum-Of-Products result of **Out=AB+CD**.

Fold up the corners of the map below like it is a napkin to make the four cells physically adjacent.

Out= $\overline{A}\overline{B}\overline{C}\overline{D} + \overline{A}B\overline{C}\overline{D} + A\overline{B}\overline{C}\overline{D} + AB\overline{C}\overline{D}$

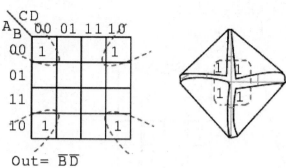

Out= $\overline{B}\overline{D}$

The four cells above are a group of four because they all have the Boolean variables **B'** and **D'** in common. In other words, B=0 for the four cells, and D=0 for the four cells. The other variables (A, B) are 0 in some cases, 1 in other cases with respect to the four corner cells. Thus, these variables (A, B) are not involved with this group of four. This single group comes out of the map as one product term for the simplified result: **Out=B'C'**

For the K-map below, roll the top and bottom edges into a cylinder forming eight adjacent cells.

Out= $\overline{A}\overline{B}\overline{C}\overline{D} + \overline{A}\overline{B}\overline{C}D + \overline{A}\overline{B}CD + \overline{A}\overline{B}C\overline{D}$
    $+ A\overline{B}\overline{C}\overline{D} + A\overline{B}\overline{C}D + A\overline{B}CD + A\overline{B}C\overline{D}$

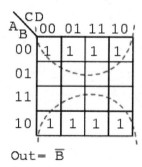

Out= $\overline{B}$

The above group of eight has one Boolean variable in common: **B=0**. Therefore, the one group of eight is covered by one p-term: **B'**. The original eight term Boolean expression simplifies to **Out=B'**

The Boolean expression below has nine p-terms, three of which have three Booleans instead of four. The difference is that while four Boolean variable product terms cover one cell, the three Boolean p-terms cover a pair of cells each.

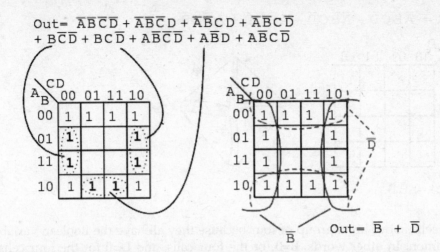

The six product terms of four Boolean variables map in the usual manner above as single cells. The three Boolean variable terms (three each) map as cell pairs, which is shown above. Note that we are mapping p-terms into the K-map, not pulling them out at this point.

For the simplification, we form two groups of eight. Cells in the corners are shared with both groups. This is fine. In fact, this leads to a better solution than forming a group of eight and a group of four without sharing any cells. Final Solution is **Out=B'+D'**

Below we map the unsimplified Boolean expression to the Karnaugh map.

$$\text{Out} = \overline{A}\,\overline{B}\,\overline{C}\,\overline{D} + \overline{A}\,\overline{B}\,C\,\overline{D} + A\,\overline{B}\,\overline{C}\,\overline{D} + ABCD$$

$$\text{Out} = \overline{B}\,\overline{C}\,\overline{D} + \overline{A}\,\overline{B}\,\overline{D} + ABCD$$

Above, three of the cells form into a groups of two cells. A fourth cell cannot be combined with anything, which often happens in "real world" problems. In this case, the Boolean p-term **ABCD** is unchanged in the simplification process. Result: **Out= B'C'D'+A'B'D'+ ABCD**

Often times there is more than one minimum cost solution to a simplification problem. Such is the case illustrated below.

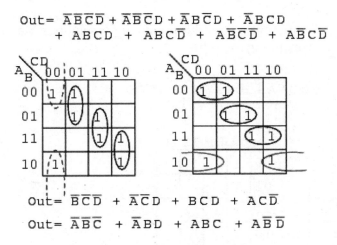

Both results above have four product terms of three Boolean variable each. Both are equally valid *minimal cost* solutions. The difference in the final solution is due to how the cells are grouped as shown above. A minimal cost solution is a valid logic design with the minimum number of gates with the minimum number of inputs.

Below we map the unsimplified Boolean equation as usual and form a group of four as a first simplification step. It may not be obvious how to pick up the remaining cells.

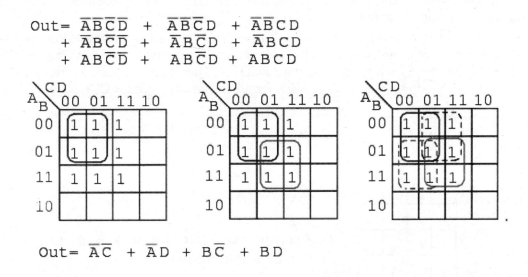

Pick up three more cells in a group of four, center above. There are still two cells remaining. the minimal cost method to pick up those is to group them with neighbouring cells as groups of four as at above right.

On a cautionary note, do not attempt to form groups of three. Groupings must be powers of 2, that is, 1, 2, 4, 8 ...

Below we have another example of two possible minimal cost solutions. Start by forming a couple of groups of four after mapping the cells.

$$\text{Out} = \overline{A}\overline{B}\overline{C}\overline{D} + \overline{A}\overline{B}CD + \overline{A}B\overline{C}\overline{D} + \overline{A}BCD + AB\overline{C}\overline{D}$$
$$+ AB\overline{C}D + ABCD + A\overline{B}\overline{C}\overline{D} + A\overline{B}CD$$

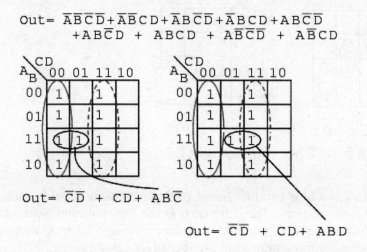

The two solutions depend on whether the single remaining cell is grouped with the first or the second group of four as a group of two cells. That cell either comes out as either **ABC'** or **ABD**, your choice. Either way, this cell is covered by either Boolean product term. Final results are shown above.

Below we have an example of a simplification using the Karnaugh map at left or Boolean algebra at right. Plot **C'** on the map as the area of all cells covered by address **C=0**, the 8-cells on the left of the map. Then, plot the single **ABCD** cell. That single cell forms a group of 2-cell as shown, which simplifies to P-term **ABD**, for an end result of **Out = C' + ABD**.

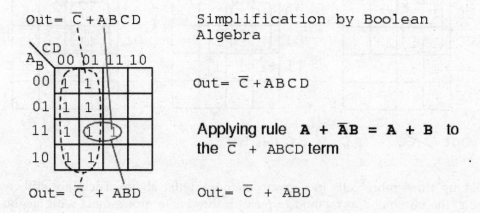

This (above) is a rare example of a four variables problem that can be reduced with Boolean algebra without a lot of work, assuming that you remember the theorems.

13.8 MINTERM VS MAXTERM SOLUTION

So far we have been finding Sum-Of-Product (SOP) solutions to logic reduction problems. For each of these SOP solutions, there is also a Product-Of-Sums solution (POS), which could be more useful, depending on the application. Before working a Product-Of-Sums solution, we need to introduce some new terminology. The procedure below for mapping product terms is not new to this chapter. We just want to establish a formal procedure for minterms for comparison to the new procedure for maxterms.

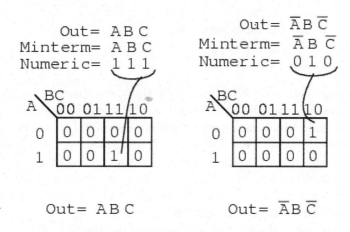

A *minterm* is a Boolean expression resulting in **1** for the output of a single cell, and **0**s for all other cells in a Karnaugh map, or truth table. If a minterm has a single **1** and the remaining cells as **0**s, it would appear to cover a minimum area of **1**s. The illustration above left shows the minterm **ABC**, a single product term, as a single **1** in a map that is otherwise **0**s. We have not shown the **0**s in our Karnaugh maps up to this point, as it is customary to omit them unless specifically needed. Another minterm **A'BC'** is shown above right. The point to review is that the address of the cell corresponds directly to the minterm being mapped. That is, the cell **111** corresponds to the minterm **ABC** above left. Above right we see that the minterm **A'BC'** corresponds directly to the cell **010**. A Boolean expression or map may have multiple minterms.

Referring to the above figure, Let's summarize the procedure for placing a minterm in a K-map:

- Identify the minterm (product term) term to be mapped.
- Write the corresponding binary numeric value.
- Use binary value as an address to place a **1** in the K-map
- Repeat steps for other minterms (P-terms within a Sum-Of-Products).

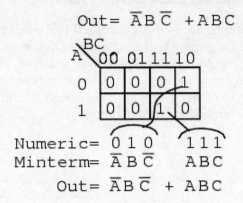

A Boolean expression will more often than not consist of multiple minterms corresponding to multiple cells in a Karnaugh map as shown above. The multiple minterms in this map are the individual minterms which we examined in the previous figure above. The point we review for reference is that the 1s come out of the K-map as a binary cell address which converts directly to one or more product terms. By directly we mean that a **0** corresponds to a complemented variable, and a **1** corresponds to a true variable. Example: **010** converts directly to **A'BC'**. There was no reduction in this example. Though, we do have a Sum-Of-Products result from the minterms.

Referring to the above figure, Let's summarize the procedure for writing the Sum-Of-Products reduced Boolean equation from a K-map:
- Form largest groups of **1**s possible covering all minterms. Groups must be a power of 2.
- Write binary numeric value for groups.
- Convert binary value to a product term.
- Repeat steps for other groups. Each group yields a p-terms within a Sum-Of-Products.

Nothing new so far, a formal procedure has been written down for dealing with minterms. This serves as a pattern for dealing with maxterms.

Next we attack the Boolean function which is **0** for a single cell and **1**s for all others.

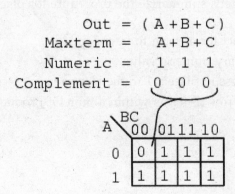

A *maxterm* is a Boolean expression resulting in a **0** for the output of a single cell expression, and **1**s for all other cells in the Karnaugh map, or truth table. The illustration above left shows the maxterm **(A+B+C)**, a single sum term, as a single **0** in a map that is otherwise **1**s. If a maxterm has a single **0** and the remaining cells as **1**s, it would appear to cover a maximum area of **1**s.

There are some differences now that we are dealing with something new, maxterms. The maxterm is a **0**, not a **1** in the Karnaugh map. A maxterm is a sum term, **(A+B+C)** in our example, not a product term.

It also looks strange that **(A+B+C)** is mapped into the cell **000**. For the equation **Out=(A+B+C)=0**, all three variables **(A, B, C)** must individually be equal to **0**. Only **(0+0+0)=0** will equal **0**. Thus we place our sole **0** for minterm **(A+B+C)** in cell **A,B,C=000** in the K-map, where the inputs are all **0**. This is the only case which will give us a **0** for our maxterm. All other cells contain **1**s because any input values other than **((0,0,0)** for **(A+B+C)** yields **1**s upon evaluation.

Referring to the above figure, the procedure for placing a maxterm in the K-map is:
- Identify the Sum term to be mapped.
- Write corresponding binary numeric value.
- Form the complement
- Use the complement as an address to place a **0** in the K-map
- Repeat for other maxterms (Sum terms within Product-of-Sums expression).

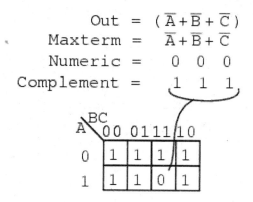

Another maxterm **A'+B'+C'** is shown above. Numeric **000** corresponds to **A'+B'+C'**. The complement is **111**. Place a **0** for maxterm **(A'+B'+C')** in this cell **(1,1,1)** of the K-map as shown above.

Why should **(A'+B'+C')** cause a **0** to be in cell **111**? When **A'+B'+C'** is **(1'+1'+1')**, all **1**s in, which is **(0+0+0)** after taking complements, we have the only condition that will give us a **0**. All the **1**s are complemented to all **0**s, which is **0** when **OR**ed.

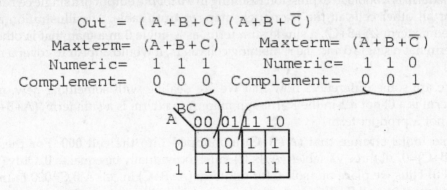

A Boolean Product-Of-Sums expression or map may have multiple maxterms as shown above. Maxterm **(A+B+C)** yields numeric **111** which complements to **000**, placing a **0** in cell (0,0,0). Maxterm **(A+B+C')** yields numeric **110** which complements to **001**, placing a **0** in cell (0,0,1).

Now that we have the k-map set-up, what we are really interested in is showing how to write a Product-Of-Sums reduction. Form the **0**s into groups. That would be a group of two below. Write the binary value corresponding to the sum-term which is **(0,0,X)**. Both A and B are **0** for the group. But, C is both **0** and **1** so we write an **X** as a place holder for C. Form the complement **(1,1,X)**. Write the Sum-term **(A+B)** discarding the C and the X which held its' place. In general, expect to have more sum-terms multiplied together in the Product-Of-Sums result. Though, we have a simple example here.

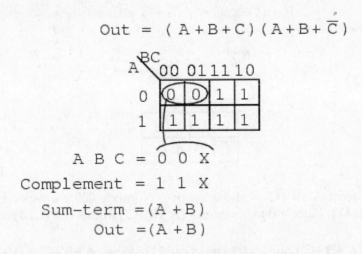

Let's summarize the procedure for writing the Product-Of-Sums Boolean reduction for a K-map:

· Form largest groups of **0**s possible, covering all maxterms. Groups must be a power of 2.

- Write binary numeric value for group.
- Complement binary numeric value for group.
- Convert complement value to a sum-term.
- Repeat steps for other groups. Each group yields a sum-term within a Product-Of-Sums result.

Example 15: Simplify the Product-Of-Sums Boolean expression below, providing a result in POS form.

$$Out = (A+B+C+\overline{D})(A+B+\overline{C}+D)(A+\overline{B}+C+\overline{D})(A+\overline{B}+\overline{C}+D)$$
$$(\overline{A}+\overline{B}+\overline{C}+D)(\overline{A}+B+C+\overline{D})(\overline{A}+B+\overline{C}+D)$$

Solution: Transfer the seven maxterms to the map below as **0s**. Be sure to complement the input variables in finding the proper cell location.

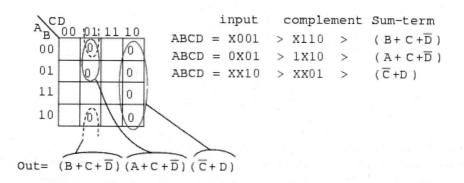

We map the **0s** as they appear left to right top to bottom on the map above. We locate the last three maxterms with leader lines..

Once the cells are in place above, form groups of cells as shown below. Larger groups will give a sum-term with fewer inputs. Fewer groups will yield fewer sum-terms in the result.

	input	complement	Sum-term
ABCD	= X001	> X110 >	$(B+C+\overline{D})$
ABCD	= 0X01	> 1X10 >	$(A+C+\overline{D})$
ABCD	= XX10	> XX01 >	$(\overline{C}+D)$

$$Out = (B+C+\overline{D})(A+C+\overline{D})(\overline{C}+D)$$

We have three groups, so we expect to have three sum-terms in our POS result above. The group of 4-cells yields a 2-variable sum-term. The two groups of 2-cells give us two 3-variable sum-terms. Details are shown for how we arrived at the Sum-terms above. For a group, write the binary group input address, then complement it, converting that to the Boolean sum-term. The final result is product of the three sums.

Example 16: Simplify the Product-Of-Sums Boolean expression below, providing a result in SOP form.

$$\text{Out} = (A+B+C+\overline{D})(A+B+\overline{C}+D)(A+\overline{B}+C+\overline{D})(A+\overline{B}+\overline{C}+D)$$
$$(\overline{A}+\overline{B}+\overline{C}+D)(\overline{A}+B+C+\overline{D})(\overline{A}+B+\overline{C}+D)$$

Solution: This looks like a repeat of the last problem. It is except that we ask for a Sum-Of-Products Solution instead of the Product-Of-Sums which we just finished. Map the maxterm 0s from the Product-Of-Sums given as in the previous problem, below left.

$$\text{Out} = (A+B+C+\overline{D})(A+B+\overline{C}+D)(A+\overline{B}+C+\overline{D})(A+\overline{B}+\overline{C}+D)$$
$$(\overline{A}+\overline{B}+\overline{C}+D)(\overline{A}+B+C+\overline{D})(\overline{A}+B+\overline{C}+D)$$

AB\CD	00	01	11	10
00		0		0
01		0		0
11			0	
10		0		0

AB\CD	00	01	11	10
00	1	0	1	0
01	1	0	1	0
11	1	1	1	0
10	1	0	1	0

Then fill in the implied 1s in the remaining cells of the map above right.

AB\CD	00	01	11	10
00	1	0	1	0
01	1	0	1	0
11	1	1	1	0
10	1	0	1	0

$$\text{Out} = \overline{C}\overline{D} + CD + ABD$$

Form groups of 1s to cover all 1s. Then write the Sum-Of-Products simplified result as in the previous section of this chapter. This is identical to a previous problem.

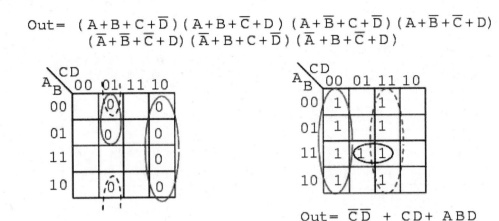

Out= $(A+B+C+\bar{D})(A+B+\bar{C}+D)(A+\bar{B}+C+\bar{D})(A+\bar{B}+\bar{C}+D)$
$(\bar{A}+\bar{B}+\bar{C}+D)(\bar{A}+B+C+\bar{D})(\bar{A}+B+\bar{C}+D)$

Out= $(B+C+\bar{D})(A+C+\bar{D})(\bar{C}+D)$

Out= $\bar{C}\bar{D} + CD + ABD$

Above we show both the Product-Of-Sums solution, from the previous example, and the Sum-Of-Products solution from the current problem for comparison. Which is the simpler solution? The POS uses 3-OR gates and 1-AND gate, while the SOP uses 3-AND gates and 1-OR gate. Both use four gates each. Taking a closer look, we count the number of gate inputs. The POS uses 8-inputs; the SOP uses 7-inputs. By the definition of minimal cost solution, the SOP solution is simpler. This is an example of a technically correct answer that is of little use in the real world.

The better solution depends on complexity and the logic family being used. The SOP solution is usually better if using the TTL logic family, as NAND gates are the basic building block, which works well with SOP implementations. On the other hand, A POS solution would be acceptable when using the CMOS logic family since all sizes of NOR gates are available.

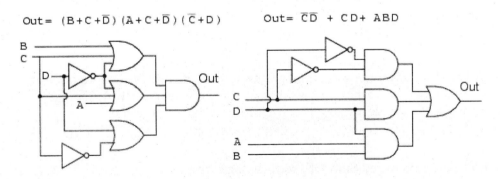

The gate diagrams for both cases are shown above, Product-Of-Sums left, and Sum-Of-Products right.

Below, we take a closer look at the Sum-Of-Products version of our example logic, which is repeated at left.

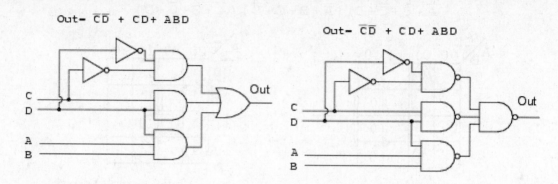

Above all AND gates at left have been replaced by NAND gates at right. The OR gate at the output is replaced by a NAND gate. To prove that AND-OR logic is equivalent to NAND-NAND logic, move the inverter invert bubbles at the output of the 3-NAND gates to the input of the final NAND as shown in going from above right to below left.

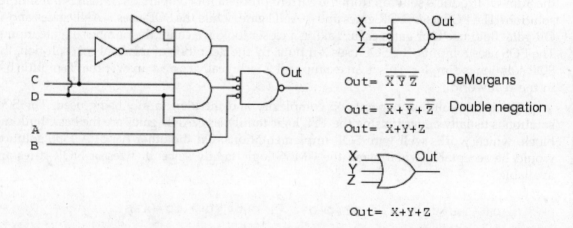

Above right we see that the output NAND gate with inverted inputs is logically equivalent to an OR gate by DeMorgan's theorem and double negation. This information is useful in building digital logic in a laboratory setting where TTL logic family NAND gates are more readily available in a wide variety of configurations than other types.

The Procedure for constructing NAND-NAND logic, in place of AND-OR logic is as follows:

- Produce a reduced Sum-Of-Products logic design.
- When drawing the wiring diagram of the SOP, replace all gates (both AND and OR) with NAND gates.
- Unused inputs should be tied to logic High.

- In case of troubleshooting, internal nodes at the first level of NAND gate outputs do NOT match AND-OR diagram logic levels, but are inverted. Use the NAND-NAND logic diagram. Inputs and final output are identical, though.
- Label any multiple packages U1, U2,.. etc.
- Use data sheet to assign pin numbers to inputs and outputs of all gates.

Example 17: Let us revisit a previous problem involving an SOP minimization. Produce a Product-Of-Sums solution. Compare the POS solution to the previous SOP.

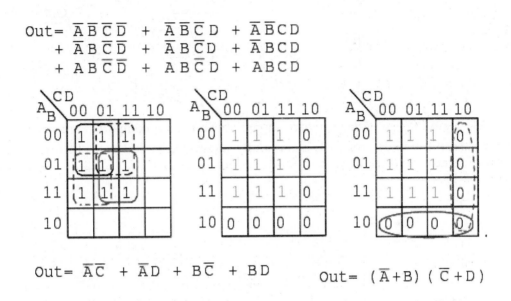

Solution: Above left we have the original problem starting with a 9-minterm Boolean unsimplified expression. Reviewing, we formed four groups of 4-cells to yield a 4-product-term SOP result, lower left.

In the middle figure, above, we fill in the empty spaces with the implied 0s. The 0s form two groups of 4-cells. The solid red group is **(A'+B)**, the dashed red group is **(C'+D)**. This yields two sum-terms in the Product-Of-Sums result, above right **Out = (A'+B)(C'+D)**

Comparing the previous SOP simplification, left, to the POS simplification, right, shows that the POS is the least cost solution. The SOP uses 5-gates total, the POS uses only 3-gates. This POS solution even looks attractive when using TTL logic due to simplicity of the result. We can find AND gates and an OR gate with 2-inputs.

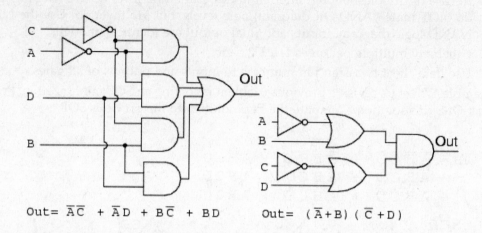

The SOP and POS gate diagrams are shown above for our comparison problem.

Given the pin-outs for the TTL logic family integrated circuit gates below, label the maxterm diagram above right with Circuit designators (U1-a, U1-b, U2-a, etc), and pin numbers.

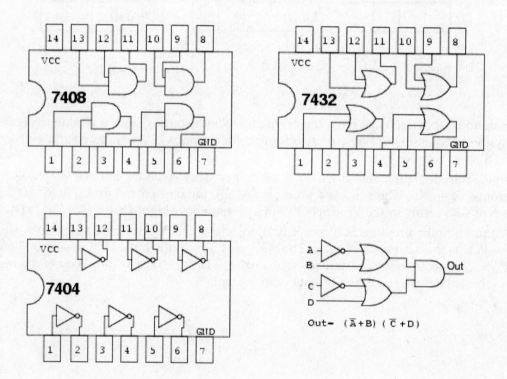

Each integrated circuit package that we use will receive a circuit designator: U1, U2, U3. To distinguish between the individual gates within the package, they are identified as a, b, c, d, etc. The 7404 hex-inverter package is U1. The individual inverters in it are are U1-a, U1-b, U1-c, etc. U2 is assigned to the 7432 quad OR gate. U3 is assigned to the 7408 quad AND gate. With reference to the pin numbers on the package diagram above, we assign pin numbers to all gate inputs and outputs on the schematic diagram below.

We can now build this circuit in a laboratory setting. Or, we could design a *printed circuit board* for it. A printed circuit board contains copper foil "wiring" backed by a non conductive substrate of phenolic, or epoxy-fiberglass. Printed circuit boards are used to mass produce electronic circuits. Ground the inputs of unused gates.

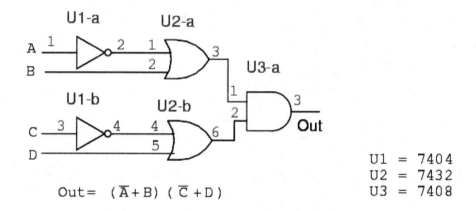

Label the previous POS solution diagram above left (third figure back) with Circuit designators and pin numbers. This will be similar to what we just did.

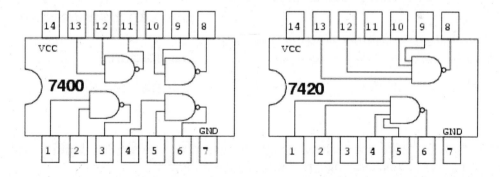

We can find 2-input AND gates, 7408 in the previous example. However, we have trouble finding a 4-input OR gate in our TTL catalog. The only kind of gate with 4-inputs is the 7420 NAND gate shown above right.

We can make the 4-input NAND gate into a 4-input OR gate by inverting the inputs to the NAND gate as shown below. So we will use the 7420 4-input NAND gate as an OR gate by inverting the inputs.

$\overline{Y} = \overline{A}\,\overline{B} = \overline{A+B}$ DeMorgan's
$Y = A+B$ Double negation

We will not use discrete inverters to invert the inputs to the 7420 4-input NAND gate, but will drive it with 2-input NAND gates in place of the AND gates called for in the SOP, minterm, solution. The inversion at the output of the 2-input NAND gates supply the inversion for the 4-input OR gate.

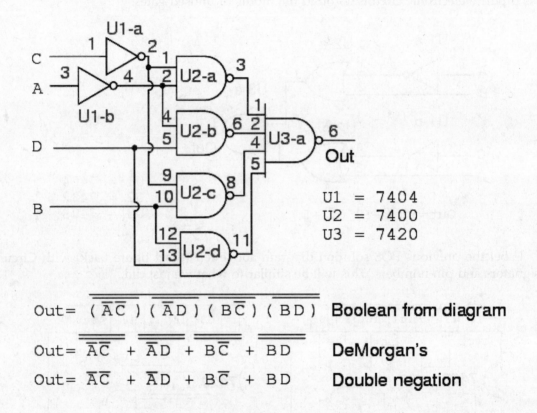

$Out = \overline{(\overline{AC})(\overline{AD})(\overline{BC})(\overline{BD})}$ **Boolean from diagram**

$Out = \overline{\overline{AC}} + \overline{\overline{AD}} + \overline{\overline{BC}} + \overline{\overline{BD}}$ **DeMorgan's**

$Out = \overline{A}C + \overline{A}D + B\overline{C} + BD$ **Double negation**

- The result is shown above. It is the only practical way to actually build it with TTL gates by using NAND-NAND logic replacing AND-OR logic.

13.9 SUM AND PRODUCT NOTATION

For reference, this section introduces the terminology used in some texts to describe the minterms and maxterms assigned to a Karnaugh map. Otherwise, there is no new material here.

Σ (sigma) indicates sum and lower case "m" indicates minterms. Σm indicates sum of minterms. The following example is revisited to illustrate our point. Instead of a Boolean equation description of unsimplified logic, we list the minterms.

f(A,B,C,D) = Σm(1, 2, 3, 4, 5, 7, 8, 9, 11, 12, 13, 15)

or

f(A,B,C,D) = Σ($m_1, m_2, m_3, m_4, m_5, m_7, m_8, m_9, m_{11}, m_{12}, m_{13}, m_{15}$)

The numbers indicate cell location, or address, within a Karnaugh map as shown below right. This is certainly a compact means of describing a list of minterms or cells in a K-map.

$$\text{Out} = \bar{A}\bar{B}\bar{C}\bar{D} + \bar{A}\bar{B}\bar{C}D + \bar{A}\bar{B}CD$$
$$+ \bar{A}B\bar{C}\bar{D} + \bar{A}B\bar{C}D + \bar{A}BCD$$
$$+ AB\bar{C}\bar{D} + AB\bar{C}D + ABCD$$

$$f(A,B,C,D) = \Sigma m(0,1,3,4,5,7,12,13,15)$$

A\B\CD	00	01	11	10
00	0	1	3	2
01	4	5	7	6
11	12	13	15	14
10	8	9	11	10

A\B\CD	00	01	11	10
00	1	1	1	0
01	1	1	1	0
11	1	1	1	0
10	0	0	0	0

A\B\CD	00	01	11	10
00	1	1	1	0
01	1	1	1	0
11	1	1	1	0
10	0	0	0	0

$$f(A,B,C,D) = \bar{A}\bar{C} + \bar{A}D + B\bar{C} + BD$$

The Sum-Of-Products solution is not affected by the new terminology. The minterms, 1s, in the map have been grouped as usual and a Sum-OF-Products solution written.

Below, we show the terminology for describing a list of maxterms. Product is indicated by the Greek (pi), and upper case "M" indicates maxterms. M indicates product of maxterms. The same example illustrates our point. The Boolean equation description of unsimplified logic, is replaced by a list of maxterms.

f(A,B,C,D) = ΠM(2, 6, 8, 9, 10, 11, 14)

or

f(A,B,C,D) = Π($M_2, M_6, M_8, M_9, M_{10}, M_{11}, M_{14}$)

Once again, the numbers indicate K-map cell address locations. For maxterms this is the location of 0s, as shown below. A Product-OF-Sums solution is completed in the usual manner.

412   Discrete Mathematics

$$\text{Out} = \overline{(\overline{A}+\overline{B}+\overline{C}+D)}\,(\overline{A}+\overline{B}+\overline{C}+D)\,(\overline{A}+\overline{B}+\overline{C}+D)\,(\overline{A}+B+C+D)$$
$$(\overline{A}+B+\overline{C}+D)\,(\overline{A}+B+\overline{C}+\overline{D})\,(\overline{A}+B+\overline{C}+D)$$

$$f(A,B,C,D) = \prod M(2,6,8,9,10,11,14)$$

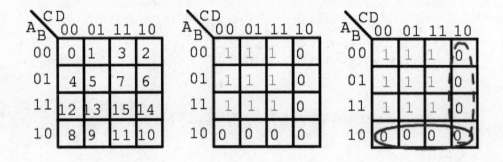

$$f(A,B,C,D) = \overline{(\overline{A}+B)}\,(\overline{C}+D)$$

13.10 DON'T CARE CELLS IN THE KARNAUGH MAP

Up to this point we have considered logic reduction problems where the input conditions were completely specified. That is, a 3-variable truth table or Karnaugh map had $2^n = 2^3$ or 8-entries, a full table or map. It is not always necessary to fill in the complete truth table for some real-world problems. We may have a choice to not fill in the complete table.

For example, when dealing with BCD (Binary Coded Decimal) numbers encoded as four bits, we may not care about any codes above the BCD range of (0, 1, 2...9). The 4-bit binary codes for the hexadecimal numbers (Ah, Bh, Ch, Eh, Fh) are not valid BCD codes. Thus, we do not have to fill in those codes at the end of a truth table, or K-map, if we do not care to. We would not normally care to fill in those codes because those codes (1010, 1011, 1100, 1101, 1110, 1111) will never exist as long as we are dealing only with BCD encoded numbers. These six invalid codes are *don't cares* as far as we are concerned. That is, we do not care what output our logic circuit produces for these don't cares.

Don't cares in a Karnaugh map, or truth table, may be either 1s or 0s, as long as we don't care what the output is for an input condition we never expect to see. We plot these cells with an asterisk, *, among the normal 1s and 0s. When forming groups of cells, treat the don't care cell as either a 1 or a 0, or ignore the don't cares. This is helpful if it allows us to form a larger group than would otherwise be possible without the don't cares. There is no requirement to group all or any of the don't cares. Only use them in a group if it simplifies the logic.

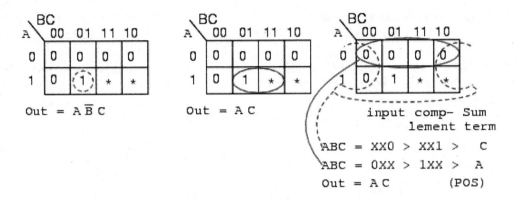

Above is an example of a logic function where the desired output is **1** for input **ABC = 101** over the range from **000 to 101**. We do not care what the output is for the other possible inputs (**110, 111**). Map those two as don't cares. We show two solutions. The solution on the right Out = AB'C is the more complex solution since we did not use the don't care cells. The solution in the middle, Out=AC, is less complex because we grouped a don't care cell with the single 1 to form a group of two. The third solution, a Product-Of-Sums on the right, results from grouping a don't care with three zeros forming a group of four 0s. This is the same, less complex, **Out=AC**. We have illustrated that the don't care cells may be used as either **1**s or **0**s, whichever is useful.

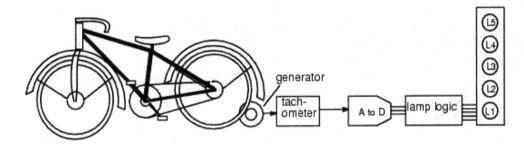

The electronics class of Lightning State College has been asked to build the lamp logic for a stationary bicycle exhibit at the local science museum. As a rider increases his pedaling speed, lamps will light on a bar graph display. No lamps will light for no motion. As speed increases, the lower lamp, L1 lights, then L1 and L2, then, L1, L2, and L3, until all lamps light at the highest speed. Once all the lamps illuminate, no further increase in speed will have any effect on the display.

A small DC generator coupled to the bicycle tire outputs a voltage proportional to speed. It drives a tachometer board which limits the voltage at the high end of speed where all lamps light. No further increase in speed can increase the voltage beyond this level. This is crucial because the downstream A to D (Analog to Digital) converter puts out a 3-bit code, **ABC**, 2^3 or 8-codes, but we only have five lamps. **A** is the most significant bit, **C** the least significant bit.

The lamp logic needs to respond to the six codes out of the A to D. For **ABC=000**, no motion, no lamps light. For the five codes **(001 to 101)** lamps L1, L1&L2, L1&L2&L3, up to all lamps will light, as speed, voltage, and the A to D code (ABC) increases. We do not care about the response to input codes **(110, 111)** because these codes will never come out of the A to D due to the limiting in the tachometer block. We need to design five logic circuits to drive the five lamps.

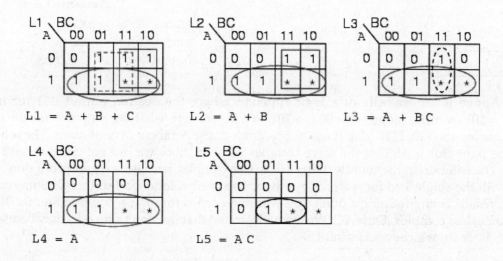

Since, none of the lamps light for **ABC=000** out of the A to D, enter a **0** in all K-maps for cell **ABC=000**. Since we don't care about the never to be encountered codes **(110, 111)**, enter asterisks into those two cells in all five K-maps.

Lamp L5 will only light for code **ABC=101**. Enter a **1** in that cell and five **0**s into the remaining empty cells of L5 K-map.

L4 will light initially for code **ABC=100**, and will remain illuminated for any code greater, **ABC=101**, because all lamps below L5 will light when L5 lights. Enter 1s into cells **100** and **101** of the L4 map so that it will light for those codes. Four **0**'s fill the remaining L4 cells

L3 will initially light for code **ABC=011**. It will also light whenever L5 and L4 illuminate. Enter three 1s into cells **011, 100, 101** for L3 map. Fill three 0s into the remaining L3 cells.

L2 lights for **ABC=010** and codes greater. Fill 1s into cells **010, 011, 100, 101**, and two 0s in the remaining cells.

The only time L1 is not lighted is for no motion. There is already a **0** in cell **ABC=000**. All the other five cells receive **1**s.

Group the **1**'s as shown above, using don't cares whenever a larger group results. The L1 map shows three product terms, corresponding to three groups of 4-cells. We used both don't cares in two of the groups and one don't care on the third group. The don't cares allowed us to form groups of four.

In a similar manner, the L2 and L4 maps both produce groups of 4-cells with the aid of the don't care cells. The L4 reduction is striking in that the L4 lamp is controlled by the most significant bit from the A to D converter, **L5=A**. No logic gates are required for lamp L4. In the L3 and L5 maps, single cells form groups of two with don't care cells. In all five maps, the reduced Boolean equation is less complex than without the don't cares.

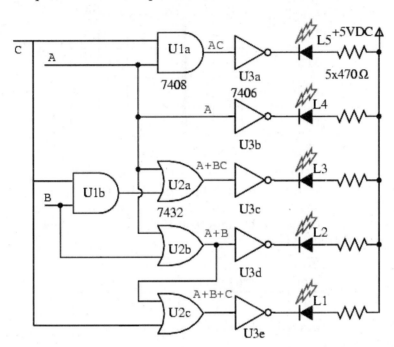

The gate diagram for the circuit is above. The outputs of the five K-map equations drive inverters. Note that the L1 **OR** gate is not a 3-input gate but a 2-input gate having inputs **(A+B), C**, outputting **A+B+C** The *open collector* inverters, **7406**, are desirable for driving LEDs, though, not part of the K-map logic design. The output of an open collecter gate or inverter is open circuited at the collector internal to the integrated circuit package so that all collector current may flow through an external load. An active high into any of the inverters pulls the output low, drawing current through the LED and the current limiting resistor. The LEDs would likely be part of a solid state relay driving 120VAC lamps for a museum exhibit, not shown here.

13.11 LARGER 5 AND 6-VARIABLES KARNAUGH MAPS

Larger Karnaugh maps reduce larger logic designs. How large is large enough? That depends on the number of inputs, *fan-ins*, to the logic circuit under consideration. One of the large programmable logic companies has an answer.

Altera's own data, extracted from its library of customer designs, supports the value of heterogeneity. By examining logic cones, mapping them onto LUT-based nodes and sorting them by the number of inputs that would be best at each node, Altera found that the distribution of fan-ins was nearly flat between two and six inputs, with a nice peak at five.

The answer is no more than six inputs for most all designs, and five inputs for the average logic design. The five variable Karnaugh map follows.

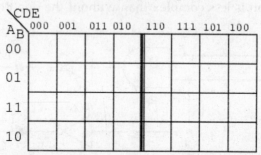

5-variable Karnaugh map (Gray code)

The older version of the five variable K-map, a Gray Code map or reflection map, is shown above. The top (and side for a 6-variable map) of the map is numbered in full Gray code. The Gray code reflects about the middle of the code. This style map is found in older texts. The newer preferred style is below.

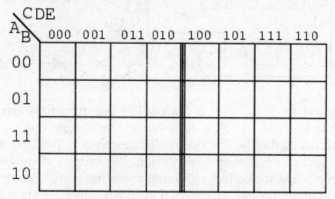

5-variable Karnaugh map (overlay)

The overlay version of the Karnaugh map, shown above, is simply two (four for a 6-variable map) identical maps except for the most significant bit of the 3-bit address across the top. If we look at the top of the map, we will see that the numbering is different from the previous Gray code map. If we ignore the most significant digit of the 3-digit numbers, the sequence **00, 01, 11, 10** is at the heading of both sub maps of the overlay map. The sequence of eight 3-digit numbers is not Gray code. Though the sequence of four of the least significant two bits is.

Let's put our 5-variable Karnaugh Map to use. Design a circuit which has a 5-bit binary input (A, B, C, D, E), with A being the MSB (Most Significant Bit). It must produce an output logic High for any prime number detected in the input data.

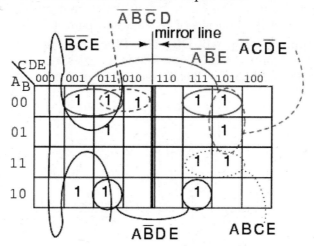

5-variable Karnaugh map (Gray code)

We show the solution above on the older Gray code (reflection) map for reference. The prime numbers are (1,2,3,5,7,11,13,17,19,23,29,31). Plot a 1 in each corresponding cell. Then, proceed with grouping of the cells. Finish by writing the simplified result. Note that 4-cell group A'B'E consists of two pairs of cell on both sides of the mirror line. The same is true of the 2-cell group AB'DE. It is a group of 2-cells by being reflected about the mirror line. When using this version of the K-map look for mirror images in the other half of the map.

Out = A'B'E + B'C'E + A'C'DE + A'CD'E + ABCE + AB'DE + A'B'C'D

Below we show the more common version of the 5-variable map, the overlay map.

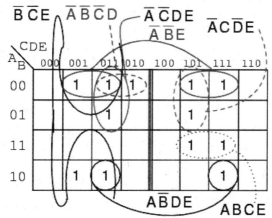

5-variable Karnaugh map (overlay)

If we compare the patterns in the two maps, some of the cells in the right half of the map are moved around since the addressing across the top of the map is different. We also need to take a different approach at spotting commonality between the two halves of the map. Overlay one half of the map atop the other half. Any overlap from the top map to the lower map is a potential group. The figure below shows that group AB'DE is composed of two stacked cells. Group A'B'E consists of two stacked pairs of cells.

For the **A'B'E** group of 4-cells **ABCDE = 00xx1** for the group. That is A,B,E are the same **001** respectively for the group. And, **CD=xx** that is it varies, no commonality in **CD=xx** for the group of 4-cells. Since **ABCDE = 00xx1**, the group of 4-cells is covered by **A'B'XXE = A'B'E**.

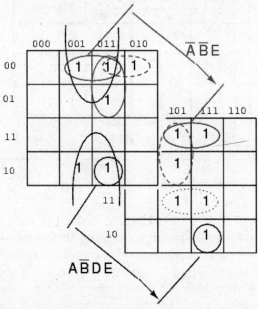

The above 5-variable overlay map is shown stacked.

An example of a six variable Karnaugh map follows. We have mentally stacked the four sub maps to see the group of 4-cells corresponding to **Out = C'F'**

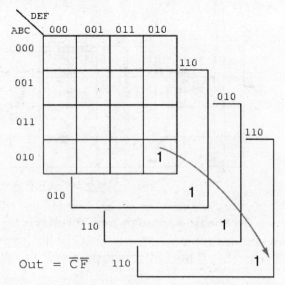

A magnitude comparator (used to illustrate a 6-variable K-map) compares two binary numbers, indicating if they are equal, greater than, or less than each other on three respective outputs. A three bit magnitude comparator has two inputs $A_2A_1A_0$ and $B_2B_1B_0$. An integrated circuit magnitude comparator (7485) would actually have four inputs, But, the Karnaugh map below needs to be kept to a reasonable size. We will only solve for the **A>B** output.

Below, a 6-variable Karnaugh map aids simplification of the logic for a 3-bit magnitude comparator. This is an overlay type of map. The binary address code across the top and down the left side of the map is not a full 3-bit Gray code. Though the 2-bit address codes of the four sub maps is Gray code. Find redundant expressions by stacking the four sub maps atop one another (shown above). There could be cells common to all four maps, though not in the example below. It does have cells common to pairs of sub maps.

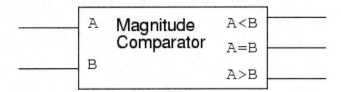

The A>B output above is ABC>XYZ on the map below.

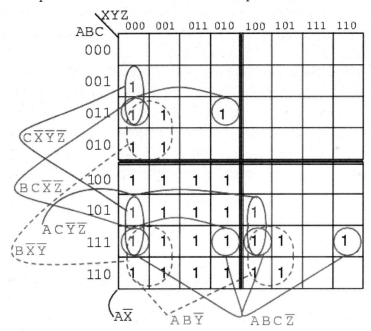

Out = $A\overline{X} + AB\overline{Y} + B\overline{X}\overline{Y} + ABC\overline{Z} + AC\overline{Y}\overline{Z} + BC\overline{X}\overline{Z} + C\overline{X}\overline{Y}\overline{Z}$

6-variable Karnaugh map (overlay)

Where ever **ABC** is greater than **XYZ**, a 1 is plotted. In the first line **ABC=000** cannot be greater than any of the values of **XYZ**. No 1s in this line. In the second line, **ABC=001**, only the first cell **ABCXYZ= 001000** is **ABC** greater than **XYZ**. A single **1** is entered in the first cell of the second line. The fourth line, **ABC=010**, has a pair of **1**s. The third line, **ABC=011** has three **1**s. Thus, the map is filled with **1**s in any cells where **ABC** is greater than **XXZ**.

In grouping cells, form groups with adjacent sub maps if possible. All but one group of 16-cells involves cells from pairs of the sub maps. Look for the following groups:

- 1 group of 16-cells
- 2 groups of 8-cells
- 4 groups of 4-cells

The group of 16-cells, **AX'** occupies all of the lower right sub map; though, we don't circle it on the figure above.

One group of 8-cells is composed of a group of 4-cells in the upper sub map overlaying a similar group in the lower left map. The second group of 8-cells is composed of a similar group of 4-cells in the right sub map overlaying the same group of 4-cells in the lower left map.

The four groups of 4-cells are shown on the Karnaugh map above with the associated product terms. Along with the product terms for the two groups of 8-cells and the group of 16-cells, the final Sum-Of-Products reduction is shown, all seven terms. Counting the **1**s in the map, there is a total of 16+6+6=28 ones. Before the K-map logic reduction there would have been 28 product terms in our SOP output, each with 6-inputs. The Karnaugh map yielded seven product terms of four or less inputs. This is really what Karnaugh maps are all about!

The wiring diagram is not shown. However, here is the parts list for the 3-bit magnitude comparator for ABC>XYZ using 4 TTL logic family parts:

- 1 ea 7410 triple 3-input NAND gate AX', ABY', BX'Y'
- 2 ea 7420 dual 4-input NAND gate ABCZ', ACY'Z', BCX'Z', CX'Y'Z'
- 1 ea 7430 8-input NAND gate for output of 7-P-terms

CHAPTER 14

GRAPH COLOURING

14.1 INTRODUCTION

In graph theory, **graph coloring** is a special case of graph labeling; it is an assignment of labels traditionally called "colors" to elements of a graph subject to certain constraints. In its simplest form, it is a way of coloring the vertices of a graph such that no two adjacent vertices share the same color; this is called a **vertex coloring**. Similarly, an **edge coloring** assigns a color to each edge so that no two adjacent edges share the same color, and a **face coloring** of a planar graph assigns a color to each face or region so that no two faces that share a boundary have the same color.

Vertex coloring is the starting point of the subject, and other coloring problems can be transformed into a vertex version. For example, an edge coloring of a graph is just a vertex coloring of its line graph, and a face coloring of a planar graph is just a vertex coloring of its planar dual. However, non-vertex coloring problems are often stated and studied *as is*. That is partly for perspective, and partly because some problems are best studied in non-vertex form, as for instance is edge coloring.

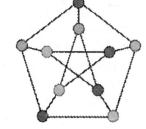

Figure 14.1. A proper vertex colouring of the Petersen graph with 3 colors, the minimum number possible.

The convention of using colors originates from coloring the countries of a map, where each face is literally colored. This was generalized to coloring the faces of a graph embedded in the plane. By planar duality it became coloring the vertices, and in this form it generalizes to all graphs. In mathematical and computer representations it is typical to use the first few positive or nonnegative integers as the "colors". In general one can use any finite set as the "color set". The nature of the coloring problem depends on the number of colors but not on what they are.

Graph coloring enjoys many practical applications as well as theoretical challenges. Beside the classical types of problems, different limitations can also be set on the graph, or on the way a color is assigned, or even on the color itself. It has even reached popularity with the general public in the form of the popular number puzzle Sudoku. Graph coloring is still a very active field of research.

14.2 DEFINITION AND TERMINOLOGY

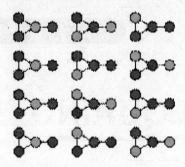

Figure 14.2. This graph can be 3-colored in 12 different ways.

Vertex Coloring

When used without any qualification, a **coloring** of a graph is almost always a proper vertex coloring, namely a labelling of the graph's vertices with colors such that no two vertices sharing the same edge have the same color. Since a vertex with a loop could never be properly coloured, it is understood that graphs in this context are loopless.

The terminology of using *colors* for vertex labels goes back to map coloring. Labels like *red* and *blue* are only used when the number of colors is small, and normally it is understood that the labels are drawn from the integers {1, 2, 3, ...}

A coloring using at most k colors is called a (proper) **k-coloring**. The smallest number of colors needed to color a graph G is called its **chromatic number,** $\chi(G)$. A graph that can be assigned a (proper) k-coloring is **k-colorable,** and it is **k-chromatic** if its chromatic number is exactly k. A subset of vertices assigned to the same color is called a *color class*, every such class forms an independent set. Thus, a k-coloring is the same as a partition of the vertex set into k independent sets, and the terms *k-partite* and *k-colorable* have the same meaning.

Chromatic Polynomial

In the below given figure all nonisomorphic graphs on 3 vertices and their chromatic polynomials. The empty graph E_3 (red) admits a 1-coloring, the others admit no such colorings. The green graph admits 12 colorings with 3 colors.

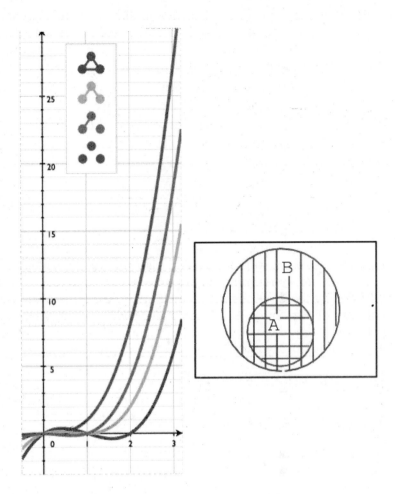

Figure 14.3. Chromatic polynomial

The **chromatic polynomial** counts the number of ways a graph can be colored using no more than a given number of colors. For example, using three colors, the graph in the image to the right can be colored in 12 ways. With only two colors, it cannot be colored at all. With four colors, it can be colored in $24 + 4 Å" 12 = 72$ ways: using all four colors, there are $4! = 24$ valid colorings (*every* assignment of four colors to *any* 4-vertex graph is a proper coloring); and for every choice of three of the four colors, there are 12 valid 3-colorings. So, for the graph in the example, a table of the number of valid colorings would start like this:

| Available colors | 1 | 2 | 3 | 4 | ... |
| Number of colorings | 0 | 0 | 12 | 72 | ... |

The chromatic polynomial is a function $P(G, t)$ that counts the number of t-colorings of G. As the name indicates, for a given G the function is indeed a polynomial in t. For the example graph, $P(G, t) = t(t-1)^2(t-2)$, and indeed $P(G, 4) = 72$.

The chromatic polynomial includes at least as much information about the colorability of G as does the chromatic number. Indeed, χ is the smallest positive integer that is not a root of the chromatic polynomial

$\chi(G) = \min\{k : P(G,k) > 0\}$.

Chromatic polynomials for certain graphs

Triangle K_3	$t(t-1)(t-2)$
Complete graph K_n	
Tree with n vertices	$t(t-1)^{n-1}$
Cycle C_n	$(t-1)^n + (-1)^n(t-1)$
Petersen graph	$t(t-1)(t-2)(t^7 - 12t^6 + 67t^5 - 230t^4 + 529t^3 - 814t^2 + 775t - 352)$

Edge coloring

An **edge coloring** of a graph, is a proper coloring of the *edges*, meaning an assignment of colors to edges so that no vertex is incident to two edges of the same color. An edge coloring with k colors is called a k-edge-coloring and is equivalent to the problem of partitioning the edge set into k matchings. The smallest number of colors needed for an edge coloring of a graph G is the **chromatic index,** or **edge chromatic number,** $\chi_2(G)$. A **Tait coloring** is a 3-edge coloring of a cubic graph. The four color theorem is equivalent to the assertion that every planar cubic bridgeless graph admits a Tait coloring.

14.3 PROPERTIES

Bounds on the Chromatic Number

Assigning distinct colors to distinct vertices always yields a proper coloring, so

$$1 \leq \chi(G) \leq n.$$

The only graphs that can be 1-colored are edgeless graphs, and the complete graph K_n of n vertices requires $\chi(K_n) = n$ colors. In an optimal coloring there must be at least one of the graph's m edges between every pair of color classes, so

$$\chi(G)(\chi(G) - 1) \leq 2m.$$

If G contains a clique of size k, then at least k colors are needed to color that clique; in other words, the chromatic number is at least the clique number:

$$\chi(G) \geq \omega(G).$$

For interval graphs this bound is tight.

The 2-colorable graphs are exactly the bipartite graphs, including trees and forests. By the four color theorem, every planar graph can be 4-colored.

A greedy coloring shows that every graph can be colored with one more color than the maximum vertex degree,

$$\chi(G) \leq \Delta(G) + 1.$$

Complete graphs have $\chi(G) = n$ and $\Delta(G) = n-1$, and odd cycles have $\chi(G) = 3$ and $\Delta(G) = 2$, so for these graphs this bound is best possible. In all other cases, the bound can be slightly improved; Brooks' theorem states that

Brooks' theorem: $\chi(G) \leq \Delta(G)$ for a connected, simple graph G, unless G is a complete graph or an odd cycle.

Graphs with High Chromatic Number

Graphs with large cliques have high chromatic number, but the opposite is not true. The Grötzsch graph is an example of a 4-chromatic graph without a triangle, and the example can be generalised to the Mycielskians.

Mycielski's Theorem: There exist triangle-free graphs with arbitrarily high chromatic number.

From Brooks's theorem, graphs with high chromatic number must have high maximum degree. Another local property that leads to high chromatic number is the presence of a large clique. But colorability is not an entirely local phenomenon: A graph with high girth looks locally like a tree, because all cycles are long, but its chromatic number need not be 2:

Theorem (Erdős): There exist graphs of arbitrarily high girth and chromatic number.

Bounds on the Chromatic Index

An edge coloring of G is a vertex coloring of its line graph $L(G)$, and vice versa. Thus,

$$\chi(G) = \chi'(L(G)).$$

There is a strong relationship between edge colorability and the graph's maximum degree $\Delta(G)$. Since all edges incident to the same vertex need their own color, we have

$$\chi'(G) \geq \Delta(G).$$

Moreover,

König's theorem: $\chi'(G) = \Delta(G)$ if G is bipartite.

In general, the relationship is even stronger than what Brooks's theorem gives for vertex coloring:

Vizing's Theorem: A graph of maximal degree Δ has edge-chromatic number Δ or $\Delta + 1$.

Other Properties

For planar graphs, vertex colorings are essentially dual to nowhere-zero flows.

About infinite graphs, much less is known. The following is one of the few results about infinite graph coloring:

If all finite subgraphs of an infinite graph G are k-colorable, then so is G, under the assumption of the axiom of choice.

Open Problems

The chromatic number of the plane, where two points are adjacent if they have unit distance, is unknown, although it is one of 4, 5, 6, or 7. Other open problems concerning the chromatic number of graphs include the Hadwiger conjecture stating that every graph with chromatic number k has a complete graph on k vertices as a minor, the Erdõs–Faber–Lovász conjecture bounding the chromatic number of unions of complete graphs that have at exactly one vertex in common to each pair, and the Albertson conjecture that among k-chromatic graphs the complete graphs are the ones with smallest crossing number.

When Birkhoff and Lewis introduced the chromatic polynomial in their attack on the four-color theorem, they conjectured that for planar graphs G, the polymomial $P(G,t)$ has no zeros in the region $[4, \infty]$. Although it is known that such a chromatic polynomial has no zeros in the region $[5, \infty]$ and that $P(G, 4) \neq 0$, their conjecture is still unresolved. It also remains an unsolved problem to characterize graphs which have the same chromatic polynomial and to determine which polynomials are chromatic.

14.4 ALGORITHMS

Efficient Algorithms

Determining if a graph can be coloured with 2 colors is equivalent to determining whether or not the graph is bipartite, and thus computable in linear time using breadth-first search. More generally, the chromatic number and a corresponding coloring of perfect graphs can be computed in polynomial time using semidefinite programming. Closed formulas for chromatic polynomial are known for many classes of graphs, such as forest, chordal graphs, cycles, wheels, and ladders, so these can be evaluated in polynomial time.

Brute-force Search

Brute-force search for a k-coloring considers every of the k^n assignments of k colors to n vertices and checks for each if it is legal. To compute the chromatic number and the chromatic polynomial, this procedure is used for every $k = 1, ..., n - 1$, impractical for all but the smallest input graphs.

Contraction

The contraction G/uv of graph G is the graph obtained by identifying the vertices u and v, removing any edges between them, and replacing them with a single vertex w where any edges that were incident on u or v are redirected to w. This operation plays a major role in the analysis of graph coloring.

The chromatic number satisfies the recurrence relation:

$$\chi(G) = \min\{\chi(G + uv), \chi(G / uv)\}$$

due to Zykov, where u and v are nonadjacent vertices, $G + uv$ is the graph with the edge uv added. Several algorithms are based on evaluating this recurrence, the resulting computation tree is sometimes called a Zykov tree. The running time is based on the heuristic for choosing the vertices u and v.

The chromatic polynomial satisfies following recurrence relation

$$P(G" uv, k) = P(G / uv, k) + P(G, k)$$

where u and v are adjacent vertices and $G"uv$ is the graph with the edge uv removed. $P(G"uv, k)$ represents the number of possible proper colourings of the graph, when the vertices may have same or different colors. The number of proper colorings therefore come from the sum of two graphs. If the vertices u and v have different colours, then we can as well consider a graph, where u and v are adjacent. If u and v have the same colors, we may as well consider a graph, where u and v are contracted. Tutte's curiosity about which other graph properties satisfied this recurrence led him to discover a bivariate generalization of the chromatic polynomial, the Tutte polynomial.

The expressions give rise to a recursive procedure, called the *deletion–contraction algorithm*, which forms the basis of many algorithms for graph coloring. The running time satisfies the same recurrence relation as the Fibonacci numbers, so in the worst case, the algorithm runs in time within a polynomial factor of $((1+\sqrt{5})/2)^{n+m} = O(1.6180)^{n+m}$. The analysis can be improved to within a polynomial factor of the number $t(G)$ of spanning trees of the input graph. In practice, branch and bound strategies and graph isomorphism rejection are employed to avoid some recursive calls, the running time depends on the heuristic used to pick the vertex pair.

Greedy Colouring

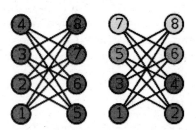

Figure 14.4

In the above figure two greedy colourings of the same graph using different vertex orders. The right example generalises to 2-colourable graphs with n vertices, where the greedy algorithm expends $n/2$ colors.

The greedy algorithm considers the vertices in a specific order $v_1,...,v_n$ and assigns to v_i the smallest available color not used by v_i's neighbours among $v_1,...,v_{i"1}$, adding a fresh color if needed. The quality of the resulting coloring depends on the chosen ordering. There exists an ordering that leads to a greedy coloring with the optimal number of $\div(G)$ colors. On the other hand, greedy colorings can be arbitrarily bad; for example, the crown graph on n vertices can be 2-colored, but has an ordering that leads to a greedy coloring with $n/2$ colors.

If the vertices are ordered according to their degrees, the resulting greedy coloring uses at most $\max_i \min\{d(x_i)+1, i\}$ colors, at most one more than the graph's maximum degree. This heuristic is sometimes called the Welsh–Powell algorithm. Another heuristic establishes the ordering dynamically while the algorithm proceeds, choosing next the vertex adjacent to the largest number of different colours. Many other graph coloring heuristics are similarly based on greedy coloring for a specific static or dynamic strategy of ordering the vertices, these algorithms are sometimes called **sequential coloring** algorithms.

Computational Complexity

Graph coloring is computationally hard. It is NP-complete to decide if a given graph admits a k-coloring for a given k except for the cases $k = 1$ and $k = 2$. Especially, it is NP-hard to compute the chromatic number. Graph coloring remains NP-complete even on planar graphs of degree 4.

The best known approximation algorithm computes a colouring of size at most within a factor $O(n(\log n)^{-3}(\log \log n)^2)$ of the chromatic number. For all $\varepsilon > 0$, approximating the chromatic number within $n^{1-\varepsilon}$ is NP-hard.

It is also NP-hard to color a 3-colorable graph with 4 colors and a k-colorable graph with $k^{(\log k)/25}$ colours for sufficiently large constant k.

Computing the coefficients of the chromatic polynomial is #P-hard. In fact, even computing the value of $\chi(G,k)$ is #P-hard at any rational point k except for $k = 1$ and $k = 2$. There is no FPRAS for evaluating the chromatic polynomial at any rational point $k \geq 1.5$ except for $k = 2$ unless NP = RP.

For edge colouring, the proof of Vizing's result gives an algorithm that uses at most $\Delta+1$ colors. However, deciding between the two candidate values for the edge chromatic number is NP-complete. In terms of approximation algorithms, Vizing's algorithm shows that the edge chromatic number can be approximated within $4/3$, and the hardness result shows that no $(4/3 - \varepsilon)$-algorithm exists for any $\varepsilon > 0$ unless P = NP. These are among the oldest results in the literature of approximation algorithms, even though neither paper makes explicit use of that notion.

Parallel and Distributed Algorithms

In the field of distributed algorithms, graph coloring is closely related to the problem of *symmetry breaking*. In a symmetric graph, a deterministic distributed algorithm cannot find a proper vertex coloring. Some auxiliary information is needed in order to break symmetry. A standard assumption is that initially each node has a *unique identifier*, for example, from the set $\{1, 2, ..., n\}$ where n is the number of nodes in the graph. Put otherwise, we assume that we are given an n-coloring. The challenge is to *reduce* the number of colors from n to, e.g., $\Delta + 1$.

A straightforward distributed version of the greedy algorithm for $(\Delta + 1)$-coloring requires $\theta(n)$ communication rounds in the worst case – information may need to be propagated from one side of the network to another side. However, much faster algorithms exist, at least if the maximum degree Δ is small.

The simplest interesting case is an n-cycle. Richard Cole and Uzi Vishkin show that there is a distributed algorithm that reduces the number of colors from n to $O(\log n)$ in one synchronous communication step. By iterating the same procedure, it is possible to obtain a 3-colouring of an n-cycle in $O(\log^* n)$ communication steps (assuming that we have unique node identifiers).

The function log*, iterated logarithm, is an extremely slowly growing function, "almost constant". Hence the result by Cole and Vishkin raised the question of whether there is a *constant-time* distribute algorithm for 3-coloring an *n*-cycle. Linial showed that this is not possible: any deterministic distributed algorithm requires $\Omega(\log^* n)$ communication steps to reduce an *n*-coloring to a 3-coloring in an *n*-cycle.

The technique by Cole and Vishkin can be applied in arbitrary bounded-degree graphs as well; the running time is $\text{poly}(\Delta) + O(\log^* n)$. The current fastest known algorithm for $(\Delta + 1)$-coloring is due to Leonid Barenboim and Michael Elkin, which runs in time $O(\Delta) + \log^*(n)/2$, which is optimal in terms of *n* since the constant factor $1/2$ cannot be improved due to Linial's lower bound.

The problem of edge coloring has also been studied in the distributed model. We can achieve a $(2\Delta - 1)$-coloring in $O(\Delta + \log^* n)$ time in this model. Linial's lower bound for distributed vertex coloring applies to the distributed edge coloring problem as well.

14.5 APPLICATIONS

Scheduling

Vertex coloring models to a number of scheduling problems. In the cleanest form, a given set of jobs need to be assigned to time slots, each job requires one such slot. Jobs can be scheduled in any order, but pairs of jobs may be in *conflict* in the sense that they may not be assigned to the same time slot, for example because they both rely on a shared resource. The corresponding graph contains a vertex for every job and an edge for every conflicting pair of jobs. The chromatic number of the graph is exactly the minimum *makespan*, the optimal time to finish all jobs without conflicts.

Details of the scheduling problem define the structure of the graph. For example, when assigning aircrafts to flights, the resulting conflict graph is an interval graph, so the coloring problem can be solved efficiently. In bandwidth allocation to radio stations, the resulting conflict graph is a unit disk graph, so the coloring problem is 3-approximable.

Register Allocation

A compiler is a computer program that translates one computer language into another. To improve the execution time of the resulting code, one of the techniques of compiler optimization is register allocation, where the most frequently used values of the compiled program are kept in the fast processor registers. Ideally, values are assigned to registers so that they can all reside in the registers when they are used.

The textbook approach to this problem is to model it as a graph coloring problem. The compiler constructs an *interference graph*, where vertices are symbolic registers and an edge connects two nodes if they are needed at the same time. If the graph can be colored with *k* colors then the variables can be stored in *k* registers.

Other Applications

The problem of coloring a graph has found a number of applications, including pattern matching.

The recreational puzzle Sudoku can be seen as completing a 9-coloring on given specific graph with 81 vertices.

EXERCISE 14.1

1. What do you understand by coloring of graphs?
2. Describe vertex coloring.
3. What do you mean by edge coloring?
4. What are the main properties of coloring of a graph?
5. Discuss applications of coloring of a graph.
6. What is Pólya enumeration theorem?
7. Give the formal statement of Pólya enumeration theorem and its proof.

CHAPTER 15

POLYA COUNTING THEOREM

15.1 PÓLYA ENUMERATION THEOREM

The **Pólya enumeration theorem** (PET), also known as **Redfield–Pólya's Theorem**, is a theorem in combinatorics, generalizing Burnside's lemma about number of orbits. This theorem was first discovered and published by John Howard Redfield in 1927 but its importance was overlooked and Redfield's publication was not noticed by most of mathematical community. Independently the same result was proved in 1937 by George Pólya, who also demonstrated a number of its applications, in particular to enumeration of chemical compounds.

The PET gave rise to symbolic operators and symbolic methods in enumerative combinatorics and was generalized to the fundamental theorem of combinatorial enumeration.

15.2 INFORMAL PET STATEMENT

Suppose you have a set of n slots and a set of objects being distributed into these slots and a generating function $f(a, b, ...)$ of the objects by weight. Furthermore there is a permutation group A acting on the slots that creates equivalence classes of filled slot configurations (two configurations are equivalent if one may be obtained from the other by a permutation from A). Then the generating function of the equivalence classes by weight, where the weight of a configuration is the sum of the weights of the objects in the slots, is obtained by evaluating the cycle index $Z(A)$ of A i.e.

$$Z(A)(t_1, t_2, \ldots, t_n) = \frac{1}{|A|} \sum_{g \in A} t_1^{j_1(g)} t_2^{j_2(g)} \cdots t_n^{j_n(g)}$$

at

$$t_1 = f(a, b, \ldots),\ t_2 = f(a^2, b^2, \ldots),\ t_3 = f(a^3, b^3, \ldots),\ \ldots\ t_n = f(a^n, b^n, \ldots).$$

15.3 DEFINITIONS FOR THE UNIVARIATE CASE

We have two finite sets X and Y, as well as a weight function $w: Y \to \mathbb{N}$. If $n = |X|$, without loss of generality we can assume that $X = \{1, 2, \ldots, n\}$.

Consider the set of all mappings $F = \{f \mid f: X \to Y\}$. We can define the weight of a function $f \in F$ to be

$$w(f) = \sum_{x \in X} w(f(x)).$$

Every subgroup of the symmetric group on n elements, S_n, acts on through permutations. If is one such subgroup, an equivalence relation on is defined as

$$f \sim_A g \iff f = g \circ a \text{ for some } a \in A$$

Denote by $[f] = \{g \in F \mid f \sim_A g\}$ the equivalence class of f with respect to this equivalence relation. $[f]$ is also called the orbit of . Since each acts bijectively on X, then

$$w(g) = \sum_{x \in X} w(g(x)) = \sum_{x \in X} w(g(a \circ x))) = \sum_{x \in X} w(f(x)) = w(f)$$

Therefore we can safely define . In other words, permuting the summands of a sum does not change the value of the sum.

Generating function by weight of the objects being distributed into the slots
Let
$c_k = |\{y \in Y \mid w(y) = k\}|$- the number of elements of of weight ;
The generating function by weight of the source objects is

Generating function by weight of the filled slot configurations (orbits)
Let
$C_k = |\{[f]\} w([f]) = k\}|$ -the number of *orbits* of weight k;
The generating function of the filled slot configurations is

$$C(t) = \sum_k C_k \cdot t^k.$$

15.4 THEOREM STATEMENT

Given all the above definitions, **Pólya's enumeration theorem** asserts that
$$C(t) = Z(A)(c(t), c(t^2), \ldots, c(t^n))$$
where is the cycle index of A.

15.5 EXAMPLE COMPUTATIONS

Enumerating graphs on three and four vertices by the number of edges
The graphs on three vertices without taking symmetries into account are shown at right. There are $2^{\binom{3}{2}}$ i.e. 8 of them ($\binom{3}{2}$ gives the number of pairs of vertices, i.e. edges, chosen from among three vertices).

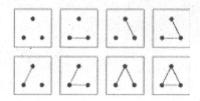

Figure 15.1: All graphs on three vertices.

We want to enumerate these graphs taking symmetries into account. There are only four nonisomorphic graphs, also shown at right.

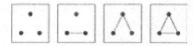

Figure 15.2: Nonisomorphic graphs on three vertices.

In this problem X is the set of all edges between vertices and Y is $\{0, 1\}$. Each mapping $X \to Y$ defines a graph on the m vertices. If we define $w(y) = y$, $\forall\ y \in Y$, then $w(f)$ is the number of edges in the graph resulting from f. Clearly, $c(t) = t + 1$ there are 2 elements in Y, one of weight 0 and one of weight 1.

The graph preserving edge permutations are directly generated by permutations of the vertices. Therefore, the subgroup A of $S_{\binom{m}{2}}$ acting on the edges (the edge permutation group of the graph) is of size $m!$.

The cycle index of the permutation group of the edges is

$$Z_A = \frac{1}{6}\left(t_1^3 + 3t_1 t_2 + 2t_3\right).$$

It follows from the enumeration theorem that the generating function of the non-isomorphic graphs on 3 vertices $C(t)$ is

$$C(t) = \frac{1}{6}\left((t+1)^3 + 3(t+1)(t^2+1) + 2(t^3+1)\right)$$

or

$$t^3 + t^2 + t + 1$$

which says that there is one graph with three edges, one with two, one with one edge, and one with no edges.

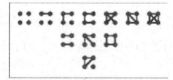

Figure 15.3: Nonisomorphic graphs on four vertices.

The cycle index of the edge permutation group for graphs on four vertices, which has degree six (there are six edges) and order twenty-four (each vertex permutation of the four vertices induces an edge permutation) is:

$$Z_A = \frac{1}{24}\left(t_1^6 + 9t_1^2 t_2^2 + 8t_3^2 + 6t_2 t_4\right).$$

Hence

$$C(t) = \frac{1}{24}\left((t+1)^6 + 9(t+1)^2(t^2+1)^2 + 8(t^3+1)^2 + 6(t^2+1)(t^4+1)\right).$$

or

$t^6 + t^5 + 2t^4 + 3t^3 + 2t^2 + t + 1$.

These graphs are shown at right.

A program that computes the generating function of the nonisomorphic graphs on vertices as well as a detailed description of the algorithm used can be found in the *GNUstep cookbook*.

Enumerating Rooted Ternary Trees

The set T3 of rooted ternary trees consists of rooted trees where every node has exactly three children (leaves or subtrees). Small ternary trees are shown at right. Note that there is a direct bijection between ternary trees with N non-leaf vertices and arbitrary trees with N vertices and degree at most 3.

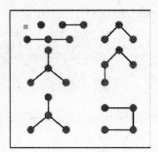

Figure 15.4: Ternary trees on 0, 1, 2, 3 and 4 vertices.

Two trees that can be obtained from one another by repeatedly permuting the children of some node are considered equivalent. In other words, the group that acts on the children of a node is the symmetric group S_3.

We use the following recursive decomposition of T3: an element of T3 is either a leaf of size zero, or a node with three children, where the order of the children is not important. The slots in this problem are the three slots where the children are attached to their parent node, and the objects that go into them are the elements of T3 itself. The group that acts on the slots is the symmetric group S_3 with cycle index

$$Z(S_3) = \frac{1}{6}\left(a_1^3 + 3a_1 a_2 + 2a_3\right).$$

It follows that the functional equation for the generating function $T(z)$ of the set T3 of rooted ternary trees is

$$T(z) = 1 + \frac{1}{6}z\left(T(z)^3 + 3T(z)T(z^2) + 2T(z^3)\right).$$

This translates into the following recurrence relation for the number t_n of rooted ternary trees on n nodes:

$t_0 = 1$
and

$$t_n = \frac{1}{6}\left(\sum_{a+b+c=n-1} t_a t_b t_c + 3 \sum_{a+2b=n-1} t_a t_b + 2 \sum_{3a=n-1} t_a\right) \text{ when } n > 0$$

where a, b and c are nonnegative integers.
The first few values of t_i are
1, 1, 1, 2, 4, 8, 17, 39, 89, 211, 507, 1238, 3057, 7639, 19241 (sequence A000598 in OEIS)

Enumerating Necklaces with Double Colors

Suppose you have a necklace containing $2n$ beads of n different colors, where every color is present exactly twice. How many different necklaces are there where two necklaces are considered equivalent if there exists a sequence of rotations and/or reflections that transforms one into the other?

In this problem the slots are the $2n$ locations where beads may be placed on the necklace and the objects that go into them are the $2n$ beads. The group that acts on the slots is the dihedral group D_{2n}. The three types of symmetries that may occur are illustrated at right for the case $n = 4$.

They are: rotations, reflections in an axis passing through opposite beads and reflections in an axis passing through opposite links.

The cycle index of the dihedral group is D_{2n} is

$$Z(D_{2n}) = \frac{1}{4n}\left(\sum_{d|2n} \varphi(d) a_d^{2n/d} + n a_1^2 a_2^{n-1} + n a_2^n\right),$$

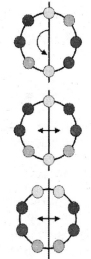

Figure 15.5: Three types of dihedral symmetries on necklaces with eight heads and four colors with two heads each.

where the sum represents the rotations, the second term the reflections in an axis passing through opposite beads, and the third term the reflections in an axis passing through opposite links.

Let the variable c_1 represent the first color, c_2 the second, c_3 the third and so on, up to c_n. The number of necklaces is then given by

$$[c_1^2 c_2^2 c_3^2 \cdots c_n^2] Z(D_{2n})(c_1, c_2, c_3, \ldots c_n).$$

Considering the rotations first, we see that only $d = 1$ and $d = 2$ contribute, namely through

$$[c_1^2 c_2^2 \cdots c_n^2](c_1 + c_2 + \cdots + c_n)^{2n} = \binom{2n}{2,2,\ldots 2} = \frac{(2n)!}{2^n}$$

and
$$[c_1^2 c_2^2 \cdots c_n^2](c_1^2 + c_2^2 + \cdots + c_n^2)^n = \binom{n}{1,1,\ldots 1} = n!.$$

The reflections in an axis passing through opposite beads contribute
$$n[c_1^2 c_2^2 \cdots c_n^2](c_1 + c_2 + \cdots + c_n)^2 (c_1^2 + c_2^2 + \cdots + c_n^2)^{n-1}$$
or
$$n n (n-1)! = n n!,$$
because there are n ways to choose a color from the first term, and $(n-1)!$ ways to choose the remaining colors from the second term (sum of squares raised to the $n-1$th power).

Finally, the reflections in an axis passing through opposite links contribute
$$n[c_1^2 c_2^2 \cdots c_n^2](c_1^2 + c_2^2 + \cdots + c_n^2)^n = n \binom{n}{1,1,\ldots 1} = n n!.$$

This yields the closed form expression for the number of necklaces containing n different colors exactly twice:
$$\frac{1}{4n} \left(\frac{(2n)!}{2^n} + (2n+1) n! \right).$$

The first few terms are
1, 2, 11, 171, 5736, 312240, 24327000, 2554072920 (sequence A120445 in OEIS)

Enumerating Coloured Cubes

Suppose you have an ordinary cube in three-space whose faces may take on one of three colors and are being permuted by the automorphisms of the cube (rotations in three-space). Here the slots are the six faces and the objects that go into them are the three colors. The object generating function by weight is
$$f(x, y, z) = x + y + z$$
which indicates that there are three colors and every color has weight one.

Figure 15.6: Cube

The cycle index of the permutation group C of the faces is
$$Z(C) = \frac{1}{24} \left(a_1^6 + 6a_1^2 a_4 + 3a_1^2 a_2^2 + 8a_3^2 + 6a_2^3 \right).$$

It follows that the generating function of the equivalence classes i.e. colored cubes taking symmetries into account is

$$g(x,y,z) = \frac{1}{24}(x+y+z)^6$$
$$+ \frac{1}{4}(x+y+z)^2(x^4+y^4+z^4)$$
$$+ \frac{1}{8}(x+y+z)^2(x^2+y^2+z^2)^2$$
$$+ \frac{1}{3}(x^3+y^3+z^3)^2$$
$$+ \frac{1}{4}(x^2+y^2+z^2)^3$$

or

$$g(x,y,z) = z^6 + yz^5 + xz^5 + 2y^2z^4 + 2xyz^4 + 2x^2z^4 + 2y^3z^3$$
$$+ 3xy^2z^3 + 3x^2yz^3 + 2x^3z^3 + 2y^4z^2 + 3xy^3z^2$$
$$+ 6x^2y^2z^2 + 3x^3yz^2 + 2x^4z^2 + y^5z$$
$$+ 2xy^4z + 3x^2y^3z + 3x^3y^2z + 2x^4yz + x^5z + y^6 + xy^5$$
$$+ 2x^2y^4 + 2x^3y^3 + 2x^4y^2 + x^5y + x^6.$$

This says e.g. that there is one cube using color *x* on five faces and color *z* on the sixth, and there are six cubes using *x*, *y*, and *z* twice:

- the cube where opposite faces have the same color
- the three cubes where two opposite faces have the same color and the remaining two pairs of opposite faces do not
- the two cubes where opposite faces always have different colors.

It also says that there are three cubes using color *x* on three faces, color *y* on two faces and color *z* on the remaining face:

- the cube where all faces of color *x* share a vertex
- the cube where the two faces of color *y* are adjacent to all three faces of color *x*
- the cube where one face of color *y* is adjacent to only two faces of color *x*.

Note that $g(1, 1, 1) = 57$ and there are 57 distinct colored cubes in total when there are three colors.

15.6 FORMAL STATEMENT OF THE THEOREM

The following statement of the theorem is for the general multivariate case as in the example of the colored necklaces and cubes.

Let A be a group acting on a set X (the "slots") and consider the set Y^X of all functions from a set X to a weighted set Y (the objects) with weight function ù (the "filled slot configurations"), where the weight of a function *f* is the sum of the weights of its range.

The Pólya enumeration theorem states that the sum of the weights of the orbits of A on Y^X (the equivalence classes of configurations induced by X) is given by

$$Z(A)\left(\sum_{y\in Y}\omega(y), \sum_{y\in Y}\omega(y)^2, \ldots, \sum_{y\in Y}\omega(y)^n\right).$$

When $\grave{u}(y)$ is a monomial in the variables (including constants) we have

$$\sum_{y\in Y}\omega(y) = f(a,b,c,\ldots)$$

where
$f(a, b, c, \ldots)$
is the generating function of the set Y by weight. Hence

$$\sum_{y\in Y}\omega(y)^k = f(a^k, b^k, c^k, \ldots)$$

and we obtain the alternate form

$$Z(A)(f(a,b,c,\ldots), f(a^2,b^2,c^2,\ldots), f(a^3,b^3,c^3,\ldots), \ldots f(a^n,b^n,c^n,\ldots)).$$

Proof of Theorem

The Pólya enumeration theorem follows from Burnside's lemma, which says that the number of orbits (equivalence classes of filled slot configurations) is the average of the number of elements of Y^X fixed by the permutation g of A over all permutations g. This value is a number and yields the count of the orbits, whereas PET enumerates orbits by weight, a more detailed classification, from which the count may be recovered by setting all variables of the weight function to one.

Applying the lemma to orbits of weight $\grave{u}$, the number of orbits of this weight is

$$\frac{1}{|A|}\sum_{g\in A}|\text{Fix}_\omega(g)|$$

where $\text{Fix}_{\grave{u}}(g)$ is the set of functions of weight $\grave{u}$ fixed by g.

Summing over all possible weights, we have

$$\sum_\omega \omega \frac{1}{|A|} \sum_{g\in A}|\text{Fix}_\omega(g)| = \frac{1}{|A|}\sum_{g\in A}\sum_\omega \omega|\text{Fix}_\omega(g)|.$$

Let the cycle structure of g be represented by

$t_1^{j_1(g)} t_2^{j_2(g)} \ldots t_n^{j_n(g)}$.

If g fixes an element of Y^X then it must be constant on every cycle q of g. The generating function by weight of a cycle q of $|q|$ identical elements from the set of objects enumerated by $f(a, b, c, \ldots)$ is

$f(a^{|q|}, b^{|q|}, c^{|q|}, \ldots)$.

It follows that the generating function by weight of the points fixed by g is the product of the above term over all cycles of g, i.e.

$$\sum_\omega \omega|\text{Fix}_\omega(g)| = \prod_{q\in g} f(a^{|q|}, b^{|q|}, c^{|q|}, \ldots)$$

or
$$f(a,b,c,\ldots)^{j_1(g)} f(a^2,b^2,c^2,\ldots)^{j_2(g)} \cdots f(a^n,b^n,c^n)^{j_n(g)}.$$

Substituting this for $\sum_\omega \omega|\text{Fix}_\omega(g)|$ in the sum over all g yields the substituted cycle index as claimed.

EXERCISE 15.1

1. What is Pólya enumeration theorem?
2. Give the formal statement of Pólya enumeration theorem and its proof.

SOLVED EXERCISES

MULTIPLE CHOICE QUESTIONS

In each case there is one correct answer (given at the end of the problem set). Try to work the problem first without looking at the answer. Understand both why the correct answer is correct and why the other answers are wrong.

1. Which of the following statements is **FALSE**?
   (a) $2 \in A \cup B$ implies that if $2 \notin A$ then $2 \in B$.
   (b) $\{2, 3\} \subseteq A$ implies that $2 \in A$ and $3 \in A$.
   (c) $A \cap B \supseteq \{2, 3\}$ implies that $\{2, 3\} \subseteq A$ and $\{2,3\} \subseteq B$.
   (d) $A - B \supseteq \{3\}$ and $\{2\} \subseteq B$ implies that $\{2, 3\} \subseteq A \cup B$.
   (e) $\{2\} \in A$ and $\{3\} \in A$ implies that $\{2, 3\} \subseteq A$.

2. Let $A = \{0, 1\} \times \{0, 1\}$ and $B = \{a, b, c\}$. Suppose A is listed in lexicographic order based on $0 < 1$ and B is in alphabetic order. If $A \times B \times A$ is listed in lexicographic order, then the next element after $((1, 0), c, (1, 1))$ is
   (a) $((1, 0), a, (0, 0))$ (b) $((1, 1), c, (0, 0))$
   (c) $((1, 1), a, (0, 0))$ (d) $((1, 1), a, (1, 1))$
   (e) $((1, 1), b, (1, 1))$

3. Which of the following statements is **TRUE**?
   (a) For all sets A, B, and C, $A - (B - C) = (A - B) - C$.
   (b) For all sets A, B, and C, $(A - B) \cap (C - B) = (A \cap C) - B$.
   (c) For all sets A, B, and C, $(A - B) \cap (C - B) = A - (B \cup C)$.
   (d) For all sets A, B, and C, if $A \cap C = B \cap C$ then $A = B$.
   (e) For all sets A, B, and C, if $A \cup C = B \cup C$ then $A = B$.

4. Which of the following statements is **FALSE**?
   (a) $C - (B \cup A) = (C - B) - A$
   (b) $A - (C \cup B) = (A - B) - C$
   (c) $B - (A \cup C) = (B - C) - A$
   (d) $A - (B \cup C) = (B - C) - A$
   (e) $A - (B \cup C) = (A - C) - B$

5. Consider the true theorem, "For all sets A and B, if $A \subseteq B$ then $A \cap B^c = \emptyset$." Which of the following statements is **NOT** equivalent to this statement:
   (a) For all sets A^c and B, if $A \subseteq B$ then $A^c \cap B^c = \emptyset$.
   (b) For all sets A and B, if $A^c \subseteq B$ then $A^c \cap B^c = \emptyset$.
   (c) For all sets A^c and B^c, if $A \subseteq B^c$ then $A^c \cap B = \emptyset$.
   (d) For all sets A^c and B^c, if $A^c \subseteq B^c$ then $A^c \cap B = \emptyset$.
   (e) For all sets A and B, if $A^c \supseteq B$ then $A \cap B = \emptyset$.

6. The power set $\mathcal{P}((A \times B) \cup (B \times A))$ has the same number of elements as the power set $\mathcal{P}((A \times B) \cup (A \times B))$ if and only if
    (a) $A = B$
    (b) $A = \theta$ or $B = \theta$
    (c) $B = \theta$ or $A = B$
    (d) $A = \theta$ or $B = \theta$ or $A = B$
    (e) $A = \theta$ or $B = \theta$ or $A \cap B = \theta$

7. Let $\sigma = 452631$ be a permutation on $\{1, 2, 3, 4, 5, 6\}$ in one-line notation (based on the usual order on integers). Which of the following is **NOT** a correct cycle notation for σ?
    (a) (614) (532)
    (b) (461) (352)
    (c) (253) (146)
    (d) (325)(614)
    (e) (614)(253)

8. Let $f: X \to Y$. Consider the statement, "For all subsets C and D of Y, $f^{-1}(C \cap D^c) = f^{-1}(C) \cap [f^{-1}(D)]^c$. This statement is
    (a) True and equivalent to:
        For all subsets C and D of Y, $f^{-1}(C - D) = f^{-1}(C) - f^{-1}(D)$.
    (b) False and equivalent to:
        For all subsets C and D of Y, $f^{-1}(C - D) = f^{-1}(C) - f^{-1}(D)$.
    (c) True and equivalent to:
        For all subsets C and D of Y, $f^{-1}(C - D) = f^{-1}(C) - [f^{-1}(D)]^c$.
    (d) False and equivalent to:
        For all subsets C and D of Y, $f^{-1}(C - D) = f^{-1}(C) - [f^{-1}(D)]^c$.
    (e) True and equivalent to:
        For all subsets C and D of Y, $f^{-1}(C - D) = [f^{-1}(C)]^c - f^{-1}(D)$.

9. Define $f(n) = \dfrac{n}{2} + \dfrac{1 - (-1)^n}{4}$ for all $n \in \mathbb{Z}$. Thus, $f: \mathbb{Z} \to \mathbb{Z}$, $\mathbb{Z}$ the set of all integers. Which is correct?
    (a) f is not a function from $\mathbb{Z} \to \mathbb{Z}$ because $\dfrac{n}{2} \notin \mathbb{Z}$.
    (b) f is a function and is onto and one-to-one.
    (c) f is a function and is not onto but is one-to-one.
    (d) f is a function and is not onto and not one-to-one
    (e) f is a function and is onto but not one-to-one.

10. The number of partitions of $\{1, 2, 3, 4, 5\}$ into three blocks is $S(5, 3) = 25$. The total number of functions $f: \{1, 2, 3, 4, 5\} \to \{1, 2, 3, 4\}$ with $|\text{Image}(f)| = 3$ is
    (a) 4×6
    (b) 4×25
    (c) 25×6
    (d) $4 \times 25 \times 6$
    (e) $3 \times 25 \times 6$

11. Let $f: X \to Y$ and $g: Y \to Z$. Let $h = g \circ f: X \to Z$. Suppose g is one-to-one and onto. Which of the following is **FALSE**?

(a) If f is one-to-one then h is one-to-one and onto.
(b) If f is not onto then h is not onto.
(c) If f is not one-to-one then h is not one-to-one.
(d) If f is one-to-one then h is one-to-one.
(e) If f is onto then h is onto.

12. Which of the following statements is **FALSE**?
    (a) $\{2, 3, 4\} \subset A$ implies that $2 \in A$ and $\{3, 4\} \subseteq A$
    (b) $\{2, 3, 4\} \in A$ and $\{2, 3\} \in B$ implies that $\{4\} \subseteq A - B$.
    (c) $A \cap B \subseteq \{2, 3, 4\}$ implies that $\{2, 3, 4\} \subseteq A$ and $\{2, 3, 4\} \subseteq B$.
    (d) $A - B \supseteq \{3, 4\}$ and $\{1, 2\} \subseteq B$ implies that $\{1, 2, 3, 4\} \subseteq A \cup B$.
    (e) $\{2, 3\} \subseteq A \cup B$ implies that if $\{2, 3\} \cap A = 0$ then $\{2, 3\} \subseteq B$.

13. Let $A = \{0, 1\} \times \{0, 1\} \times \{0, 1\}$ and $B = \{a, b, c\} \times \{a, b, c\} \times \{a, b, c\}$. Suppose A is listed in lexicographic order based on $0 < 1$ and B is listed in lexicographic order based on $a < b < c$. If $A \times B \times A$ is listed in lexicographic order, then the next element after $((0, 1, 1), (c, c, c), (1, 1, 1))$ is
    (a) $((1, 0, 1), (a, a, b), (0, 0, 0))$ (b) $((1, 0, 0), (b, a, a), (0, 0, 0))$
    (c) $((1, 0, 0), (a, a, a), (0, 0, 1))$ (d) $((1, 0, 0), (a, a, a), (1, 0, 0))$
    (e) $((1, 0, 0), (a, a, a), (0, 0, 0))$

14. Consider the true theorem, "For all sets A, B, and C if $A \subseteq B \subseteq C$ then $C^c \subseteq B^c \subseteq A^c$." Which of the following statements is **NOT** equivalent to this statement:
    (a) For all sets A^c, B^c, and C^c, if $A^c \subseteq B^c \subseteq C^c$ then $C \subseteq B \subseteq A$
    (b) For all sets A^c, B, and C^c, if $A^c \subseteq B \subseteq C^c$ then $C \subseteq B^c \subseteq A$
    (c) For all sets A, B, and C^c, if $A^c \subseteq B \subseteq C$ then $C^c \subseteq B^c \subseteq A$
    (d) For all sets A^c, B^c, and C^c, if $A^c \subseteq B^c \subseteq C$ then $C^c \subseteq B^c \subseteq A$
    (e) For all sets A^c, B^c, and C^c, if $A^c \subseteq B^c \subseteq C$ then $C^c \subseteq B \subseteq A$

15. Let $\mathcal{P}(A)$ denote the power set of A If $\mathcal{P}(A) \subseteq B$ then
    (a) $2^{|A|} \leq |B|$ (b) $2^{|A|} \geq |B|$
    (c) $2^{|A|} < |B|$ (d) $|A| + 2 \leq |B|$
    (e) $2^{|A|} \geq 2^{|B|}$

16. Let $f: \{1, 2, 3, 4, 5, 6, 7, 8, 9\} \to \{a, b, c, d, e\}$. In one-line notation, $f = (e, a, b, b, a, c, c, a, c)$ (use number order on the domain). Which is correct?
    (a) Image(f) = $\{a, b, c, d, e\}$, Coimage(f) = $\{\{6, 7, 9\}, \{2, 5, 8\}, \{3, 4\}, \{1\}\}$
    (b) Image(f) = $\{a, b, c, e\}$, Coimage(f) = $\{\{6, 7, 9\}, \{2, 5, 8\}, \{3, 4\}\}$
    (c) Image(f) = $\{a, b, c, e\}$, Coimage(f) = $\{\{6, 7, 9\}, \{2, 5, 8\}, \{3, 4\}, \{1\}\}$
    (d) Image(f) = $\{a, b, c, e\}$, Coimage(f) = $\{\{6, 7, 9, 2, 5, 8\}, \{3, 4\}, \{1\}\}$
    (e) Image(f) = $\{a, b, c, d, e\}$, Coimage(f) = $\{\{1\}, \{3, 4\}, \{2, 5, 8\}, \{6, 7, 9\}\}$

17. Let $\Sigma = \{x, y\}$ be an alphabet. The strings of length seven over S are listed in dictionary (lex) order. What is the first string after $xxxxyxx$ that is a palindrome (same read forwards and backwards)?
    (a) $xxxxyxy$ (b) $xxxyxxx$
    (c) $xxyxyxx$ (d) $xxyyyxx$
    (e) $xyxxxyx$

18. Let σ = 681235947 and τ = 627184593 be permutations on {1,2,3,4,5,6,7,8,9} in one-line notation (based on the usual order on integers). Which of the following is a correct cycle notation for τ o σ?
    (a) (124957368)
    (b) (142597368)
    (c) (142953768)
    (d) (142957368)
    (e) (142957386)

19. Let S = {0, 1, 2, 3, 4, 5, 6, 7, 8, 9}. What is the smallest integer K such that any subset of S of size K contains two disjoint subsets of size two, {x_1, x_2} and {y_1, y_2}, such that $x_1 + x_2 = y_1 + y_2 = 9$?
    (a) 8
    (b) 9
    (c) 7
    (d) 6
    (e) 5

20. There are K people in a room, each person picks a day of the year to get a free dinner at a fancy restaurant. K is such that there must be at least one group of six people who select the same day. What is the smallest such K if the year is a leap year (366 days)?
    (a) 1829
    (b) 1831
    (c) 1830
    (d) 1832
    (e) 1833

21. A mineral collection contains twelve samples of Calomel, seven samples of Magnesite, and N samples of Siderite. Suppose that the smallest K such that choosing K samples from the collection guarantees that you have six samples of the same type of mineral is K = 15. What is N?
    (a) 6
    (b) 2
    (c) 3
    (d) 5
    (e) 4

22. What is the smallest N > 0 such that any set of N nonnegative integers must have two distinct integers whose sum or difference is divisible by 1000?
    (a) 502
    (b) 520
    (c) 5002
    (d) 5020
    (e) 52002

23. Let S = {1, 3, 5, 7, 9, 11, 13, 15, 17, 19, 21}. What is the smallest integer N > 0 such that for any set of N integers, chosen from S, there must be two distinct integers that divide each other?
    (a) 10
    (b) 7
    (c) 9
    (d) 8
    (e) 11

24. The binary relation R= {(0, 0), (1, 1)} on A = {0, 1, 2, 3,} is
    (a) Reflexive, Not Symmetric, Transitive
    (b) Not Reflexive, Symmetric, Transitive
    (c) Reflexive, Symmetric, Not Transitive
    (d) Reflexive, Not Symmetric, Not Transitive
    (e) Not Reflexive, Not Symmetric, Not Transitive

25. Define a binary relation R = {(0, 1), (1, 2), (2, 3), (3, 2), (2, 0)} on A = {0, 1, 2, 3}. The directed graph (including loops) of the transitive closure of this relation has
    (a) 16 arrows
    (b) 12 arrows
    (c) 8 arrows
    (d) 6 arrows
    (e) 4 arrows

26. Let $\mathbb{N}^+$ denote the nonzero natural numbers. Define a binary relation R on $\mathbb{N}^+ \times \mathbb{N}^+$ by $(m, n)R(s, t)$ if $gcd(m, n) = gcd(s, t)$. The binary relation R is
    (a) Reflexive, Not Symmetric, Transitive
    (b) Reflexive, Symmetric, Transitive
    (c) Reflexive, Symmetric, Not Transitive
    (d) Reflexive, Not Symmetric, Not Transitive
    (e) Not Reflexive, Not Symmetric, Not Transitive

27. Let $\mathbb{N}_2^+$ denote the natural number's greater than or equal to 2. Let mRn if $gcd(m, n) > 1$. The binary relation R on $\mathbb{N}_2$ is
    (a) Reflexive, Symmetric, Not Transitive
    (b) Reflexive, Not Symmetric, Transitive
    (c) Reflexive, Symmetric, Transitive
    (d) Reflexive, Not Symmetric, Not Transitive
    (e) Not Reflexive, Symmetric, Not Transitive

28. Define a binary relation R on a set A to be *antireflexive* if xRx doesn't hold for any $x \in A$. The number of symmetric, antireflexive binary relations on a set of ten elements is
    (a) 2^{10}
    (b) 2^{50}
    (c) 2^{45}
    (d) 2^{90}
    (e) 2^{55}

29. Let R and S be binary relations on a set A. Suppose that R is reflexive, symmetric, and transitive and that S is symmetric, and transitive but is **not** reflexive. Which statement is always true for any such R and 3I
    (a) R ∪ S is symmetric but not reflexive and not transitive.
    (b) R ∪ S is symmetric but not reflexive.
    (c) R ∪ S is transitive and symmetric but not reflexive
    (d) R ∪ S is reflexive and symmetric.
    (e) R ∪ S is symmetric but not transitive.

30. Define an equivalence relation R on the positive integers A = {2, 3, 4, ... , 20} by m R n if the largest prime divisor of m is the same as the largest prime divisor of n. The number of equivalence classes of R is
    (a) 8
    (b) 10
    (c) 9
    (d) 11
    (e) 7

31. Let R = nar {(a, a), (a, b), (b, b), (a, c), (c, c)} be a partial order relation on Σ = {a, b, c}. Let $\preceq$ be the corresponding lexicographic order on Σ*. Which of the following is true?
   (a) $bc \preceq ba$
   (b) $abbaaacc \preceq abbaab$
   (c) $abbac \preceq abb$
   (d) $abbac \preceq abbab$
   (e) $abbac \preceq abbaac$

32. Consider the divides relation, $m \mid n$, on the set A = {2, 3, 4, 5, 6, 7, 8, 9, 10}. The cardinality of the covering relation for this partial order relation (i.e., the number of edges in the Hasse diagram) is
   (a) 4
   (b) 6
   (c) 5
   (d) 8
   (e) 7

33. Consider the divides relation, $m \mid n$, on the set A = {2, 3, 4, 5, 6, 7, 8, 9, 10}. Which of the following permutations of A is not a topological sort of this partial order relation?
   (a) 7, 2, 3, 6, 9, 5, 4, 10, 8
   (b) 2, 3, 7, 6, 9, 5, 4, 10, 8
   (c) 2, 6, 3, 9, 5, 7, 4, 10, 8
   (d) 3, 7, 2, 9, 5, 4, 10, 8, 6
   (e) 3, 2, 6, 9, 5, 7, 4, 10, 8

34. Let A = {2, 3, 4, 5, 6, 7, 8, 9, 10, 11, 12, 13, 14, 15, 16} and consider the divides relation on A. Let C denote the length of the maximal chain, M the number of maximal elements, and m the number of minimal elements. Which is true?
   (a) C = 3, M = 8, m = 6
   (b) C = 4, M = 8, m = 6
   (c) C = 3, M = 6, m = 6
   (d) C = 4, M = 6, m = 4
   (e) C = 3, M = 6, m = 4

35. Let
   m = "Juan is a math major,"
   c = "Juan is a computer science major,"
   g = "Juan's girlfriend is a literature major,"
   h = "Juan's girlfriend has read Hamlet," and
   t = "Juan's girlfriend has read The Tempest."
   Which of the following expresses the statement "Juan is a computer science major and a math major, but his girlfriend is a literature major who hasn't read both The Tempest and Hamlet."
   (a) $c \wedge m \wedge (g \vee (\sim h \vee \sim t))$
   (b) $c \wedge m \wedge g \wedge (\sim h \wedge \sim t)$
   (c) $c \wedge m \wedge g \wedge (\sim h \vee \sim t)$
   (d) $c \wedge m \wedge (g \vee (\sim h \wedge \sim t))$
   (e) $c \wedge m \wedge g \wedge (h \vee t)$

36. The function $((p \vee (r \vee q)) \wedge \sim(\sim q \wedge \sim r)$ is equal to the function
   (a) $q \vee r$
   (b) $((p \vee r) \vee q)) \wedge (p \vee r)$
   (c) $(p \wedge q) \vee (p \wedge r)$
   (d) $(p \vee q) \wedge \sim (p \vee r)$
   (e) $(p \wedge r) \vee (p \wedge q)$

37. The truth table for $(p \vee q) \vee (p \wedge r)$ is the same as the truth table for
    (a) $(p \vee q) \wedge (p \vee r)$
    (b) $(p \vee g) \wedge r$
    (c) $(p \vee q) \wedge r$
    (d) $p \vee q$
    (e) $(p \wedge q) \vee p$

38. The Boolean function $[\sim(\sim p \wedge q) \wedge \sim (\sim p \wedge \sim g)] \vee (p \wedge r)$ is equal to the Boolean function
    (a) q
    (b) $p \wedge r$
    (c) $p \vee q$
    (d) r
    (e) p

39. Which of the following functions is the constant 1 function?
    (a) $\sim p \wedge (p \wedge q)$
    (b) $(p \wedge q) \vee (\sim p \vee (p \wedge \sim q))$
    (c) $(p \wedge \sim q) \wedge (\sim p \vee q)$
    (d) $((\sim p \wedge q) \wedge (q \wedge r)) \wedge \sim q$
    (e) $(\sim p \wedge q) \vee (p \wedge q)$

40. Consider the statement, "Either $-2 \leq x \leq -1$ or $1 \leq x \leq 2$." The negation of this statement is
    (a) $x < -2$ or $2 < x$ or $-1 < x < 1$
    (b) $x < -2$ or $2 < x$
    (c) $-1 < x < 1$
    (d) $-2 < x < 2$
    (e) $x \leq -2$ or $2 \leq x$ or $-1 < x < 1$

41. The truth table for a Boolean expression is specified by the correspondence $(P, Q, R) \to S$ where $(0, 0, 0) \to 0$, $(0, 0, 1) \to 1$, $(0, 1, 0) \to 0$, $(0, 1, 1) \to 0$, $(1, 0, 0) \to 0$, $(1, 0, 1) \to 0$, $(1, 1, 0) \to 0$, $(1, 1, 1) \to 1$. A Boolean expression having this truth table is
    (a) $[(\sim P \wedge \sim Q) \vee Q] \vee R$
    (b) $[(\sim P \wedge \sim Q) \wedge Q] \wedge R$
    (c) $[(\sim P \wedge \sim Q) \vee \sim Q] \wedge R$
    (d) $[(\sim P \wedge \sim Q) \vee Q] \wedge R$
    (e) $[(\sim P \vee \sim Q) \wedge Q] \wedge R$

42. Which of the following statements is **FALSE:**
    (a) $(P \wedge Q) \vee (\sim P \wedge Q) \vee (P \wedge \sim Q)$ is equal to $\sim Q \wedge \sim P$
    (b) $(P \wedge Q) \vee (\sim P \wedge Q) \vee (P \wedge \sim Q)$ is equal to $Q \vee P$
    (c) $(P \wedge Q) \vee (\sim P \wedge Q) \vee (P \wedge \sim Q)$ is equal to $Q \vee (P \wedge \sim Q)$
    (d) $(P \wedge Q) \vee (\sim P \wedge Q) \vee (P \wedge \sim Q)$ is equal to $[(P \vee \sim P) \wedge Q] \vee (P \wedge \sim Q)$
    (e) $(P \wedge Q) \vee (\sim P \wedge Q) \vee (P \wedge \sim Q)$ is equal to $P \vee (Q \wedge \sim P)$.

43. To show that the circuit corresponding to the Boolean expression $(P \wedge Q) \vee (\sim P \wedge Q) \vee (\sim P \wedge \sim Q)$ can be represented using two' logical gates, one shows that this Boolean expression is equal to $\sim P \vee Q$. The circuit corresponding to $(P \wedge Q \wedge R) \vee (\sim P \wedge Q \wedge R) \vee (\sim P \wedge (\sim Q \wedge \sim R))$ computes the same function as the circuit corresponding to
    (a) $(P \wedge Q) \wedge \sim R$
    (b) $P \vee (Q \wedge R)$
    (c) $\sim P \vee (Q \wedge R)$
    (d) $(P \wedge \sim Q) \vee R$
    (e) $\sim P \vee Q \vee R$

44. Using binary arithmetic, a number y is computed by taking the n-bit two's complement of $x - c$. If n is eleven, $x = 10100001001_2$ and $c = 10101_2$ then $y =$
    (a) 01100001111_2
    (b) 01100001100_2
    (c) 01100011100_2
    (d) 01000111100_2
    (e) 01100000000_2

45. In binary, the sixteen-bit two's complement of the hexadecimal number DEAF_{16} is
  (a) 0010000101010111_2
  (b) 1101111010101111_2
  (c) 0010000101010011_2
  (d) 0010000101010001_2
  (e) 0010000101000001_2

46. In octal, the twelve-bit two's complement of the hexadecimal number 2AF_{16} is
  (a) 6522_8
  (b) 6251_8
  (c) 5261_8
  (d) 6512_8
  (e) 6521_8

47. Consider the statement form $p \Rightarrow q$ where p = "If Tom is Jane's father then Jane is Bill's niece" and q = "Bill is Tom's brother." Which of the following statements is equivalent to this statement?
  (a) If Bill is Tom's Brother, then Tom is Jane's father and Jane is not Bill's niece.
  (b) If Bill is not Tom's Brother, then Tom is Jane's father and Jane is not Bill's niece.
  (c) If Bill is not Tom's Brother, then Tom is Jane's father or Jane is Bill's niece.
  (d) If Bill is Tom's Brother, then Tom is Jane's father and Jane is Bill's niece.
  (e) If Bill is not Tom's Brother, then Tom is not Jane's father and Jane is Bill's niece.

48. Consider the statement, "If n .is*divisible by 30 then n is divisible by 2 and by 3 and by 5." Which of the following "statements is equivalent to this statement?
  (a) If n is not divisible by 30 then n is divisible by 2 or divisible by 3 or divisible by 5.
  (b) If n is not divisible by 30 then n is not divisible by 2 or not divisible by 3 or not divisible by 5.
  (c) If n is divisible by 2-divisible by 3 and divisible by 5 then n is divisible by 30.
  (d) If n is not divisible by 2 or not divisible by 3 or not divisible by 5 then n is not divisible by 30.
  (e) If n is divisible by 2 or divisible by 3 or divisible by 5 then n is divisible by 30.

49. Which of the following statements is the contrapositive of the statement, "You win the game if you know the rules but are not overconfident."
  (a) If you lose the game then you don't know the rules or you are overconfident.
  (b) A sufficient condition that you win the game is that you know the rules or you are not overconfident.
  (c) If you don't know the rules or are overconfident you lose the game.
  (d) If you know the rules and are overconfident then you win the game.
  (e) A necessary condition that you know the rules or you are not overconfident is that you win the game.

50. The statement form $(p \Leftrightarrow r) \Leftrightarrow (q \Leftrightarrow r)$ is equivalent to
  (a) $[(\sim p \vee r) \wedge (p \vee \sim r)] \vee \sim[(\sim q \vee r) \wedge (q \vee \sim r)]$
  (b) $\sim[(\sim p \vee r) \wedge (p \vee \sim r)] \wedge [(\sim q \vee r) \wedge (q \vee \sim r)]$
  (c) $[(\sim p \vee r) \wedge (p \vee \sim r)] \wedge [(\sim q \vee r) \wedge (q \vee \sim r)]$

(d) $[(\sim p \vee r) \wedge (p \vee \sim r)] \vee [(\sim q \vee r) \wedge (q \vee \sim r)]$
(e) $\sim[(\sim p \vee r) \wedge (q \vee \sim r)] \vee [(\sim q \vee r) \wedge (q \vee \sim r)]$

51. Consider the statement, "Given that people who are in need of refuge and consolation are apt to do odd things, it is clear that people who are apt to do odd things are in need of refuge and consolation." This statement, of the form $(P \Rightarrow Q) \Rightarrow (Q \Rightarrow P)$, is logically equivalent to
    (a) People who are in need of refuge and consolation are not apt to do odd things.
    (b) People are apt to do odd things if and only if they are in need of refuge and consolation.
    (c) People who are apt to do odd things are in need of refuge and consolation.
    (d) People who are in need of refuge and consolation are apt to do odd things.
    (e) People who aren't apt to do odd things are not in need of refuge and consolation.

52. A sufficient condition that a triangle T be a right triangle is that $a^2 + b^2 = c^2$. An equivalent statement is
    (a) If T is a right triangle then $a^2 + b^2 = c^2$.
    (b) If $a^2 + b^2 = c^2$ then T is a right triangle.
    (c) If $a^2 + b^2 = c^2$ then T is not a right triangle.
    (d) T is a right triangle only if $a^2 + b^2 = c^2$.
    (e) T is a right triangle unless $a^2 + b^2 = c^2$.

53. Which of the following statements is **NOT** equivalent to the statement, "There exists either a computer scientist or a mathematician who knows both Discrete Math and Java."
    (a) There exists a person who is a computer scientist and who knows both discrete Math and Java or there exists a person who is a mathematician and who knows both discrete Math and Java.
    (b) There exists a person who is a computer scientist or there exists a person who is a mathematician who knows Discrete Math or who knows Java.
    (c) There exists a person who is a computer scientist and who knows both discrete Math and Java or there exists a mathematician who knows both discrete math and Java.
    (d) There exists a computer scientist who knows both Discrete Math and Java or there exists a person who is a mathematician who knows both Discrete Math and Java.
    (e) There exists a person who is a computer scientist or a mathematician who knows both Discrete Math and Java.

54. Which of the following is the negation of the statement, "For all odd primes $p < q$ there exists positive non-primes $r < s$ such that $p^2 + q^2 = r^2 + s^2$."
    (a) For all odd primes $p < q$ there exists positive non-primes $r < s$ such that $p^2 + q^2 \neq r^2 + s^2$.

(b) There exists odd primes $p < q$ such that for all positive non-primes $r < s$, $p^2 + q^2 = r^2 + s^2$.
(c) There exists odd primes $p > q$ such that for all positive non-primes $r < s$, $p^2 + q^2 \neq r^2 + s^2$.
(d) For all odd primes $p < q$ and for all positive non-primes $r < s$, $p^2 + q^2 \neq r^2 + s^2$.
(e) There exists odd primes $p < q$ and there exists positive non-primes $r < s$ such that $p^2 + q^2 \neq r^2 + s^2$.

55. Consider the following assertion: "The two statements
    (1) $\exists x \in D, (P(x) \wedge Q(x))$ and   (2) $(\exists x \in D, P(x)) \wedge (\exists x \in D, Q(x))$
have the same truth value." Which of the following is correct?
   (a) This assertion is false. A counterexample is $D = \mathbb{N}$, $P(x) = $ "x is divisible by 6," $Q(x) = $ "x is divisible by 3."
   (b) This assertion is true. The proof follows from the distributive law for $\wedge$.
   (c) This assertion is false. A counterexample is $D = \mathbb{Z}$, $P(x) - $ "$x < 0$," $Q(x) = $ "$x \geq 0$."
   (d) This assertion is true. To see why, let $D = \mathbb{N}$, $P(x) = $ "x is divisible by 6," $Q(x) = $ "x is divisible by 3." If $x = 6$, then x is divisible by both 3 and 6 so both statements in the assertion have the same truth value for this x.
   (e) This assertion is false. A counterexample is $D = \mathbb{N}$, $P(x) = $ "x is a square," $Q(x) = $ "x is odd."

56. Which of the following is an unsolved conjecture?
   (a) $\exists n \in \mathbb{N}, 2^{2^n} + 1 \notin \mathbb{P}$
   (b) $\exists K \in \mathbb{N}, \forall n > K, n$ odd, $\exists p, q, r \in \mathbb{P}, n = p + q + r$
   (c) $(\exists x, y, z, n \in \mathbb{N}^+, x^n + y^n = z^n) \Leftrightarrow (n = 1, 2)$
   (d) $\forall m \in \mathbb{N}, \exists n > m, n$ even, $\exists p, q \in \mathbb{P}, n = p + q$
   (e) $\forall m \in \mathbb{N}, \exists n > m, n \in \mathbb{P}$ and $n + 2 \in \mathbb{P}$

57. Which of the following is a solved conjecture?
   (a) $\forall m \in \mathbb{N}, \exists n \geq m, n$ odd, $\exists p, q \in P, n = p + q$
   (b) $\forall m \in \mathbb{N}, \exists n \geq m, n \in \mathbb{P}$ and $n + 2 \in \mathbb{P}$
   (c) $\forall m \in \mathbb{N}, \exists n \geq m, 2^{2^n} + 1 \in \mathbb{P}$
   (d) $\forall k \in \mathbb{N}, \exists p \in P, p > k, 2^p - 1 \in \mathbb{P}$
   (e) $\forall n \geq 4, n$ even, $\exists p, q \in \mathbb{P}, n = p + q$

58. What logic function corresponds to the following arrangement?

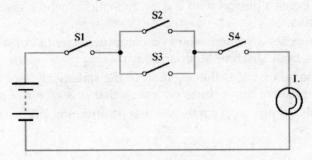

(a) L = (S1 OR S2) AND (S3 OR S4).
(b) L = S1 OR (S2 AND S3) OR S4.
(c) L = (S1 AND S2) OR (S3 AND S4).
(d) L = S1 AND (S2 OR S3) AND S4.

59. Which logic gate has the following truth table?

A	B	C
0	0	0
0	1	1
1	0	1
1	1	1

(a) An exclusive NOR gate.   (b) A two-input AND gate.
(c) An exclusive OR gate.   (d) A two-input OR gate.

60. In Boolean algebra the AND function is represented by the '+' sign.
   (a) True   (b) False

61. What Boolean expression describes the output X of this arrangement?

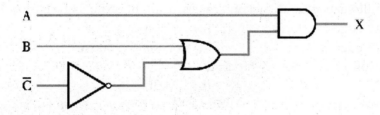

(a) X = A + (B.C)   (b) X = A + B + C
(c) X = (A.B) + C   (d) X = A.(B + C)

62. In the Karnaugh map shown below, which of the loops shown represents a legal grouping?

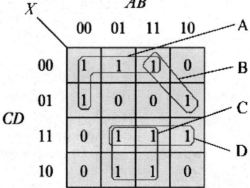

(a) A. (b) B.
(c) C. (d) D

63. What is the most widely used method for the automated simplification of Boolean expressions?
    (a) Binary reduction.   (b) Quine-McCluskey minimisation.
    (c) Fast Fourier transforms.   (d) Karnaugh maps.
64. Physical logic gates take a finite time to respond to changes in their input signals. What name is given to this time?
    (a) Propagation delay time.   (b) Hold time.
    (c) Rise time.   (d) Set-up time.
65. Express the binary number 1001 in decimal.
    (a) 9   (b) 11
    (c) 13   (d) 15
66. Express the decimal number 57 in binary.
    (a) 110011   (b) 111010
    (c) 111001   (d) 111101
67. Convert the hexadecimal number 59 into binary.
    (a) 1010101   (b) 1101101
    (c) 1001101   (d) 1011001
68. A Hamiltonian cycle in a Hamiltonian graph of order 24 has
    (a) 12 edges.   (b) 24 edges.
    (c) 23 edges.   (d) none of the above
69. The graph given below is bipartite.
    (a) TRUE.   (b) FALSE.
70. A simple graph G with 13 vertices has 4 vertices of degree 3, 3 vertices of degree 4 and 6 vertices of degree 1.
    The graph G must be a tree:
    (a) TRUE.   (b) FALSE.
71. A spanning tree for a simple graph of order 24 has
    (a) 12 edges.   (b) 6 edges,
    (c) 23 edges   (d) none of the above.
72. The number of spanning trees in the complete graph K_8
    (a) 48.   (b) 6^8.
    (c) 8^6 (d) none of the above.
73. The order of a forest, F, with 17 vertices and 4 components is
    (a) 17   (b) 4.
    (c) 16.   (d) none of the above.
74. The size of a forest, F, with 17 vertices and 4 components is
    (a) 14.   (b) 4.
    (c) 16.   (d) none of the above.

75. A forest, F, with 17 vertices and 4 components has at least one 3-clique.
    (a) TRUE.   (b) FALSE
76. The number of different labelled trees of order n is
    (a) n^n (b) $(n-2)^n$
    (c) n^{n-2} (d) none of the above
77. Consider the Prüfer sequence, S = (1,7,1,5,2,5).
    Let T be the labelled tree corresponding to S. The degree of the vertex 7 in T has degree
    (a) 2.   (b) 1.
    (c) 5.   (d) 7.
    (e) none of the above.
78. Consider the Prüfer sequences S = (1,7,1,5,2,5).
    Let T be the labelled tree corresponding to S.
    The size of T is
    (a) 2.   (b) 6.
    (c) 5.   (d) 7
    (e) none of the above.
79. The number of walks of length 3 through the vertices A, B and C in the graph, G, drawn above, is
    (a) 3.   (b) 6.
    (c) 12.   (d) 1.
    (e) none of the above.
80. The cube graph Q_5 is planar,
    (a) TRUE.   (b) FALSE
81. The complete graph K_7 is non-planar.
    (a) TRUE.   (b) FALSE.
82. The complete bipartite graph $K_{4,3}$ is non-planar.
    (a) TRUE.   (b) FALSE.
83. If G is a simple connected 3-regular planar graph where every face is bounded by exactly 3 edges, then the size of G is
    (a) 3.   (b) 4.
    (c) 6   (d) 5.
    (e) none of the above.

    Let G be a graph. The next three questions refer to A which is an adjacency matrix for G.

$$A = \begin{bmatrix} 0 & 1 & 1 & 2 & 0 \\ 1 & 0 & 0 & 0 & 1 \\ 1 & 0 & 0 & 1 & 1 \\ 2 & 0 & 1 & 0 & 0 \\ 0 & 1 & 1 & 0 & 0 \end{bmatrix}$$

84. The degree sequence of G is
    (a) (2,2,3,3,4).
    (b) (0,1,1,2,0).
    (c) (0,0,1,1,2)
    (d) (2,3,4).
    (e) none of the above.
85. The size of the graph G is
    (a) 2.
    (b) 3.
    (c) 7.
    (d) none of the above.
86. The entry at position (4,2) in the matrix A^2 is
    (a) 2
    (b) 1
    (d) 4.
    (e) none of the above.

87. Let H be the plane drawing of the planar graph (7, drawn above. The graph H has
    (a) 10 faces.
    (b) 5 faces.
    (c) 11 faces,
    (d) 6 faces.
    (e) none of the above.

The next two questions refer to the graph H drawn below

88. The size of H*, the dual of H is
    (a) 4.
    (b) 3.
    (c) 6
    (d) none of the above
89. The order of H* is
    (a) 4.
    (b) 3
    (c) 6.
    (d) none of the above.
90. If G is a connected plane graph of order v, size e and with f faces, then
    (a) $v - e + f = 2$
    (b) $e - v + f = 2$
    (c) $v + e - f = 2$
    (d) $e + v - f = 2$
    (e) none of the above.
91. The chromatic number of the cyclic graph C_{15} is
    (a) 3.
    (b) 2.
    (c) 6
    (d) 15.
    (e) none of the above.
92. The chromatic number of the complete graph $K\backslash \$$ is
    (a) 3;
    (b) 2;
    (c) 6;
    (d) 15,
    (e) none of the above.

93. The chromatic polynomial $P_G(t)$, where G is the complete graph K_3 is
    (a) $t(t-1)^2$
    (b) $t(t-1)(t-2)$
    (c) t^3
    (d) $t(t-1)(t+1)$
    (e) none of the above.

94. The polynomial $t^4 - 4t^3 + 6t^2 - 3t$ *is* the chromatic polynomial for a graph G. Thus the size of G is
    (a) 4
    (b) 6
    (c) 3.
    (d) 2.
    (e) none of the above.

95. The polynomial $t4 - 4t^3 + 6t^2 - 3t$ is the chromatic polynomial for a graph G. Thus the order of G is
    (a) 4
    (b) 6.
    (c) 3
    (d) 2
    (e) none of the above.

96. The edge chromatic number (or chromatic index) of the complete graph K_{15} is
    (a) 3.
    (b) 2.
    (c) 6.
    (d) 15.
    (e) none of the above.

Questions 97 to 102 refer to the graph G drawn below.

97. The degree sequence for G is:
    (a) (2,2,3,3,3,4,4);
    (b) (2,3,3,3,3,4,4);
    (c) (2, 3, 3, 3, 3, 4, 4);
    (d) (2,3,3,3,4,4,4).

98. The size of G is:
    (a) 11;
    (b) 4;
    (c) 7;
    (d) 2

99. The order of G is :
    (a) 11;
    (b) 4;
    (c) 7;
    (d) 2.

100. The graph G is best described as
    (a) a multigraph;
    (b) a pseudograph;
    (c) a complete graph;
    (d) a simple graph

101. The graph G has
    (a) a 3~clique as a subgraph;
    (b) a 4-clique as a subgraph;
    (c) a 4-clique and a 3-clique as subgraphs;
    (d) neither a 4-clique nor a 3-clique *as* subgraphs.

102. The order of every spanning subgraph of G is:
    (a) 5;
    (b) 9;
    (c) 7;
    (d) 4.

Questions 103 to 110. refer to the graph G drawn below :

G :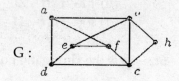

103. The graph G is not a regular graph because:
    (a) not all edges are the same length;
    (b) it is a complete graph
    (c) not all vertices have the same degree;
    (d) it has a vertex of degree 3.
104. The graph G is a bipartite graph:
    (a) TRUE;   (b) False
105. The order in G - h is :
    (a) 4;   (b) 6;
    (c) 5;   (d) 7.
106. bh is an edge in G. The order in G — bh is:
    (a) 4;   (b) 6;
    (c) 5;   (d) 7
107. dacbeb is
    (a) a walk in G;   (b) a walk but not a trail in
    (c) a trail but not a path in G; (d) not a walk in G.
108. befcdef is
    (a) a trail in G;   (b) a walk but not a trail in G;
    (c) a cycle in G;   (d) a path but not a cycle in G,
109. The edge ef is not a bridge of G because;
    (a) the degree of e is 3;   (b) the edge ef is parallel to the edge do;
    (c) G is 3-regular;   (d) G - ef is not connected.
110. The graph G is:
    (a) semi-Hamiltonian but not Eulerian;
    (b) Hamiltonian but not Eulerian;
    (c) Hamiltonian and semi-Eulerian;
    (d) not Hamiltonian nor Eulerian.
111. The maximum size of a simple graph of order 15 is
    (a) 105;   (b) 210;
    (c) 21;   (d) 45
    H = (V.E) is a graph, where V = $\{v_1, v_2, v_3, v_4, v_5\}$
    and E = $\{v_1v_1, v_1v_2, v_5v_3, v_2v_4, v_3v_4, v_4v_5, v_5v_4\}$

Questions 112 and 113 refer to H = (V, E).

112. The degree of the vertex v_1 in H = (V, E) is:
    (a) 2;   (b) 1;
    (c) 3;   (d) 4

113. The size of H = (V, E) is:
    (a) 5;   (b) 8;
    (c) 7;   (d) 6.

114. Exactly one of the following sequences is graphic. Which one is it?
    (a) (1, 3, 3, 4, 5, 6, 6);   (b) (1, 2, 2, 3, 4, 4);
    (c) (2, 3, 3, 4, 4, 5);   (d) (2, 3, 4, 4, 5)

115. The size of the path graph P_7 is
    (a) 5;   (b) 6;
    (c) 4;   (d) 7

116. Let H be a simple graph of order 8 and with 3 components. Then the number of the maximum size of H is:
    (a) 15;   (b) 10;
    (c) 11;   (d) 24.

117. Let G be the complete graph K_{12}. Then G is:
    (a) Semi-Eulerian and semi-Hamiltonian;
    (b) Eulerian and Hamiltonian;
    (c) Neither Eulerian but Hamiltonian;
    (d) Eulerian and not Hamiltonian,

118. The size of the complete bipartite graph $K_{5,4}$ is:
    (a) 10;   (b) 9;
    (c) 25;   (d) 20.

119. The order of the complete bipartite graph $K_{5,4}$ is:
    (a) 10;   (b) 9;
    (c) 25;   (d) 20

120. The order of the 4-cube graph Q_4 is:
    (a) 16;   (b) 4;
    (c) 8;   (d) 32

121. The size of the 4-cube graph Q_4 is:
    (a) 16;   (b) 4;
    (c) 8;   (d) 32

122. The order of the Petersen graph is
    (a) 16;   (b) 4;
    (c) 10;   (d) 6

123. The size of the complement of the cycle graph C5 is :
    (a) 25;   (b) 5;
    (c) 10;   (d) 15

Questions 124 and 126, refer to the graph G which is drawn below :

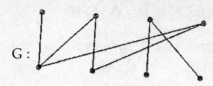

124. The graph G is bipartite
    (a) True
    (b) FALSE.
125. The graph G has
    (a) 2 components;
    (b) 1 component;
    (c) 8 components;
    (d) 7 components.
126. H = (V, E) is a graph, where V = {a, b, c, d, e, f}
    (a) {ab, be, de, be, cf, cd}
    (b) {af, ad, ac, bc, be, cd, cf, ae, df}
    (c) {af, fb, bd, ce, ef, ea}
    (d) {ac, fb, ba, dc, fe, ea}
127. The minimum number of elements in a Boolean algebra is
    (a) 1
    (b) 2
    (c) 3
    (d) 4
128. Idempotent law in Boolean algebra is
    (a) (a')' = a
    (b) a + a • b = a
    (c) a + a = a
    (d) a + 1 = a
129. In Boolean algebra, if $a, b \in B$, then absorption law is
    (a) a + (a • b) = a
    (b) a • a = a
    (c) a + a = a
    (d) none of these
130. In Boolean algebra a • a = a is known as
    (a) De-Morgan's law
    (b) Absorption law
    (c) Idempotent law
    (d) none of these
131. In Boolean algebra a • (a + b) = a is known as
    (a) Idempotent law
    (b) Absorption law
    (c) De-Morgan's law
    (d) none of these
132. In Boolean algebra a + (b • c) = (a + b) • (a + c) follows from
    (a) Distributive law
    (b) Associative law
    (c) Idempotent law
    (d) none of these
133. For any two elements a and b in Boolean algebra (a + b)' = a' • b' and (a • b)' = a' + b' are known
    (a) Idempotent law
    (b) Absorption law
    (c) De-Morgan's law
    (d) none of these
134. In Boolean algebra, the dual of a • 0 = 0 is
    (a) a + 1 = 1
    (b) a • 0 = 1
    (c) 0 • a = 0
    (d) none of these

135. In Boolean algebra, the dual of a • (b + c) = (a • b) + (a • c) is
    (a) a + (b • c) = (a + b) • (a + c)
    (b) (b + c) • a = (a • c) + (a • b)
    (c) a • b + a • c = (a • b) + (a • c)
    (d) none of these
136. In Boolean algebra, the dual of a • a'- 0 is
    (a) a + a' = 1
    (b) a' • a = 0
    (c) a • a' = 1
    (d) none of these
137. In Boolean algebra, which of the following statements is true for x, y ∈ B
    (a) (x • y)' = x' – y'
    (b) (x • y) = x' + y'
    (c) (x • y)' = x' • y'
    (d) none of these
138. In Boolean algebra, if a, b ∈ B, then a + a' b is equal to
    (a) a + b
    (b) a' + b'
    (c) a + ab'
    (d) none of these
139. In Boolean algebra, if a, b ∈ B, then value of a' + a • b is equal to
    (a) a + b'
    (b) a' + b'
    (c) a' + b'
    (d) none of these
140. In Boolean algebra, if a, b ∈ B, then value of a • (a • b) is
    (a) a • a
    (b) b + a
    (c) a + b
    (d) none of these
141. Which of the following statement is true in Boolean algebta, where a, b ∈ B
    (a) (x + y)' = x' • y'
    (b) (x + y)' = x' + y'
    (c) (x + y)' = x • y
    (d) none of these
142. In Boolean algbera if a ∈ B, then
    (a) (a')' = a'
    (b) (a') = a • a
    (c) (a')' = a
    (d) none of these
143. Which of the following statements is false in Boolean algebra, where a, b ∈ B
    (a) a • (a + b) = b
    (b) (a')' = a
    (c) a + a = a
    (d) a • a = a
144. Which of the following statements are true in Boolean algebra if a ∈ B
    (a) a + 1 = a
    (b) a + 1 = 1
    (c) a • 0 = a
    (d) a • 0 = 1
145. Which of the following statement is true in Boolean algebra if x ∈ B
    (a) x + x' = 1
    (b) x + 0 = 0
    (c) x • 1 = 1
    (d) x • x' = 1
146. A Boolean algebra can not have
    (a) 2 elements
    (b) 3 elements
    (c) 4 elements
    (d) 5 elements

147. Simplified form of the switching function $F(x, y) = x + x \cdot y$ is
   (a) $x \cdot y$
   (b) x
   (c) y
   (d) $x + y$

148. Simplified form of the switching function $F(x, y, z) = x \cdot y + y \cdot z + y \cdot z'$ is
   (a) y
   (b) x
   (c) $x \cdot y$
   (d) none of these

149. The Boolean expression $x \cdot y' + x \cdot y + x \cdot y^x + x' y$ is equivalent to
   (a) 0
   (b) 1
   (c) $x y$
   (d) none of these

150. l.u.b. of the elements a and b of a Boolean algebra B is
   (a) $a + b$
   (b) 1
   (c) 0
   (d) $a \cdot b$

151. g.l.b. {a, b} of a Boolean algebra B is
   (a) $a + b$
   (b) $a \cdot b$
   (c) 1
   (d) 0

152. Complete D.N.R of a Boolean function in two variables p and q is
   (a) $p' \cdot q + p' \cdot q + p \cdot q' + p' \cdot q'$
   (b) $p \cdot q' + p'q$
   (c) $p \cdot q$
   (d) $P \cdot p + q \cdot q + p' \cdot q + p \cdot q'$

153. The complement of Boolean function $F(x, y) = x' y + x y' + x' y'$ is
   (a) $x \cdot y$
   (b) $x + y$
   (c) $(x \cdot y)'$
   (d) none of these

154. Negation of the statement "He is neither intelligent nor a player" is
   (a) He is intelligent or a player.
   (b) He is intelligent and a player.
   (c) He is intelligent but not a player.
   (d) He is not intelligent but a player.

155. Fill in the blank spaces in each of the following statements by choosing the most appropriate word out of the 4 alternatives given below each statement.
   4-alternatives: (a) necessary (b) sufficient (c) necessary and sufficient (d) neither necessary nor sufficient
   Statements:
   (i) The .... condition that the figure PQRS is a rectangle is that it is a quadrilateral
   (ii) The .... condition that I am more than 40 years old is that I am more than 20 years old.
   (iii) The .... condition that p.q is odd is that p is odd and q is odd
   (iv) The .... condition that I am more than 40 years old is that I am more than 60 years old.

156. The number of rows in the truth table of a compound statement having n distinct primary or atomic statements is
    (a) 2^n (b) n^2
    (c) n   (d) 2

157. If S(x) = x is a teacher, W(x) = x is intelligent, Aen symbolic representation of the statement "Every teacher is not intelligent" is
    (a) $\forall x\{S(x)\}$ (b) $\exists x\{S(x) \wedge \sim W(x)\}$
    (c) $\exists x\{\sim S(x) \wedge \sim W(x)\}$ (d) $\exists x\{W(x) \wedge S(x)\}$

158. If F(x) ≡ x is female, T(x) ≡ x is a teacher, P(x) ≡ x is a player, then symbolic representation of the statement "There are some female teachers who are not players" is
    (a) $\exists x\{F(x) \wedge T(x) \wedge \sim P(x)\}$ (b) $\exists x\{F(x) \wedge T(x)\}$
    (c) $\forall x\{F(x) \vee T(x) \wedge P(x)\}$ (d) $\forall x\{\sim P(x) \vee \sim T(x)\}$

159. If T(x) ≡ x is a teacher, M(x) ≡ x is a male then symbolic representation of the statement "All teachers are male" is
    (a) $\forall x\{T(x) \Rightarrow M(x)\}$ (b) $\forall x\{T(x) \vee M(x)|$
    (c) $\exists x\{T(x) \wedge M(x)\}$ (d) $\exists x\{T(x) \wedge \sim M(x)\}$

160. Which of the following is a tautology?
    (a) $\sim\{p \wedge (\sim p)\}$ (b) $p \wedge p$
    (c) $\sim p \wedge \sim p$ (d) $p \vee p$

161. Which of the following is not a tautology?
    (a) $\sim\{p \wedge (\sim p)\}$ (b) $\{p \vee (\sim p)\}$
    (c) $p \wedge (\sim p)$ (d) $\{q \vee (\text{-}q)\}$

162. $p \vee q$ is false when
    (a) p is true, q is false   (b) p is true, q is true
    (c) p is false, q is true   (d) p is false, q is false

163. $p \wedge q$ is true when
    (a) p is true, q is false   (b) p is true, q is true
    (c) p is false, q is true   (d) p is false, q is false

164. $p \Rightarrow q$ is false when
    (a) p is true, q is false   (b) p is false, q is true
    (c) p is true, q is false   (d) p is false, q is false

163. If p is true and q is false, then $p \Leftrightarrow q$ is
    (a) false   (b) true
    (c) neither true nor false   (d) p is false, q is false

166. Which of the following are tautologies?
    (a) $((p \vee q) \wedge q) \Leftrightarrow q$ (b) $((p \vee q) \wedge (\sim p)) \Rightarrow q$
    (c) $(p \vee (p \rightarrow q)) \rightarrow p$ (d) $((p \vee q) \wedge p) \Rightarrow q$

ANSWERS OF MULTIPLE CHOICE QUESTIONS

1. e	2. c	3. b	4. d	5. a	6. d
7. b	8. a	9. e	10. d	11. a	12. b
13. e	14. d	15. a	16. c	17. b	18. d
19. c	20. b	21. e	22. a	23. d	24. b
25. a	26. b	27. a	28. c	29. d	30. a
31. b	32. e	33. c	34. a	35. b	36. d
37. a	38. e	39. c	40. b	41. b	42. c
43. c	44. e	45. a	46. c	47. a	48. d
49. e	50. b	51. a	52. d	53. a	54. c
55. b	56. d	57. e	58. d	59. d	60. b
61. d	62. c	63. b	64. a	65. a	66. c
67. d	68. b	69. a	70. b	71. c	72. c
73. a	74. d	75. b	76. c	77. a	78. d
79. b	80. b	81. a	82. a	83. c	84. a
85. c	86. a	87. e	88. c	89. b	90. a
91. a	92. d	93. b	94. a	95. a	96. d
97. b	98. a	99. c	100. d	101. a	102. c
103. c	104. b	105. b	106. d	107. d	108. d
109. d	110. b	111. a	112. c	113. b	114. b
115. b	116. a	117. c	118. d	119. b	120. a
121. d	122. c	123. b	124. a	125. a	126. c;
127. b	128. c	129. a	130. c	131. b	132. a
133. c	134. a	135. a	136. a	137. b	138. c
139. b	140. c	141. a	142. c	143. a	144. b
145. a	146. b	147. b	148. a	149. b	150. a
151. b	152. a	153. a	154. a		

155 (i) necessary, (ii) necessary; (iii) necessary and sufficient; (iv) sufficient;

156. a	157. b	158. a	159. a	160. a	161. c
162. d	163. b	164. a	165. a	166. a and b.

FILL IN THE BLANKS

1. The set of integers between $\frac{9}{4}$ and $\frac{11}{4}$ is ....
2. Number of elements in a singlet on set is ....
3. The union of set {3, 4, 7} and {3, 7, 9} is ....
4. A ∪ A′ is equal to ....
5. A ∩ A′ is equal to ....
6. A − B is equal to ....

7. $A \cap \phi$ is equal to ....
8. $x \notin A$ and $x \notin B \Leftrightarrow$ ....
9. $x \notin A$ or $x \notin B \Leftrightarrow$ ....
10. $x \in A$ and $x \in B \Leftrightarrow$ ....
11. If $x \in A$ and $x \notin B$ then ....
12. If $x \in A$ or $x \in A$ then ....
13. If $A = \{4, 5\}$, $B = \{5, 6\}$ then $A \times B$ is equal to ....
14. If $A = \{3, 5\}$, then $A \times A$ is equal to ....
15. If two sets P and Q have 3 elements in common and $n(P) = 7$, $n(Q) = 5$, then $n(P \times Q)$ is equal to ....
16. The number of elements in the power set of the set $\{\{\phi\}, 1 \{4, 5\}\}$ is ....
17. If $f: R \to R$ and $f(x) = 3x - 7$, then $f^{-1}(14)$ is equal to ....
18. R is transitive if ....
19. The function $f: Q \to Q$ defined by $f(x) = 3x + 5$, $x \in Q$ is ....
20. The range of the function $f(x) = \dfrac{x^2}{1+x^2}$ is ....
21. The function $f: A \to B$ is a bijection if and only if ....
22. If the operation • distributes over the operation +, then $a \bullet (b + c)$ is equal to ....
23. If + is a binary operation on a set A and e is the identity for this binary operation, then $\forall \ a \in A$.
14. If $A = \{3, 5\}$, then $A \times A$ is equal to ....
15. If two sets P and Q have 3 elements in common and $n(P) = 7$, $n(Q) = 5$, then $n(P \times Q)$ is equal to ....
16. The number of elements in the power set of the set $\{\{\phi\}, 1 \{4, 5\}\}$ is ....
17. If $f: R \to R$ and $f(x) = 3x - 7$, then $f^{-1}(14)$ is equal to ....
18. R is transitive if ....
19. The function $f: Q \to Q$ defined by $f(x) = 3x + 5$, $x \in Q$ is ....
20. The range of the function $f(x) = \dfrac{x^2}{1+x^2}$ is ....
21. The function $f: A \to B$ is a bijection if and only if ....
22. If the operation • distributes over the operation +, then $a \bullet (b + c)$ is equal to ....
23. If + is a binary operation on a set A and e is the identity element for this binary operation, then $\forall \ a \in A$, ...
24. A group $(G, *)$ is abelian group if $\forall \ a, b \in G$ the operation satisfy ....
25. The number of identity element in a group $(G, *)$ is ....
26. The number of inverse of any element $a \in G$ in the group $(G, *)$ is ....
27. If $a, b \in G$ in the group $(G, *)$ then ....
28. In a group $(G, *)$, if $a, b, c, \in G$ then ....

29. In a group (G, *), if $a, b, c \in G$ then $c * a = c * b$ implies that ....
30. A group homomorphism is called isomorphism if the mapping is ....
31. If a semi-group (M, *) has an identity element, then (M, *) is a ....
32. If for some element $a \in G$ of a group (G, *) every element $x \in G$ is of form a^n, where n is an integer then (G, *) is called a ....
33. If G = {– 1, – 1, i, – i} then operation × denotes ordinary multiplication H = {1, –1} is a sub-group of G, then distinct right cosets of H in G are ....
34. If (G, *) is an abelian group, then $\forall\ x, y \in G$ ....
35. Algebraic structure (R, +, •) is a ring if R is an abelian group with respect to the operation + and the second operation • is ....
36. If a ring contains an identity element for the second operation it is called ....
37. If cancellation laws hold in a ring R, then it is ....
38. A ring which (i) is communicate (ii) has unit element and is without zero division is called ....
39. (S, $\subseteq$) is ....
40. The maximum number of zero element and unit element each in a poset is ....
41. The lattice (P(S), $\subseteq$) is bounded with greatest and least element equal to ....
42. The number of compliment of an element in a bounded lattice if it exists, is ....
43. The poset (Z^+, $\leq$) is ....
44. Generating function for the sequence 1, 2, 4, 8, 16, 32, ... is ....
45. The function $A(z) = e^{2z}$ generates the sequence ....
46. The number of non-negative interal solutions of the equation $x + y + z = 9$ is ....
47. The number of non-negative integral solutions of the equation $x + y + z = 7$ is ....
48. The recurrence relation $u_n = 7\ u_{n-1} + n^3$ is ....
49. The recurrence relation $u_n = u^2_{n-1} + u_{n-2}\ u_{n-3}\ u_{n-4}$ is ....
50. The characteristic roots of the characteristic equation $(x – 2) (x – 3)^4 = 0$ are ....
51. If $G_1 = (V_1, E_1)$ and $G_2 = (V_2, E_2)$ are two graphs, then union of G_1 and G_2 is given by $G = G_1 \cup G_2$ where $G = (V, E)$ in which ....
52. Deletion of a vertex implies ....
53. Number of vertices in isomorphic graphs ....
54. Number of edges in isomorphic graphs ....
55. The number of vertices with a given degree in two isomorphic graphs ....

ANSWERS OF FILL IN THE BLANKS

1. a null set 2. 1 3. {3, 4, 7, 9} 4. $\cup$ 5. ϕ 6. $A \cap B'$
7. ϕ 8. $x \notin A \cup B$ 9. $x \notin A \cap B$ 10. $x \notin A \cap B$ 11. $x \in U$ 12. $x \in \phi$
13. {(4, 5) (4, 6, (5, 5), (5, 6))} 14. {(3, 3), (3, 5), (5, 3), (5, 5)} 15. 35
16. 8 17. 7 18. $a\,R\,b$ and $b\,R\,c \Rightarrow a\,R\,c$ 19. one-one onto
20. {0, 1} 21. $f(a) = f(b) \Rightarrow a = b\ \forall\ a, b$ 22. $a • b + a • c$ 23. $a + e = a = e + a$

24. $a*b = b*a$ 25. 1   26. 1   27. $(a*b)^{-1} = b^{-1} * a^{-1}$
28. $a*c = b*c \Rightarrow a = b$ 29. $a = b$ 30. one-one onto   31. Monoid
32. cyclic group   33. H and Hi 34. $(x*y)^2 = x^2 * y^2$
35. associative and distributes over the first operation
36. ring with unity   37. a ring without zero divisions
38. integral domain   39. a poset   40. 1   41. δ and ϕ respectively
42. 1   43. a lattice   44. $(1-2z)^{-1}$ 45. $1, \dfrac{2}{\lfloor 1}, \dfrac{4}{\lfloor 2}, \dfrac{8}{\lfloor 3}, ...$ 46. $^8C_2$
47. $^9C_2$ 48. non-homogenous of order 1 and degree 1.
49. homogeneous of order 4 and degree 3
50. $x = 2$ of multiplicity 1 and $x = 3$ of multiplicity 4
51. $V = V_1 \cup V_2$ and $E = E_1 \cup E_2$
52. removal fo all edges incident on that vertex.
53. is always the same   54. is always the same
55. is always the same

TRUE/FALSE QUESTIONS

1. A tree is always a connecte graph.
2. A tree has self loop as well as parallel edges.
3. A tree is an acyclic connected graph.
4. A tree with n vertices has n edges.
5. Number of paths between any two vertices of a tree is exactly 1.
6. An arborescence is a directed tree in which in degree of every vertex other than the root is zero.
7. In a tree with two on more vertices the number pendant vertices is 1.
8. The eccentricity of centre in a tree is called its diameter.
9. In a binary tree each vertex can have at most 2 children.
10. If every vertex of a binary tree has either two children on no child, it is called full or complete binary tree.
11. Number of vertices in a binary tree can be even on odd.
12. The number of pendant vertices in a binary trees with n vertices is 0.
13. The number of labelled tree with 4 vertices is one.
14. Each chromatic number of a cycle have 8 edges is zero.
15. Edge chromatic number of a cyclic having 9 edges is zero.
16. Chromatic nmber of a bipantite graph with edge set non-empty is zero.
17. Chromatic number of a tree with 10 vertices is zero.
18. The vertices of every planner graph can be property coloured with one colour.
19. If a set A = {2, 4, 7, 9, 10, 12}, then the number of subsets of A having less than 4 elements are 26.
20. The number of numers from 1 to 100 that are divisible by 2 or 3 is 67.

Answer of True/False

1. True
2. False
3. True
4. False
5. True
6. False
7. False
8. False
9. True
10. True
11. False
12. False
13. False
14. False
15. False
16. False
17. False
18. False
19. True
20. True

APPENDIX

AUTOMATA, GRAMMARS AND LANGUAGES

FINITE STATE MACHINES

Finite-State Machines. Combinatorial circuits have no memory or internal states, their output depends only on the current values of their inputs. Finite state machines on the other hand have internal states, so their output may depend not only on its current inputs but also on the past history of those inputs.

A *finite-state machine* consists of the following :
1. A finite set of states S.
2. A finite set of input symbols $\mathcal{J}$.
3. A finite set of output symbols $\mathcal{O}$.
4. A next-state or transition function $f: S \times \mathcal{J} \to S$.
5. An output function $g: S \times \mathcal{J} \to \mathcal{O}$.
6. An initial state $\sigma \in S$.

We represent the machine $M = (S, \mathcal{J}, \mathcal{O}, f, g, \sigma)$

Example 1: We describe a finite state machine with two input symbols $\mathcal{J} = \{a, b\}$ and two output symbols $\mathcal{O} = \{0, 1\}$ that accepts any string from $\mathcal{J}^*$ and outputs as many 1's as a's there are at the beginning of the string, then it outputs only 0's. The internal states are $S = \{\sigma_0, \sigma_1\}$, where σ_0 is the initial state—we interpret it as not having seeing any "b" yet; then the machine will switch to σ_1 as soon as the first "b" arrives. The next-state and output functions are as follows :

	$\mathcal{J}$	f		g	
S		a	b	a	b
σ_0		σ_0	σ_1	1	0
σ_1		σ_1	σ_1	0	0

This finite-state machine also can be represented with the following *transition diagram*:

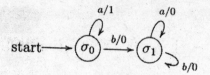

The vertices of the diagram are the states. If in state σ an input *i* causes the machine to output *o* and go to state σ′ then we draw an arrow from σ to σ′ labeled *i/o* or *i, o*.

Example 2: The following example is similar to the previous one but the machine outputs 1 only after a change of input symbol, otherwise it outputs 0:

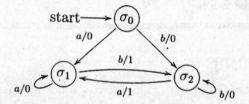

Example 3: A Serial-Adder. A serial adder accepts two bits and out-puts its sum. So the input set is $\mathcal{I} = \{00, 01, 10, 11\}$. The output set is $\mathcal{O} = \{0, 1\}$. The set of states is $S = \{NC, C\}$, which stands for "no carry" and "carry" respectively. The transition diagram is the following:

A.2 FINITE-STATE AUTOMATA

A *finite-state automaton* is similar to a finite-state machine but with no output, and with a set of states called *accepting* or *final states*. More specifically, finite-state automaton consists of:

1. A finite set of *states* S.
2. A finite set of *input symbols* $\mathcal{I}$.
3. A *next-state* or *transition function* $f: S \times \mathcal{I} \to S$.
4. An *initial state* $\sigma \in S$.
5. A subset $\mathcal{F} \subseteq S$ of *accepting* or *final states*.

We represent the automaton $A = (S, \mathcal{I}, f, \sigma, \mathcal{F})$. We say that an automaton *accepts* or *recognizes* a given string of input symbols if that strings takes the automaton from the start state to a final state.

Example 1: The following transition diagrams represent an automaton accepting any string of *a*'s and *b*'s ending with an *a*. The first diagram uses the same scheme as with finite-state machines, with 1 representing "accept" or "recognize", and "0" representing "not accept":

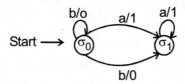

The second kind of diagram omits the outputs and represents the accepting states with double circles:

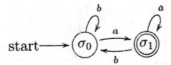

Two finite-state automata that accept exactly the same set of strings are said to be *equivalent*. For instance the following automaton also accepts precisely strings of *a*'s and *b*'s that end with an *a*, so it is equivalent to the automaton shown above:

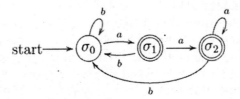

Example 2: The following automaton accepts strings of *a*'s and *b*'s with exactly an even number of *a*'s:

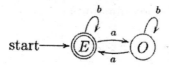

Example 3: The following automaton accepts strings starting with one a followed by any number of *b*'s :

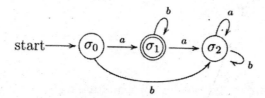

Example 4: The following automaton accepts strings ending with *aba* :

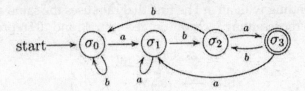

Languages and Grammars

Formal Languages. Consider algebraic expressions written with the symbols A = $\{x, y, z, +, *, (,)\}$. The following are some of them: "$x + y * y$", "$y + (x * y + y) * x$", "$(x + y) * x + z$", etc. There are however some strings of symbols that are not legitimate algebraic expressions, because they have some sort of syntax error, e.g.: "$(x + y$", "$z + + y * x$", "$x(*y) + z$", etc. So syntactically correct algebraic expressions are a subset of the whole set A^* of possible strings over A.

In general, given a finite set A (the *alphabet*), a (*formal*) *language* over A is a subset of A^* (set of strings of A).

Although in principle any subset of A^* is a formal language, we are interested only in languages with certain structure. For instance: let A — $\{a, b\}$. The set of strings over A with an even number of a's is a language over A.

Grammars. A way to determine the structure of a language is with a *grammar*. In order to define a grammar we need two kinds of symbols: *non-terminal*, used to represent given subsets of the language, and *terminal*, the final symbols that occur in the strings of the language. For instance in the example about algebraic expressions mentioned above, the final symbols are the elements of the set $A = \{x, y, z, +, *, (,)\}$. The non-terminal symbols can be chosen to represent a complete algebraic expression (E), or terms (T) consisting of product of factors (F). Then we can say that an algebraic expression E consists of a single term

$$E \to T,$$

or the sum of an algebraic expression and a term

$$E \to E + T.$$

A term may consist of a factor or a product of a term and a factor

$$T \to F$$
$$T \to T * F$$

A factor may consist of an algebraic expression between parenthesis

$$F \to (E),$$

or an isolated terminal symbol

$$F \to x,$$
$$F \to y,$$
$$F \to z.$$

Those expressions are called *productions*, and tell us how we can generate syntactically correct algebraic expressions by replacing successively the symbols on the left by the expressions on the right. For instance the algebraic expression "$y + (x * y + y) * x$" can be generated like this:

$$E \Rightarrow E + T \Rightarrow T + T \Rightarrow F + T \Rightarrow y + T \Rightarrow y + T * F \Rightarrow y + F * F \Rightarrow$$
$$y + (E) * F \Rightarrow y + (E + T) * F \Rightarrow y + (T + T) * F \Rightarrow y + (T * F + T) * F \Rightarrow$$
$$y + (F * F + T) * F \Rightarrow y + (x * T + T) * F \Rightarrow y + (x * F + T) * F \Rightarrow$$
$$y + (x * y + T) * F \Rightarrow y + (x * y + F) * T \Rightarrow y + (x * y + y) * F \Rightarrow$$
$$y + (x * y + y) * x.$$

In general a *phrase-structure grammar* (or simply, *grammar*) G consists of:
1. A finite set V of symbols called *vocabulary* or *alphabet*.
2. A subset $T \subseteq V$ of *terminal symbols*. . The elements of $N = V - T$ are called *nonterminal symbols* or *'nonterminals*.
3. A start symbol $\sigma \in N$.
4. A finite subset P of $(V^* - T^*) \times V^*$ called the set of *productions*.

We write $G = (V, T, \sigma, P)$.

A production $(A, B) \in P$ is written:
$$A \to B.$$

The right hand side of a production can be any combination of terminal and nonterminal symbols. The left hand side must contain at least one nonterminal symbol.

If $\alpha \to \beta$ is a production and $x\alpha y \in V^*$, we say that $x\beta y$ is *directly derivable* from $x\alpha y$, and we write
$$x\alpha y \Rightarrow x\beta y.$$

If we have $\alpha_1 \Rightarrow \alpha_2 \Rightarrow \ldots \Rightarrow \alpha_n$ ($n \geq 0$), we say that α_n is *derivable* from σ_1, and we write $\alpha_1 \stackrel{*}{\Rightarrow} \alpha_n$ (by convention also $\alpha_1 \stackrel{*}{\Rightarrow} \alpha_1$).

Given a grammar G, the language L(G) associated to this grammar is the subset of T^* consisting of all strings derivable from σ.

Backus Normal Form. The *Backus Normal Form* or *BNF* is an alternative way to represent productions. The production $S \to T$ is written $S ::= T$. Productions of the form $S ::= T_1, S ::= T_2, \ldots, S ::= T_n$ can be combined as
$$S ::= T_1 \mid T_2 \mid \ldots \mid T_n.$$

So, for instance, the grammar of algebraic expressions defined above can be written in BNF as follows :
$$E ::= T \mid E + T$$
$$T ::= F \mid T * F$$
$$F ::= (E) \mid x \mid y \mid z$$

Combining Grammars. Let $G_1 = (V_1, T_1, \sigma_1, P_1)$ and $G_2 = (V_2, T_2, \sigma_2, P_2)$ be two grammars, where $N_1 = V_1 - T_1$ and $N_2 = V_2 - T_2$ are disjoint (rename nonterminal symbols if necessary). Let $L_1 = L(G_1)$ and $L_2 = L(G_2)$ be the languages associated respectively to G_1 and G_2. Also assume that σ is a new symbol not in $V_1 \cup V_2$. Then

1. *Union Rule*: the language union of L_1 and L_1
$$L_1 \cup L_2 = \{\alpha \mid \alpha \in L_1 \text{ or } \alpha \in L_1\}$$
starts with the two productions
$$\sigma \to \sigma_1, \sigma \to \sigma_2$$

2. *Product Rule:* the language *product of* L_1 *and* L_2
$$L_1 L_2 = \{\alpha\beta \mid \alpha \in L_1, \beta \in L_1\}$$
where $\alpha\beta$ = string concatenation of α and β, starts with the production
$$\sigma \to \sigma_1 \sigma_2$$
3. *Closure Rule:* the language *closure of* L_1
$$L_1^* = L_1^0 \cup L_1^1 \cup L_1^2 \cup \ldots$$
where $L_1^0 = \{\lambda\}$ and $L_1^n = \{\alpha_1 \alpha_2 \ldots \alpha_n \mid \alpha_\kappa \in L_1, k = 1, 2, \ldots, n\}$
($n = 1, 2, \ldots$), starts with the two productions
$$\sigma \to \sigma_1 \sigma, \quad \sigma \to \lambda$$

Types of Grammars (Chomsky's Classification). Let G be a grammar and let λ denote the null string.

0. G is a *phrase-structure* (or type 0) *grammar* if every production is of the form:
$$\alpha \to \delta,$$
where $\alpha \in V^* - T^*, \delta \in V^*$.

1. G is a *context-sensitive* (or type 1) *grammar* if every production is of the form:
$$\alpha A \beta \to \alpha \delta \beta$$
(*i.e.*: we may replace A with δ in the context of α and β),
where $\alpha, \beta \in V^*, A \in N \; \delta \in V^* - \{\lambda\}$

2. G is a *context-free* (or type 2) *grammar* if every production is of the form:
$$A \to \delta$$
where $A \in N, \delta \in V^*$.

3. G is a *regular* (or type 3) *grammar* if every production is of the form:
$A \to a$ or $A \to aB$ or $A \to \lambda$
where $A, B \in N, a \in T$.

A language L is *context-sensitive* (respectively *context-free, regular)* if there is a context-sensitive (respectively context-free, regular) grammar G such that $L = L(G)$.

The following examples show that these grammars define different kinds of languages.

Example 1: The following language is type 3 (regular):
$$L = \{a^n b^m \mid n = 1, 2, 3, \ldots m = 1, 2, 3, \ldots\}.$$
A type 3 grammar for that language is the following: $T = \{a, b\}$, $N = \{\sigma, S\}$, with start symbol σ, and productions:
$$\sigma \to a\sigma, \quad \sigma \to aS, \quad S \to bS, \quad S \to b.$$

Example 2: The following language is type 2 (context-free) but not type 3:
$$L = \{a^n b^n \mid n = 1, 2, 3, \ldots\}.$$
A type 2 grammar for that language is the following:
$T = \{a, b\}$, $N = \{\sigma\}$, with start symbol σ, and productions
$$\sigma \to a\sigma b, \quad \sigma \to ab$$

Example 3: The following language is type 1 (context-sensitive) but not type 2:
$$L = \{a^n b^n c^n \mid n = 1, 2, 3, \ldots\}.$$

A type 1 grammar for that language is the following :
T = {a, b, c}, N = {σ, A, C}, with start symbol σ, and productions

$$\sigma \to abc, \quad \sigma \to aAbc,$$
$$A \to abC, \quad A \to aAbC$$
$$Cb \to bC, \quad Cc \to cc.$$

.There are also type 0 languages that are not type 1, but they are harder to describe.

Equivalent Grammars. Two grammars G and G" are equivalent if L(G) = L(G').

Example: The grammar of algebraic expressions defined at the beginning of the section is equivalent to the following one :

Terminal symbols = {x, y, z, +, *, (,)}, nonterminal symbols = {E, T, F, L}, with start symbol E, and productions

$$E \to T, \quad E \to E + T$$
$$T \to F, \quad T \to T * F$$
$$F \to (E), \quad F \to L,$$
$$L \to x, L \to y, L \to z$$

Context-Free Interactive Lindenmayer Grammar. A context-free interactive Lindenmayer grammar is similar to a usual context-free grammar with the difference that it allows productions of the form A → B where A ∈ N ∪ T (in a context free grammar A must be nonterminal). Its rules for deriving strings also are different. In a context-free interactive Lindenmayer grammar, to derive string β from string α, all symbols in α must be replaced *simultaneously*.

Example: The von Koch Snowflake. The von Koch Snowfiake is a fractal curve obtained by start with a line segment and then at each stage transforming all segments of the" figure into a four segment polygonal line, as shown below. The von Koch Snowflake fractal is the limit of the sequence of curves defined by that process.

Figure A.1: Von Koch Snowflake, stages 1-3.

Figure A.2: Von Koch Snowflake, stages 4-5

A way to represent an intermediate stage of the making of the fractal is by representing it as a sequence of movements of three kinds: 'd' = draw a straight line (of a fix length) in the current direction, r = turn right by 60°, l = turn left by 60°. For instance we start with a single horizontal line d, which we then transform into the polygonal *dldrrdld,* then each segment is transformed into a polygonal according to the rule $d \to dldrrdld,$ so we get *dldrrdldldldrrdldrrdldrrdldldldrrdld*

If we represent by D a segment that may no be final yet, then the sequences of commands used to build any intermediate stage of the curve can be defined with the following grammar:

N = {D}, T = {d, r, l}, with start symbol D, and productions:

$$D \to DlDrrDlD, \quad D \to d, \quad r \to r, \quad l \to l.$$

Example: The Peano curve. The Peano curve is a space filling curve, *i.e.*, a function $f: [0, 1] \to [0, 1]^2$ such that the range of f is the whole square $[0, 1]^2$, defined as the limit of the sequence of curves shown in the figures below.

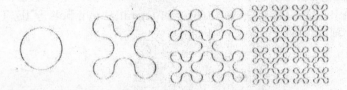

Figure A.3: Peano curve, stages 1-4.

Each element of that sequence of curves can be described as a sequence of 90° arcs drawn either anticlockwise ('*l*') or clockwise ('*r*'). The corresponding grammar is as follows:

T = {*l, r*}, N = {C, L, R}, with and start symbol C, and productions

C → LLLL,

L → RLLLR, R → RLR,

L → *l*, R → *r*, *l* → *l*, *r* → *r*.

Language Recognition

Regular Languages. Recall that a regular language is the language associated to a regular grammar, *i.e.*, a grammar G = (V, T, σ, P) in which every production is of the form:

A → *a* or A → *a*B or A → λ,

where A, B ∈ N = V − T, *a* ∈ T.

Regular languages over an alphabet T have the following properties (recall that λ = 'empty string', αβ = 'concatenation of α and β', α^n = 'α concatenated with itself *n* times'):

1. ϕ, {λ}, and {*a*} are regular languages for all *a* ∈ T.
2. If L_1 and L_2 are regular languages over T the following languages also are regular:

$$L_1 \cup L_2 = \{\alpha \mid \alpha \in L_1 \text{ or } \alpha \in L_2\}$$

$$L_1 L_2 = \{\alpha\beta \mid \alpha \in L_1 \; \beta \in L_2\}$$

$$L_1^* = \{\alpha_1 \ldots \alpha_n / \alpha_k \in L_1 \; n \in \mathbb{N}\}$$

$$T^* - L_1 = \{\alpha \in T^* \mid \alpha \notin L_1\}$$

$$L_1 \cap L_2 = \{\alpha \mid \alpha \in L_1 \text{ and } \alpha \in L_2\}$$

We justify the above claims about $L_1 \cup L_2$, $L_1 L_2$ and L_1^* as follows. We already know how to combine two grammars L_1 and L_2 to obtain $L_1 \cup L_2$, $L_1 L_2$ and L_1^*, the only problem

is that the rules do no have the form of a regular grammar, so we need to modify them slightly.
1. *Union Rule:* Instead of adding $\sigma . \to \sigma_1$ and $\sigma \to \sigma_2$, add all productions of the form $\sigma \to RHS$, where RHS is the right hand side of some production $(\sigma_1 \to RHS) \in P_1$ or $(\sigma_2 \to RHS) \in P_2$.
2. *Product Rule:* Instead of adding $\sigma \to \sigma_1 \sigma_2$, use σ_1 as start symbol and replace each production $(A \to a) \in P_1$ with $A \to a\sigma_2$ and $(A \to \lambda) \in P_1$ with $A \to \sigma_2$.
3. *Closure Rule:* Instead of adding $\sigma \to \sigma_1 \sigma$ and $\sigma \to \lambda$, use σ_1 as start symbol, add $\sigma_1 \to \lambda$, and replace each production $(A \to a) \in P_1$ with $A \to a\sigma_1$ and $(A \to \lambda) \in P_1$ with $A \to \sigma_1$.

GLOSSARY

Adjacent Vertices, Adjacency Matrix

Two vertices in a graph are said to be adjacent if they are connected directly by an edge.

An adjacency matrix is a means of representing a graph in the form of a matrix. In the adjacency matrix, both the rows and columns represent the vertices. If If two vertices are adjacent, then a 1 is place in the corresponding position in the matrix, otherwise a zero appears. Note this applies to a simple graph, where there are no multiple edges connecting different vertices, nor any loops. In the case of such multigraphs, the entries of the matrix will be integers larger than one. If the graph is not a digraph, the adjacency matrix will be symmetric.

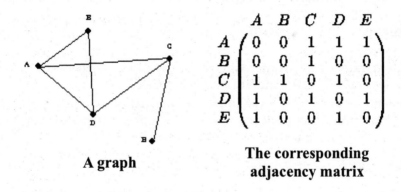

A graph · The corresponding adjacency matrix

Algorithm: In essence, an algorithm is a sequence of instructions that, if followed for an operation (whether mathematical or not), will always lead to a result.

Apportionment: Apportionment deals with the issue of dividing items of value among a group of people who desire them. Apportionment may include division of discrete objects, as in the division of seats of a representative government among districts, or may involve division of continuous objects such as dividing a cake among a group of people. In addition, topics such as estate division combine division of both discrete and continuous objects. Apportionment represents an aspect of both fair division as well as an aspect of election theory.

Apportionment Algorithms: Various apportionment algorithms are used to allocate seats in a representative government system. The need for apportionment algorithms arise when the number of seats given to each group (possibly states, or student classes) cannot exactly represent their relative sizes in terms of the ideal ratio. Apportionment algorithms include the Hamilton method, the Hill method, the Jefferson method and the Webster method.

In addition, various apportionment algorithms are used for dividing estates as well as continuous objects, such as cakes using the cut and choose method.

Approval Voting: In approval voting, you may vote for as many choices as you like, but you do not rank them. You mark all those of which you approve. Approval voting violates Arrow's criteria less than any other known method.

Arithmetic Series: Arrow's Conditions, Arrow's Criteria and Arrow's Paradox

Arrow's conditions (or criteria) are a list of five conditions which Kenneth Arrow selected as necessary for a fair determination of a group ranking. These conditions are:

1. **Nondictatorship:** The preferences of a single individual should not become the group ranking without considering the other individuals' preferences.
2. **Individual sovereignty:** Each individual is allowed to order the choices in any way and can indicate ties.
3. **Unanimity:** If every individual prefers one choice to another, the group ranking should do the same.
4. **Freedom from irrelevant alternatives:** The group ranking between any pair of choices does not depend on the individuals' preferences for the remaining choices.
5. **Uniqueness of the group ranking:** The method of producing the group ranking should give the same result whenever it is applied to a given set of preferences. The group ranking must also be transitive.

Binary Tree: A binary tree is a tree in which each vertex has at most two children.

Binomial Theorem: For all possible integers *n* and numbers *x* and *y*,

$$(x+y)^n = \binom{n}{0}x\sqrt{n} + \binom{n}{1}x^{n-1}y + \binom{n}{2}x^{n-2}y^2 + ... + \binom{n}{k}x^{n-k}y^k + ... + \binom{n}{n}y^n$$

or, equivalently,

$$(x+y)^n = \sum_{k=0}^{n} \binom{n}{k} x^{n-k} y^k$$

Borda Count, Borda Method:

The Borda method is a group ranking method which assigns weights to individual preference schedules by giving a number of points for each first place vote another number for each second place vote and so on. The most common way of assigning values in a group ranking of *n* choices is to assign *n* points to first place, *n-1* points to second, and so on. The group ranking is determined by totalling each choice's points

Child, Children: A child of a vertex (called the parent) is an adjacent vertex one level lower on the tree.

Chromatic Number: The chromatic number of a graph is the minimum number of colors needed to color the graph.

Circuit: A circuit is a path which begins and ends at the same vertex. Note that a circuit does not necessarily have to pass by every vertex or edge, nor is a circuit restricted to passing vertices and edges only once. Compare with a cycle.

Combination: A situation in which the order of selecting various items does not matter. The combination of *n* things taken *m* at a time is written formally as $C(n,m)$ or $_nC_m$ or $\binom{n}{m}$ In general, $C(n,m)$ is calculated by evaluating the expression $P(n,m)/m!$, But

$$P(m,n) = \frac{n!}{(n-m)!} \text{ SO } C(n,m) = \frac{n!}{(n-m)!m!}.$$

Complete Graph: A complete graph is a graph where each vertex is connected to every other with a single edge.

Conditional Probability: A conditional probability is the probability of one event occurring given that another event has already occurred. For instance, if we are rolling two dice, the probability that their sum will be greater than seven is 11/36 (since of the 36 possible rolls, 11 are eight or greater), however the conditional probability that their sum will be greater than seven given that the first die rolled is a 4 is 2/6 or 1/2 (since we must roll a five or a six).

Condorcet Method, Condorcet Paradox: The Condorcet method is a means of determining a winner of an election using group rankings. Condorcet's method requires that the winning choice be able to defeat each of the other choice in a one-on-one contest. Unfortunately, for a given a set of preference schedules, it may be impossible to generate a Condorcet winner.

What this means is that in a set of head-to-head elections, it would be possible for choice A to beat choice B, choice B would beat choice C, yet choice C would beat choice A. This represents the Condorcet paradox. For instance, in this set of preference schedules, A would beat B by 10 votes to 9, B would beat C by 10 votes to 9 yet C would beat A by 10 votes to 9 as well!

Continuous Mathematics: A branch of Mathematics which deals with problems that are based on a continuous set of numbers. As opposed to Discrete Mathematics.

Counting Problems, Counting Techniques: A category of discrete Mathematics which investigates how many solutions exist for problems with known solutions.

Critical Path: When a graph is used to model project scheduling, the critical path is a path from the start to the finish which passes through all of the tasks which are critical to completing the project in the shortest amount of time.

Cycle: A cycle is a path that begins and ends at the same vertex and does not use any edge or vertex more than once. Note that a cycle is necessarily a curcuit but a curcuit is not necessarily a cycle.

Degree: In a graph, the degree of a vertex is the number of edges which have that vertex as an endpoint.

Dimension: The dimenstion of a square matrix is the number of rows (or columns in that matrix.

Directed Graph: A directed graph is a graph which has edges which can only be traversed in one direction.

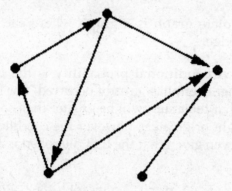

A directed graph

Discrete Mathematics: Discrete mathematics problems can be classified in three broad categories, existence problems, counting problems and optimization problems.

Earliest Start Time: The earliest start time (EST) is the earliest that an activity can begin if all the activities preceding it begin as early as possible.

Edge: An edge connects a pair of vertices.

Election Theory: Branch of social decision making which deals with making decisions based on preferences of a group of individuals.

Estate Division: Estate division is a category of fair division which deals with determining a fair method for dividing an estate among a number heirs.

Euler Circuit: An Euler circuit is a circuit which traverses each edge of a graph exactly once. (Since this is a circuit, it must start and end at the same vertex.)

Euler Path: An Euler path is a path which traverses each edge of a graph exactly once. (Since this is a path, the route need not start and end at the same vertex.)

Existence Problems: A category of discrete mathematics which deals with whether a given problem has a solution or not.

Fair Division, Fairness: Fair division deals with dividing an object among a number of people in such a way as to treat each person fairly. Fair division issues include topics such as apportionment, divide continuous items such as food, or estate division. In the continuous case, we call a division among n people *fair* if each person feels that he or she has received at least $1/n$th of the object.

Four Colour Theorem: The four Colour Theorem, proved in 1976 by Appel and Haken, states that all planar graphs can be coloured with only four colors

Graph: Informally, a graph consists of a finite set of vertices and edges which connect them. Graphs are usually depicted pictorially as a set of points representing the vertices with lines (usually straight, but not necessarily so) connecting them to represent the edges. Types of graphs are: simple, directed or digraphs, multigraphs or planar.

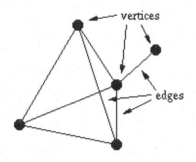

A Graph

Graph Colouring: Graph colouring consists of of assigning colours to the vertices of a graph so that adjacent vertices are given different colors.

Group Ranking: A group ranking is a determination of the preferences of a group towards a set of choices. Often the preferences of the individuals are represented in the form of preference schedules. Some group ranking methods include the Borda method, the Condorcet Method or plurality and sequential runoff winners, among others.

Hamilton Method: The Hamilton method is an apportionment algorithm which assigns additional representative seats based on looking at the decimal portion of the quotas for the various groups. The group with the largest decimal remainder to their qouta is given the first additional extra seat.

Hamiltonian cycle, path or curcuit: In a graph a Hamiltonian path is a path that contains each vertex once and only once. A Hamiltonian cycle is a cycle that includes each vertex.

Hill Method of Apportionment: The Hill method is an apportionment algorithm which is used to assign seats in a representative body. The Hill method requires adjusting the ideal ratio until the quota of one of the groups exceeds the geometric mean of two consecutive integers. While seemingly complicated, the only difference between the the Hill method, the Jefferson method and the Webster method is how the quotas are rounded.

Ideal Ratio</>: When representative seats are being apportioned amongst various groups, the ideal ratio represents the number of people each seat should ideally represent. For example, if the total population of a school is 400 and there are 20 voting seats in a student council, then each seat should ideally represent 20 students. Usually, however, it is impossible to allocate the seats in such a manner so that each group has the ideal number of seats represented so one of various apportionment algorithms must be implemented.

Independent Random Events: Two events A and B are said to be independent if the probability of both events occurring is the same as the product of each of the events occurring on their own. Symbolically,

Insincere Voting: Insincere voting occurs when people vote differently than the way they actually feel. Insincere voting usually occurs when people have some knowledge beforehand of the possible outcomes of an election. Insincere voting has been seeen in plurality elections, such as three-party presidential elections as well as sports ranking polls, such as the AP football poll.

Jefferson Method: The Jefferson method is an apportionment algorithm which assigns representative seats by decreasing the ideal ratio until the quota for one of the groups exceeds an integer value. This group is then given that number of seats.

Latest Start Time: In a project scheduling situation, the latest that a task can begin without delaying the project's completion is known as the latest start time (LST) for the task.

Leaf, leaves: A leaf is a vertex in a tree which has no children.

Loop: In a graph, a loop is an edge which connects a vertex to itself.

Majority Winner: A choice which is placed first on over half of a group's preference schedules is the majority winner of that group's election.

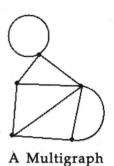

A Multigraph

    Markov Chain

    Matrix

Multigraph: A multigraph is a graph in which there are multiple edges connecting the same pair of vertices. If an edge connects a vertex to itself, then that edge forms a loop.

Multiplication Principle: The multiplication principle states that if events A and B can occur in *a* and *b* ways, respectively, then events A *and* B can occur together in *a times b* ways.

    The word *and* in the description of an event often indicates that the multiplication principle should be used.

Nearest Neighbour Algorithm: The nearest neighbor algorithm is a method for solving the traveling salesperson problem. The algorithm requires following a path which always travels along the edge with the lowest weight connecting the current vertex to another which has not yet been visited. Note that the nearest neighbor algorithm will find a circuit, however, there is no guarantee that this solution will represent the minimal possible route.

Odds: The odds of an event occuring is the ratio of the probability of that event occurring to the probability of the event not occuring. Note that this is significantly different from just the probability of the event occurring. For example, the probability of rolling a six on one fair die is 1/6, however, the *odds* of rolling a six is 1:5.

Optimization Problems: A category of discrete mathematics which focuses on finding a best solution for a particular problem.

Pascal's Triangle: Pascal's triangle is an array of numbers that is constructed by beginning with a row of two ones: 1 1. Each new row after the starts and ends with a one, and the other numbers are formed by adding the numbers above and on either side of them:

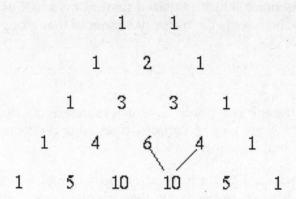

Path: A path is a sequence of adjacent vertices. In other words, a path is a route through a graph which travels along the edges from vertex to vertex. If a path starts and ends at the same vertex, then it is a circuit.

Permutation: A situation in which order matters. In general, the number of permutations of *n* things taken *m* at a time is denoted as *P(n, m)* and is calculated by evaluating the expression

$$P(m,n) = \frac{nl}{(n-m)l}$$

Planar Graph: A planar graph is a graph which can be drawn in the plane with no edges crossing.

Plurality Winner: In a election, the choice which receives the most first place votes is determined to be the plurality winner.

Power Index: In a situation where weighted voting is used, the number of winning coalitions to which a group is essential is that group's power index.

Preference Schedule: A list of alternatives and individual rankings, sometimes presented as a graph with higher-ranked preferences listed above lower-ranked ones. For instance, this preference schedule indicates that the individual prefers option E best, followed by option C, then A, then B. Option D would be this individual's least preferred choice.

Prerequisite: In a project scheduling situation, any tasks which must be completed before a given task can be started are that task's prerequisites.

Probability: When an event can occur in a finite number of discrete outcomes, the probability of an event is the ratio of the number of ways in which the event can occur to the total number of possibilities, assuming that each of them is equally likely.

Quota: In a representative body, the quota is defined as the ratio of the size of the popluation of a group being represented to the ideal ratio. The quota represents the ideal number of representatives that that specific group should have.

Root: A root of a tree is the vertex in a tree such that there is a uniques route from the root to any other vertex in the tree. In a digraph, if a root exists, it is unique. If the tree is undirected, the any vertex can be a root.

Runoff Elections: In a runoff election, all choices but the two receiving the most votes are eliminated and then these two choices are again voted upon by the group. In terms of a typical election, this would require that the population vote twice, once for the initial decision of which two choices would run off, and then again to determine the winner. Using preference schedules, we can model both votes by examining which two choices receive the most first place votes and then comparing the head-to-head preferences of the group for those two choices. See the <u>sequential runoff</u> method.

Sequential Runoff: A sequential runoff election is similar to a basic runoff election except that instead of eliminating all choices but two and then deciding between those two, choices are eliminated one at a time, starting with the one that receives the least number of first place votes.

Shortest Path Algorithm: The shortest path algorithm is a method for finding the shortest route from one vertex to another in a weighted graph. The shortest path algorithm requires extending a tree from the starting vertex until it reaches the finishing vertex. At each step, the tree is extended in a direction such that the new leaf is at the shortest possible distance from the root.

Simple Graph: A simple graph is a graph which is not a multigraph. In other words, a simple graph is a graph which has at most one edge connecting each pair of vertices.

Social Decision Making: Social Decision Making includes the topics of Election Theory, Fair Division and Apportionment.

Spanning Tree: A spanning tree is a tree which is a subgraph of a graph which contains all of the vertices of the graph.

Subgraph: A subgraph is of a graph is itself a graph in which all of the edges and vertices are contained within the original graph.

Task Table: A task table is a list of tasks, the time needed to complete them and the prerequisite tasks which must be completed before each can be started. A task table is used to organize information in a project scheduling situation.

Travelling Salesperson Problem: "The Travelling Salesperson Problem calls for a minimal-cost Hamilton circuit in a complete graph having an associated cost matrix C. Entry c_{ij} in C is the cost of using the edge from the ith vertex to the jth vertex. Minimal-cost means minimizing the sum of the costs of the edges." In other words, given a complete graph with weighted edges, the Traveling Salesperson Problem asks for the shortest route which visits all of the edges.

Tree: A tree is a graph which contains no cycles. We can visualize a tree by drawing it with a root at the top with the vertices below leading to the leaves at the lowest. If the vertices are placed on levels, higher level vertices are referred to the parents of the vertices directly below them, while the lower vertices are similarly referred to as their children.

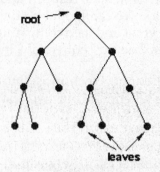

A Tree

Vertex: Informally, a vertex is located at the end of an edge a
nd is usually represented pictorially by a dot.

Webster Method: The Webster method is an apportionment algorithm used to determine how seats in a representative body are assigned to various groups. The Webster method involves decreasing the ideal ratio until the quota of one of the groups passes the arithmetic mean of two consecutive integers (or, in other words, until the quota rounds up instead of down.)

Weighted Digraph: A weighted digraph is a weighted graph which is also a directed graph.

Weighted Graph: A weighted graph is a graph in which a number (often representing a distance) is assigned to each edge.

Weighted Voting: Weighted voting occurs when different numbers of votes are allotted to different members of a voting body. This can occur when voting members represent a group of people, such as on a school's student council where each representative is given a different number of votes depending on the size of the class that they represent.

Winning Coalition: A winning coalition is a collection of representatives voting under a weighted voting scheme which together has enough votes to pass an issue.

INDEX

A
Abelian Group 127
Ackermann Function 96
And Gate 233
Antisymmetric Relation 43
Asymmetric Relation 43

B
'Big-Omega' 93
'Big-oh' 93
Binary Operation 121
Binary Trees 288
Binomial Coefficients 340
Bipartite Graph 254
Boolean Function 242
Boolean Rings 175
Brute-force Search 426

C
Cauchy Product 354
Cayley's Theorem 159
Chromatic Polynomial 422
Closed Walk 253
Closure of Relations 68
Co-domain 84
Combinations 339
Commutative Law 174
Comparability 209
Complement of A Set 5
Complete Bipartite Graph 254
Complete Graph 253
Composition Table 124
Conjunction Operation 314
Connected Graph 253
Constant Function 85
Cosets 150
Counting Techniques 337
Cyclic Groups 149

D
Demorgan's Laws 13
Digraph 251
Dihedral Groups 166
Direct Sum 176
Disconnected Graph 253
Disjoint Sets 4
Disjunction Operation 314
Distinct 1
Domain 84

E
Edge Coloring 421
Empty Set 2
Equal Set 2
Equivalence Classes 54
Equivalence Relation 46
Euclidean Domains 198
Eulerian Graphs 261
Even Integers 1
Exclusive or Operation 314
Existential Quantification 318

F
Face Colouring 421
Fallacies 110
Fibonacci Sequence 96
Field of Fractions 204
Finite Fields 203
Finite Group 127
Finite Set 2

G
George Boole 229
Graph Colouring 421

H
Hamiltonian Cycle 264
Hamiltonian Graph 261, 262
Handshaking Lemma 252
Hasse Diagrams 212
Homomorphic 126
Homomorphisms 268

I
Ï 29
Ideal Domains 196
Identity Law 173
Identity Mapping 85
Identity Relation 40
Inclusion Mapping 85
Inference Theory 320
Infimum 211
Infinite Group 127
Infinite Set 2
Injective Mapping 83
Integers 1
Intersection Of Sets 4, 7
Inverse Law 171
Irreflexive Relation 45

Isomorphic 126
Isomorphism 265

K

Karnaugh Maps 371
Konigsberg Bridges 265
Kruskal's Algorithm 302

L

Lagrange's Theorem 151
Lattice 215
Logic 233
Lower Bound 60

M

Many to Many 82
Many to One 82
Matrix Rings 172, 177
Maximal Ideals 201
Minimum Connector 302
Modular Arithmetic 177
Modular Lattices 216
Multisets 6

N

Nand Gate 245, 406
National Numbers: 1
Negation Operator 313
Non-negative Integers 1
Non-reflexive Relation 45
Normal Subgroup 155, 163
Not Gate 233
Notation 104

O

One to Many 82
One to One 81
Or Gate 233
Order of A Group 127
Ordered Pairs 33

P

Pair Set 2
Partition 6
Partition of a Set 55
Path 246
Path Graph 254
Peano Axioms 106
Permutation Group 158
Permutations 338, 339
Pigeonhole Principle 342, 343
Pólya Enumeration Theorem 431
Polyhedral Groups 169
Polynomial Rings 175, 181
Poset 209
Power Sets 210

Prim's Algorithm 303
Proper Subset 3

Q

Quotient Groups 155
Quotient Set 56

R

Range of Mapping 84
Recurrence Relation 346
Reflexive Closure 68
Regular Graph 253
Rings of Functions 177
Rings of Sets 176
Root of A Polynomial 202
Rooted Ternary Trees 434
Roster Method 3
Rubik's Cube Puzzle 159

S

Set Builder Method 3
Singleton Set 2
Spanning Tree 286
Status of Zero 103
Supremum 211
Surjective Mapping 83
Symmetric 42
Symmetric Closure 68
Symmetric Difference 5

T

'Theta' 93
Trail 270
Transformation 85
Transitive Closure 69
Transitive Relation 43
Traversal of A Binary Tree 289
Tree Structures 284

U

Union of Sets 4
Universal Relation 41
Universal Set 4
Upper Bound 60

V

Venn Diagrams 21
Vertex Coloring 422
Void Relation 40

W

Walk 246
Well Ordered Set 59, 210

Z

Zero Divisors 191
Zero Rings 176